Early Stages of
Oxygen Precipitation in Silicon

NATO ASI Series

Advanced Science Institutes Series

A Series presenting the results of activities sponsored by the NATO Science Committee, which aims at the dissemination of advanced scientific and technological knowledge, with a view to strengthening links between scientific communities.

The Series is published by an international board of publishers in conjunction with the NATO Scientific Affairs Division

A	Life Sciences	Plenum Publishing Corporation
B	Physics	London and New York
C	Mathematical and Physical Sciences	Kluwer Academic Publishers
D	Behavioural and Social Sciences	Dordrecht, Boston and London
E	Applied Sciences	
F	Computer and Systems Sciences	Springer-Verlag
G	Ecological Sciences	Berlin, Heidelberg, New York, London,
H	Cell Biology	Paris and Tokyo
I	Global Environmental Change	

PARTNERSHIP SUB-SERIES

1.	Disarmament Technologies	Kluwer Academic Publishers
2.	Environment	Springer-Verlag / Kluwer Academic Publishers
3.	High Technology	Kluwer Academic Publishers
4.	Science and Technology Policy	Kluwer Academic Publishers
5.	Computer Networking	Kluwer Academic Publishers

The Partnership Sub-Series incorporates activities undertaken in collaboration with NATO's Cooperation Partners, the countries of the CIS and Central and Eastern Europe, in Priority Areas of concern to those countries.

NATO-PCO-DATA BASE

The electronic index to the NATO ASI Series provides full bibliographical references (with keywords and/or abstracts) to more than 50000 contributions from international scientists published in all sections of the NATO ASI Series.
Access to the NATO-PCO-DATA BASE is possible in two ways:

– via online FILE 128 (NATO-PCO-DATA BASE) hosted by ESRIN,
Via Galileo Galilei, I-00044 Frascati, Italy.

– via CD-ROM "NATO-PCO-DATA BASE" with user-friendly retrieval software in English, French and German (© WTV GmbH and DATAWARE Technologies Inc. 1989).

The CD-ROM can be ordered through any member of the Board of Publishers or through NATO-PCO, Overijse, Belgium.

Series 3. High Technology – Vol. 17

Early Stages of Oxygen Precipitation in Silicon

edited by

R. Jones

**Department of Physics,
University of Exeter,
Exeter, U.K.**

Kluwer Academic Publishers

Dordrecht / Boston / London

Published in cooperation with NATO Scientific Affairs Division

Proceedings of the NATO Advanced Research Workshop on
Early Stages of Oxygen Precipitation in Silicon
Exeter, U.K.
March 26-29, 1996

A C.I.P. Catalogue record for this book is available from the Library of Congress.

ISBN-13: 978-94-010-6645-7   e-ISBN-13: 978-94-009-0355-5
DOI: 10.1007/978-94-009-0355-5

Published by Kluwer Academic Publishers,
P.O. Box 17, 3300 AA Dordrecht, The Netherlands.

Kluwer Academic Publishers incorporates the publishing programmes of
D. Reidel, Martinus Nijhoff, Dr W. Junk and MTP Press.

Sold and distributed in the U.S.A. and Canada
by Kluwer Academic Publishers,
101 Philip Drive, Norwell, MA 02061, U.S.A.

In all other countries, sold and distributed
by Kluwer Academic Publishers Group,
P.O. Box 322, 3300 AH Dordrecht, The Netherlands.

Printed on acid-free paper

Softcover reprint of the hardcover 1st edition 1996

Contents

Preface

It was found as long ago as 1954 that heating oxygen rich silicon to around 450°C produced electrical active defects — the so called *thermal donors*. The inference was that the donors were created by some defect produced by the aggregation of oxygen. Since then, there has been an enormous amount of work carried out to elucidate the detailed mechanism by which they, and other defects, are generated. This task has been made all the more relevant as silicon is one of the most important technological materials in everyday use and oxygen is its most common impurity. However, even after forty years, the details of the processes by which the donors and other defects are generated are still obscure.

The difficulty of the problem is made more apparent when it is realised that there is only one oxygen atom in about ten thousand silicon atoms and so it is difficult to devise experiments to 'see' what happens during the early stages of oxygen precipitation when complexes of two, three or four oxygen atoms are formed. However, new important new findings have emerged from experiments such as the careful monitoring of the changes in the infrared lattice absorption spectra over long durations, the observation of the growth of new bands which are correlated with electronic infra-red data, and high resolution ENDOR studies. In addition, progress has been made in the improved control of samples containing oxygen, carbon, nitrogen and hydrogen.

In parallel with these developments, there have been great advances in modelling techniques. A large amount of modelling had already been carried out and the central role of an *overcoordinated* oxygen atom in donors had already been emphasised. Recent progress in first principles methods has, above all else, given greater confidence in predictive theory, and it is thus natural that the subject has excited the interest of so many theoreticians.

The time seemed ripe for a workshop to bring together leading workers in the field. A key aim of the workshop was to promote collaboration between experimental and modelling groups.

The workshop was made possible by funding both from NATO and the University of Exeter and both organisations are thanked for their support. I also thank my co-organisers Peter Deák and Stefan Estreicher for support and encouragement. The bulk of the daily administrative work fell locally on the shoulders of Chris Ewels, Jon Goss, Paul Leary, Chris Latham, Paul Rawlins and Sailesh Patel, all of whom worked very hard to ensure the workshop was a success. They were keen to ensure that details of itineraries, the abstracts for the talks and posters and the excursions were available on the internet. This proved to be very useful for those young enough – or with enough time – to browse these pages, but the responses made it clear that this service was of great use. The time might come when we all will stay at home and have a complete, but virtual, conference over the internet. That will, however, remove the pleasure of meeting directly with the people who are daily involved with problems regarding the fate of oxygen in silicon.

R. Jones

Contributers

C. A. J. Ammerlaan, Van der Waals-Zeeman Institute, University of Amsterdam, Valckenierstraat 65-7, NL-1018 XE Amsterdam, The Netherlands

B. A. Andreev, Institute of Chemistry of High-Purity Substances, Russian Academy of Sciences, 49 Tropinin St., GSP-445, Nizhnyi Novgorod, 603600 Russia

D. Ballutaud, Laboratoire de Physique des Solides, 1 Place Aristide Briand 92195, Meudon Cedex, France

F. Berg Rasmussen, Van der Waals-Zeeman Laboratorium, Universiteit van Amsterdam, Valckenierstraat 65-67, NL-1018 XE Amsterdam, The Netherlands

J. von Boehm, Laboratory of Computational Dynamics, Helsinki University of Technology, 02150 Espoo, Finland

D. I. Bohne, Phoenix Contact GmbH, Flachsmarktstr. 8-28, D-32825 Blomberg, Germany

A. Boutry-Forveille, Laboratoire d" Electrochimie Interfaciale, 1 Place Aristide Briand 92195, Meudon Cedex, France

Yu. A. Bumai, Belarussian State Polytechnical Academy, F. Skorina 65 Ave., Minsk 220027, Belarus

P. Clauws, RUG, Krijgslaan 281-S1, B-9000 Gent, Belgium

A. Correia, Laboratoire de Physique des Solides, 1 Place Aristide Briand 92195, Meudon Cedex, France

S. E. Daly, School of Physical Sciences, Dublin City University, Dublin 9, Ireland

G. Davies, Physics Department, King's College London, Strand, London WC2R 2LS, UK

P. Deák, Department of Atomic Physics, Technical University of Budapest, Budafoki út 8., Budapest, H-1111, Hungary

M. Depas, IMEC, Kapeldreef 75, B-3001 Leuven, Belgium

V. V. Emtsev, Ioffe Physicotechnical Institute, Russian Academy of Sciences, 194021 St. Petersburg, Russia

K.L. Enisherlova, Research and Production Company, ELLINA-NT, 38 Vavilov Street, Moscow, 117942, Russia

S. K. Estreicher, Physics Department, Texas Tech University, Lubbock, TX 79409, USA

C. P. Ewels, Department of Physics, University of Exeter, Exeter, EX4 4QL, UK,

W. R. Fahrner, Fern University of Hagen, Haldener Str. 182, Hagen 588084, Germany

P. A. Fedders, Physics Dept., Washington University, St. Louis, MO 63130, USA

M. Fujinami, Advanced Technology Research Laboratories, Nippon Steel Corporation 1618 Ida, Nakahara-ku, Kawasaki 211, Japan

L. G. Fytros, University of Athens, Department of Physics, Solid State Section, Panepistimiopolis, Zografos, Athens 157 84, Greece

U. Gösele, Max-Planck-Institute of Microstructure Physics, Weinberg 2, D-06120 Halle, Germany

A. Gali, Department of Atomic Physics, Technical University of Budapest, Budafoki út. 8, H-1111 Budapest, Hungary

R. D. Goldberg, Department of Physics, University of Western Ontario, London, Ontario, Canada N6A 3K7

J. Goss, Department of Physics, University of Exeter, Exeter, EX4 4QL, UK

J. E. Gower, Physics Department, King's College London, Strand, London WC2R 2LS, UK

D. Gräf, Wacker-Siltronic GmbH, PO Box 1140, D-84479 Burghausen, Germany

P. J. Grönberg, Laboratory of Physics, Helsinki University of Technology, 02150 Espoo, Finland

T. Gregorkiewicz, Van der Waals-Zeeman Institute, University of Amsterdam, Valckenierstraat 65-7, NL-1018 XE Amsterdam, The Netherlands

R. Habu, SKY Aluminium Co., Ltd, 1351 Uenodai, Fukaya 366, Japan

T. Hallberg, Department of Physics and Measurement, Linköping University, Linköping, S-58183, Sweden

H. Harada, Advanced Technology Research Labs., Nippon Steel Co. 3434 Shimata, Hikari 743, Japan

M. O. Henry, School of Physical Sciences, Dublin City University, Dublin 9, Ireland

V. Higgs, Bio-Rad Micromeasurements, Hemel Hempstead, HP2 7TD, UK

G. Hobler, Institut für Festkörperelektronik, TU Wien, Gusshausstr. 27-9/E362, A-1040 Wien,

N.A. Iasamanov, Museum of Earth of the Moscow State University, Vorobyovy Gory, Moscow, Russia

T. Ikarashi, Microelectronics Research Laboratories, NEC Corporation, 34 Miyukigaoka, Tsukuba, Ibaraki 305, Japan

A. I. Ivanov, Belarussian State Polytechnical Academy, F. Skorina 65 Ave., Minsk 220027, Belarus

T. Iwasaki, Advanced Technology Research Labs., Nippon Steel Co. 3434 Shimata, Hikari 743, Japan

R. Jones, Department of Physics, University of Exeter, Exeter, EX4 4QL, UK,

K. Kawakami, Advanced Technology Research Labs., Nippon Steel Co. 3434 Shimata, Hikari 743, Japan

N. M. Kazychits, Belorussian State University, F. Skorina 4 Ave., Minsk, 220028 Belarus

K. Kenis, IMEC, Kapeldreef 75, B-3001 Leuven, Belgium

L. I. Khirunenko, Institute of Physics of the National Academy of Sciences of Ukraine, Pr. Nauki, 46, 252650, Kiev-22, Ukraine

S. Kimura, Microelectronics Research Laboratories, NEC Corporation, 34 Miyukigaoka, Tsukuba, Ibaraki 305, Japan

G. Kissinger, Institute of Semiconductor Physics, D-15204 Frankfurt (Oder), Germany

A. P. Knights, Department of Physics, University of Western Ontario, London, Ontario, Canada N6A 3K7

U. Lambert, Wacker-Siltronic GmbH, PO Box 1140, D-84479 Burghausen, Germany

G. Langouche, KUL, Celestijnenlaan 200D, B-3001 Leuven, Belgium

E. C. Lightowlers, King's College, Strand, London, WC2R 2LS, UK

J. L. Lindström, Department of Physics and Measurement, Linköping University, Linköping, S-58183, Sweden

C. A. Londos, University of Athens, Department of Physics, Solid State Section, Panepistimiopolis, Zografos, Athens 157 84, Greece

H. E. Maes, KUL, Celestijnenlaan 200D, B-3001 Leuven, Belgium

V. P. Markevich, Institute of Solid State and Semiconductor Physics, P. Brovki str. 17, Minsk 220072, Belarus

Yu. V. Martynov, Van der Waals-Zeeman Institute, University of Amsterdam, Valckenierstraat 65-7, NL-1018 XE Amsterdam, The Netherlands

E. McGlynn, School of Physics Sciences, Dublin City University, Collins Avenue, Dublin 9, Ireland

K. G. McGuigan, Department of Physics, Royal College of Surgeons in Ireland, Dublin 2, Ireland

S. A. McQuaid, Laboratorio de Microelectrónica, Departamento de Física Aplicada, Universidad Autónoma de Madrid, Cantoblanco, 28049 Madrid, Spain

I. F. Medvedeva, Institute of Solid State and Semiconductor Physics, P. Brovki str. 17, Minsk 220072, Belarus

S. Messoloras, J. J. Thomson Physical Laboratory, University of Reading, Whiteknights, Reading RG6 6 AF, UK

J. Miro, Department of Atomic Physics, Technical University of Budapest, Budafoki út 8, H-1111 Budapest, Hungary

A. Misiuk, Institute of Electron Technology, S8 Al. Lotników 32/46, 02-668 Warsaw, Poland

L. I. Murin, Institute of Solid State and Semiconductor Physics, P. Brovki str. 17, Minsk 220072, Belarus

U. Myler, Department of Physics, University of Western Ontario, London, Ontario, Canada N6A 3K7

R. C. Newman, IRC Semiconductor Materials, The Blackett Laboratory, Imperial College of Science and Technology and Medicine, London SW7 2BZ, UK

R. M. Nieminen, Laboratory of Physics, Helsinki University of Technology, 02150 Espoo, Finland

S. Öberg, Department of Mathematics, University of Luleå, Luleå, S95187, Sweden

H. Ono, Microelectronics Research Laboratories, NEC Corporation, 34 Miyuki-gaoka, Tsukuba, Ibaraki 305, Japan

B. Pajot, Groupe de Physique des Solides, Tour 23, Université Denis Diderot, 2 place Jussieu, Paris, France

S. T. Pantelides, Department of Physics and Astronomy, Vanderbilt University, Nashville, TN 37235, USA

Y. K. Park, Physics Department, University of California, Irvine, CA 92717, USA,

G.N. Petrov, Research and Production Company, ELLINA-NT, 38 Vavilov Street, Moscow, 117942, Russia

M. Ramamoorthy, Department of Physics and Astronomy, Vanderbilt University, Nashville, TN 37235, USA

S. Rycroft, J. J. Thomson Physical Laboratory, University of Reading, Whiteknights, Reading RG6 6 AF, UK

A. N. Safonov, Physics Department, King's College London, Strand, London WC2R 2LS, UK

N. V. Sarlis, University of Athens, Department of Physics, Solid State Section, Panepistimiopolis, Zografos, Athens 157 84, Greece

K. Schmalz, Institute of Semiconductor Physics, Walter-Korsing-Str. 2, D-15230 Frankfurt (Oder), Germany

E. Schroer, Max-Planck-Institute of Microstructure Physics, Weinberg 2, D-06120 Halle, Germany

S. Senkader, Institut für Festkörperelektronik, TU Wien, Gusshausstr. 27-9/E362, A-1040 Wien, Austria

V. I. Shakhovtsov, Institute of Physics of the National Academy of Sciences of Ukraine, Pr. Nauki, 46, 252650, Kiev-22, Ukraine

Y. Shirakawa, Process Development Div., Fujitsu Ltd., 1015 Kamikodanaka, Nakahara-ku, Kawasaki 211, Japan

V. V. Shumov, Institute of Physics of the National Academy of Sciences of Ukraine, Pr. Nauki, 46, 252650, Kiev-22, Ukraine

P. J. Simpson, Department of Physics, University of Western Ontario, London, Ontario, Canada N6A 3K7

L. C. Snyder, The University at Albeny, 1400 Washington Avenue, Albany, New York 12222, USA

J. -M. Spaeth, Fachbereich Physik, Universität-GH Paderborn, 33095 Paderborn, Germany

M. Stavola, Department of Physics, Lehigh University, Bethlehem, PA 18015, USA

A. Stesmans, KUL, Celestijnenlaan 200D, B-3001 Leuven, Belgium

R. J. Stewart, J. J. Thomson Phyiscal Laboratory, University of Reading, Whiteknights, Reading RG6 6 AF, UK

M. Suezawa, Institute for Materials Research, Tohoku University, 2-1-1 Katahira, Aoba-ku, Sendai, 980-77 Japan

K. Sumino, Nippon Steel Corporation, 20-1 Shintomi, Futtsu City, Chiba Prefacture 293, Japan

T. Y. Tan, School of Engineering, Duke University, Durham, NC 27706, USA

A. Tanikawa, Microelectronics Research Laboratories, NEC Corporation, 34 Miyukigaoka, Tsukuba, Ibaraki 305, Japan

K. Terashima, Microelectronics Research Laboratories, NEC Corporation, 34 Miyukigaoka, Tsukuba, Ibaraki 305, Japan

T. M. Tkacheva, Institute of Metallurgy of the Russian Academy of Sciences, 49 Leninsky prospkt, Moscow, 117333 Russia

M.-A. Trauwaert, IMEC, Kapeldreef 75, B-3001 Leuven, Belgium

A. G. Ulyashin, Belarussian State Polytechnical Academy, F. Skorina 65 Ave., Minsk 220027, Belarus

J. Vanhellemont, IMEC, Kapeldreef 75, B-3001 Leuven, Belgium

V. S. Varichenko, Belorussian State University, F. Skorina 4 Ave., Minsk, 220028 Belarus

P. Wagner, Wacker-Siltronic GmbH, PO Box 1140, D-84479 Burghausen, Germany

G. D. Watkins, Department of Physics, Lehigh University, Bethlehem, PA 18015, USA

J. Weber, Max-Planck-Institut für Festkörperforschung, Postfach 800665, D-70506 Stuttgart, Germany

P. Werner, Max-Planck-Institute of Microstructure Physics, Weinberg 2, D-06120 Halle, Germany

R. Wu, The University at Albeny, 1400 Washington Avenue, Albany, New York 12222, USA

C. Yamada-Kaneta, Fujitsu Laboratories Ltd., 10-1, Morinosato-wakamiya, Atsugi 243-01, Japan

H. Yamada-Kaneta, Process Development Div., Fujitsu Ltd., 1015 Kamiko-danaka, Nakahara-ku, Kawasaki 211, Japan

V. I. Yashnik, Institute of Physics of the National Academy of Sciences of Ukraine, Pr. Nauki, 46, 252650, Kiev-22, Ukraine

A. M. Zaitsev, Fern University of Hagen, Haldener Str. 182, Hagen 588084, Germany

I. S. Zevenbergen, Van der Waals-Zeeman Institute, University of Amsterdam, Valckenierstraat 65, NL-1018 XE Amsterdam, The Netherlands

A.-M. van Bavel, KUL, Celestijnenlaan 200D, B-3001 Leuven, Belgium

OXYGEN-RELATED DEFECTS IN SILICON: STUDIES USING STRESS-INDUCED ALIGNMENT

G. D. WATKINS
Department of Physics, Lehigh University,
Bethlehem, PA 18015, USA

1. Introduction

Although the primary concern relating to a defect in a semiconductor is apt to be the effect it has on the electrical properties of the material, the experimental techniques of choice to identify it, to determine its structure, and unravel its role in various processes occuring in the material are primarily the spectroscopic ones, such as electron paramagnetic resonance (EPR), electron-nuclear double resonance (ENDOR), local vibrational mode (LVM) spectroscopy, or absorption and luminescence associated with electronic transitions involving the defect. In the course of this workshop, we will have the chance to learn in detail how each of these have contributed to our understanding of oxygen and its complexes in silicon.

In this paper, I will concentrate on a simple auxiliary technique which has been used successfully with each of the above for oxygen-related defects in silicon — the application of uniaxial stress. Uniaxial stress can do two things: (1) It can split and shift the spectroscopic transitions, which, if resolved, can be used to determine the symmetry of the defect and the polarization properties of the transitions. (2) If the defect is anisotropic and is free to reorient, it may align preferentially with respect to the applied stress, revealing the magnitude and sense of the local strain it produces in the lattice. Combined with studies *vs* temperature, the kinetics of its reorientational motion can also be determined. It is this second phenomenon – stress-induced alignment – that I will concentrate on here. The important information that can result from the first will be discussed in some of the other papers to follow.

The outline of the present paper is as follows: In section 2, I present a brief outline of the theory of stress-induced alignment, and the experimental method for measuring the relevant alignment parameters and the reorien-

1

R. Jones (ed.), Early Stages of Oxygen Precipitation in Silicon, 1–18.
© 1996 *Kluwer Academic Publishers.*

tation kinetics. In section 3, I review the results for three examples, (i) isolated interstitial oxygen, (ii) isolated substitutional oxygen, and (iii) the interstitial-carbon–interstitial-oxygen pair. In these cases, well established models have emerged for the structure of the defects, the stress-induced studies having played an important role in the process. In section 4, I discuss results for the thermal donor. We will see that, in this case, although the structure of the core is still uncertain, the stress-induced alignment studies define interesting properties that any final model must somehow accomodate.

2. Stress-Alignment Technique

For an anisotropic defect in a cubic crystal, the change in its total energy due to externally applied stress $\boldsymbol{\sigma}$ (or strain $\boldsymbol{\varepsilon}$) can be written [1]

$$\Delta E_i = Tr[\mathbf{A} \cdot \boldsymbol{\sigma}_i] = Tr[\mathbf{B} \cdot \boldsymbol{\varepsilon}_i], \tag{1}$$

where the subscript i refers to a specific orientation for the defect, the $\boldsymbol{\sigma}_i$ (or $\boldsymbol{\varepsilon}_i$) tensor is the applied stress (or strain) resolved into the principal axis system of defect i, and $\mathbf{A}$ (or $\mathbf{B}$) is the "piezo-spectroscopic tensor" [1], which completely characterizes the coupling properties of the defect to the stress (or strain) in its principal axis system. At a given temperature T, if the defect is free to reorient, the relative populations n_i of the individual defect orientations will assume a Boltzmann distribution

$$\frac{n_i}{n_j} = \exp[(\Delta E_j - \Delta E_i)/kT], \tag{2}$$

from which, by applying stress along a few specific crystallographic directions, the complete set of traceless components for $\mathbf{A}$ (or $\mathbf{B}$) can be determined using Eq. (1), if the individual defect orientations can be resolved. Since $\mathbf{A}$ (or $\mathbf{B}$) is symmetric it can always be diagonalized, leading, for example, to the following form for $\mathbf{B}$

$$\begin{pmatrix} B_1 & & \\ & B_2 & \\ & & B_3 \end{pmatrix}, \tag{3}$$

where the 1,2,3 coordinates define the principle axis system of the defect, reflecting its point group symmetry. [This convention differs from that in the tables of Kaplyanskii [1], which provide the nondiagonal tensor components in the crystal cubic axis system. The two are related by the appropriate rotation of coordinates for each symmetry class. Here I will use the principal axis system, and $\mathbf{B}$, to take advantage of the more direct intuitive physical insight they contain.]

In the case of EPR, the anisotropy of the g-tensor or fine structure terms generally allows easy spectral separation of the individual defect orientations, and their populations can be monitored directly from their relative signal amplitudes. In the case of optical transitions, if stress-induced splittings due to the individual orientations can be resolved, the same is true. If they are not resolved, less information is available, but dichroism (difference in absorbed or emitted light polarized parallel vs perpendicular to the stress direction) still provides a quantitative measure of alignment, which may be sufficient to determine the tensor components for high symmetry defects. Used in conjunction with EPR studies, it serves to unambiguously identify an optical transition with a specific defect seen in EPR, and to establish the dipole moment direction for the transition.

Usually in these experiments, the stress is applied at an elevated temperature T where the defects are able to reorient. After waiting a sufficient time to establish thermal equilibrium, the sample is cooled rapidly with the stress on to a lower temperature (selected as optimum for the particular spectroscopic technique involved), and the quenched-in alignment monitored with stress removed. The kinetics for reorientation can subsequently be determined by successive anneals vs time at different temperatures, monitoring the recovery after each annealing step again at the lower observation temperature. The results are then matched to a simple first order recovery process

$$n_j(t) - n_{j0} = [n_j(0) - n_{j0}] \exp(-t/\tau), \qquad (4)$$

where n_{j0} in the equilibrium value of n_j, and the time constant τ is given by

$$\tau^{-1} = \nu_0 \exp(-U/kT). \qquad (5)$$

Here, the preexponential factor ν_o is a suitably weighted "attempt frequency", and U is the energy barrier that must be surmounted for the defect to reorient.

3. Simple oxygen-related defects

3.1. ISOLATED INTERSTITIAL OXYGEN

One of the first uses of this technique was reported 35 years ago for isolated interstitial oxygen [2, 3]. At that time, an LVM band at 9μ (1100 cm^{-1}) had been established to arise from dissolved oxygen in silicon and the now accepted model shown in Fig. 1(a) of an isolated interstitial oxygen atom squeezed and bonding between two normal nearest neighbors had already been proposed to explain the LVM properties [4, 5, 6]. Quenched-in alignment achieved by 265 MPa stress along a $\langle 111 \rangle$ direction at 400°C produced dichroism, as shown in Fig. 1(b). Combined with results for stress

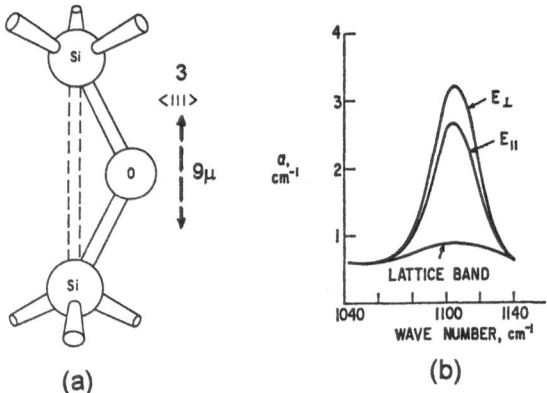

Figure 1. (a) Model of interstitial oxygen in silicon. (b) Dichroism at room temperature in the 9μ LVM band after 265 MPa $\langle 111 \rangle$ stress at $400°$C [3].

along $\langle 100 \rangle$ (no dichroism), and $\langle 110 \rangle$, the predictions of the model were confirmed in detail, the dipole moment for the 9μ band being along the Si-O-Si $\langle 111 \rangle$ 3-axis direction with axially symmetric "piezo-spectroscopic" tensor components $B_3 = -2B_2 = -2B_1 = -15.2$ eV. The large negative value for B_3 confirmed the squeezed nature of the Si-O-Si bond and its tendency to avoid applied compressional strain (negative) along its axis.

The biggest bonus of this study, however, has come from the recovery kinetics. At that time, it was also recognized that reorientation required oxygen to hop from one Si-Si bond to another, and that in measuring its kinetics, one was therefore also measuring the single jump diffusion process. Combining this with internal friction studies done earlier by Southgate at 100-300 kHz [7], and using the relation for the diffusion constant [8]

$$D = a^2/12\tau, \qquad (6)$$

where a is the nearest neighbor distance in the lattice, a remarkably accurate estimate of the oxygen diffusion constant was possible [3]. Since that time, mass transport diffusion studies at elevated temperatures [9, 10, 11, 12, 13] and extended studies of stress-induced dichroism recovery [14, 15] have been performed as shown in Fig. 2, confirming the early results (black triangles in the figure), and leading to the accurate determination of its diffusion now over eleven decades [16], and the remarkable establishment that it proceeds over the full range by a single well defined process.

It is interesting to note that first principle theoretical calculations of the diffusion barrier as the total energy difference between stable and saddle point configurations, have often found much lower values (1.2 -1.8 eV) [17,

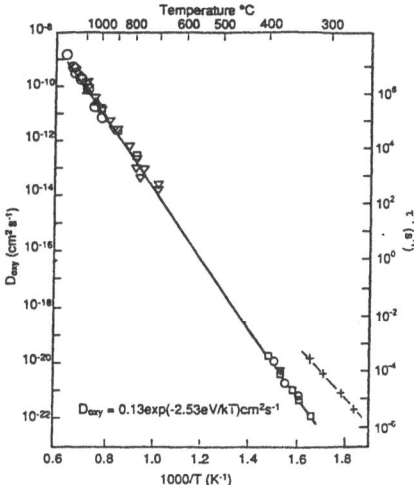

Figure 2. Diffusion constant of oxygen in silicon deduced from stress-induced alignment, internal friction, and mass transport studies [14, 16].

18]. An explanation for this disturbing discrepancy has only very recently been proposed [19, 20], and will be the subject also of one of the papers in this conference.

Before we leave isolated oxygen, we should point out another important discovery, unexpected and surprising at the time, which found that after certain heat treatments, a much more rapid recovery from stress-induced alignment can sometimes be observed, indicating a diffusional activation energy of only 1.9 eV [14]. This is indicated by the dashed line in Fig. 2. The origin of this phenomenon has been the subject of a number of studies which have variously attributed it to the presence of trace metallic impurities [21] and most recently to hydrogen [22]. We will hear more about this also in later chapters.

3.2. SUBSTITUTIONAL OXYGEN

Substitutional oxygen is formed when a mobile lattice vacancy produced by radiation damage is trapped by interstitial oxygen. It introduces an acceptor level at $E_C - 0.16$ eV and has been extensively studied by EPR [23] in its negatively charged state and by LVM spectroscopy in both its neutral [24] and negative [25] states, leading to the well established off-center oxygen model in Fig. 3(a). From the EPR studies alone, it could be established from hyperfine interactions with ^{29}Si that the unpaired electron was localized primarily in a bond between two normally next nearest sili-

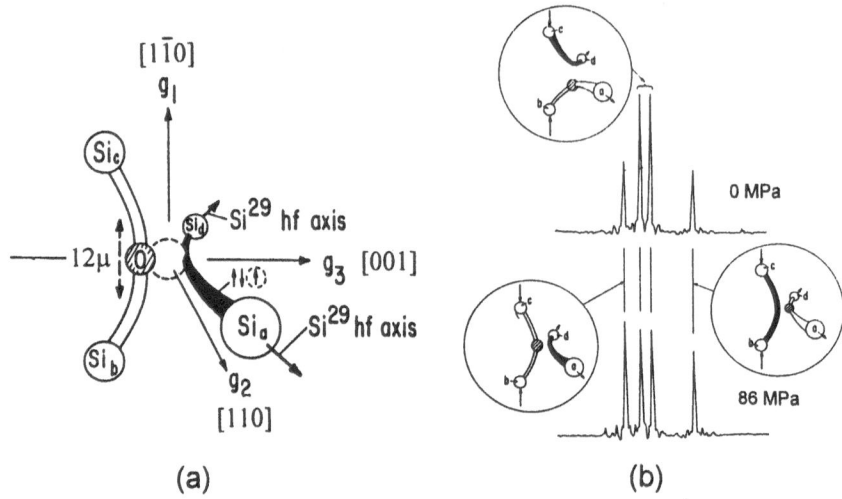

Figure 3. (a) Model of substitutional oxygen deduced from EPR and LVM studies. (b) Change in the EPR spectrum produced by ⟨110⟩ stress at 125°C, observation at 77°C [23].

con atoms, and there was strong suspicion that oxygen was also somehow involved. However, it was the stress-induced alignment studies that served to construct the details.

An example of the EPR stress-induced alignment results is illustrated in Fig. 3(b). From these studies, the values $B_1 = +8.8$, $B_2 = -0.4$, $B_3 = -8.4$ eV were determined for the EPR active negatively charged state, the C_{2v} principal axes being indicated in Fig. 3(a). From the EPR studies it was also possible to deduce values for the neutral charge state of $B_1 = +6.1$, $B_2 = +4.9$, $B_3 = -11.1$ eV [26]. From these, and their change vs charge state, it could be deduced that both pairs of dangling silicon bonds had collapsed together as shown, and further that the orbital containing the unpaired electron was an *antibonding* one, its collapse being subtantially reduced when occupied.

At the same time, an oxygen-associated LVM band at 12μ was also detected in the material after radiation damage [24]. Detailed correlation between the dichroism induced in this band and the alignment observed in the EPR confirmed unambiguously that the oxygen atom is part of the defect and determination of the LVM dipole moment direction established the oxygen position as bonded between the other two silicon neighbors, as shown. In effect, EPR looks at one side of the defect, the LVM studies at the other, and it was the stress-induced alignment studies that supplied the

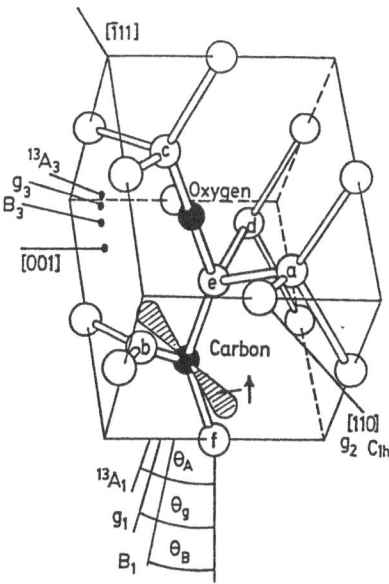

Figure 4. Model deduced for the interstitial-carbon–interstitial-oxygen pair in silicon, showing the various defect axes [28]. The oxygen is shown on bond-center, but is probably distorted out to one side or the other in the C_{1h} plane.

essential connection.

3.3. INTERSTITIAL-CARBON–INTERSTITIAL-OXYGEN PAIR

Interstitial oxygen interacts strongly with interstitial carbon. The resulting complex is readily produced by radiation damage when interstitial carbon, formed by trapping a radiation-produced mobile interstitial silicon atom, migrates and is trapped by interstitial oxygen. It introduces a deep donor level at $E_V + 0.38$ eV and is EPR active in its unoccupied positive charge state. A model deduced from the EPR studies [27, 28] is shown in Fig. 4. From the EPR alone, the symmetry was established to be C_{1h}, with its **g** and ^{13}C hyperfine interactions very similar to those of positively charged isolated interstitial carbon, but with evidence of a nearby perturbation. Confirmation of this was also provided by stress-induced alignment in the EPR studies giving $B_1 = 8.6$, $B_2 = 0.2$, and $B_3 = -8.8$ eV, where the principal axes are indicated in the figure. These are very close to those determined for isolated interstitial carbon [29], except for the $\theta_B = 15°$ tilt in the axes which reduces the symmetry from C_{2v} for isolated interstitial carbon, to C_{1h}, resulting from the perturbing defect nearby.

The identification of the perturbing defect as oxygen came from two independent stress-induced alignment experiments [27, 28]. In one experiment, it was demonstrated that stress-induced dichroism in a much studied radiation-produced luminescence band, labeled the C-line, with zero phonon line at 0.79 eV, has a one-to-one relationship to the alignment of the EPR center both in its sense and magnitude, and in its kinetics for recovery. The C-line had previously been established to contain both carbon (from an isotope shift of its zero phonon line) and oxygen (from isotope shifts of two localized mode phonon replicas) [30, 31], and this one-to-one correlation with the EPR center served to establish unambiguously that they both arise from the same center.

In a second experiment, a quenched-in alignment of interstitial oxygen was first produced in the sample before radiation to produce the complex [27, 28]. The result of this novel approach was that the newly formed complexes observed by EPR grew in with a partial alignment reflecting the original preferential $\langle 111 \rangle$ axis of the interstitial oxygen. This clearly established that the trapping defect was *isolated interstitial* oxygen and served to suggest at the same time the necessary spatial relationship between the oxygen and the interstitial carbon.

Putting this all together, the model of Fig. 4 was constructed. Of course, any other model which can account for all of this detailed information must also be considered. It is important to point this out because recent theoretical modeling of the center has come up with a different bonding arrangement of the atoms [32]. Hopefully this will be discussed later in the workshop.

4. NL8 Thermal donors

As we will see in later presentations in this workshop, what was originally thought to be a conceptually simple process of oxygen aggregation at $\sim$ 450°C to form "thermal donors" in silicon, has evolved into a very a complex and poorly understood phenomenon. There are, for example, now several identified groups of thermal donors (TD's), labeled NL8 [33, 34], NL10 [33, 35], STD [36], and there are also more recent similar contenders related to hydrogen [37]. Here, I will confine my considerations to the family of TD's denoted NL8, which have been the most studied through the years, and are often, but not always, the dominant ones.

NL8 denotes a family of, at the latest count [38], at least sixteen double donors (which we will denote TDD1, TDD2, .., TDDn) that grow in progressively *vs* time at temperatures $\sim$450°C in oxygen-containing Czochralski-grown silicon. They can be monitored individually by their optical excitation spectra to neutral or singly ionized excited hydrogenic effective mass

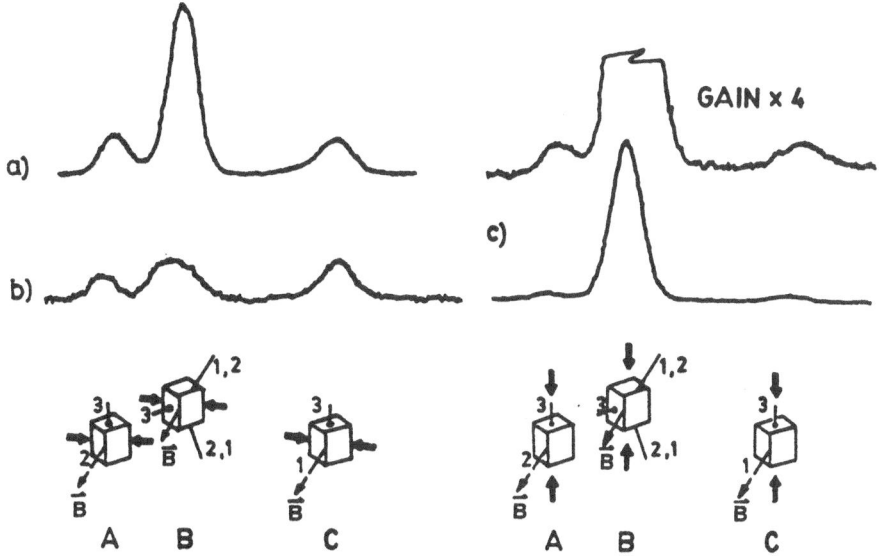

Figure 5. NL8 EPR spectra with **B** ∥ [110] for TDD's grown in with and without compressive stress for 90 min. at 460°C. (a) $\sigma = 0$; (b) $\sigma \parallel [\bar{1}10]$, 600 MPa; (c) $\sigma \parallel [001]$, 600 MPa [40]. Insets show defect orientations associated with each of the resolved EPR lines along with the stress and magnetic field directions.

states, and are observed to progressively become shallower, approaching the binding energies expected for shallow effective mass double donor ground states. They can also be monitored by a C_{2v} symmetry EPR signal, originally labeled NL8 [33] (and hence the name), arising from their paramagnetic positive charge state, and whose g-values slowly shift *vs* annealing time, reflecting also the changing relative contributions of the different species. (The proof that the EPR NL8 spectrum arises from the singly ionized donor state of the same double donors observed optically was achieved also by the application of uniaxial stress [39], and will be discussed later by Stavola, when he describes other important additional information concerning the electronic structure of the TDD's obtained from such studies. It was not a stress-*alignment* experiment, however, and will not be treated here.)

To explore the effect of stress-induced alignment for the NL8 TDD's, uniaxial stress has been applied during the heat treatment production of the TDD's [40, 41, 42, 43]. A large preferential alignment resulted, as is shown in Fig. 5 for the EPR spectrum in samples heat treated 90 min. at 460°C under 600 MPa compressional stress. This alignment is clearly demonstrated also by dichroism in the individual optical excitation spectra,

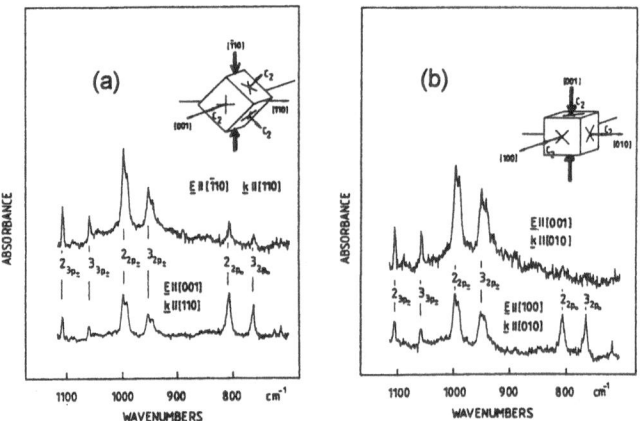

Figure 6. IR spectra of the individual TDD$^+$'s for the samples of Fig. 5 [40].

as shown in Fig. 6.

The proper interpretation of the above results rests heavily upon the mechanism for the alignment. It could represent the result of alignment of an early precursor to the TDD series which for the most part grows in retaining its alignment as it progresses from one TDD to the next. Or it could result from reorientation of each of the individual species. It is only in the latter case that the arguments of thermal equilibrium would be valid, and from which therefore a "piezospectroscopic tensor" could be derived for the individual TDD's. To test this question, several additional samples were prepared and isothermal annealing experiments performed at several temperatures to determine the recovery kinetics [43].

4.1. TDD RECOVERY MECHANISM AND KINETICS

In Fig. 7, we show the recovery at 400°C as measured by the NL8 EPR spectrum for the $\langle 110 \rangle$ stressed sample of Fig. 5. It will be seen later that this recovery time of ~ 5 hrs. is probably faster than the TDD growth rate at this low temperature, suggesting that the defects are actually reorienting. However, the results in the figure contain an even stronger indication of reorientation: Note that the recovery of component B of the spectrum is uniformly $1.5\times$ faster throughout the decay than the recovery between A and C. If the recovery were due to the growth in of unaligned defects, all components should decay together. Instead, it is straightforward to show that the factor of 1.5 is a unique indicator that the defects must be *reorienting*, the conversion between A and C requiring the intermediate step to B [43].

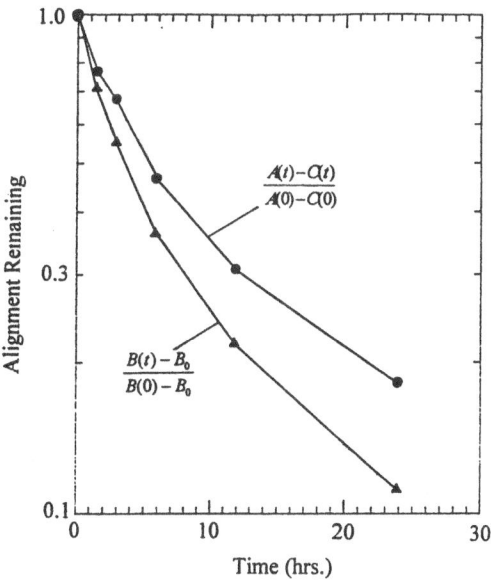

Figure 7. Recovery at 400°C of the NL8 EPR spectrum after ⟨110⟩ stress-induced alignment [43].

The non-exponential decay seen in Fig. 7 has a natural explanation in that the NL8 spectrum is composed of contributions from several TDD^+'s with different reorientation times. A confirmation of this is apparent by comparison to the 400°C alignment decay observed for the individual re-solved TDD's via their optical excitation spectra in a similarly prepared sample [41]. There, roughly exponential decay times were reported for each species, which appeared to increase a factor of ∼10 from TDD2 to TDD4. The decay time for TDD3 most closely matched the NL8 value for half recovery at that temperature, suggesting its domination in the EPR spectrum for that sample. (The optical excitation spectrum for this sample indicates comparable concentrations of TDD2 and TDD3, see Fig. 6. Due to partial compensation by the initial boron concentration, the TDD^+ EPR signal will tend to be dominated by the shallower higher number TDD's, the deeper ones being neutral.)

In Fig. 8, the results of isothermal anneals of the NL8 EPR spectrum for a set of similarly prepared samples are presented [43]. To cover this range of temperatures, overlapping (for temperature calibration) furnace and Rapid Thermal Anneal techniques were combined, and the value for the time constant $\tau_{1/2}$ was determined by the time for 50% recovery of the B signal. [This procedure was dictated by the nonexponential recovery behavior as seen in Fig. 7. In the higher temperature anneals, the decays

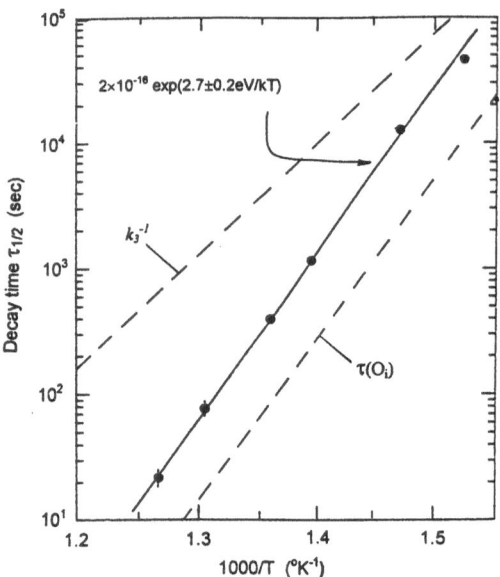

Figure 8. Reorientation kinetics for NL8 [43]. Shown for comparison is the reorientation time constant for isolated oxygen and k_3, an estimate of the lifetime of TDD3 (before conversion to TDD4) from the studies of Claybourn and Newman.

were observed to be more exponential.] The straight line on a semi-log plot over $3\frac{1}{2}$ decades reveals a single thermally activated process, giving with high precision,

$$\tau_{1/2} = 2 \times 10^{-16} \exp(2.7 \pm 0.2eV/kT). \tag{7}$$

This presumably reflects the property of TDD3, the primary contributor to the NL8 EPR signal in the samples.

This is an interesting result. This activation energy, within experimental accuracy, is identical to that for the single hop diffusional motion of isolated oxygen (2.53 eV, Fig. 2). There is a difference in the preexponential factor, however. To illustrate this, the reorientation time constant for isolated oxygen is also plotted in the figure, which is uniformly a factor ∼5 shorter. This has led to a very simple interpretation [43]. It suggests that reorientation is being limited by the motion of interstitial oxygen, presumably aggregated in the vicinity of the TDD3 core, and that ∼5 jumps are required to complete the process.

The reorientation rate becomes slower as the TDD's progress through the series, as already noted above in the optical dichroism recovery of TDD2, TDD3, and TDD4 at 400°C. The results of a detailed study at several additional temperatures from 375°C to 490°C [42], is shown in Fig.

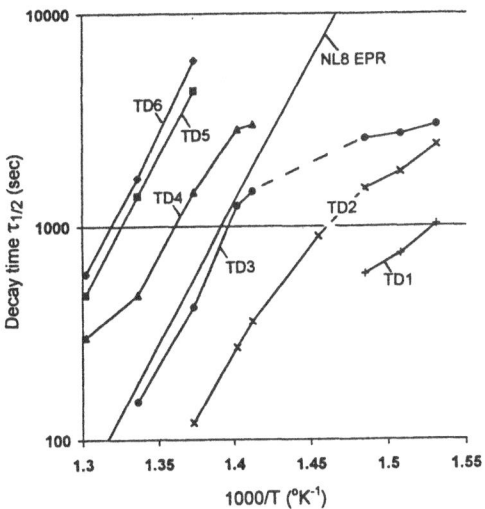

Figure 9. Recovery from ⟨100⟩ stress growth alignment observed for the individual TDD's by their IR excitation spectra [42].

9 for two sets of samples - one, heat treated for 90 min. to optimize the earlier TDD's, the other for 480 min. to optimize the later ones. These more difficult measurements are not as precise as those for the NL8 EPR studies but clearly demonstrate the trend, the TDD3 values matching well, as expected, in the temperature range above ~435°C those deduced from the NL8 EPR studies, as given in Fig. 8. The values increase uniformly as the TDD's progress, with each displaying approximately the same activation energy.

Below ~430°C, the results reveal the onset of a lower activation energy process. A puzzle at the time these were first obtained [42], we can conclude now that this reflects the onset again of the low temperature Stavola "anomalous" oxygen reorientation rate, as shown in Fig. 2, which for some reason happens to have been operative in this series of anneals. This rapid recovery was not observed in the 400°C furnace anneal described earlier which matched the much longer anneal times of the NL8 EPR results [41]. [The set of anneals in Fig. 9 was performed as a series of consecutive short 5 min. anneals, with optical observation between, perhaps enhancing the opportunity for hydrogen and/or trace metallic impurity incorporation.]

Finally, one more independent check was made as to whether the conclusion of rapid reorientation on the time scale of growth is reasonable. For this, the results of Claybourn and Newman [44] were used, in which a detailed study was performed of the growth of the individual TDD's. (Since growth rates can depend sensitively upon the actual material studied due to differing oxygen concentrations, homogeneity of distribution, the presence of hydrogen, etc., this necessarily serves only as as a rough guide for the rates to be expected in our samples.)

According to their analysis,

$$\frac{d[TDD_{n+1}]}{dt} = k_n[TDD_n] - k_{n+1}[TDD_{n+1}], \tag{8}$$

where $[TDD_n]$ is the concentration of the n-th TDD. We estimated each k_n by evaluating $[TDD_n]$ and $d[TDD_{n+1}]/dt$ from their published figures and extrapolating to $t = 0$, to avoid contributions from the second term on the right. This gave a characteristic capture time for TDD3 to go to TDD4 (i.e., the lifetime of TDD3) of

$$1/k_3 \approx 2 \times 10^{-9} \exp\left(1.8eV/kT\right) sec. \tag{9}$$

The values for $1/k_2$ and $1/k_4$ have the same activation energy but with preexponential factors a factor of $\sim$3 smaller or larger, respectively.

This result is also plotted in Fig. 8, allowing a direct comparison to the reorientation time constant of TDD3, as measured from the NL8 EPR recovery. The fact that this estimated lifetime for TDD3 substantially exceeds its measured reorientation time in the temperature range of the EPR measurements provides additional confirmation that the conclusion of reorientation as the dominant process is consistent with expected TDD growth rates.

4.2. TDD PIEZOSPECTROSCOPIC TENSOR

Having established that each of the individual TDD species can reorient, and that its alignment therefore represents thermal equilibrium under the applied stress at the treatment temperature, one can analyze the alignment to determine its "piezospectroscopic coupling tensor". The NL8 alignment shown in the samples of Fig. 5 gives an average effective C_{2v} strain coupling tensor $B_1 = 1.9$, $B_2 = 10.3$, $B_3 = -12.2$ eV, where the labeling of the principal axes are as shown in Fig.5, and correspond to the convention for the g-value component labeling [33].

The samples used for the above results were heat treated under 600 MPa stress for a total of 90 min. at 460°C, and, as discussed above, primarily reflect the properties of TDD3. A separate sample, heat treated at 460°C for 360 min., was found in the optical studies to be dominated by TDD3 and TDD4, and for it the NL8 EPR alignment was measurably less, giving $B_1 = 1.1$, $B_2 = 8.5$, $B_3 = -9.6$ eV. A third sample, similarly treated for 600 min., was dominated by TDD5 and gave $B_1 = 1.1$, $B_2 = 6.6$, $B_3 = -7.7$ eV. In all cases, the negative sign for B_3 indicates a large net compressive strain

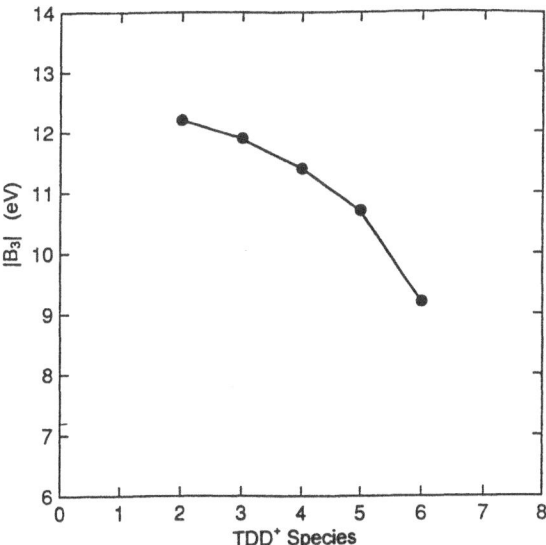

Figure 10. The B_3 coupling coefficient estimated for the individual TDD's from their IR excitation spectra [43].

produced by the TDD in the lattice along its $\langle 100 \rangle$ C_{2v} 3-axis which causes the defect to align with this axis perpendicular to the applied compressional strain. In the competition between the 1- and 2-axes, the 2-axis prefers the applied compressional strain the most.

The decrease in the strength of the coupling as the TDD's progress through their series was also monitored in the optical studies for samples that had been stressed along the $\langle 100 \rangle$ direction during TDD formation. From these studies, B_3 could be estimated and the results are plotted in Fig. 10. We note the close agreement between the value for TDD3 and the corresponding NL8 EPR value for the 90 min. 460°C treated sample, as expected. It is clear here also that the EPR NL8 TDD$^+$ results tend, as expected, to reflect the properties of the higher TDD's than indicated by their relative total concentrations to explain their somewhat larger measured reductions.

4.3. CONCLUSIONS

The various TDD's can reorient readily on the time scale of their lifetime and their alignment therefore reflects the strain fields that they each individually introduce into the lattice. Subtracting out any component of hydrostatic ("breathing") strain, which our experiments cannot reveal, the core produces a large net compressional strain field along its C_{2v} $\langle 100 \rangle$ axis.

16

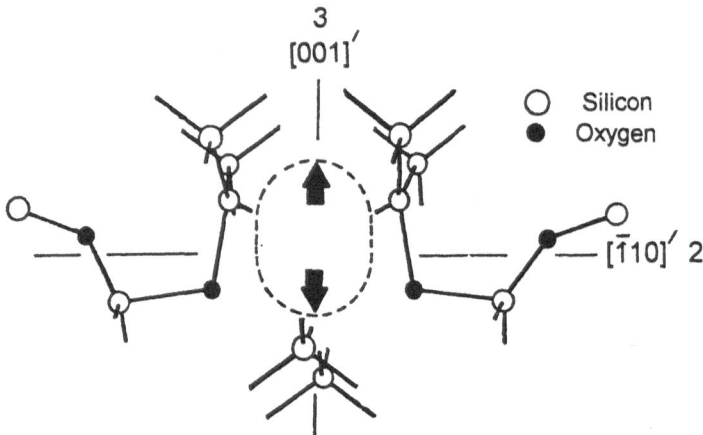

Figure 11. The thermal donor core produces a strong net compression along its [001]′ C_{2v} 3-axis. The model here depicts the relief of the strain by the accumulation of bond-centered oxygens in chains along the perpendicular [$\bar{1}$10]′ 2-axis direction.

There is considerable anisotropy in the net tension perpendicular to this direction, being substantially greater along one of the ⟨110⟩ principal axes (that of g_2 for the NL8 EPR) than that along the other. This is summarized in Fig. 11, where specific (unknown) details of the core have been appropriately omitted. The magnitude of the strains decrease as the TD-D's advance through the series, suggesting strain relief as a driving force for the aggregation of the diffusing species. In the figure, we illustrate how the accumulation of bond-centered interstitial oxygen can serve this purpose [45].

The reorientation kinetics suggest that the process may be occuring by the diffusive motion of interstitial oxygen, but with ~5 jumps required for TDD3 and progressively more for the subsequent ones. According to Fig. 11, this suggests that the limiting process is diffusion and reassembly of the surrounding aggregated oxygen atoms, the core supplying the driving force, but not the major barrier.

Any model for the TDD's will have to address and accomodate these properties.

5. Summary

When combined with one or more of the various spectroscopic techniques, stress-induced defect alignment serves as a powerful auxiliary tool for revealing important additional energetic and kinetic information concerning defects. In the example of isolated interstitial oxygen in silicon, it provided an early confirmation of its atomic configuration and at the same time provided a high precision method of determining both the mechanism and the magnitude of its diffusion in the lattice. In the case of isolated substitutional oxygen in silicon, it provided a unique method for correlating information obtained from its EPR and LVM infrared spectra to come up with a highly detailed structural model. In the case of the interstitial-carbon–interstitial-oxygen pair in this material, it served to confirm its structure as that of an electrically active interstitial carbon atom (from EPR) perturbed by a nearby interstitial oxygen atom (from a correlated luminescence band) and provided information concerning the relative atomic arrangements of the two constituents (from oxygen stress-induced alignment prior to the pair formation).

Finally, we have discussed what has been learned about the NL8 thermal donor system using these techniques: (1) Reorientation of each is limited by what appears to be single diffusive hops of the aggregated interstitial oxygen atoms surrounding the core. (2) Strong dihedral strain is produced in the lattice by the core, which is compressive along a $\langle 100 \rangle$ axis with substantial tensile anisotropy between the two perpendicular $\langle 110 \rangle$ directions. (3) The strain decreases as the TDD's grow, consistent with strain relief as the driving mechanism for interstitial oxygen accumulation in the tensile $\langle 110 \rangle$ directions.

References

1. Kaplyanskii, A.A. (1964) *Opt. Spektrosk.*, **16**, p. 606 [*Opt. Spectrosc. (USSR)*, **16**, p. 329].
2. Corbett, J.W. and Watkins, G.D. (1961) *J. Phys. Chem. Solids* **20**, 319.
3. Corbett, J.W., McDonald, R.S. and Watkins, G.D. (1964) *J. Phys. Chem. Solids* **25**, p. 873.
4. Kaiser, W., Keck, W.P.H. and Lange, C.F. (1956) *Phys. Rev.* **101**, 1264.
5. Bond, W.L. and Kaiser, W. (1960) *J. Phys. Chem. Solids* **16**, 44, and references therein.
6. Hrostowski, H.J. and Kaiser, R.H. (1957) *Phys. Rev.* **107**, 966.
7. Southgate, P.D. (1960) *Proc. Phys. Soc. Lond.* **76**, 385, 398.
8. Haas, C. (1960) *J. Phys. Chem. Solids* **15**, 108.
9. Mikkelsen Jr., J.C. (1982) *Appl. Phys. Lett.* **40**, 336.
10. Lee, S.-T. and Nichols, D. (1985) *Appl. Phys. Lett.* **47**, 1001.
11. Lee, S.-T. and Nichols, D. (1986) in J.C. Nikkelsen, Jr., S.J. Pearton, J.W. Corbett and S.J. Pennycook (eds.) *Oxygen, Carbon, Hydrogen and Nitrogen in Crystalline Silicon*, Mater. Res. Symp. Proc. **59**, Pittsburgh, p. 31.
12. Takano, Y. and Maki, M. (1973) in H.R. Huff and R.R. Burgers (eds.) *Semiconductor Silicon 1973*, Electrochem. Soc., Pennington, p. 469.
13. Gass, J., Muller, H.H., Stussi, H. and Schweitzer, S. (1980) *J. Appl. Phys.* **51**, 2030.
14. Stavola, M., Patel, J.R., Kimerling, L.C. and Freeland, P.E. (1983) *Appl. Phys. Lett.* **42**, 73.
15. Newman, R.C., Tucker, J.H. and Livingston, F.M. (1983) *J. Phys. C: Solid State Phys.* **16**, L151.
16. Newman, R.C. and Jones, R. (1994) in F. Shimura (ed.) *Oxygen in Silicon*, Vol.

18

42 of Willardson, R.K., Beer, A.C. and Weber, E.R. (eds.) *Semiconductors and Semimetals*, Academic Press, San Diego, Chapter 8.

17. Saito, M. and Oshiyama, A. (1988) *Phys. Rev. B* **38**, 10711.
18. Needels, M., Joannopoulos, J.D., Bar-Yam, Y., Pantelides, S.T. and Wolfe, R.H. (1991) in P. D. Bristowe, J. E. Epperson, J. E. Griffith and Z. Lilienthal-Weber (eds.) *Defects in Materials*, Mater. Res. Soc. Symp. Proc. **209**, Pittsburgh, p. 103.
19. Jiang, Z. and Brown, R.A. (1995) *Phys. Rev. Lett.* **74**, 2046.
20. Ramamoorthy, M. and Pantelides, S.T. (1996) *Phys. Rev. Lett.* **76**, 267.
21. Tipping, A.K., Newman, R.C., Newton, D.C. and Tucker, J.H. (1986) *Mater. Sc. Forum* **10-12**,887.
22. Newman, R.C., Tucker, J.H., Brown, A.R. and McQuaid, S.A. (1991) *J. Appl. Phys.* **70**, 3061.
23. Watkins, G.D. and Corbett, J.W. (1961) *Phys. Rev.* **121**, 1001.
24. J. W. Corbett, G. D. Watkins, R. M. Chrenko and R. S. McDonald (1961) *Phys. Rev.* **121**, 1015.
25. Bean, A.R. and Newman, R.C. (1971) *Sol. St. Commun.* **9**, 271.
26. In reference [23], the results were expressed in terms of three parameters, M, M^*, and N. We have converted these to the corresponding components of **B** by $B_1 = N - Tr\mathbf{B}$, $B_3 = -Tr\mathbf{B}$, and $B_2 = M - Tr\mathbf{B}$ for the neutral state, and $B_2 = M + M^* - Tr\mathbf{B}$ for the negative state.
27. Trombetta, J.M. and Watkins, G.D. (1987) *Appl. Phys. Lett.* **51**, 1103.
28. Trombetta, J.M. and Watkins, G.D. (1988) in M. Stavola, S.J. Pearton and G. Davies (eds.), *Defects in Electronic Materials*, Mater. Res. Soc. Symp. Proc. **104**, Pittsburgh, p. 93.
29. Watkins, G.D. and Brower, K.L. (1976) *Phys. Rev. Lett.* **36**, 1329.
30. Thonke, K., Watkins, G., and Sauer, R. (1984) *Sol. State Commun.* **51**, 127.
31. Davies, G., Lightowlers, E.C., Wooley, R., Newman, R. and Oates (1984) *J. Phys. C* **17**, L499.
32. Jones, R. and Öberg, S. (1992) *Phys. Rev. Lett.* **68**, 86.
33. Müller, S.N., Sprenger,M., Sieverts, E.G. and Ammerlaan, C.A.J. (1978) *Sol. St. Commun.* **25**, 987.
34. Wagner, P. and Hage, J. (1989) *Appl. Phys. A* **49**, 123.
35. Gregorkiewicz, T., van Wezep, D.A., Bekman, H.H.P.Th. and Ammerlaan, C.A.J. (1987) *Phys. Rev. B* **35**, 3810.
36. Navarro, H., Griffin, J., Weber, J. and Genzel, L. (1986) *Solid State Commun.* **58**, 151.
37. Hartung, J. and Weber, J. (1993) *Phys. Rev. B* **48**, 14161.
38. Götz, W., Pensl, G. and Zulehner, W. (1992) *Phys. Rev. B* **46**, 4312.
39. Lee, K.M., Trombetta, J.M., and Watkins, G.D. (1985) in N.M. Johnson, S.G. Bishop and G.D.Watkins (eds.), *Microscopic Identification of Electronic Defects in Semiconductors*, Mater. Res. Soc. Symp. Proc. **46**, Pittsburgh, p. 263.
40. Wagner, P., Gottschalk, H., Trombetta, J. and Watkins, G.D. (1987) *J. Appl. Phys.* **61**, 346.
41. Wagner, P., Gottschalk, H., Trombetta, J. and Watkins, G.D. (1986) *Mater. Sci. Forum* **10-12**, 961.
42. Wagner, P., Hage, J., Trombetta, J.M. and Watkins, G.D. (1992) *Mater. Sci. Forum* **83-87**, 401.
43. Trombetta, J., Watkins, G.D., Hage, J. and Wagner, P. 1996, to be published.
44. Claybourn, M. and Newman, R.C. (1987) *Appl. Phys. Lett.* **51**, 2197.
45. Ourmazd, A., Schröter, W. and Bourret, A. (1984) *J. Appl. Phys.* **56**, 1670.

THE INITIAL STAGES OF OXYGEN AGGREGATION IN SILICON: DIMERS, HYDROGEN AND SELF-INTERSTITIALS

R. C. NEWMAN

IRC Semiconductor Materials, The Blackett Laboratory, Imperial College of Science, Technology and Medicine, London SW7 2BZ, UK.

Abstract

Oxygen precipitation in CZ Si at T > 500°C is reviewed briefly. Oxygen clustering at T ≤ 500°C is then discussed in terms of dimer formation, dissociation effects and the apparent requirement for rapid dimer diffusion. Interactions of O_i atoms with vacancies and self-interstitials (I-atoms) are re-examined and it is shown that the presence of hydrogen leads to enhanced O_i diffusion that is the rate limiting step in the aggregation process. Self-interstitial generation is analysed and it is concluded that TD-defects cannot be O_2I_n clusters with a large number of I-atoms. It has been proposed that TD defects, partially passivated with H (or D) atoms, give rise to the STD and NL10 spectra. Different spectra are generated in Al-doped Si.

1. Introduction

Silicon crystals grown by the Czochralski (CZ) method contain oxygen impurities at a concentration $[O_i]$ close to 10^{18} cm^{-3}. The isolated atoms occupy off-axis bond-centred sites and are electrically neutral. They give rise to an infrared (IR) absorption band at 9 μm (full width at half maximum $\Delta \sim 34$ cm^{-1} at 300 K) and $[O_i]$ is determined from measurements of the peak absorption coefficient (α_{MAX}) of the band together with established calibrations : using the calibration [1] for which $[O_i] = 3.14 \times \alpha_{MAX} \times 10^{17}$ cm^{-3}, we can alternatively write $[O_i] = 0.94 \times 10^{16} \times IA$, where IA is the integrated absorption coefficient of the band.

When CZ material is annealed (T > 300°C), isolated O_i atoms diffuse and form clusters since the grown-in $[O_i]$ constitutes a supersaturated solution. There are two complementary methods of following the precipitation process: firstly, the loss of O_i from solution may be determined at any stage from the reduction in the strength of the 9 μm band although measurements at 4.2 K may be necessary to separate this absorption (Δ is then only ~ 0.6 cm^{-1}) from the broad absorption of precipitated SiO_2 particles and secondly, direct observations may be made of the size and number density of the

R. Jones (ed.), Early Stages of Oxygen Precipitation in Silicon, 19–39.
© 1996 *Kluwer Academic Publishers.*

TABLE 1. Oxygen precipitation at 750°C

Time (h)	Particle Size (Å)		Number Density (cm^{-3})	
	SANS	TEM	SANS	TEM
48	56	49	1.4 x 10^{13}	1.1 - 1.6 x 10^{13}
96	93	82	0.9	1.0 - 2.0
431	123	106	0.48	0.5 - 1.0

oxygen aggregates. For T ≥ 650°C precipitates have been measured by transmission electron microscopy (TEM) and correlations have been established with small angle neutron scattering (SANS) [2], as well as defect etching. The rate of growth of precipitates once nucleated is then controlled by O_i diffusion with the "normal" value of $D_{oxy} = 0.13 \exp(-2.53 \text{ eV}/kT)$ cm^2 s^{-1} [3] determined by measurements of internal friction, the relaxation of stress-induced dichroism and measurements of diffused profiles [4].

To accommodate the local volume change around growing SiO_2 particles [5], there is formation of self-interstitials (I-atoms) [6] that are detected by TEM as stacking faults or punched-out dislocation loops but they do not give rise to detectable SANS since there is no change in the local average nuclear scattering factor. An important result is that extended anneals lead to reductions in the number density N of particles nucleated in the early stages of the process due to the redissolution of smaller SiO_2 particles and a concomitant increase in the size of the larger particles [2] (Table 1). This Ostwald ripening shows that samples are in a state of dynamic quasi-equilibrium and demonstrates the need to consider dissociation of small aggregates in any modelling. Models proposed for the precipitation process without prior assumptions about the value of D_{oxy} can be verified only if simultaneous measurements are made of the oxygen loss and the number density of SiO_2 particles (the combined measurements yield the average particle size, Fig. 1), while I_n clusters must be clearly distinguished from O_n clusters.

To study the initial stages of oxygen clustering, it is necessary to make measurements at low T so that small O_2, O_3, etc clusters form on a timescale compatible with that required to carry out a controlled anneal. There are, however, basic problems relating to anneals at T < 650°C since definitive identifications of the aggregates that form are unclear, although the loss of O_i atoms can still be determined from careful measurements using a double beam grating spectrometer (e.g. a Perkin Elmer, PE983 instrument) [7]. This lower temperature range has to be split into two parts. In the first region (650 > T > 500°C), TEM measurements have shown the formation of relatively large ribbon-like defects (RLDs) that were originally attributed to coesite, a high pressure phase of SiO_2, leading to a requirement that D_{oxy} be enhanced by a factor of ~ 10^4 although the enhancement mechanism was not identified [8]. For T ≤ 500°C, there is sequential formation, with overlapping, of a family of up to sixteen double thermal donors TD(N) (1 ≤ N ≤ 16) [9] and traditionally these defects

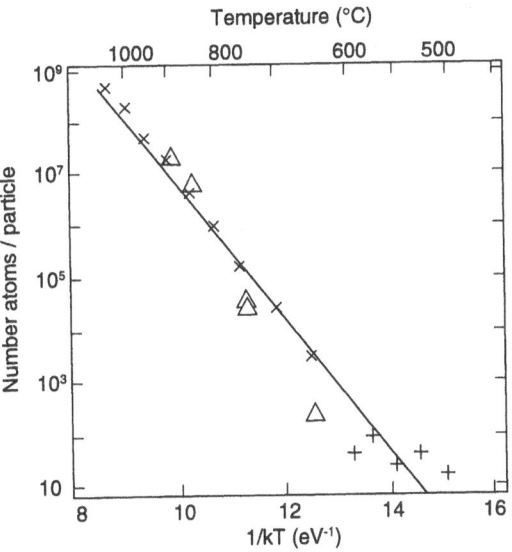

Figure 1. Number of oxygen atoms per precipitate particle (after extended anneals) determined from IR measurements of O_i loss, together with SANS and TEM data for $T \geq 650°C$: for $T \leq 600°C$, O_i, loss measurements were made to find the particle sizes assuming that D_{oxy} had normal values (see Reference [4] and Section 2).

have also been attributed to O_i clusters in which case there would have to be incorporation of more than 16 atoms in TD16. Enhanced O_i diffusion would again have to be invoked to sustain this interpretation if isolated O_i atoms were the only mobile species and an enhancement is required even if clusters contain only 4 O_i atoms, as thought in early work [10].

It is now apparent that the presence of hydrogen in a Si crystal does lead to enhanced values of D_{oxy} [11] and the consequences of this will be discussed in Section 4.3. Nevertheless, hydrogen is not normally present at a high concentration in as-grown material. Other mechanisms that might lead to enhancements of D_{oxy} have also been studied, involving interactions of O_i atoms with I-atoms [12] and lattice vacancies, (V). In addition it has been proposed that O_2 dimers diffuse much more rapidly than isolated O_i atoms (by a factor of $\sim 10^7$) at low temperatures [13]. There are further uncertainties since it is necessary to establish whether or not I-atoms (and/or vacancies) are generated during anneals at $T < 500°C$ and, if so, there is a further requirement to determine the minimum size of an $(O_i)_n$ cluster that would generate these defects. It is also essential to determine the atomic composition of RLDs and TDs, to discover whether they are oxygen clusters, I-atom clusters or combinations of the two species and to identify the way in which vacancies are accommodated when there is I-atom emission from a very small cluster.

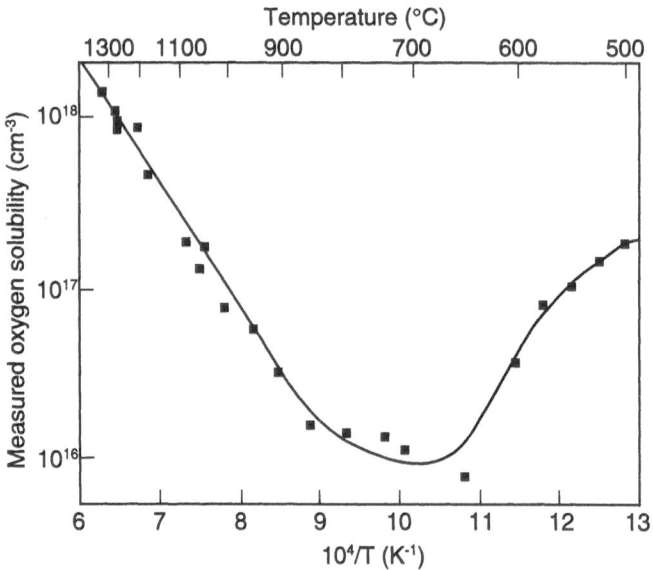

Figure 2. Values of the solubility of O_i atoms in silicon determined from IR measurements of the 9 μm band at 4.2 K, following extended sample anneals. The increase in the quasi-equilibrium values for $T \leq$ 700°C is attributed to the small precipitate size [14].

2. Heat treatments in the range 500 < T < 650°C.

We have argued previously from measurements of the temperature dependence of oxygen loss that there is a change from macroscopic to microscopic (involving relatively few oxygen atoms) precipitation in the temperature range 500 < T < 650°C and that "black dot" contrast observed by TEM should be identified with very small oxygen precipitate particles [14]. This led to a re-appraisal of the origin of the TEM contrast from the much larger RLDs with the conclusion that they were not due primarily to oxygen aggregates (coesite) [15]. The most recent high resolution TEM data support this view and the defects have been re-assigned to silicon structures nucleated by the incorporation of I-atoms [16,17]. This interpretation is then consistent with the observations of similar structures in float zone (FZ) silicon (essentially oxygen-free) following high energy electron irradiation. The change also explains the previous lack of detection of SANS from CZ Si heated in this temperature range. Weak SANS has however now been observed (using the latest ILL (Grenoble) neutron beam facilities) from large Si samples annealed at 600°C and 550°C and it is concluded that very small oxygen clusters are indeed present [18]. At 500°C we expect the clusters to contain only some 10-20 O_i atoms (Fig. 1).

Dissociation of SiO_2 precipitates formed at higher temperatures (T = 750°C) had already been demonstrated and this process would also be expected for much smaller particles. In that case, they would have to be in dynamic equilibrium with the O_i atoms

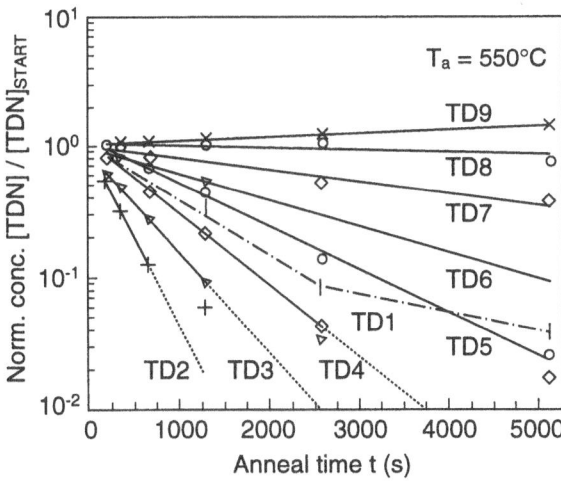

Figure 3. The destruction of TD(N) defects by dissociation during anneals at 550°C, showing the loss of defects with small N at shorter times than those with larger N (for TD(1), see text). The measurements indicated an activation energy ~ 2.5 eV, the same as that for D_{oxy} [19].

remaining in solution leading to a concentration $[O_i]_s$ much greater than the true solubility because of the small radius of curvature of the particles and consequently a high interface energy. Data, again derived from the kinetics of O_i loss, support this view (Fig. 2) since $[O_i]_s$ first decreases with decreasing T, corresponding to a heat of solution of 1.4 eV, passes through a broad minimum and then increases as T is lowered further to 500°C [14]. It is implied that dissociation of very small particles is of paramount importance and should not be ignored in modelling of the initial stages of O_i aggregation.

If a sample first annealed at 450°C to produce TD(N) defects (with a range of N up to 8) is subsequently heated for short periods at 550°C, the donors are destroyed progressively [19] so that, as expected, the smaller clusters (small N) are lost before the larger ones (Fig. 3), except for TD1 which has a longer lifetime possibly related to its known bistability [20]. These observations do not resolve the question of whether TD(N) defects should be identified with clusters predominantly of oxygen atoms or predominantly of I-atoms [21]. However, extrapolation of the data shown in Fig. 1 to lower temperatures would imply that O_2 dimers would be the predominant defects formed initially for T < 450°C unless D_{oxy} were enhanced or O_2 dimers diffused much more rapidly than O_i atoms.

3. Oxygen-loss resulting from anneals at T < 500°C

In a crystal containing only dispersed O_i atoms the first stage of oxygen aggregation has to be the formation of O_2 dimers [22]. This is a second order process described by the

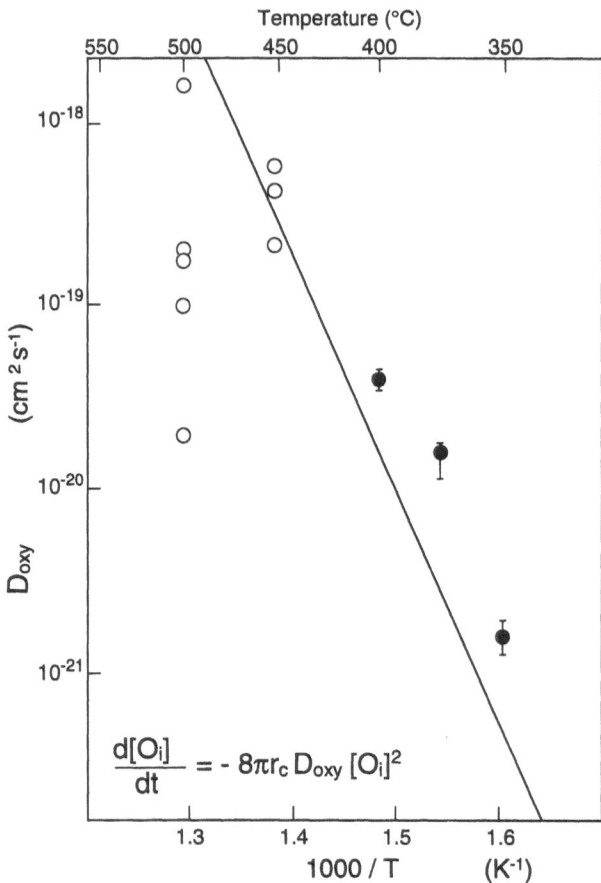

Figure 4. Values of the diffusion coefficient of O_i atoms determined from the rates of O_i loss from solution and deduced from second order kinetics for CZ crystals with various values of $[O_i]_0$. D_{oxy} is independent of $[O_i]_0$ for $T \leq 400°C$ but larger than the "normal" values (the solid line). For $T \geq 450°C$ the apparent values of D_{oxy} decrease with decreasing values of $[O_i]_0$ and fall below the"normal" values [7].

equation $d[O_i]/dt = -8\pi r_c D_{oxy}[O_i]^2$ so that a plot of the reciprocal of $[O_i]_t$ versus time (t) should yield a straight line with a slope equal to $8\pi r_c D_{oxy}$, where r_c is the separation at which the two atoms bind together (the capture radius). Extensive data (Fig. 4) for oxygen loss show that the choice of $r_c = 5$ Å leads to values of D_{oxy} close to the "normal" values for $T \sim 420°C$ [4,7]. At higher temperatures ($500°C$), too small a value is always determined but this is expected since a fraction of the dimers once formed would later dissociate. The dissociation rate depends only on the properties of a small cluster and would be a constant at a fixed T but the rate of formation of dimers

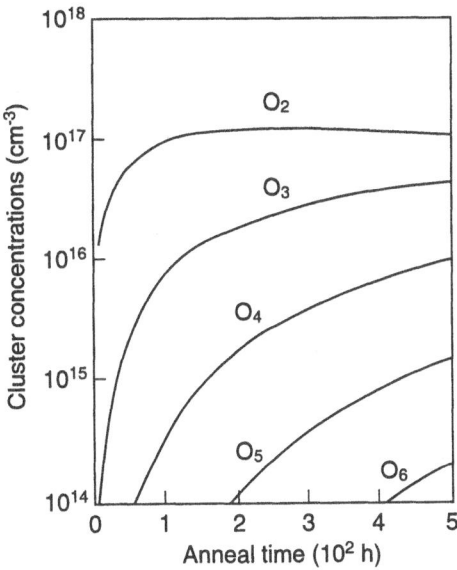

Figure 5. The serial growth of oxygen clusters O2, O3, etc. as a function of annealing time for a simulated anneal at 450°C with a normal value of D_{oxy}, assuming O2 dimers are immobile [23].

will decrease as the grown-in oxygen concentration decreases so that larger discrepancies are then expected. The real situation, as discussed later, is more complicated. For T < 400°C, the values of D_{oxy} are too large by a factor of between 3 and 10 indicating that an enhancement does occur : possible reasons for this will be discussed in Section 4.

More detailed calculations, allowing for the formation of O_3, O_4 etc clusters at 450°C with a normal value of D_{oxy} show that $[O_2]$ reaches ~ 10^{17} cm^{-3} after 200 h, $[O_3]$ ~ 10^{16} cm^{-3}, with reductions of about a decade for each additional O atom that is added to the cluster (Fig. 5) [23]. Thus, it would be impossible for clusters of 10 O_i atoms to form (to produce say TD9) unless D_{oxy} were greatly enhanced (by a factor much greater than 10^4), but then the calculated rate of loss of O_i from solution would be correspondingly greater than that measured. We therefore have to reject all models of TD-formation involving large oxygen clusters that invoke such large enhancements of D_{oxy} at low temperatures but still have to explain the much smaller enhancements found for anneals at 350 - 400°C. Finally, it is important to note that $d[O_i]/dt$ assumed to vary as $[O_i]^n$ leads to values of the exponent n ranging from 9 at 500°C to 2 at T ≤ 400°C [7] : there is, however, no discrepancy with early data since a value of n close to 4 is found at 450°C [10]. These results suggest that dimers are stable (do not dissociate) for T ≤ 400°C and that clusters of about 10 O_i atoms are the smallest stable species at T = 500°C. The latter result clearly links to the data shown in Fig. 1, but the indicated high stability of the dimers has to be explained.

4. Enhancements of O_i diffusion

4.1. OXYGEN VACANCY PAIRS

Possible mechanisms for the enhancement of D_{oxy} must involve interactions with an intrinsic defect (V or I-atom) or a second impurity atom. It is well known that there is formation of O-V pairs (A-centres) during high energy irradiation of Si at T ~ 300 K that leads to the displacement of Si atoms from their lattice sites. The resulting vacancies are mobile at 300 K and are trapped by O_i atoms. Reorientation of the A-centre bonding occurs at low temperatures [24] but this process does not constitute a diffusion jump that requires dissociation of the defect followed by renewed pairing. Evidence apparently in support of the dissociation process was obtained by annealing irradiated samples at T ≥ 300°C [25]. There was a loss of A-centres determined from reductions in the strength of the associated IR localized vibrational mode (LVM) at 836 cm^{-1} (77 K) and an increase in the strength of the 9 μm (O_i) band. However, more detailed measurements showed that the latter changes ceased at some stage in the anneal and there was instead formation of the 889cm^{-1} (300 K) oxygen complex attributed to O_2V centres (Fig. 6) [4]. It was therefore proposed that I-atom clusters dissociated in the early stages of the anneal and that mobile I-atoms were captured by A-centres to regenerate isolated O_i atoms: I atom release would also inhibit the formation of O_2V defects It was inferred that A-centres do not dissociate so that the presence of vacancies would not lead directly to enhancements of D_{oxy}. In the course of this work [25] it was deduced that the calibration factor to convert the integrated absorption of the IR A-centre band to a concentration is the same as that for the 9 μm (O_i) band, both for [O-V]° (mode at 836 cm^{-1}, 77K) and [O-V]$^-$ (mode at 884 cm^{-1}, 77K) [26].

4.2. OXYGEN I-ATOM PAIRS

There is only limited experimental information about O_i-I defects, although *ab initio* theory predicts that they act as donors and exhibit bistability with negative-U behaviour [27]. O_i-I and $2O_i$-I defects are proposed structures for the cores of TD1 and TD2 defects respectively, and ENDOR measurements support the view that an I-atom is present in the core [28]. Secondly, as stated above, it has been proposed [12] that the presence of I-atoms leads to enhancements of D_{oxy} (see also Ref. 27) (Fig. 7).

Available IR LVM data are due to Brelot [29, 30] who studied the effects of low temperature (120 K) electron irradiation (up to 3 MeV) on CZ Si doped with Ge (up to 3×10^{20} cm^{-3}) or Sn (up to 10^{19} cm^{-3}). Vacancies are trapped by these isoelectronic impurities while I-atoms were inferred to be trapped by O_i atoms, leading to defects that give LVMs at 935 (B), 944 (B^1) and 956 (B^{11}) cm^{-1} : these lines are not produced in FZ Si. In samples containing a high carbon concentration the strengths of the B lines are reduced due to the competition for trapping of diffusing I-atoms. The relative strengths of the B lines changed when the samples were illuminated at low temperature, indicating changes in the charge state of the defect responsible or possible bistability. On heating samples to 230 K there was dissociation of Ge-V pairs but the strengths of the lines B, B^1 and B^{11} did not change, although A-centres were produced as vacancies

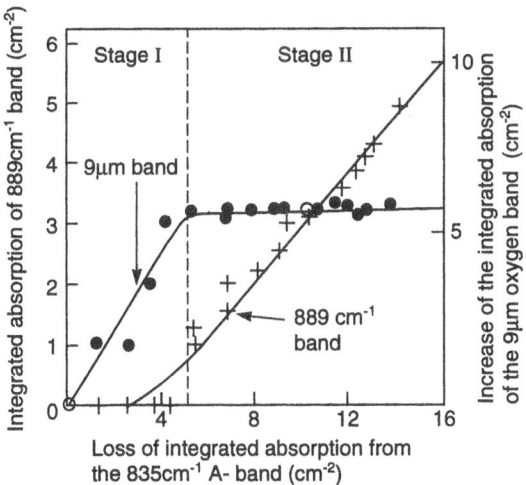

Figure 6. The loss in the integrated absorption coefficient (IA) of the 835 cm^{-1} IR line from O_i-V pairs (A-centres) during an isothermal anneal of a sample at 300°C following 2 MeV electron irradiation at room temperature to a dose of 6 x 10^{18} cm^{-3}. In stage I, the lost absorption reappears as an increase in the IR absorption of the 9μm O_i band. At a later stage (II), this process terminates and there is correlated growth of the 889 cm^{-1} (O_2-V) centre [4].

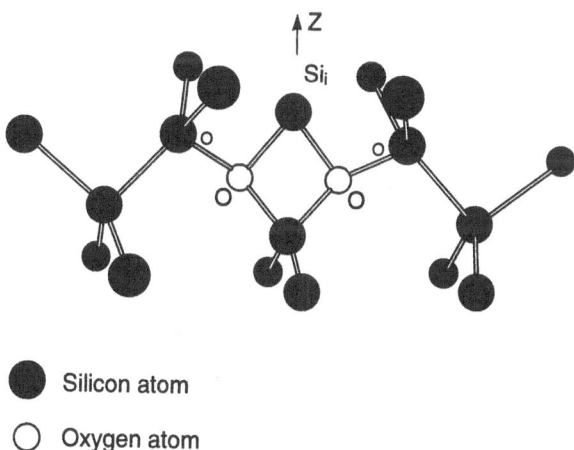

● Silicon atom

○ Oxygen atom

Figure 7. O_2-I model for the core of a double thermal donor [27].

were captured by O_i atoms. The B defects anneal at ~150°C. Finally, correlations of the strength (IA) of the 9 μm band with the B band (935 cm^{-1}), established from either irradiation or annealing treatments, led to the conclusion that the calibration factors were the same for the two centres [30].

These results link to recent electron paramagnetic resonance (EPR) studies of the AA13 and AA14 defects produced in proton irradiated Si . The measurements were sensitive to illumination and it has been proposed that a bistable O_i-I complex is responsible [31].

The conclusions may be summarised as follows : (a) O_i-I pairs do not trap mobile vacancies released from Ge-V pairs; (b) I-atoms released from O_i-I pairs are however trapped by O_i-V pairs so that an oxygen diffusion jump can occur; (c) the low concentrations of O_i-I defects produced in undoped Si (no Ge or Sn) must be a result of A-centre formation followed by the rapid trapping of an I-atom, (d) it is possible that transient trapping of I atoms with O_i at intermediate temperatures $150 \leq T \leq 500°C$ followed by dissociation may lead to enhancements in D_{oxy}, providing an explanation for the enhancements revealed in Fig. 4 for T ~ 350°C; (e) the O_i-I pair cannot be identified with TD1 because of its lack of stability. These conclusions are entirely consistent with those deduced from a study that demonstrated enhancements of D_{oxy} by the relaxation of stress-induced dichroism in undoped Si induced by 2 MeV electron irradiation (T < 300°C) and the inhibition of this enhancement in silicon containing tin impurities in a concentration of $10^{19} cm^{-3}$ [32]. It would be valuable to have corresponding experimental data for O_2I interactions to compare with theory [27].

4.3. ENHANCEMENTS OF D_{oxy} DUE TO HYDROGEN.

Hydrogen may be introduced into CZ Si by heating it to a high temperature (T > 600°C) in H_2 gas and then quenching the sample to room temperature, or by exposing samples to a hydrogen plasma with T < 500°C [4]. Values of D_{oxy} determined from the relaxation of stress-induced dichroism in material quenched from a temperature in the range $800 \leq T \leq 1300°C$ are enhanced and we have D_{oxy} (enh) = 3.2 x 10^{-4} exp (-1.96 eV/kT) cm^2 s^{-1} [7] (Fig. 8). There are corresponding enhancements in the rates of O_i loss, TD-formation and other reactions involving carbon atoms by factors of 5, 30 and 200 at 450, 400 and 350°C respectively (see Ref. 4). The concentration of rapidly diffusing atomic hydrogen present during these anneals is not known but it must be less than the solubility, given by $[H]_s = 9.1$ x 10^{21} exp (-1.8 eV/kT) cm^{-3}, leading to $[H]_s <$ 10^{14} cm^{-3} at 900°C. The presence of a small concentration of grown-in hydrogen could account for the relatively small enhancements of D_{oxy} in as-received material annealed at the lowest temperatures indicated in Fig. 4. The enhancement mechanism must involve a catalytic process so that the hydrogen is not consumed significantly during the formation of oxygen clusters. Two *ab initio* models have been proposed to explain these observations. In the first model [33] it was argued that if a H-atom occupies a bond-centred site in a bond that has a common Si apex with a bond containing an O_i atom, the diffusion barrier of an O_i diffusion jump is lowered. In the second model [34] an enhancement was predicted when the H-atom occupied an anti-bonding site in line

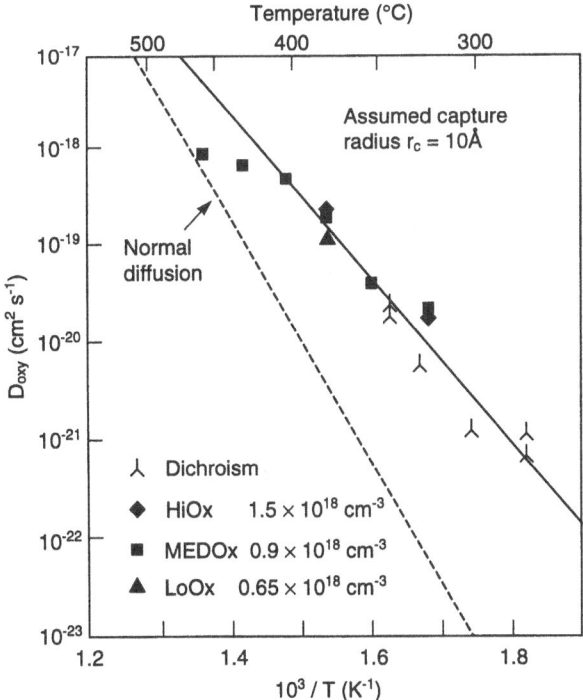

Figure 8. Arrhenius plot of D_{oxy} (enh), determined from a model of stable dimer formation with a capture radius of 10 Å for samples pre-heated at 1300°C and quenched in H_2 gas. There is a clear linkage to values of D_{oxy} (enh) determined from the relaxation of stress-induced dichroism (solid line) [4].

with the Si_i-O_i-Si_i bond. It is possible that a similar lowering of the barrier could occur when some atom other than hydrogen is involved; this could be a second O_i atom.

Recently an IR absorption line at 1075.1 cm^{-1} has been observed in CZ (but not FZ) Si that had been heated in H_2 gas and quenched to room temperature [35]. The frequency of the line increased to 1076.6 cm^{-1} when the pre-treatment was alternatively in D_2 gas rather than H_2. This shift demonstrates that the defect responsible incorporates a H atom and oxygen is also likely to be involved. It was therefore proposed that the IR line is an LVM corresponding to a perturbed oxygen mode. However, an increase in the frequency resulting from an increase in the mass of an atom could then occur only if there is strong anharmonicity associated with the H-mode (not yet detected).

In earlier work [4], it was implied that the presence of Cu or Fe also led to enhancements of D_{oxy}. However, these rapidly diffusing impurities that occupy interstitial sites were introduced into Si from hydrated salts (nitrates) in quartz ampoules that were heated and then quenched. There would certainly have been introduction of hydrogen into the samples and so effects due to the presence of these fast diffusing interstitial metals still have to be determined. There is a complication since there is a possibility that the presence of transition metal atoms could catalyse the dissociation of H_2 molecules or larger clusters to generate more hydrogen atoms than would otherwise be present; such effects have been proposed previously for hydrogen in Ge crystals [36].

4.4. RAPIDLY DIFFUSING OXYGEN DIMERS.

If TD(N) centres are to be identified with large oxygen clusters, there is a need for a rapidly diffusing oxygen-related species to be generated as aggregation proceeds. It does not appear feasible that an O_3, or a larger cluster, could be mobile and hence the defect responsible would have to involve a dimer that is formed at a rate consistent with essentially "normal" values of D_{oxy}. It follows that O_2 formation would be the rate-limiting step in the overall process. Indeed, it was proposed some years ago that molecular-like O_2 dimers did diffuse more rapidly than isolated O_i atoms [13] in order to explain the formation of O_4 clusters at a time when TD-defects were still thought to have only four O_i atoms in their cores because at 450°C measurements of $d[O_i]/dt$ were proportional to the fourth power of the grown-in $[O_i]$ concentration [10]. This latter result was then explained self-consistently by two dimers coming together to form an O_4 complex but the role of O_3 defects was not clear and dissociation of small clusters was ignored. Subsequently, models of dimers (adjacent bond-centred O_i atoms) have been examined using *ab initio* theory but calculated binding energies range from 0.1 eV [37] to 1.0 eV [38] : a clarification of this energy is clearly needed. It has also been reported that the migration energy of O_2 is only 1.4 eV [37] (see also Ref. 27) and it was argued that one O_i atom effectively pulled the second atom through the diffusion barrier. Superficially, this process would appear to be similar to that proposed for the enhancement of oxygen diffusion due to interactions with H-atoms. Rapid diffusion of dimers has also been implied in reference 39. Finally, we comment that it is unknown whether or not the formation of a dimer would lead to the ejection of an I-atom to produce an O_2V centre, although theory implies that this is an endothermic process. On the other hand, a binding energy of 1.1eV has been calculated for the O_2I complex that is predicted to be a negative-U centre [27].

Irrespective of any uncertainties, modelling of the growth of oxygen clusters has been carried out with the assumption that $D(O_2)$ is much greater than D_{oxy} [7]. Because the ratio $D(O_2)/D_{oxy}$ may be as large as 10^7 [13], diffusing dimers would be trapped rapidly by remaining O_i atoms to form O_3 complexes so that after a short transient the concentration $[O_2]$ becomes essentially constant at a low value. The consequence is that dissociation of dimers is unimportant at low temperatures (T < 450°C) since they exist for only a very short time. However, the trimers are immobile and they would be expected rapidly to re-generate O_2 and O_i during the much longer period of the anneal. The latter process can therefore lead to an effective diffusion jump of the O_i atom and this enhancement might explain the low T effects shown in Fig. 4, as an alternative explanation to postulating an enhancement in D_{oxy} due to H-O_i or O_i-I interactions. Clearly, two dimers could also interact to form O_4 and if this cluster dissociates, again with emission of dimers, even larger clusters can be formed (e.g. $O_2 + O_4 \rightarrow O_6$), etc until the clusters reach a size when they are stable. This whole process could therefore account for large cluster formation controlled by the rate of dimer formation. A previous objection to the process of converting TD1 to TD2 by the capture of a fast diffusion species identified with dimers may not be valid even though measurements indicated that the rate was proportional to the grown-in $[O_i]$ concentration and not on the square of this quantity [40]. Because a pseudo-equilibrium of $[O_2]$ is formed, modelling shows that the concentration of dimers may be proportional to $[O_i]$ or even a lower power of $[O_i]$ rather than $[O_i]^2$, provided that the dissociation rate of O_2 is small [7].

At higher T > 400°C, O_2 dissociation would be expected on a short time scale and the order of the kinetics then tends to $[O_i]^n$ where n is the number of atoms in the smallest stable cluster. Further details are presented in a separate paper [41] but we note that the "third order kinetics" (D_{oxy} apparently proportional to $[O_i]$ measured for T = 450°C) can be explained [23].

So far, emission of I-atoms has been ignored. The maximum concentration that could be produced could not exceed $[O_2]$ (Fig. 5) that reaches $\sim 10^{17}$ cm^{-3}, (without rapid dimer diffusion), according to modelling for an anneal at 450°C for 200 h. At this point, we note that extensive measurements show that the total concentration of donor centres $\Sigma TD(N)$ is always $\sim$ 10 times smaller than the concentration of $[O_i]$ lost, irrespective of whether or not there is enhanced O_i diffusion due to hydrogen, the anneal temperature (T $\leq$ 500°C) or the anneal time up to the stage when donors start to be destroyed. This ratio implies that on average only five I atoms could be generated from 5 dimers that would be converted to O_2V during the period when there is formation of one donor. If, however, an O_3 complex is the smallest cluster that could lead to I-atom ejection, there would be only 3 I-atoms available per donor. It therefore seems impossible to envisage a process that could lead to the formation of O_2I_n defects with a large value of n that might be identified with the family of thermal donor defects contrary to our previous proposal [21,22]. If, in addition, the core of a TD-centre has a structure with a bonded Si, as proposed by Deák et al [27] (Fig. 7) not all the dimers (or trimers) will contribute to I-atom formation. We could then reasonably write $[(O_3V + I) + O_2] \rightarrow O_3V + O_2I$, so that 5 O_i atoms are consumed per donor, but there is still a discrepancy of a factor of two with the measurements.

5. Further experimental data.

It is necessary to seek further evidence that O_2 dimers diffuse rapidly and that there is I-atom generation during anneals of CZ Si at T $\leq$ 500°C. In early work, it was demonstrated that the presence of carbon in CZ Si inhibited the formation of large concentrations of TD(N) centres during anneals at 430°C [42]. Recent measurements of $\Delta[O_i]/\Delta\Sigma TD(N)$ have then shown much increased values in such material, as would be expected, because there was no evidence for enhanced (or retarded) O_i loss. It has also been shown that during anneals up to 800 h at 450°C there were monotonic decreases in $[O_i]$ and substitutional carbon $[C_s]$ (measured from the IA of the 16.5 μm LVM) with a ratio close to 2:1 when the grown-in value of $[C_s]$ was close to 7 x 10^{17}

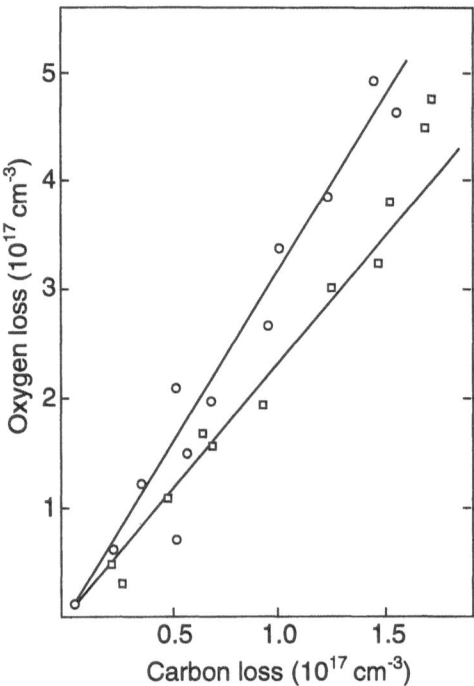

Figure 9. The correlated loss of O_i and C_s atoms from solution in Si heated at 450°C : □, sample with $[C_s] = 7 \times 10^{17}$ cm^{-3} and $[O_i] = 10^{18}$ cm^{-3} ; O, sample with $[C_s] = 4 \times 10^{17}$cm^{-3} and $[O_i] = 10^{18}$ cm^{-3}. Note $\Delta[O_i]/\Delta[C_s]$ is ~ 2 for the former sample [43].

cm^{-3} ($[O_i]$ was ~ 9.5 x 10^{17} cm^{-3}) (Fig. 9) [43]. It is important to recall that there was no significant growth of O_i-C_s pairs but other IR lines were detected and attributed to $(O_i)_2C_i$ complexes. A current version of the original interpretation is that formation of an O_2 cluster (the kinetic model described in these early papers is incorrect) led to the emission of an I-atom : the I atom diffused and was trapped by a C_s atom and following an exchange of sites the mobile C_i was trapped by another O_i pair that had not ejected an I-atom to produce an $(O_i)_2C_i$ complex. This process would correspond to the formation of O_iC_i complexes (C(3)-centre) during 2 MeV electron irradiation. For the annealed CZ samples we should therefore expect 4 O_i atoms to be lost per C_s atom lost, contrary to the observations. On the other hand, if an O_2 dimer diffused rapidly to a C_s atom the loss ratio would be only 2:1. The carbon would not be a true interstitial, but that might not be inconsistent with the unusual structure of the related C_iO_i pair [44]. Photoluminescence measurements led to a proposal from isotopic shifts ^{16}O, ^{18}O, ^{12}C, ^{13}C that the P-centre produced by the anneals is a $(O_i)_2C$ complex [45]. If I-atoms are consumed in this way, they would not be available to form core structures of TDs and the low rates of formation of these defects would be explained.

Recently, these ideas have been questioned as a result of further measurements that imply that TD centres are still formed at the usual rate when carbon is present but that their rate of destruction is much enhanced. One set of electrical measurements relates to the formation of low concentrations of the bistable TD1 and TD2 defects and it was proposed that I-atoms dissociate from these defects due to strain effects resulting from the presence of the carbon atoms at remote lattice sites [46]. The second set refers to IR measurements of the TD centres with the interpretation that the I-atom originally present is displaced at a later stage of the anneal by the capture of a C_i atom [47]. If carbon complexes with associated deep electron traps were formed there would also be a reduction in the donor activity, even if these defects were at remote sites but IR transitions from $TD(N)^+$ would be expected if the samples were n-type.

It is also known from irradiation studies of Si that substitutional boron and aluminium impurities are displaced from their lattice sites by the capture of an I-atom [48]. Samples of silicon doped with $[O_i] = 7.5 \times 10^{17}$ cm^{-3} and $[B] = 7 \times 10^{16}$ cm^{-3} were therefore annealed at 450°C for periods up to 2700 h [49]. The rate of TD-formation was deduced from resistivity measurements that gave $\Delta|N_A-N_D|$, while the boron concentration was monitored from the associated changes in the strength of its LVM and electronic absorption that could be measured after anneals for $t \geq 150$ h. The TD-formation was the same as that found in undoped samples but at the same time there was a loss of B atoms from substitutional sites that was almost exponential with time. There was no evidence that boron was incorporated in the donor defects and the initial value of $\Delta[O_i]/\Delta[B_s]$ was ~ 16, indicating that O_i - B_s pairs may have formed (a new IR LVM appeared at 1146 cm^{-1}).

Results for a sample containing Al in a similar concentration of 8×10^{16} cm^{-3} were remarkably different. There was very rapid donor formation and the sample became n-type but not due to the presence of the usual TDs [49]. Instead, the IR electronic absorption was concentrated in the spectral region from 180-260°C, corresponding to shallow delocalised donors. The ratio of the change in the carrier concentration (resistivity) to the loss of O_i was close to unity (Fig. 10). It had been proposed previously [22] that Al_s atoms captured I atoms generated by O_2 formation and that Al_i atoms would be formed. This species is also mobile, even at 300 K, and Al_i^{2+}-Al_s^- pairs form in irradiated silicon [48,50] : there is therefore the likelihood that O_2 dimers would trap Al_i defects and by comparison with the trapping of I-atoms, single shallow donors should be formed rather than double donors. So the reaction $O_2 + Al_s \rightarrow O_2Al$ would cause one O_i to be lost for each hole lost in p-type material. More recent work has revealed the IR electronic donor spectrum more clearly [51] but further studies are required.

We point out that Al_i is not necessarily expected to behave in the same manner as B_i, since this isolated defect has deep levels and is a negative-U centre [52]. At temperatures above ~ 200 K B-B pairs form and at higher temperatures B_i clusters appear to be produced [53]. The data for Si:Al are complemented by PL analyses that were interpreted in terms of the emission from isoelectronic complexes formed between a series of TDs and captured Al_i atoms [54]. Irrespective of the details, these results provide a basis for explanations for (a) early resistivity data [55] and b) the existence of

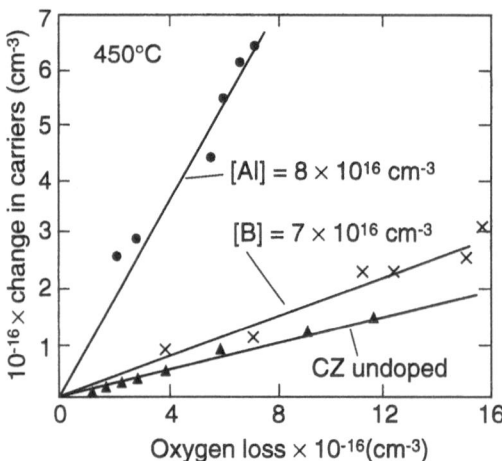

Figure 10. Comparisons of the dependence of $\Delta n/\Delta[O_i]$ for CZ undoped Si with B and Al doped crystals (~8 x 10^{16}cm^{-3}) Si, where n is the concentration of carriers lost during anneals at 450°C. N.B. TD(N) defects are double donors but are not all fully ionized at 300 K in undoped Si (measurement temperature) at high concentrations [49]. However they would be ionized in the p-type crystals.

two NL10 electron paramagnetic resonance spectra NL10 itself and NL10Al, distinguished by their ENDOR spectra [56].

6. Further observations relating to TD defects

There are two further topic areas relating to TD defects that have not been discussed : the first concerns hydrogen passivation and the second direct IR vibrational absorption.

It is now well known that defects or impurities with either deep or shallow levels may trap a hydrogen (or deuterium) atom, leading to partially or completely passivated complexes. Thus, various defects incorporating carbon and/or oxygen impurities form complexes that give rise to photoluminescence transitions that show small shifts in their frequencies when an H atom is replaced by D [57]. This is due to the large difference in the zero point energies of the H oscillator and anharmonic effects that produce small changes in the electronic levels. Single shallow donors and acceptors can be completely passivated but it was then shown that sulphur double donors could be partially passivated to produce single donors [58]. It follows that there could also be partial passivation of TD(N) double donors as indicated previously from measurements of deep level transient spectroscopy [59]. The replacement of H by D in the core of a donor would be expected to produce small changes in the ground state energy, for the reasons already stated, and indeed small shifts (~ 0.1 cm^{-1}) were detected in hydrogen related donors produced in n°- irradiated Si that was then annealed in a hydrogen plasma [60]. We then showed that similar shifts occurred in the frequencies of shallow

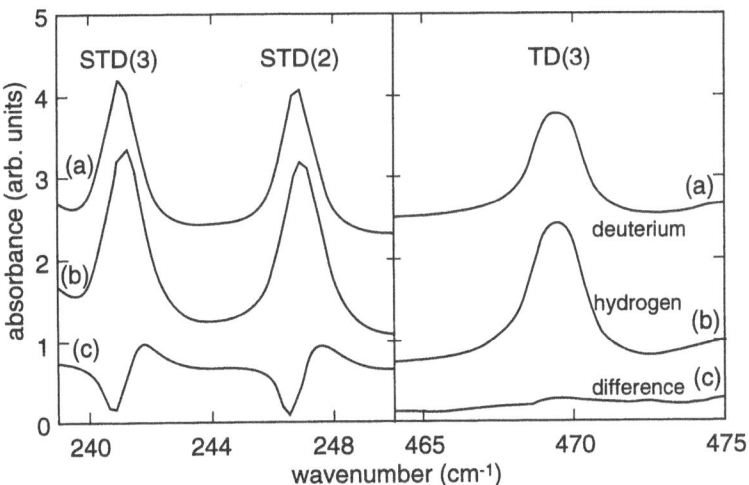

Figure 11. Spectra showing STD electronic transitions (1s → 2p±) in samples heated at 470°C in an RF hydrogen or deuterium plasma (13.56 MHz, 40 W, 2 mbar, t = 70 h) : the latter lines reveal an isotopic shift of ~ 0.1 cm^{-1} to lower frequencies compared with the former, (difference spectrum), whereas there is no shift in the TD(N) lines [61].

donor transitions formed in hydrogenated (deuterated) CZ Si [61]. These measurements have now been extended to the family of STD centres (Fig. 11) and the various IR transitions show shifts of ~ 0.1 cm^{-1} to lower energies when H is replaced by D [62]. It is therefore proposed that STDs are simply partially passivated TD(N) centres It then seems likely that the NL10 EPR spectrum found in undoped Si is another signature of the STD(N) centres since recent electron nuclear double resonance (ENDOR) measurements reveal the presence of a hydrogen (deuterium) atom in the core [56]. Attempts to establish a more definitive correlation are in progress [62].

There have been recent publications [63,64] describing a series of IR absorption lines at frequencies close to 1000 cm^{-1} and there have been correlations with various TD(N) centres and possible precursors to these defects. These LVMs with halfwidths of ~ 3 cm^{-1} (300 K) have therefore been attributed to oxygen clusters and it has been stated that the total absorption gained in these lines is equal to that lost from the 9 μm band as a result of the anneals. However, this conclusion was based on measurements of the peak absorption coefficients of the lines and the difference in linewidth was not taken into account. When this is done, the new absorption is only about 10% of that lost. This would mean that the dipole moments of the new lines are a factor of ~ 3 smaller than that of O_i atoms, that have a very large moment corresponding to a charge of η ~ 3e per unit displacement in the mode. However, a much reduced moment would imply a much modified bonding arrangement. The shift in frequency does not provide that evidence as there are no significant changes in η for either charge state or the A-centre or the O_i-I complex (Sections 4.1 and 4.2)

7. Conclusions

Progress continues to be made in understanding the oxygen aggregation process both from experiment and theory. Models involving rapid dimer diffusion provide mechanisms for the formation of large clusters at low temperatures without leading to disagreements with the rate of O_i loss. However, it appears that dissociation of clusters with the release of a dimer is an essential step. The rate of oxygen loss is correlated with the rate of $\Sigma TD(N)$ formation even when D_{oxy} is enhanced by the presence of hydrogen. A model of TD centres involving say two O_i atoms and a large number of I-atoms does not seem likely since we have $\Delta[O_i]/\Delta\Sigma[TD(N)] \sim 10$ and there would, on average, be only two I-atoms/donor if O_3 were the smallest cluster that could release an I-atom and the Deák model of the core is correct (this removes one of the two I-atoms). The Deák model has been favoured in recent EPR/ENDOR measurements [28] that suggest that there is aggregation of O_i atoms in BC-sites plus a central I-atom. It is not clear however whether there is addition of one or two extra O_i atoms, when TD(N) is converted to TD(N+1), and there is still a problem in understanding the strength of the IR absorption from oxygen present as TD-centres. It has become apparent that hydrogen plays an important role in relation to the observations of STD and NL10 spectra that are likely to be due to the same defects. The NL10Al centre may then be related to a different IR spectrum arising from donors incorporating an Al_i atom rather than an I-atom.

Acknowledgements

The author thanks The Engineering and Physical Sciences Research Council, U.K. for their financial support of this work (grant number GR/J 97540).

References

1. Baghdadi, A., Bulles, W.M., Croarkin, M. C., Li, Yue-Zhen, Scace, R.I., Series, R.W., Stallhofer, P. and Watanabe, M., (1989) Interlaboratory determination of the calibration factor for the measurement of the interstitial oxygen content of silicon by infrared absorption, *J. Electrochem. Soc.*, **136**, 2015-2024.
2. Bergholz, W., Binns, M.J., Booker, G.R., Hutchison, J.C., Kinder, S. H., Messoloras, S., Newman, R.C., Stewart, R.J. and Wilkes, J.G., (1989). A study of oxygen precipitation in silicon using high resolution transmission electron microscopy, small angle neutron scattering and infrared absorption, *Phil. Mag.* **B59**, 499-522.
3. Mikkelsen, J.C. Jr., (1986). The diffusivity and solubility of oxygen in silicon. *Mat. Res. Soc. Symp. Proc.* **59**, 19-30
4. Newman, R.C. and Jones, R., (1994) Diffusion of oxygen in silicon, F. Shimura (ed.), *Semicond. and Semimetals* (Academic, San Diego) **42**, 289-352.

5. Newman, R.C. and Bullough, R., (1970) The interactions of impurities with dislocations in silicon and germanium, *Rep. Prog. in Phys.* **77**, 101-134.

6. Gösele, U.M., (1986) The role of carbon and point defects in silicon. *Mater. Res. Soc. Symp. Proc.* **59**, 419-431.

7. McQuaid, S.A., Binns, M.J., Londos, C.A., Tucker, J.H., Brown, A.R. and Newman, R.C., (1995) Oxygen loss during thermal donor formation in Czochralski silicon : new insights into oxygen diffusion mechanisms, *J. Appl. Phys.* **77**, 1427-1442.

8. Bourret, A., Thibault-Desseaux , J. and Seidman,D.N., (1984) Early stages of oxygen segregation and precipitation in silicon *J. Appl. Phys.* **55**, 825-836.

9. Götz, W., Pensl, G and Zulehner, W., (1992) Observation of five additional thermal donor species TD12 to TD16 and of regrowth of thermal donors at initial stages of the new oxygen donor formation in Czochralski grown silicon *Phys. Rev.* B**46**, 4312-4315.

10. Kaiser, W., Frisch, H.L. and Reiss, H., (1958) Mechanism of the formation of donor states in heat treated silicon *Phys. Rev.* **112**, 1546-1554.

11. McQuaid, S.A., Newman, R.C., Tucker, J. H., Lightowlers, E.C., Kubiak, R.A.A., and Goulding, M., (1991) The concentration of atomic hydrogen diffused into silicon in the temperature range 900°C to 1300°C, *Appl. Phys. Lett.*, **58**, 2933-35.

12. Ourmazd, A., Schröter, W. and Bourret, A., (1984) Oxygen-related thermal donors in silicon: A new structural and kinetic model *J. Appl. Phys.* **56**, 1670-1681.

13. Gösele, U. M. and Tan, T.Y., (1982) Oxygen diffusion and thermal donor formation in silicon, *Appl. Phys.* A**28**, 79-92.

14. Messoloras, S., Newman, R.C., Stewart, R.J. and Tucker, J.H., (1987) Is enhanced interstital oxygen diffusion necessary to explain the kinetics of precipitation in silicon at temperatures below 650°C? *Semicond. Sci. Technol.* **2**, 14-19.

15. Bourret, A., (1987) Defects induced by oxygen precipitation in silicon : a new hypothesis involving hexagonal silicon, *Inst. Phys. Conf. Ser.* **87**, 39-48.

16. Werner, P., Reiche, M. and Heydenreich J., (1993) HREM Investigations of the agglomeration of self-interstitials in silicon, *Phys. Stat. Sol* (a) **137**, 533-541.

17. Takeda, S., Kohyama, M. and Ibe, K., (1994) Interstitial defects on {113} in Si and Ge line defect configuration incorporated with a self-interstitial chain, *Phil. Mag.* A**70**, 287-312.

18. Stewart, R.J., (1996), this conference.

19. Götz, W., Pensl, G., Zulehner, W., Addinall, R. and Newman, R.C., (1995) Annihilation studies of oxygen-related (470°C) thermal donor centres in CZ-Si, *Solid St. Commun.* **93**, 454: Götz, W., (1993) Untersuchen and Chalkogen-Donatoren Silizium und Stickstoff-Donatoren in 4-H Siliziumkarbid, Ph.D Thesis, University of Erlangen-Nürnberg.

20. Markevitch, V.P., Makarenko, L.F. and Murin, L.I., (1986) Some new features of thermal donor formation in silicon at T < 800 K *Phys. Stat. Sol.* (a) **97**, K173-176.

21. Newman, R.C., (1986) Oxygen precipitation and thermal donors in silicon, *Mater. Res. Soc. Symp. Proc.* **59**, 205-206.

22. Newman R.C., (1985) Thermal donors in silicon : oxygen clusters or self-interstitial aggregates, *J. Phys. C. : Solid St. Phys.* **18**, L967-772.

23. Tan, T.Y., Kleinhenz, R. and Schneider, C.P., (1986) On the kinetics of oxygen clustering and thermal donor formation in Czochralski silcon *Mater. Res. Soc. Symp. Proc.* **59**, 195-206.

24. Watkins, G.W. and Corbett, J.W., (1961) Defects in irradiated silicon : I. Electron spin resonance of the Si-A center *Phys. Rev.* **121**, 1001-1014.

25. Oates, A.S. and Newman, R.C., (1986) The involvement of oxygen-vacancy defects in enhancing oxygen diffusion in silicon, *Appl. Phys. Lett.* **49**, 262-264.

26. Bean A.R. and Newman, R.C., (1971) An infrared study of defects produced in n-type silicon by electron irradiation at low temperatures, *Solid St. Commun.* **9**, 271-274.

27. Deák, P., Snyder, L. and Corbett, J.W., (1992) Theoretical studies on the core structure of the 450°C oxygen thermal donors in silicon, *Phys. Rev.* B**45**, 11612-11626.

38

28. Meilwes, N, Spaeth, J.-M, Götz, W. and Pensl, G., (1994), Thermal donors in silicon : an investigation of their structure with electron nuclear double resonance, *Semicond.Sci. Technol.* **9**, 1623-1632, Gregorkiewicz, T., Bekman, H.H. P.Th. and Ammerlaan, C.A.J., (1988) Microscopic structure of the NL10 heat-treatment center in silicon : study by electron-nuclear double resonance, *Phys. Rev.* **B387**, 3998-4015.

29. Brelot, A. and Charlemagne, J., (1971) Infrared studies of low temperature electron irradiated silicon containing germanium, oxygen and carbon, (Eds.) J.W. Corbett and G.D. Watkins, *Radiation Effects in Semiconductors*, (Gordon and Breach, London) p161-170.

30. Brelot, A., (1972) Étude dans le silicium de l'interaction des impuretés et des défauts produit par irradiation, Thèse de doctorat d'état, University of Paris.

31. Abdullin, Kh.A., Mukashev, B.N. and Gorelkinskii, Yu. V., (1995) New bistable oxygen-related complex in silicon, *Mater. Sci.Forum* **196-201**, 1007-1012,

32. Oates, A.S., Binns, M.J., Newman, R.C., Tucker, J.H., Wilkes, J.G. and Wilkinson, A., (1984) The mechanism of radiation-enhanced diffusion of oxygen in silicon at room temperature, *J. Phys. C : Solid St. Phys.* **17**, 5695-5705.

33. Estreicher, S.K. (1990) Interstitial oxygen in silicon and its interaction with hydrogen, *Phys. Rev.* **B41** 9886-9891.

34. Jones, R., Öberg, S. and Umerski, A., (1991) Interaction of hydrogen with impurities in semiconductors, *Mater. Sci. Forum*. **83-87**, 551-561.

35. Markevitch, V.P., Suezawa, M. And Sumino, K., (1995) Optical absorption due to vibration of hydrogen-oxygen pairs in silicon, *Mater. Sci.Forum* **196-201**, 915-920.

36. Hansen, W.L., Haller, E.E. and Luke, P.N., (1982), Hydrogen concentration and distribution in high purity germanium crystals, *IEEE Trans. Nucl. Sci.* **NS-29**, No1, 738-744.

37. Snyder, L.C., Corbett, J.W., Deák, P. and Wu, R., (1988) On the diffusion of oxygen dimer in a silicon crystal, *Mater. Res. Symp. Proc.* **104**, 179-184.

38. Needels, M., Joannopoulos, J.D., Bar-Yam, Y., and Pantiledes, S.T., (1991) Oxygen complexes in silicon, *Phys. Rev.* **B43**, 4208-4313.

39. Jones, R., (1990) *ab initio* calculations on thermal donors in Si an over-coordinated O atom model for the NL10 and NL8 centres, *Semicond. Sci. Technol.* **5**, 255-260.

40. Markevich, V.P., Makarenko, L.F. and Murin, L.I., (1989) Thermal donor formation and mechanism of enhanced oxygen diffusion in silicon, Mater. Sci. Forum **38-41**, 589-594.

41. McQuaid, S.A., Piqueras, J. and Newman, R.C. Kinetics of oxygen loss and thermal donor formation in silicon : the rapid diffusion of oxygen clusters (1996), this conference.

42. Bean, A.R. and Newman, R.C., (1972) The effect of carbon on thermal donor formation in heat treated pulled silicon crystals, *J. Phys. Chem. Solids* **33**, 255-268.

43. Newman, R.C., Oates, A.S. and Livingston, F.M., (1983) Self-interstitials and thermal donor formation in silicon : new measurements and a model for the defects, *J. Phys. C : Solid St. Phys.* **16**, L667-674.

44. Davies, G., Oates, A.S., Newman, R.C., Woolley, R., Lightowlers, E.C., Binns, M.J. and Wilkes, J.G. (1986) Carbon-related radiation damage centres in Czochralski silicon, *J. Phys. C : Solid St. Phys.* **19** 841-855.

45. Kürner, W., Sauer, R., Dörnen, A. and Thonke, K., (1989) Structure of the 0.767 eV oxygen-carbon luminescence defect in 450°C thermally annealed Czochralski-grown silicon, *Phys. Rev.* **B39**, 13327-13337.

46. Murin, L.I., and Markevich, V.P., (1995), Effect of carbon on thermal double donor formation in silicon, *Mater. Sci. Forum*, **196-201**, 1315-1320.

47. Kamiura, Y., Maeda, T., Yamashita, Y. and Hashimoto, F., (1995) Anomalous fast annihilation of thermal donors in carbon rich silicon, *Mater. Sci. Forum* **196-201**, 1321-1326.

48. Watkins, G.D., (1965) A review of EPR studies in irradiated silicon in *Radiation Damage in Semiconductors-3,* (Dumod, Paris), pp97-113.

49. Claybourn, M. and Newman, R.C., (1989) Thermal donor formation in boron doped silicon, *Mater. Sci. Forum* **38-44**, 613-618.

50. Devine, S.D. and Newman, R.C. 1970, One phonon absorption from aluminium complexes in silicon compensated by lithium or electron irradiation *J. Phys. Chem. Solids* **31**, 685-700.

51. Kaczor, P., Godlewski, M. and Gregorkiewicz T., (1994) A New bistable shallow thermal donor in Al-doped Si, *Mater. Sci. Forum* **143-147**, 1185-1190.
52. Watkins, G.D., (1975) Defects in irradiated silicon : EPR and electron-nuclear double resonance of interstitial boron, *Phys. Rev.* **B12**, 5824-5839.
53. Tipping, A.K. and Newman, R.C., (1987) An infrared study of the production diffusion and complexing of interstitial boron in electron irradiated silicon, *Semicond. Sci. Technol.* **2**, 389-398.
54. Drakeford, A.C.T. and Lightowlers, E.C., (1988) Complex defect formation in heat treated aluminium doped CZ silicon, *Mater. Res. Soc. Symp. Proc.* **104**, 209-213.
55. Fuller, C.S., Doleiden, F.H. and Wolfstirn, K., (1960) Reactions of group III acceptors with oxygen in silicon crystals *J. Phys. Chem. Solids* **13**, 187-203.
56. Martynov, Yu. V., Gregorkiewicz, T. and Ammerlaan, C.A.J., (1995) ENDOR identification of a hydrogen-passivated thermal donor, *Mater. Sci. Forum* **196-2001**, 849-854.
57. Lightowlers, E.C., (1995) Hydrogen incorporation and interaction with impurities and defects in silicon investigated by photoluminesence spectroscopy, *Mater. Sci. Forum* **196-201**, 817-824.
58. Peale, R.E., Muro, K. and Sievers, A.J., (1990) Sulfur-hydrogen donor complexes in silicon, *Mater. Sci. Forum* **65-60**, 151-156.
59. Pearton, S.J., Chantre, A.M., Kimerling, L.C. Cummings K.D. and Dautremont-Smith, W.C., (1986) Hydrogen passivation of oxygen donors in Si, *Mater. Res. Soc. Symp. Proc.* **59**, 475-480.
60. Hartung, J. and Weber, J., (1993) Shallow hydrogen related donors in silicon, *Phys. Rev.* **B48**, 14161-14166.
61. McQuaid, S.A., Newman, R.C. and Lightowlers, E.C., (1994) Hydrogen-related shallow thermal donors in Czochralski silicon, *Semicond. Sci. Technol* **9**, 1736-1730.
62. Semaltianos, N.G., Newman, R.C., Tucker, J.H., Lightowlers, E.C. Gregorkiewicz, T. and Ammerlaan, C.A.J., (1996), (unpublished work).
63. Lindström, J.L. and Hallberg, T., (1994) Clustering of oxygen atoms in silicon at 450°C : a new approach to thermal donor formation, *Phys. Rev. Lett.* **72**, 2729-2732.
64. Lindström, J.L. and Hallberg, T., (1995) Vibrational infrared-absorption bands related to the thermal donors in silicon, *J. Appl. Phys.* **77**, 2684-2690.

INFRARED STUDIES OF THE EARLY STAGES OF OXYGEN CLUSTERING IN SILICON

J.L. LINDSTRÖM AND T. HALLBERG
*Linköping University, Department of Physics and Measurement
Technology, S-581 83 Linköping, Sweden*

Abstract

Clustering of oxygen atoms in silicon in the temperature range 350-470°C will be presented based on results from studies of infrared vibrational absorption bands. Several vibrational bands have been reported in the wavenumber range 975-1015 cm^{-1}. These bands correlate well with the formation of thermal donors (TDs). Bands at 975, 988 and 999 cm^{-1} have been found to be related to the three first appearing TDs. A band at 1006 cm^{-1} correlates with the development of the rest of the TDs. The 1012 band is suggested to originate from a different type of donor, possibly the shallow thermal donors. All these bands have a corresponding band in the 716-748 cm^{-1} range. A broad absorption band with a maximum at 1060 cm^{-1} also grows up during annealing in this temperature range. It is suggested that 90-95% of the oxygen atoms lost from the interstitial position during TD formation gives rise to this band, which originates from a type of oxygen precipitate. Different aspects of the TDs will be presented, such as results from samples doped with the O^{18} isotope, the bistability of the first two appearing TDs, the influence on the TD formation process from other impurity atoms, from electron irradiation (2 MeV) and from dispersion treatments.

1. Introduction

Oxygen is an important impurity in silicon materials used for integrated circuits due to its beneficial effects such as improving the mechanical strength of the wafers and the intrinsic gettering of other impurities. However, in a very large temperature range (up to 1200°C) the solution of oxygen remains highly supersaturated and therefore an inhomogenous distribution of oxygen can be expected in as-grown materials. Different heat treatments are used to homogenize as-grown materials in connection with circuit fabrication. However, at temperatures where oxygen becomes mobile (>350°C) different forms of aggregation of oxygen atoms will start which can cause harmful effects in circuit performance. Future processing of integrated circuits requires a deeper knowledge of the nature of these defects.

Clustering of oxygen atoms in Czochralski (CZ) silicon has been widely studied [1,2,3]. After high temperature treatments (650-1100°C) extended defects in the form of different quartz precipitates have been observed. In the temperature range 550-750°C the new donors (NDs) are formed while the thermal donors (TDs) and the shallow thermal donors (STDs) are formed in the range 350-500°C.

R. Jones (ed.), Early Stages of Oxygen Precipitation in Silicon, 41–60.
© 1996 *Kluwer Academic Publishers.*

The TDs in silicon have been extensively studied [2,4,5] using a variety of experimental techniques like resistivity measurements, infrared (IR) absorption, electron paramagnetic resonance (EPR), electron nuclear double resonance (ENDOR), Hall effect, deep level transient spectroscopy (DLTS) and photo-thermal ionization spectroscopy (PTIS).

Very detailed information about these defects comes from EPR and IR spectroscopy studies at low temperatures. Many IR absorption bands related to electronic transitions have been observed and ordered into a series corresponding to 4 [6], 6 [7], 9 [8,9], 11 [10] and 16 [11] different double donors. The structure of these defects is suggested to be a core to which oxygen atoms aggregate to form a series of closely related TD structures. Several core models of the TDs have been presented, such as the SiO_4 complex [12], the Y-lid (O_Y) complex [13], the $(O_Y)_2$ complex [14,15], the OSB-model [16] and the O_4 complex [17]. More recently a promising core model involving a Si_iO or Si_iO_2 complex has been suggested [18].

The modeling of the growth kinetics of thermal donors [1,4,5] is usually based on investigations of the loss of interstitial oxygen during TD formation. In most cases a serial formation process is used instead of parallel processes [19]. A fast diffusion mechanism of oxygen [20] and fast diffusing O_2 molecules [21] have been suggested to explain the rapid growth of the early TDs.

Experimentally it has been shown that the growth kinetics of TDs is dependent on pre heat treatments [1]. An oxygen homogenization treatment at high temperature (1300 °C) is known to delay the TD formation [22,23].

Oxygen in silicon is known to give rise to several vibrational IR absorption bands, both as an interstitial atom and combined with other impurity atoms or vacancies [2]. The oxygen atom in an interstitial position gives rise to several vibrational bands with the strongest band positioned at 1107 cm^{-1} at room temperature [24]. For the oxygen in an almost substitutional position (A-center) a corresponding band at 830 cm^{-1} has been reported [25]. Oxygen precipitates formed in the temperature range 650-1100°C give rise to IR absorption bands in the range 1050-1250 cm^{-1} [26,27]. Weak absorption bands at 1012-1013 [28-32], 1006 [31] and 728 [30] cm^{-1} have been reported over the years.

It has recently been shown that vibrational absorption bands in the wavenumber ranges 724-748 and 975-1015 cm^{-1} are developing in the temperature range 350-470°C and that these bands correlate well with the development of different TDs [33-35]. It has also been suggested that the appearance of TDs is a parallel process to the development oxygen precipitates with a corresponding band at 1060 cm^{-1} [35].

The main purpose of this paper is to present new information about the TDs and the early oxygen clustering process based on studies of these vibrational bands.

2. Experimental Procedure

In the experimental work referred to here we formed thermal donors in CZ silicon materials by heat treatments for different times in the range 350-470°C. We used as-grown P and B doped samples of 2-20 Ωcm with an interstitial oxygen concentration in the range $5-9\times10^{17}$ cm^{-3} and a carbon concentration below detection limit. Sample dimensions were 10×5×5 mm. In some experiments samples with concentrations of carbon in the range $6-50\times10^{16}$ cm^{-3} and aluminum in the range $0.9-5\times10^{16}$ cm^{-3} were included as well as samples doped with the isotope O^{18}.

The concentration of interstitial oxygen $[O_i]$ and substitutional carbon $[C_s]$ was monitored by measuring the well-known absorption bands at 1107 cm[-1] [36] and 605 cm[-1] [37] respectively. In Ref. 36 $[O_i]$ was determined by $\alpha \times 2.45 \times 10^{17}$ cm[-3], where α is the absorption coefficient of the 1107 band. The spectrometer used for the IR studies was a Bruker FTIR 113v. The heat treatment was performed in nitrogen atmosphere and was followed by a HF-dip.

The total TD concentration [TD] was determined by resistivity measurements using a four-point probe, taking into account the double donor character of the TDs and the mobility difference between electron and holes. From measurements at 10 K we studied the electronic transitions of the TDs in the spectral range 350-1200 cm[-1].

In this paper we mainly use the integrated absorption coefficient in the analysis of the TD-related vibrational bands, i.e. $I=\int\alpha(\nu)d\nu$, where ν is the wavenumber. Since the line shape of the bands is Gaussian we can approximate I by $\alpha \times \Delta$, where Δ is the full width at half maximum (FWHM).

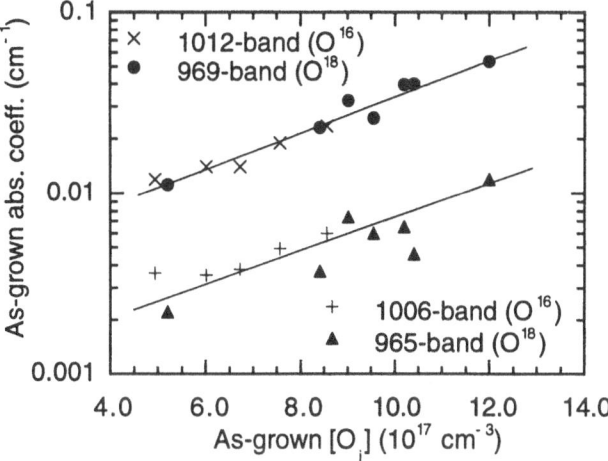

Figure 1. The strength of the bands at 1012 and 1006 cm[-1] in O^{16} as-grown material and at 969 and 965 cm[-1] in O^{18} as-grown material, plotted versus the concentration of O_i.

3. TD Vibrational Bands Related to Oxygen

Absorption spectra from room temperature measurements for the as-grown samples used in this study all showed the absorption bands at 1006 and 1012 cm[-1]. Silicon materials with different interstitial oxygen concentrations show a strong correlation between the 1107 band and the 1012 and 1006 bands [33,34], as illustrated in Fig. 1. This has also been shown for the electronic bands related to TD1, TD2 and TD3 [34]. This indicates that oxygen is included in the centers related to these bands.

In Fig. 2 we show a spectrum with the 1006 and 1012 bands together with the absorption band from interstitial oxygen at 1107 cm[-1]. The strength of the band at 1012 cm[-1] is typically 0.5-1.0% of the 1107 band. After annealing the samples more bands appear, positioned at about 975, 988 and 999 cm[-1]. When changing the temperature from room temperature to 10K the positions of the bands at 975, 988 and 999 cm[-1] increases by about 3 cm[-1]. The 1006 band increases by only 0.5-1 cm[-1] while the 1012 band shows no temperature dependence. The difference in temperature dependence of these bands should be due to different microscopic structures of the corresponding TDs.

From a very recent investigation of a silicon material doped with the oxygen isotope O^{18} results are presented in Fig. 2 for an as-grown sample. In this case the O^{18} band is as expected found at about 1058 cm^{-1}. New bands are found at about 965 and 969 cm^{-1}. These are the positions expected for the 1006 and 1012 bands in O^{18} material if these bands are due to oxygen, i.e. if we multiply the positions of the 1006 and 1012 bands with the ratio 1058/1107. The plot in Fig.1 not only shows that the 969 and 965 bands correlates with the concentration of O_i (1058 band), but it also shows a strong correlation with the 1012 and 1006 bands respectively. In annealed O^{18} material other bands appear at about 945 and 955 cm^{-1}, bands which are the equivalents to the 988 and 999 bands in O^{16} material (a more thorough analysis of these results will be published shortly). These results strongly support that the TD-related vibrational bands are indeed related to oxygen vibrations. A similar support for this assignment has also been found in oxygen doped Ge [38]. The involvement of oxygen in the core of the TDs has previously been confirmed by ENDOR investigations [17,39].

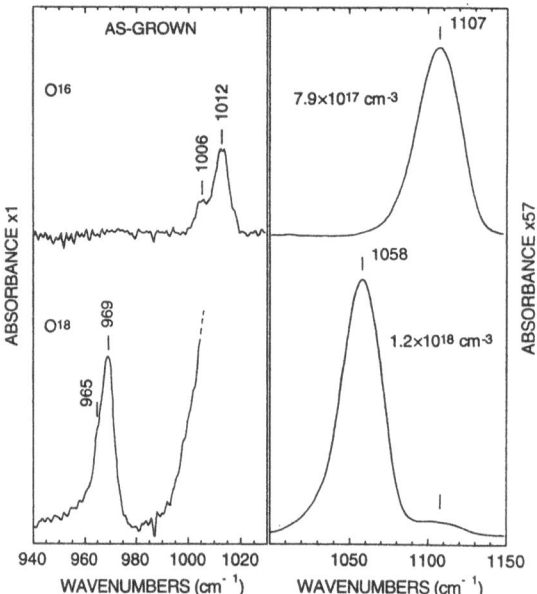

Figure 2. Spectra measured at room temperature for samples containing the oxygen isotope O^{16} (above) and O^{18} (below).

A calibration coefficient $K_{TD}=6(\pm1)\times10^{15}$ cm^{-1} for the TD-related vibrational bands was suggested in Ref. 35, estimated by fitting the different bands to the total [TD] from resistivity measurements. It was determined by using I of the bands and the best fit was obtained by using the same calibration coefficient for all the different bands. Assuming that the TD-related vibrational bands are due to oxygen atoms for which the oxygen calibration coefficient is valid, the ratio K_{O_i}/K_{TD} should give the number of oxygen atoms involved in these complexes (K_{O_i} is here the oxygen calibration coefficient based on I rather than on α). The result from such an analysis shows the involvement of 1-2 oxygen atoms.

From the results presented so far the conclusion is that the TD-related vibrational bands are due to the vibrations of oxygen atoms in a TD core of 1-2 oxygen atoms. This is in line with the ENDOR investigation results in Ref. 39 and with the core models suggested in Ref. 18, i.e. a Si_iO or Si_iO_2 complex, which were calculated to give rise to vibrational bands similar to the ones observed in our investigations. The addition of oxygen atoms to such a core should give rise to the different wavenumber positions of the TD-related vibrational bands [35].

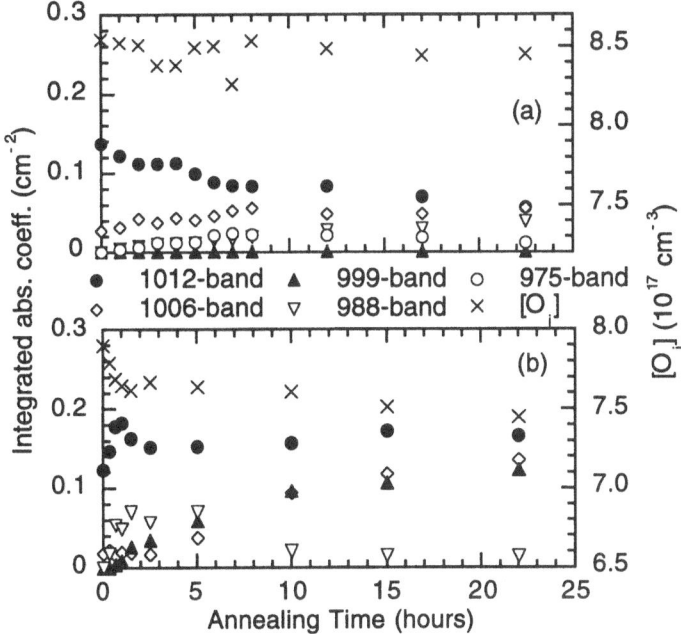

Figure 3. The initial changes of the bands 975-1012 cm^{-1} and [O_i] for annealing at 370°C (a) and 450°C (b).

4. The Thermal Donor Formation

The development of TDs in the temperature range 350-500°C is usually described as a sequential process where an increased amount of oxygen atoms are attached to oxygen clusters and a series of more and more shallow double TDs is formed. A new approach will be used here based on the behavior of the recently discovered vibrational bands related to the TDs.

Typical formation kinetics of these new vibrational bands are presented in Fig. 3 for samples annealed at 370 and 450°C. The plots show clearly that bands at 1006 and 1012 cm^{-1} are present in as-grown material and that the bands are changing in a different way at 370 and 450°C. It is found that this corresponds to two situations for TD formation, one at lower temperatures (<450°C) where pre-existing oxygen clusters are transformed and one at higher temperatures (≥450°C), where ordinary oxygen diffusion is important and where a growth of pre-existing clusters can be observed.

4.1. EARLY STAGES OF TD FORMATION.

The 370°C annealing in Fig. 3a shows that the 1012 band decreases while new bands at 975 and 988 cm^{-1} appear. The band at 999 cm^{-1} will appear after about 40 hours of annealing at 370°C. The 1006 band is initially increasing. It is found that there is a reversible correlation between the decrease of the 1012 band and the increase in the sum of the 975, 988, 999 and 1006 bands [34,35].

When comparing the growth of the TD electronic transition bands (1s-3p$_+$) for TD+1, TD+2 and TD+3 with the 975, 988 and 999 vibrational bands, a very close correlation was found as presented in Refs. 34 and 35. It was therefore concluded that TD1, TD2 and TD3 are developing in structures giving rise to the 975, 988 and 999 bands respectively.

One explanation of the behavior shown in Fig. 3a is that different clusters are "frozen in" in as-grown crystals, i.e. the clusters related to the 1006 and 1012 bands. These clusters are adjusting their size and structure to their thermodynamic equilibrium at the temperature of the post heat treatment. This explanation is supported by results from different annealing temperatures where different equilibrium values of these bands are achieved [35,40]. An initial free energy difference between the as-grown phase and another phase will occur when the material is annealed at TD formation temperatures. This is the driving force for the initial changes of the 1006 and 1012 bands. The temperature dependence of the equilibrium concentration should be related to the temperature dependence of the free energy.

It has often been pointed out that there is a difficulty to fit kinetic models to the rapid growth of TD1 and TD2 [5]. Using activation energies for normal oxygen diffusion leads to an unrealistic capture radius for the core. Different anomalous diffusion mechanisms for oxygen have also been suggested. The transformation process of pre-existing clusters suggested here gives, however, a new explanation of the early rapid donor formation.

During these transformation processes [O$_i$] stays nearly constant although most samples show some more or less stochastic variations of up to $\pm 2 \times 10^{16}$ cm^{-3}. A relatively constant value of [O$_i$] for short annealing times at low temperatures was reported in Ref. 23.

For samples annealed at temperatures $\geq$450°C the 1012 band initially increases together with the 988 band during the first annealing hour. This increase is correlated with a decrease in [O$_i$] according to Fig. 3b.

4.2 LATER STAGES OF TD FORMATION

When the initial transformation phase as described above is over for the samples annealed at <450°C or initially for these annealed at $\geq$450°C, we can observe the effects of oxygen diffusion. The loss of [O$_i$] occurs simultaneously with the growth of the bands at 988, 999, 1006 and 1012. This is illustrated in Figs. 3b and 4.

In Fig. 4, showing a sample annealed at 420°C, we also note that the growth of the 1012 band correlates in time with the decrease of the 999 and 1006 bands. In more detail we can see that the beginning of growth of the 1012 band from its initial equilibrium level is connected with the decrease in growth rate of the 999 band. The growth rate of the 1012 band is further increased as the 999 and the 1006 bands reach their maximum concentrations and start to decrease, while the saturation of the 1012 band is related to the interrupted decrease of these both bands. Fig. 4 also shows that [O$_i$] stays rather constant during the initial stages of TD formation, which was previously pointed out for annealing at 370°C (Fig. 3a). The important decrease in $[O_i]$ correlates with the growth of the 1006 band and the later formation stages of the 999 band, which we have observed at all annealing temperatures investigated. The growth of the 1012 band seems to be correlated to both the decrease in the 999 and 1006 bands and to the loss of $[O_i]$.

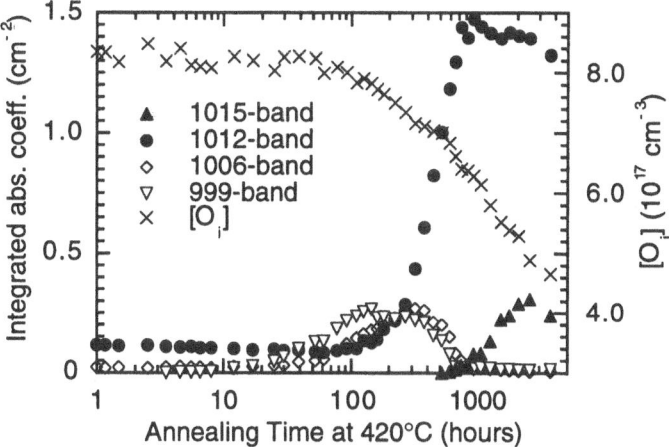

Figure 4. The formation of the 999-1015 bands and the decrease in $[O_i]$ for annealing at 420°C.

As the 1012 band is approaching its maximum level a new band at about 1015 cm⁻¹ is appearing, according to Fig. 4. The 1015 band appears as a shoulder on the broad 1012 band, as shown in Fig. 5. This band is appearing just before [TD] is starting to saturate, according to the resistivity measurements, indicating that it is the last one in the series of TD-related vibrational bands formed in this temperature range.

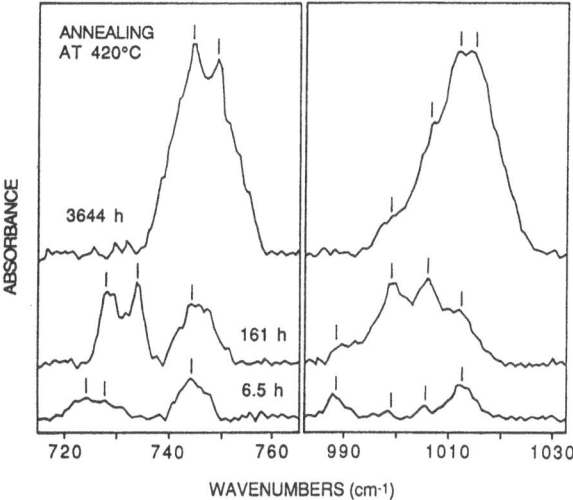

Figure 5. Room temperature spectra at different annealing times at 420°C, showing bands at about 988/724, 999/728, 1006/734, 1012/744 and 1015/748 cm⁻¹.

In comparison to any of the other TD-related vibrational bands the maximum absorption coefficient of the 1012 band becomes considerably higher and its FWHM more than twice as large. The increase of the FWHM of the 1012 band during its formation seems to be caused by an inhomogeneous broadening effect due to the locally very high density of centers related to the 1012 band. The kinetics of this band has been found to be correlated with the EPR NL10 center [35].

4.3 TD-RELATED BANDS IN THE 716-748 cm^{-1} RANGE

Previously a band at 728 cm^{-1} has been reported and suggested to possibly be related to the thermal donors [30]. Some other bands around that wavenumber were also reported in Ref. 10. Recently bands at about 724, 728, 734, 744 and 748 cm^{-1} were found to be related to the bands at 988, 999, 1006, 1012 and 1015 cm^{-1} respectively, by comparing their formation kinetics [35]. Spectra showing different annealing times of the bands in the 724-748 cm^{-1} range are shown in Fig. 5, together with the bands in the 975-1015 cm^{-1} region. Recently a band observable at 10K at about 716 cm^{-1} has been identified as a band corresponding to the 975 band.

Due to strong phonon absorption in the 724-748 cm^{-1} spectral region it is more difficult to study the bands in this region as compared with the bands in the 975-1015 cm^{-1} region.

4.4 ACTIVATION ENERGIES

Activation energies E_a for the formation of the TDs estimated by studying the TD-related vibrational bands were reported in Ref. 40. E_a for the formation of the 988, 999 and 1006 bands is according to previously reported results for TD2-TD6 [41-43], i.e. in the range 1.7±0.2 eV. The later formation kinetics of the 1012 band was found to be according to an E_a of about 1.5 eV. However, the initial changes in the temperature range 350-420°C for the 975, 1006 and 1012 bands were according to an E_a in the range 1.2-1.3 eV, showing that the initial formation process is different from the later formation of TDs.

4.5 BISTABILITY OF THERMAL DONORS

It is now well established that TD1 and TD2 are bistable, as deduced from Hall-effect measurements [44-46], DLTS studies [47] and IR spectroscopy [10,46,48]. This means that besides the configurations TD1 and TD2 there exist two other low energy configurations, here labeled X1 and X2 respectively. In order to observe the bistability the room temperature Fermi level has to be located above E_c-0.32 eV or E_c-0.25 eV, for TD1 and TD2 respectively [10]. These different configurations form an Anderson negative U system [44,45]. The conversion X→TD can be achieved by band gap illumination of the sample during cooling before the IR measurement.

In Ref. 10 it was concluded that different physical processes control the configurational transformations. X→TD was found to be due to carrier capture while TD→X should be caused by atomic displacement. Calculations in Ref. 18 show that a pair of bistable configurations can appear for both a Si_iO (TD1) and a Si_iO_2 (TD2) complex. For the Si_iO_2 complex the configurational transformation from the TD to the X configuration was obtained by moving the oxygen atoms from a more compact into a more open configuration. It was argued that this movement of the oxygen atoms should be blocked after the side chains of additional oxygen atoms are formed in higher order TDs, which could explain why only the first two TDs are showing bistable properties.

The IR results previously reported [10,46,48] on the bistability of TD1 and TD2 are based on the electronic transitions which they give rise to. So far the X1 and X2 configurations have not been detected by any spectroscopic method [10,48]. However, it has recently been found that the bistable properties of TD1 and TD2 can be observed from studying their vibrational bands at 975 and 988 cm^{-1} and that X1 and X2 as well give rise to IR absorption [49].

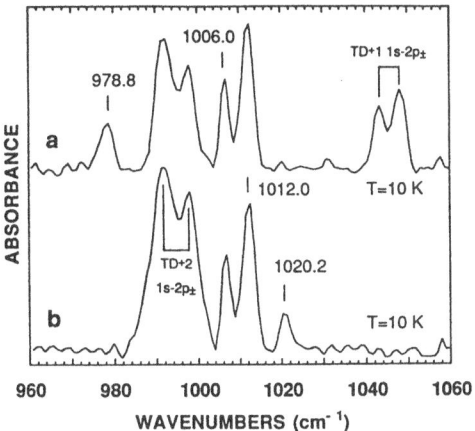

Figure 6. Spectra for a sample annealed at 350°C for 20 hours, measured at 10K. (a) was measured after cooling in band gap light while (b) was measured after cooling in darkness.

The spectra in Fig. 6 shows this bistability as illustrated by both the $1s-2p_{\pm}$ electronic transitions and the vibrational band at 978.8 cm^{-1} of TD1, for a measurement at 10K. Similar results were reported for TD2 [49]. Fig. 6a shows the spectrum of a sample cooled in band gap light while Fig. 6b shows the spectrum of a sample cooled in darkness. The sample of Fig. 6 shows no bistability of TD2 since $E_F < E_c - 0.25$ eV. An interesting observation in Figure 6b is the fact that a new band at about 1020 cm^{-1} appears when the vibrational band and the $1s-2p_{\pm}$ transition of TD1 disappear. From these results it is suggested that the 1020 band corresponds to the X1 configuration. Similar results were obtained for TD2, where the band of X2 was also suggested to give rise to IR absorption at about 1020 cm^{-1} at 10K.

A very important result from the same investigation is presented in Fig. 7. It shows results from two different samples annealed to maximum concentrations of TD1 (Fig. 7b) and TD2 (Fig. 7a). For each sample room temperature spectra with and without a Ge filter where collected and the differential spectra are presented in Fig. 7. The Ge filter will reduce the IR light to energies below the band gap light of silicon and only the X1 and X2 configurations are possible. The figure clearly shows that the TD1 and TD2 configurations change their concentrations by the same amount as the center related to the band at 1020 cm^{-1} does, which supports the identification of the 1020 band with both X1 and X2. The sharp feature on the low energy side of the 988 band is due to an instrumental error. However, while there is a temperature dependence of about 3 cm^{-1} on the peak positions of the 975 and 988 bands, the temperature dependence of the 1020 band is very small.

Since there is a common band for X1 and X2 at 1020 cm^{-1} it should mean that the microscopic configurations of these defects are identical and not different as suggested in Ref. 18.

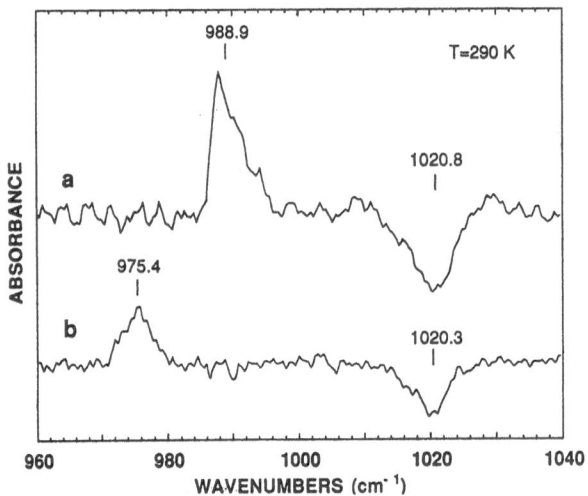

Figure 7. Differential spectra between measurements without and with a Ge filter, measured at room temperature. Spectrum (a) is for a sample annealed at 390°C for 48 hours and spectrum (b) for a sample annealed at 350°C for 20 hours.

5. The Formation of Oxygen Precipitates

Annealing of samples at 450-470°C for at least 100 hours reveals the presence of a broad (FWHM≈120 cm^{-1}) vibrational band positioned at about 1060 cm^{-1}. Fig. 8 shows this band for a sample annealed at 450°C for 8700 hours, for measurements at 10K and at room temperature. The asymmetry of the band indicates the presence of a weaker broad band in the region of about 1150 cm^{-1}. This broad structure originates from oxygen precipitates and has previously been reported by several investigators for a wide range of annealing temperatures [26,27,50-53]. The formation of such an oxygen precipitate band at TD formation temperatures was also reported in Ref. 50. It has been calculated that small SiO_2 particles shaped as either needles, prolates or spheres should give rise to a broad IR band positioned around 1100 cm^{-1} [26]. Investigations using transmission electron microscopy (TEM) also show that oxygen precipitates, mostly in the shapes of rodlike defects, are indeed formed at such low temperatures as 450°C [54].

Fig. 9 shows the formation kinetics of the 1060 band and the sum of the TD-related vibrational bands plotted together with $[O_i]$, for annealing at 470°C. There is a clear linear correlation between the loss of $[O_i]$ and the growth of both the TD-related bands and the 1060 band, showing the involvement of oxygen in these centers. Reliable data of the 1060 band for annealing times shorter than 100 hours was not possible to obtain, mainly due to its broadness and the presence of the 1107 O_i band. However, extrapolation of the 1060 band kinetics to shorter annealing times shows that oxygen precipitates should be formed from the beginning of the heat treatment, i.e. during the formation of the TDs. Spectral analysis of samples annealed in the temperature range 420-470°C show that there is a constant ratio between the concentrations of precipitates and TDs which indicates a connection between the formation of the TDs and the oxygen precipitates.

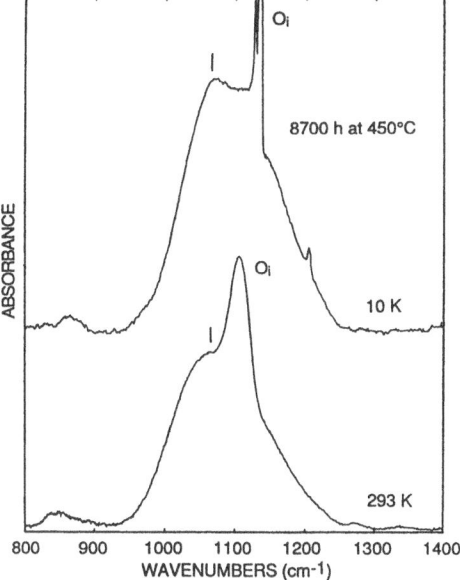

Figure 8. Spectra measured at 10K and room temperature for a sample annealed at 450°C for 8700 hours.

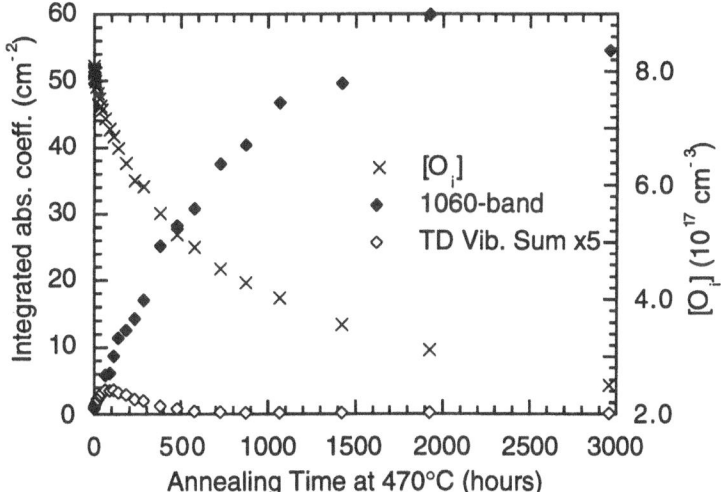

Figure 9. The loss of $[O_i]$ and the formation of the 1060 band plotted together with the development of the TD-related vibrational bands $(I_{988}+...+I_{1015})\times5$. The sample was annealed at 470°C.

According to the OSB-model [16] there is a natural link between the TDs and the rodlike defects. In that model chain-like oxygen clusters related to the TDs are embryonic forms of the much larger rodlike defects. The electrical activity is obtained by a Si_i atom bonded to two O_i atoms, which are located next to the cluster. These clusters grow by the addition of oxygen atoms in <110> directions, which above a certain critical size will terminate the electrical activity of the TDs by ejection of the Si_i atom. This should explain the annihilation of TDs as the formation of oxygen precipitates becomes increasingly important, according to Fig. 9. It has also been shown that pre-treatments at TD formation temperatures will increase the formation of oxygen precipitates or rodlike defects at higher annealing temperatures [53-55]. These observations are in line with the OSB-model. However, the involvement of oxygen in the rodlike defects has been questioned [2,55,56].

By using the calibration coefficient for interstitial oxygen K_{Oi} on the TD related vibrational bands and on the 1060 band and comparing with the loss of interstitial oxygen during TD formation, we conclude that 90-95% of the loss is involved in the formation of the precipitates.

6. The Influence of Impurities on the TD Formation Process

It has been reported that dopants like Al will enhance TD formation [57-59]. It has also been reported from EPR measurements that the formation of the TD-related NL10 center is strongly enhanced in Al doped silicon as compared with B doped [60,61] and that Al is present in some TDs. The presence of carbon is known to have the opposite effect during TD formation.

We have observed this enhancement and retardation in the TD formation process by studying the TD-related vibrational bands. A comparison was made with B and P doped silicon for which we could not observe a significant influence from the dopant.

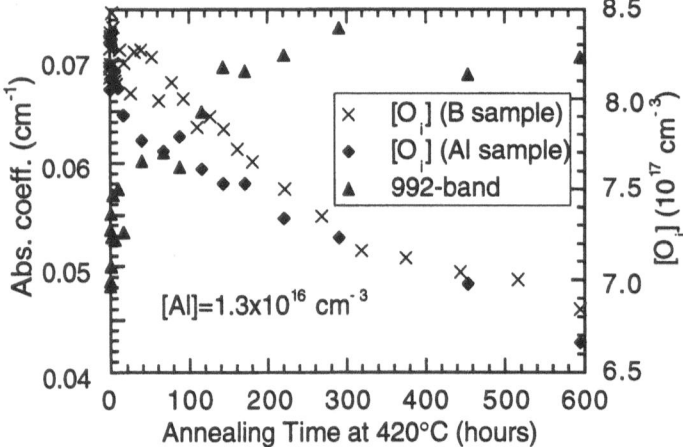

Figure 10. The loss of $[O_i]$ for Al and B doped samples and the development of the 992 band, for annealing at 420°C.

6.1. ALUMINUM DOPED SILICON

A sequential development of the TDs is typically observed during heat treatment, where the early double donors TD1 and TD2 start to grow already during the first hour of annealing at 420°C for B doped silicon. Regarding the development of TDs in Al doped samples no TDs seem to be formed during the first hours and there is a delay in the formation up to about 20 hours, according to resistivity measurements [62]. It seems like the formation of TD1 and TD2 is inhibited. There was no indication of the bands at 975 and 988 cm^{-1}. Following the development of TDs from resistivity results in different Al and B doped samples, we observed a strong influence on the TD concentration achieved at longer annealing times in low and moderately Al doped samples, while the high Al doped and the B doped samples develop in a similar way.

In as-grown material there is no significant difference between B and Al doped silicon of comparable oxygen content, when comparing the strength of the TD-related vibrational bands at 1006 and 1012 cm^{-1} [62]. However, for the Al doped samples a new band can be observed at 992 cm^{-1}. The maximum strength of this band depends linearly on the Al doping concentration as determined by resistivity measurements.

We found that the loss of $[O_i]$ during the first 200 hours at 420°C was much more pronounced in Al doped samples than in a comparable B doped sample, as shown in Fig. 10. A simultaneous growth of the 992 band occurs according to the same figure. After about 300 hours the total loss of $[O_i]$ is very similar in the two materials. At this point the growth of the 992 band has ceased, while the 1012 band continues to grow.

A possible explanation for the growth of the 992 band and the absent initial growth of the 975 and 988 bands in Al doped samples as compared with P and B doped samples, could be that a fast diffusing catalyst starts the transformation process of the centers related to the 1012 band in B-doped samples and that this catalyst is not present or takes part in a different reaction in Al doped samples. This catalyst could be Si_i atoms which in the presence of Al preferentially interacts with Al_s to form Al_i. The growth of the 992 band could then be caused by diffusing Al_i atoms forming complexes with O_i. In Ref. 62 it was suggested that this complex is $Si-O_i-Al_s$.

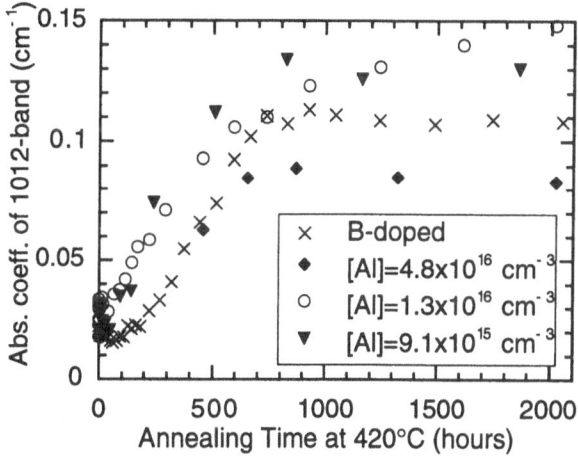

Figure 11. The growth of the 1012 band with annealing time at 420°C for Al and B doped samples.

In Al doped samples the development of the bands positioned at 999 and 1006 cm^{-1} were found to develop in a normal way with annealing time, as compared with B and P doped samples. The development of donors related to the growth of the 1012 band are clearly affected by the presence of Al, but not related to the concentration of Al in a simple way. For low and medium doped samples a clear enhancement can be seen, while for a high Al doping a suppression occurs. This is illustrated in Fig. 11. The enhancement of the 1012 band observed for low and medium Al doped samples has also been reported for the EPR NL10 center in comparable samples [63]. This supports the correlation between the 1012 band and the NL10 center [35].

6.2. CARBON RICH SILICON

High concentrations of carbon in silicon is known to strongly suppress the formation of TDs [28]. Several vibrational bands related to oxygen and carbon have been reported as well as atomic models involving different combinations of these atoms [2,28,37]. Some of these bands are often found in as-grown materials and some appear after heat treatments. Another category is found after electron irradiation (2 MeV) at room temperature due to the formation of mobile carbon interstitials.

The TD-related vibrational bands were used to reveal the suppression of the TDs by investigating samples with $[O_i]$ and $[C_s]$ of about 5×10^{17} cm^{-3}. There was no observable TD concentration even after 180 hours at 450°C, neither from resistivity measurements nor from the TD electronic transition bands. Room temperature IR results showed no sign of the bands 975-1012 cm^{-1}. Instead a different group of bands appear with a dominating band at 1026 cm^{-1} which previously has been studied [37] and suggested to be related to a carbon-oxygen complex with probably two oxygen atoms involved. Apparently oxygen is clustering with carbon in this case, giving rise to a structure which does not affect the resistivity. In samples with $[O_i]=6 \times 10^{17}$ cm^{-3} and $[C_s]=6 \times 10^{16}$ cm^{-3} the vibrational bands related to the TDs as well as the band at 1026 cm^{-1} were observed after annealing at 450°C.

These observations could be explained as follows. Self-interstitials might be ejected during the growth of oxygen clusters [2,16]. Carbon is known to eliminate mobile Si_i in silicon through the Watkins replacement mechanism, i.e. $C_s+Si_i \rightarrow C_i+Si_s$. Thus, carbon atoms efficiently can form the carbon related oxygen complexes at the same time as the concentration of self-interstitials and O_i, which both should be important in the formation of the TDs, are reduced. Also, if the TD core model discussed above [18], with a Si_i atom giving rise to the double donor character, is modified with a carbon atom replacing the Si atom the core could become an electrically neutral defect.

7. The effect of oxygen dispersion treatments on the TD formation process

Different dispersion treatments at high temperatures (≥ 1100°C) are used to improve the homogeneity of silicon materials. A "donor-kill" treatment at 650°C is another treatment which has been widely used for silicon in the industry. Some results concerning the effect on the TD-related vibrational bands discussed here have been reported [31,33,35] and are in support of the ideas that TDs are at least to some extent annihilated by such a treatment.

In Fig. 12 the result of a dispersion treatment for two hours at 1100°C shows that the integrated absorption coefficient for the 1012 band drops to nearly half of the initial value, as compared with an as-grown sample, while the band at 1006 cm^{-1} is almost completely reduced.

Concerning the TD formation in dispersed samples, the maximum concentrations of TD1 and TD2 are reduced by a factor 2.0 and 1.5 respectively, as compared with an as-grown sample [35]. This is illustrated in Fig. 12 for TD1. Apparently the initial concentration of the defects related to the 1012 band affects the formation of TD1, which can be explained by that the 1012 band is partly transformed to the 975 band. The decay of the 1012 band has been found to be according to first-order kinetics, which would suggest that these complexes initially dissociate into the first appearing donors [35]. The dispersion treatment initially also affected the formation of TD3, but after long annealing times it approached the concentration of TD3 in an as-grown sample. All these changes appeared the same for the 975, 988 and 999 bands as for the electronic transition bands for TD1, TD2 and TD3 respectively. Also, [TD] from resistivity measurements was affected in a similar way, i.e. initially delayed but equivalent to an as-grown sample after long annealing times.

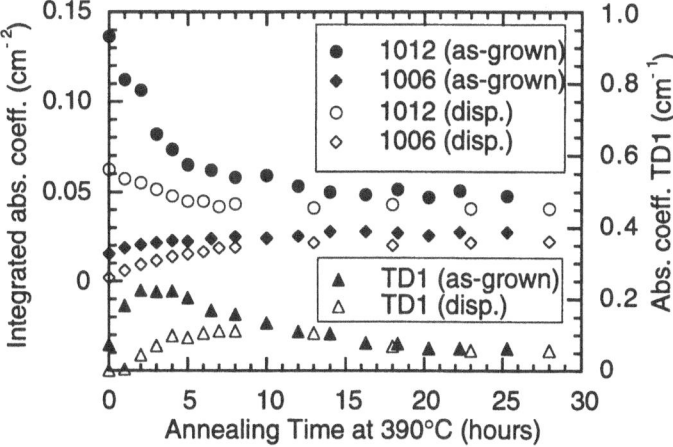

Figure 12. The initial changes of the 1006 and 1012 bands during the adjustments to their respective equilibrium levels and the growth of TD^+1 ($1s-3p_\pm$) for an as-grown sample and a dispersed sample. The samples were annealed at 390°C.

Similar differences on the formation of the first appearing donors between a dispersed and an as-grown sample have been found previously [22,64,65]. However, in our investigation we can relate the differences to a change in the starting concentration of the TDs related to the 1012 band.

8. The Effect of Electron Irradiation on the Thermal Donors

Different results have been published about the influence of electron irradiation on the formation of TDs [30,66-68]. It has been found that there is no influence on the sequential formation kinetics [66], suggesting that the vacancy is not involved in the TD core [69]. In one investigation an enhanced TD formation was observed during electron irradiation at about 200°C [30]. There is a suggestion that the dioxygen-vacancy (VO_2) center is a core for the thermal donors, but experimental investigations show that this is not the case [67]. In Ref. 68 it was suggested that the TDs do not significantly interact with vacancies and self-interstitials produced during irradiation. However, we will here report that annihilation of TDs can be done with electron irradiation. Early interpretations of the effect of electron irradiation on TDs have been complicated since there is a change of the Fermi level due to the introduction of defects during irradiation which will affect both the resistivity and the possibility to observe the electronic transitions from the double TDs. These problems are not present when studying the TD-related vibrational bands.

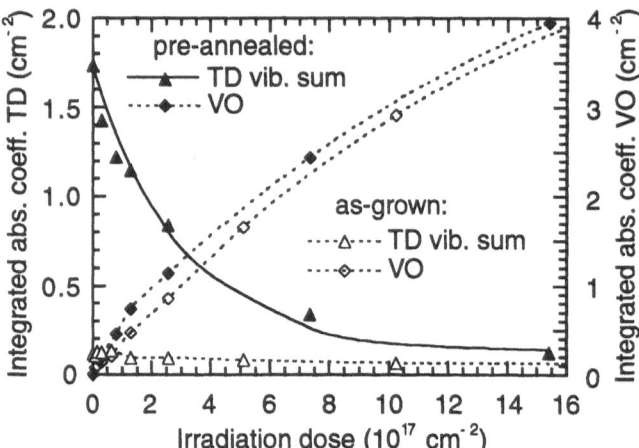

Figure 13. The effect of electron irradiation on the TDs for a pre-annealed and an as-grown sample and the formation of VO. The decay of the TD-related vibrational bands ($I_{999}+...+I_{1015}$) is fitted to first-order kinetics (solid line) for the pre-annealed sample.

The effect of electron irradiation on the TD complexes is illustrated in Fig. 13 for B doped as-grown and pre-annealed samples, where the dose dependence of the bands at 999, 1006, 1012 and 1015 cm^{-1} is shown according to the sum $I_{999}+I_{1006}+I_{1012}+I_{1015}$ for the pre-annealed sample and according to the sum $I_{1006}+I_{1012}$ for the as-grown sample. In the pre-annealed sample the bands at 999, 1006 and 1015 cm^{-1} are significantly or completely reduced while the 1012 band is more resistant against the irradiation and decays to an equilibrium level. The decay of the vibrational bands for the pre-annealed sample is in Fig. 13 fitted to first-order reaction kinetics (solid line), which indicates that the TD complexes are diminished by a dissociation process. The 1006 and 1012 bands in the as-grown sample are weakly affected by the irradiation and both are reduced to two different equilibrium levels.

Fig. 13 also illustrates that the formation of VO centers in the pre-annealed sample is slightly enhanced as compared with the as-grown sample. One possibility for enhanced VO formation is the suppression of the backward reactions, through competing reactions like $Al_s+Si_i \rightarrow Al_i$ or $C_s+Si_i \rightarrow C_i$. However, in this case the loss of $[O_i]$ is equal in the two samples, suggesting that the enhanced formation of VO requires a second source of oxygen. One possibility is that oxygen atoms involved in the TDs are transformed into VO centers.

A calibration coefficient for VO was deduced from the loss of $[O_i]$ in as-grown samples, where $[VO]=I_{VO}\times 1.3\times 10^{16}$ cm^{-3}. Using this coefficient for VO in the pre-annealed sample and comparing with the loss of $[O_i]$, the difference corresponds to the loss in strength of the TD related vibrational bands multiplied by K_{Oi}. This supports the assumption that oxygen atoms involved in the TDs are transformed to VO centers.

If the core of the TDs involves a Si_i atom giving rise to the electrical activity [16,18], the core could easily interact with vacancies produced during irradiation. This would affect the TDs in a similar way as observed in this investigation.

It can also be added that the effect of the irradiation in these experiments is similar to a thermal donor removal heat treatment at 650°C [33] or a dispersion treatment at 1100°C [35], after which only a small fraction of centers related to the 1012 and 1006 bands remains.

9. Summary

Infrared studies of vibrational bands related to oxygen atoms in different configurations in silicon have revealed new facts about the thermal donors and oxygen clusters in the temperature region 350-470°C. The results can be summarized as follows:

1. Oxygen atoms in the TD-core give rise to at least two different localized vibrational modes in the regions 975-1015 cm^{-1} and 716-748 cm^{-1}.

2. Bands at 975, 988, 999 and 1006 cm^{-1} correlate with the regular thermal double donors, while the 1012 band correlates with the EPR NL10 center.

3. Using the O_i calibration coefficient we conclude that the core of the TDs contains 1-2 oxygen atoms.

4. Transformation or dissociation of pre-existing oxygen clusters related to the 1012 band gives rise to the early TDs.

5. The bistability of TD1 and TD2 was observed from a change in the corresponding vibrational bands at 975 and 988 cm^{-1}. A band at 1020 cm^{-1} is suggested to correspond to the X1 and X2 configurations.

6. The appearance of a broad IR band at about 1060 cm^{-1} indicates that oxygen precipitates will develop in the temperature region where the TDs are formed. 90-95% of the oxygen lost from interstitial position during TD formation will appear in these precipitates.

7. The presence of Al atoms in CZ grown Si affects the TD formation by suppresses the formation of TD1 and TD2. The bands at 999 and 1006 cm^{-1} do not seem to be affected, while the centers related to the 1012 band are affected. A new IR absorption band at 992 cm^{-1} appears. The strength of the band correlates linearly with the Al doping. It is suggested that a $Si-O_i-Al_s$ center is formed.

8. In carbon rich silicon the TD-related vibrational bands are all suppressed. Other carbon related vibrational bands appear which have previously been reported.

9. The TD-related vibrational bands are significantly reduced by electron irradiation. The VO production is enhanced during the annihilation of the TDs. Oxygen atoms released from the TDs can explain the observed enhanced VO formation.

10. Acknowledgments

We appreciate discussions with Dr. B.G. Svensson. The authors are grateful to Dr. Zulehner from Wacker Chemitronic for providing the aluminum doped Si samples and to Dr. T. Carlberg for the O^{18} doped samples. Financial support was received from Teknikvetenskapliga Forskningsrådet (TFR) and from the National Defense Research Establishment.

58

11. References

1. Oehrlein, G.S. and Corbett, J.W. (1983) Mater. Res. Soc. Symp. Proc. **14**, 107.
2. *Oxygen in Silicon* (1994), Semiconductors and semimetals vol. 42, ed. F. Shimura, Academic Press, Inc..
3. Borghesi, A., Pivac, B., Sassella, A. and Stella, A. (1995) J. Appl. Phys. **77**, 4169.
4. Kimerling, L.C. (1986) in *Oxygen,Hydrogen and Nitrogen in Silicon*, eds. Pearton, S.J., Corbett, J.W., Mikkelsen Jr., J.C. and Pennycook, S.J., Materials Research Society, Pittsburg.
5. Watkins, G.D. (1986) Material Science Forum Vols. **10-12**, 953.
6. Wruck, D. and Gaworzewski, P. (1979) Phys. Stat. Sol (a) **56**, 557.
7. Schaake, H.F., Barber, S.C. and Pinizotto, R.F. (1981) in *Semiconductor Silicon*, Eds. Huff, H.R., Kriegler, R.J. and Takeishi, Y., The Electrochemical Society, Inc., Pennington, p. 273.
8. Oeder, R. and Wagner, P. (1983) Mater. Res. Soc. Symp. Proc. **14**, 171.
9. Pajot, B., Compain, H., Lerouille, J. and Clerjaud, B. (1983) Physica **117B/118B**, 110.
10. Wagner, P. and Hage, J. (1989) Appl. Phys. A **49**, 123.
11. Götz, W., Pensl, G. and Zulehner, W. (1992) Phys. Rev. B **46**, 4312.
12. Kaiser, W. (1957) Phys. Rev. **105**, 1751.
13. Stavola, M. and Snyder, L.C. (1993) in *Defects in Silicon*, eds. Kimerling, L.C. and Bullis, M.W., Electrochem. Soc., Pennington, NJ, p. 61.
14. Snyder, L.C., Deák, P., Wu and, R.Z., Corbett, J.W. (1989) Mater. Sci. Forum **38-41**, 329.
15. Jones, R. (1990) Semicond. Sci. Technol. **5**, 255.
16. Ourmazd, A., Schröter, W. and A. Bourret (1984) J. Appl. Phys. **56**, 1670.
17. Michel, J., Niklas, J.R. and Spaeth, J.-M. (1989) Phys. Rev. B **40**, 1732.
18. Deák, P., Snyder, L.C. and Corbett, J.W. (1992) Phys. Rev. B **45**, 11 612.
19. Tan, T.Y., Kleinhenz, R. and Schneider, C.P. (1986) Mat. Res. Soc. Symp. Proc. **59**, 195.
20. Stavola, M., Patel, J.R., Kimerling, L.C. and Freeland, P.E., (1983) Appl. Phys. Lett. **42**, 73.
21. Gösele, U. and Tan, T.Y. (1982) Appl. Phys. A **28**, 79.
22. Mao, B.-Y., Lagowski, J. and Gatos, H.C. (1984) J. Appl. Phys. **56**, 2729.
23. Oehrlein, G.S., Lindström, J.L. and Cohen, S.A. (1985) in *13th International Conference on Defects in Semiconductors*, eds. Kimerling, L.C. and Parsey jr., J.M., Metallurgical Society of AIME: Coronado, CA., p. 701.
24. Kaiser, W., Keck, P.H. and Lange, C.F. (1956) Phys. Rev. **101**, 1264.
25. Corbett, J.W., Watkins, G.D. and McDonald, R.S. (1964) Phys. Rev. **135**, A1381.
26. Hu, S.M. (1980) J. Appl. Phys. **51**, 5945.
27. Hallberg, T. and. Lindström, J.L (1992) J. Appl. Phys. **72**, 5130.
28. Bean, A. R. and Newman, R.C. (1972) J. Phys. Chem. Solids **33**, 255.
29. Oehrlein, G.S., Lindström, J.L and Corbett, J.W. (1982) Appl. Phys. Lett. **40**, 241.
30. Pajot, B. and von Bardeleben, J. (1985) in *13th International Conference on Defects in Semiconductors*, eds. Kimerling, L.C. and Parsey jr., J.M., Metallurgical Society of AIME: Coronado, CA., p. 685.

31. Pajot, B., Stein, H.J., Cales, B. and Naud, C. (1985) J. Electrochem. Soc. **132**, 3034.
32. Lindström, J.L., Weman, H. and Oehrlein, G.S. (1987) Phys. Status Solidi (a) **99**, 581.
33. Lindström, J.L. and Hallberg, T. (1994) Phys. Rev. Lett. **72**, 2729.
34. Lindström, J.L. and Hallberg, T. (1995) J. Appl. Phys. **77**, 2684.
35. Hallberg, T. and Lindström, J.L., to be published in J. Appl. Phys.
36. Graff, K., Grallath, E., Ades, S., Goldbach, G. and Tölg, G. (1973) Solid State Electron. **16**, 887.
37. Newman, R.C. and Willis, J.B. (1965) J. Phys. Chem. Solids **26**, 373.
38. Clauws, P., Proc. from the E-MRS 1995 Spring Meeting, Strasbourg, France, to be published in Materials Science & Engineering B.
39. van Wezep, D.A., Gregorkiewicz, T., Bekman, H.H.P.Th. and Ammerlaan, C.A.J. (1986) Mater. Sci. Forum **10-12**, 1009.
40. Hallberg, T. and Lindström, J.L. (1996) Materials Science & Engineering B **36**, 13.
41. Claybourn, M. and Newman, R.C. (1987) Appl. Phys. Lett. **51**, 2197.
42. Markevich, V.P., Makarenko, L.F. and Murin, L.I. (1986) phys. stat. sol (a) **97**, K173.
43. Stein, H.J., Hahn, S.K. and Shatas, S.C. (1986) J. Appl. Phys. **59**, 3495.
44. Tkachev, V.D., Makarenko, L.F., Markevich, V.P. and Murin, L.I. (1984) Sov. Phys. Semicond. **18**, 324.
45. Makarenko, L.F., Markevich, V.P. and Murin, L.I. (1985) Sov. Phys. Semicond. **19**, 1192.
46. Latushko, Ya.I., Makarenko, L.F., Markevich, V.P. and Murin, L.I. (1986) Phys. Stat. Sol. (a) **93**, K181.
47. Chantre, A. (1987) Appl. Phys. Lett. **50**, 1500.
48. Wruck, D. and Spiegelberg, F. (1986) Phys. Stat. Sol. (b) **133**, K39.
49. Hallberg, T. and Lindström, J.L., submitted to Appl. Phys. Lett.
50. Tempelhoff, K., Spiegelberg, F., Gleichmann, R. and Wruck, W. (1979) Phys. Stat. Sol. (a) **56**, 213.
51. Hu, S.M. (1980) Appl. Phys. Lett. **36**, 561.
52. Livingston, F.M., Messoloras, S., Newman, R.C., Pike, B.C., Stewart, R.J., Binns, M.J., Brown, W.P. and Wilkes, J.G. (1984) J. Phys. C: Solid State Phys. **17**, 6253.
53. Seres, J. and Hild, E. (1989) Materials Science Forum **38-41**, 661.
54. Bergholz, W., Pirouz, P. and Hutchison, J.L., in *13th International Conf. on Defects in Semiconductors*, eds. Kimerling, L.C. and Parsey Jr., J.M. (1985) Metallurgical Society of AIME: Coronado, CA., p. 717.
55. Reiche, M. and Reichel, J. (1989) Materials Science Forum **38-41**, 643.
56. McQuaid, S.A., Binns, M.J., Londos, C.A., Tucker, J.H., Brown, A.R. and Newman, R.C. (1995) J. Appl. Phys. **77**, 1427.
57. Fuller, C.S. and Doleiden, F.H. (1958) J. Appl. Phys. **29**, 1264.
58. Fuller, C.S., Doleiden, F.H. and Wolfstirn, K. (1960) J. Phys. Chem. Solids **13**, 187.
59. Drakeford, A.C.T. and Lightowlers, E.C. (1988) Mat. Res. Soc. Symp. Proc. **104**, 209.
60. Gegorkiewiecz, T., van Wezep, D.A., Bekman, H.H.P.Th. and Ammerlaan, C.A.J. (1987) Phys. Rev. B **35**, 3810.

61. Bekman, H.H.P.Th., Gregorkiewicz, T., van Wezep, D.A. and Ammerlaan, C.A.J. (1987) J. Appl. Phys. **62**, 4404.
62. Lindström, J.L. and Hallberg, T. (1996) Materials Science & Engineering B **36**, 150.
63. Gregorkiewiecz, T., Bekman, H.H.P.Th. and Ammerlaan, C.A.J. (1992) Phys. Rev. B **46**, 4582.
64. Fuller, C.S. and Logan, R.A. (1957) J. Appl. Phys. **28**, 1427.
65. Wagner, P. (1986) Mater. Res. Soc. Symp. Proc. **59**, 125.
66. Henry, A., Saminadayar, K., Pautrat, J.L. and Magnea, N. (1988) phys. stat. sol. (a) **107**, 101.
67. Svensson, J., Svensson, B.G. and Lindström, J.L. (1986) Appl. Phys. Lett. **49**, 1435.
68. Neimash, V.B., Sagan, T.R., Tsmots, V.M., Shakhovtsov, V.I. and Shindich, V.L. (1991) Sov. Phys. Semicond. **25**, 1117.
69. Helmreich, D. andSirtl, E. (1977) in *Semiconductor Silicon*, eds. Huff, H.R. and Sirtl, E. (1977) Electrochem. Soc., Princeton, p. 626.

MAGNETIC RESONANCE INVESTIGATIONS
OF THERMAL DONORS IN SILICON

C.A.J. AMMERLAAN, I.S. ZEVENBERGEN, Yu.V. MARTYNOV,
AND T. GREGORKIEWICZ
Van der Waals - Zeeman Institute, University of Amsterdam
Valckenierstraat 65-67, NL-1018 XE Amsterdam, The Netherlands

Abstract

Heat treatment of oxygen-rich silicon in the temperature range 450-500°C for times ranging from minutes to several hundreds of hours produces centers with donor activity. By the method of electron paramagnetic resonance (EPR) two prominent spectra, labelled Si-NL8 and Si-NL10, respectively, were observed and were associated to paramagnetic states of these centers. On the basis of angular dependent resonance patterns the centers were described with g tensors of orthorhombic-I symmetry. Introducing the ^{17}O oxygen isotope and applying electron nuclear double resonance (ENDOR) the presence of oxygen atoms in both centers was established; the oxygen atoms are located on one of the mirror planes of the 2mm symmetry group. From ENDOR on the ligand ^{29}Si nuclei the wave function describing the electron with unpaired spin was found to be quite extended; this is consistent with a shallow character of the donor state. Both centers appear to exist in a large number of configurations, up to 17 have been reported, the versions with smaller binding energy and more isotropic g tensor developing upon increase of the heat-treatment time. These very similar, but yet distinct, configurations are not resolved in EPR at standard frequencies, only a continuous shift of the overall g tensor parameters is observed. For the spectrum Si-NL8 the model of a singly ionized shallow double oxygen-related donor is generally accepted; contrary to this, the situation with spectrum Si-NL10 remained more puzzling. The two main difficulties are, first, the existence of two different families of heat-treatment centers with very similar properties and, secondly, the observation of distinct ^{27}Al ENDOR interactions in the Si-NL10 spectrum together with the observation of Si-NL10 centers in silicon with definitely no aluminum present. Recently, a prominent role of hydrogen in the formation of Si-NL10 centers has been discovered. Hydrogen hyperfine interactions were detected by ENDOR and further studied by field-scanned ENDOR. In deuterium-doped material the quadrupole interaction of the local electric field gradient with the deuteron was measured. The direction of the hydrogen

61

R. Jones (ed.), Early Stages of Oxygen Precipitation in Silicon, 61–82.
© *1996 Kluwer Academic Publishers.*

bond was found to be parallel to the <011> crystalline axis perpendicular to the oxygen-containing plane of the thermal donor. It also appeared that the true symmetry of this center is triclinic, i.e., lower than the orthorhombic symmetry as resolved in the EPR. On this basis an identification of the Si-NL10 center as a singly hydrogen-passivated thermal donor in its neutral charge state is proposed. It appears necessary to distinguish between aluminum-containing centers, Si-NL10(Al), and hydrogen-containing ones, Si-NL10(H). Infrared absorption spectra as reported in the literature, and related to so-called shallow thermal donors (STD's), are probably due to these Si-NL10 centers.

1. Introduction

Thermal donors (TD's) are heat-treatment centers of a shallow double donor character. These centers are generated in oxygen-rich silicon upon annealing in the 400-500°C temperature range and are most probably related to an oxygen aggregation process. For a recent review see [1]. Also for germanium formation of thermal donors has been observed [2]. While the effect of TD formation was reported already four decades ago the issue still lacks comprehensive understanding and constitutes perhaps the most fascinating subject in the physics of defect centers in silicon. The interest in TD's stems from the fundamental side and is closely related to such phenomena as aggregation and impurity reactions. However, it has at the same time also a considerable applied science component, since TD's can be generated in high concentration and therefore fully determine the electrical character of the material.

The most important features of TD's established so far include:

$\star$ The electrical character of a TD is that of an effective mass theory (EMT) double donor, with the first ionization level at approx. 70 meV, and the second one at approx. 150 meV below the bottom of the conduction band.

$\star$ The generation kinetics and the maximum concentration of TD's are related to the initial interstitial oxygen concentration available in the material.

$\star$ TD's constitute a family of up to 17 species which develop subsequently upon heat treatment, with the more shallow species being generated later.

$\star$ The overall symmetry of the TD is orthorhombic-I (C_{2v}).

$\star$ Short heat-treatment at a higher temperature ($T > 600°C$) destroys the electrical activity of TD's. This effect is accompanied by a coincident increase of interstitial oxygen concentration.

$\star$ Some other impurities - carbon, hydrogen and aluminum - were found to influence the TD generation process.

At the same time several questions remain open and clearly require further studies. The most prominent of these are:

$\star$ The microscopic structure of the TD core and the origin of its double donor activity.

$\star$ The growth mechanism and the related question of structural differences between different TD species.

★ The mechanism responsible for loss of electrical activity and (partial?) dissociation of the TD center.

By inspection of the above given list of unsolved problems related to TD's one concludes that whereas the general properties of these centers are well studied and, for the major part, understood, the microscopic picture is nearly completely missing. This is best illustrated by the fact that, in spite of intensive investigations, the theoretical model accounting for all the properties of TD's has yet to be developed.

In the studies of defects magnetic resonance has been frequently applied in view of its unique ability to provide information on a microscopic scale. As a result the majority of structural models of defect centers are based on data from magnetic resonance spectroscopy. In this paper we will review the results of EPR and ENDOR studies of TD's.

2. Electron paramagnetic resonance results

Among many other experimental techniques also EPR has been employed to investigate TD's. In an early study [3] several so-called *heat-treatment centers* have been reported in Czochralski-grown silicon annealed for various periods of time at a temperature of 450°C. In a later report [4] the actual number of these centers has been reduced to two - Si-NL8 and Si-NL10 and these were studied in a considerable detail. Both spectra are illustrated in Fig.1 for the three main crystallographic directions in the silicon lattice.

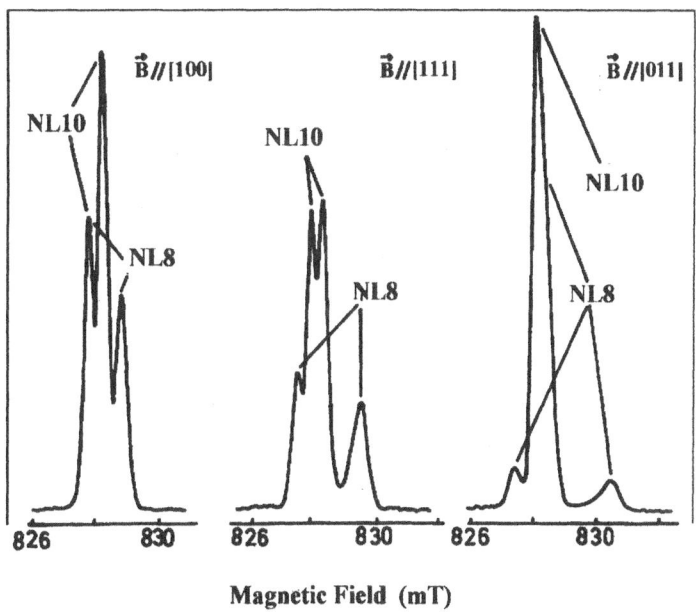

Magnetic Field (mT)

Figure 1. The TD-related EPR spectra Si-NL8 and Si-NL10 as measured in three main directions for the K microwave band of approx. 23 GHz.

64

Both spectra have spin $S=\frac{1}{2}$ and were found to be generated in oxygen-rich silicon. For p-type material the formation of Si-NL8 was practically independent of the particular acceptor (B, Al, Ga, In); formation of Si-NL10 was clearly enhanced by the presence of aluminum. Upon short annealing first the Si-NL8 center could be detected; its intensity increased with prolonged heat-treatment, reached a maximum and then decayed. The Si-NL10 spectrum appeared only for longer annealing; it could be observed also earlier but upon white-light illumination of the sample. It has been found that whenever the Si-NL8 spectrum has been detected then also the Si-NL10 would appear upon longer annealing. In n-type material [5] only Si-NL10 center could be observed.

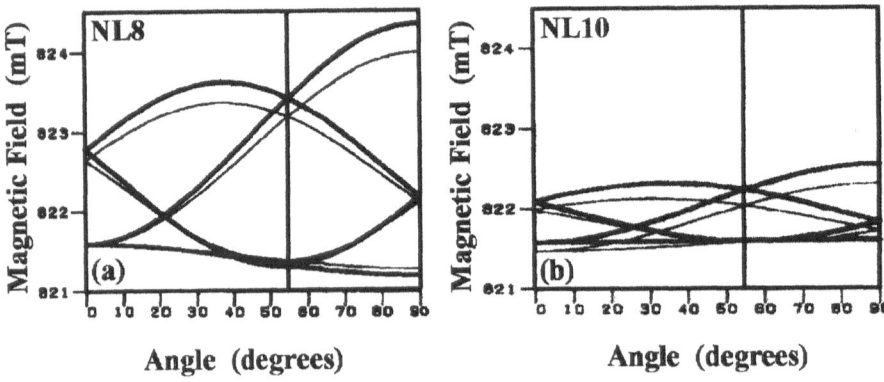

Figure 2. The angular dependence for the Si-NL8 (a) and Si-NL10 (b) EPR spectra. Thick and thin lines correspond to the annealing times of 10 and 100 hrs, respectively.

The angular dependence of both spectra as measured in the K microwave band is depicted in Fig.2. As can be concluded, both spectra are of the orthorhombic-I symmetry whose elements in the silicon lattice include 2 (perpendicular) mirror planes ((011)) and a twofold axis (<100>). The values of the g-tensors are in a close vicinity of 2 and their anisotropy is small. According to the general defects classification by Corbett and Sieverts [6], these characteristic features put both centers in the category of shallow donors. One notes further that the anisotropy of the Si-NL10 spectrum is clearly smaller than that of the Si-NL8. According to the earlier mentioned classification this indicates that the ionization energy of Si-NL10 center should be smaller than that of Si-NL8.

The observation of two TD-related EPR spectra, both with spin $S=\frac{1}{2}$, presented a problem for an interpretation. In general, a double donor DD can be paramagnetic in its singly ionized charge state DD^+, with $S=\frac{1}{2}$. Also the zero charge state DD^0 can be paramagnetic with S=1. However, in the latter case the paramagnetic state has usually higher energy than the S=0 state. Following the uniaxial stress measurements [7] the Si-NL8 spectrum has been identified with the singly ionized thermal donor TD^+. The particular identification of the other TD-related EPR spectrum - Si-NL10 - is still debated upon and several possibilities have been considered [8].

Fig.2 illustrates yet another characteristic feature of both TD-related EPR spectra. As can be seen the g-values of the spectra are not uniquely defined and change upon prolongation of the heat treatment; the longer the annealing the more isotropic the g-tensor becomes. This characteristic g-shifting effect is very unusual among defect centers in semiconductors. Its occurrence for TD-related spectra is illustrative of the multispecies character of TD's and is an EPR equivalent of the effect observed also in the IR absorption, where a somewhat different TD spectrum is found depending on the annealing stage. The g-shifting effect carries then the information about the TD transformation process and has been investigated further. It has been established that not all the components of the g-tensor were undergoing the transformation; the change was most pronounced for the off-diagonal g_{xy} element whose value is directly related to the splitting ΔT between the two spectral components as measured for the $<111>$ orientation of the magnetic field. Fig.3 illustrates the evolution of the ΔT splitting with the annealing time for the Si-NL8 and Si-NL10 spectra.

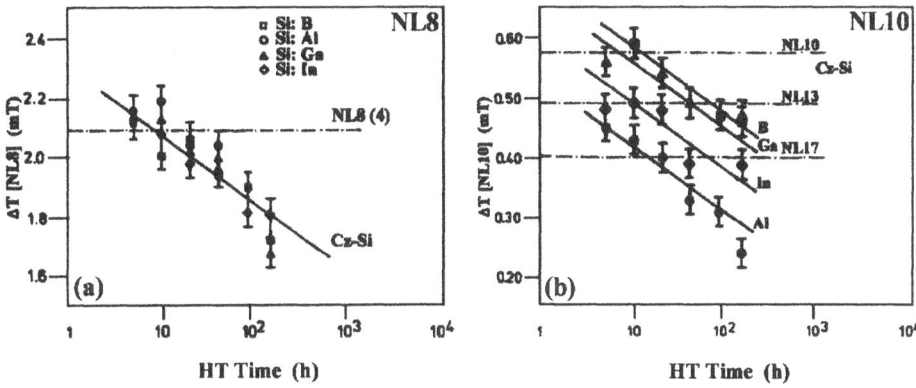

Figure 3. The annealing time dependence of the spectral splitting ΔT, as measured for the magnetic field $B \parallel <111>$, for (a) the Si-NL8, and (b) the Si-NL10 EPR spectra.

For the Si-NL8 center the g-shifting effect was found to be independent of the dopant. For the Si-NL10 such an influence can be concluded from the figure; the most isotropic spectra are observed for an aluminum-doped material. If we assume that the degree of anisotropy is related to the ionization energy of a given center then we conclude that in Al-doped silicon the most shallow NL10 species are generated. If we assume further that the g-value shift is indicative of the transformation process then we also have to acknowledge that the presence of aluminum has a catalytic effect on this process. The particular role of aluminum in the structure and generation of Si-NL10 centers still remains as one of the more puzzling questions in the EPR studies of TD's. We will come back to it in the next section. Here we only mention that the issue is being additionally complicated by the fact that while a substitutional aluminum in silicon is an acceptor, it becomes a double donor when displaced onto an interstitial site.

66

The fact that the g-shifting effect illustrates the varying composition of particular species present at a given annealing stage has been directly confirmed in a high-field EPR experiment conducted for a frequency of 350 GHz.

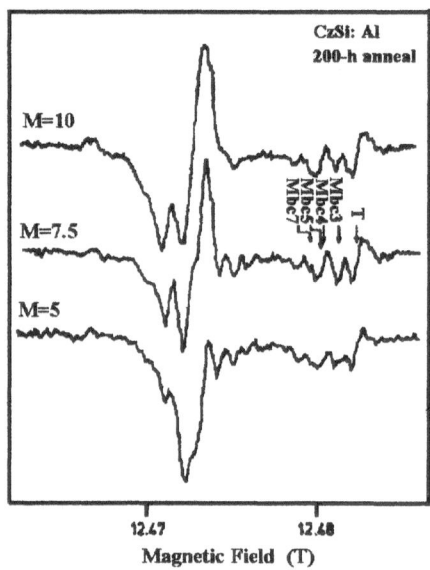

Figure 4. The Si-NL10 spectrum as measured for **B** $||<011>$ in the high-field EPR experiment - the resonance frequency ν=350 GHz

Fig.4 shows the Si-NL10 spectrum as measured for an Al-doped Cz-Si annealed for 200 hrs at 470°C, with magnetic field along the $<011>$ crystallographic direction. As can be seen, the so-called U6 point of an orthorhombic-I spectrum is now clearly split into at least 5 distinct components indicating the presence of 5 different species of the Si-NL10 center; each one characterized by a somewhat different EPR spectrum. The fact that the species-related splitting is most pronounced for the U6 point relates to the property that the individual g-tensors differ mostly in their off-diagonal element g_{xy}. In view of a rather primitive EPR detection system used in this experiment the sensitivity was relatively low. Therefore the measurements were possible, so far, only for the most intense Si-NL10 spectrum in an aluminum-doped sample annealed for 200 hrs. Experiments for differently doped and annealed materials could directly reveal the influence of these parameters on the generation kinetics of individual Si-NL10 species. These experiments require, however, a more advanced high-field EPR experimental set-up; this became recently available in our laboratory and new results might be expected. Also similar investigations of the Si-NL8 center, disclosing details of the symmetry and generation kinetics of the TD$^+$, have yet to be performed.

3. Electron nuclear double resonance results

In an ENDOR experiment hyperfine interactions between the electron spin responsible for the paramagnetism of a given center and nuclear momenta at atomic sites accessible to the spin density are probed. In this way the so-called *spin mapping* can be realized, where the spin density on sites within the center and also in a more or less direct vicinity of the defect core can be determined. For a center with an extended spin localization hyperfine interactions with several shells of neighboring atoms can usually be detected. It is customary to distinguish between the self- and ligand ENDOR; the first case concerns the interactions with the nuclear spin of an atom (or atoms, for an extended defect) which forms the paramagnetic core of the center. The second case concerns the interactions with the crystal surrounding the defect core. From ENDOR an experiment details of the electronic and atomic structure of a defect are unraveled. The ENDOR ability to deliver microscopic information is usually superior to that provided by other experimental techniques commonly used in defect studies, such as IR absorption, photoluminescence and DLTS spectroscopy. Based on the analysis of ENDOR data reliable microscopic models of many defect centers were proposed.

The power of ENDOR spectroscopy has also been applied to the studies of TD's. Unfortunately in this case the high resolution of this technique results in a low intensity of signals, as the spectra related to different TD species do not overlap and are detected separately. Consequently high concentrations of TD's are required. This effectively low sensitivity seriously hampers ENDOR studies of TD's and lowers credibility of obtained results and conclusions which are based on them. In what follows results of ENDOR investigations concerning hyperfine interactions with different magnetic nuclei are presented.

3.1 SILICON ENDOR

An observation of ^{29}Si ENDOR is (nearly) always possible for defects in silicon. This follows from a relatively high natural abundance of this magnetic isotope of approx.5%. Based on the results of ^{29}Si ENDOR one can conclude about the extension of the spin localization within the crystal rather than about the core of the center. By summing up the spin localizations determined for individual neighboring atom shells the total electron localization may be found revealing information pertinent to the general character of the electronic structure of the defect: deep $\leftrightarrow$ shallow, isotropic $\leftrightarrow$ anisotropic.

The ^{29}Si ENDOR experiments have been conducted for both TD-related EPR centers [9,10]. The results can be summarized as follows:

⋆ spin wave function character: In the LCAO (linear combination of atomic orbitals) analysis the spin wave function on an atomic site is approximated by locally available electron orbitals. For the standard sp^3 hybridization the ratio of $s-$ to $p-$orbital localizations α^2/β^2 is 1/3. For the investigated centers $\alpha^2/\beta^2 \approx 3$ was found indicating a preferred electron localization in spherically symmetric s-orbitals. In this aspect both centers resemble shallow EMT donors.

⋆ spin localization: The total localization η^2 of the electron responsible for the S=$\frac{1}{2}$ paramagnetism was found to be approx. 5.5% and 0.5% for Si-NL8 and Si-NL10, respectively. This is very small, even compared to a classical example of a shallow donor such as a substitutional phosphorus atom, for which the total localization of $\eta^2 \approx 15\%$ is measured. Such a low η^2 value indicates a very delocalized character of both centers for which the commonly applied approach of LCAO analysis fails. Since η^2 for Si-NL10 is clearly smaller than for Si-NL8, we assume that this defect is also more shallow. In this way the ENDOR experiment confirms our earlier conclusion based on the comparison of g-value anisotropy for both centers.

We note that the fact that the spin wave function is so delocalized and yet the g-tensors of both centers show measurable anisotropy is very unusual and presents a considerable challenge to any theoretical model attempting to describe the electronic structure of the TD. It is also worth pointing out that the ^{29}Si ENDOR experiments failed to unravel a single Si atom with a prominent spin localization; this result strongly disfavors numerous TD core models where the donor activity originates from a silicon interstitial atom.

3.2 OXYGEN ENDOR

As has been mentioned before, the formation process of TD's has always been related to clustering of oxygen. In EPR the presence of a particular nucleus in the structure of a center manifest itself by a hyperfine interaction, which leads to a characteristic splitting of the spectrum. However, in silicon intentionally doped with the magnetic ^{17}O isotope of oxygen no such splitting could be observed neither for Si-NL8 nor for Si-NL10 spectrum. The presence of oxygen atoms in the structure of both centers has been finally concluded from ENDOR experiments; the magnitude of the hyperfine interaction with oxygen nuclei was found much too small to be observable in EPR [11,12].

Before we summarize the information on the structure of both centers as revealed by oxygen ENDOR we start by a short explanation on how such an information can be obtained from ENDOR data. In order to be able to discuss the structure of the centers and to present the symmetry arguments we shall introduce the labelling of various symmetry classes of ENDOR tensors. The symmetry of Si-NL8 and Si-NL10, as derived from EPR, is orthorhombic-I. Such a defect has 6 equivalent orientations in the silicon lattice, labelled *ab, ac, ad, bc, bd*, and *cd* [13]. Fig.5 presents a fragment of the silicon lattice, positions of the defect's two mirror planes and the principal axes of its g-tensor. The symmetry of an ENDOR pattern due to a particular shell of magnetic nuclei is the symmetry of the wave function of the center in conjunction with one of the nuclei in the shell. Clearly, the symmetry of ENDOR is lower or equal to the symmetry of the wave function. For an orthorhombic center one can discriminate between 4 types of positions (see Fig.5): if the nucleus is on the 2-fold symmetry axis of the center (T-position), the symmetry of its ENDOR pattern is orthorhombic; if it is situated in one of the defect's symmetry planes (Mad- and Mbc-positions) the symmetry is monoclinic-I, and, finally, if it is outside both mirror planes (G-position) the symmetry will be triclinic. We note further that the shell of triclinic symmetry (G-type) will contain 4 atoms, of monoclinic symmetry (Mad and Mbc) - 2 atoms in one of the mirror planes, and, finally, the orthorhombic shell will correspond to a single atom on the 2-fold axis of the defect.

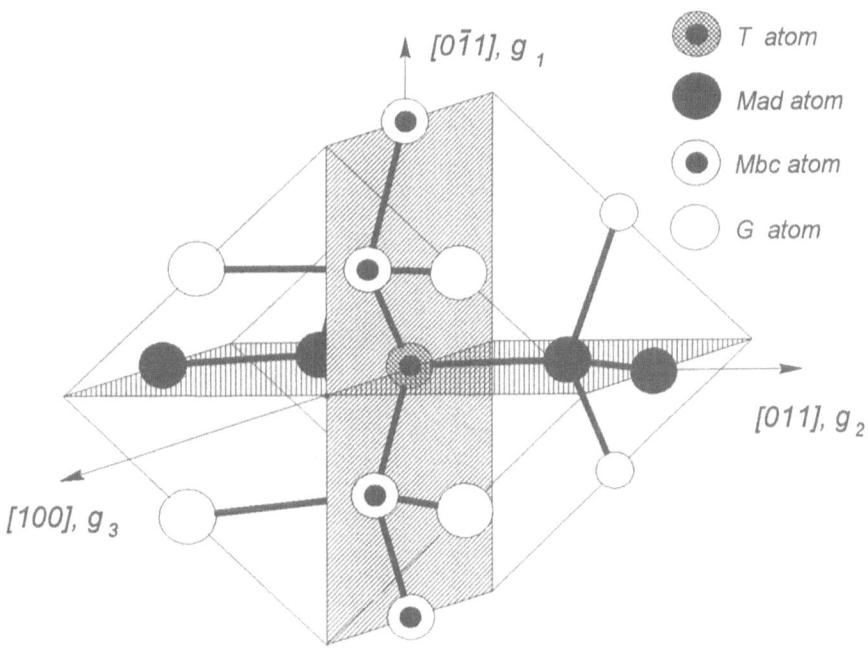

Figure 5. Principal directions of an orthorhombic-I center for *ad* orientation with respect to the silicon lattice. Two mirror planes of the defect are indicated together with atoms at four types of positions which can be distinguished in ENDOR.

Based on the above we can now use ENDOR data to understand the microscopic structure of a defect. Fig.6a shows a computer simulation of the angular dependence of the oxygen ENDOR spectrum as obtained for a single orientation of the Si-NL10 center [14]. As can be seen the pattern is rather symmetric with respect to the Zeeman frequency of ^{17}O uniquely identifying this nucleus as responsible for the hyperfine interaction. The symmetry type is Mbc and following our earlier reasoning we conclude that the relevant atomic shell contains two oxygen atoms in the mirror plane of the defect. In the experiment a series of at least 8 similar tensors have been observed - Fig.6b shows two of these. It is evident that superior resolution is necessary to discriminate between them. All the observed tensors were of the same symmetry type. Such a result indicated that oxygen atoms were always contained within a mirror plane of the same type. By the field stepped ENDOR (FSE) technique, which allows for a direct correlation between NMR and EPR transitions coupled in an ENDOR experiment, it has been shown that the oxygen tensors depicted in Fig.6b corresponded to different Si-NL10 species. In this way the oxygen ENDOR experiment directly confirmed the multispecies character of the Si-NL10 center and explained the occurrence of the g-shifting effect.

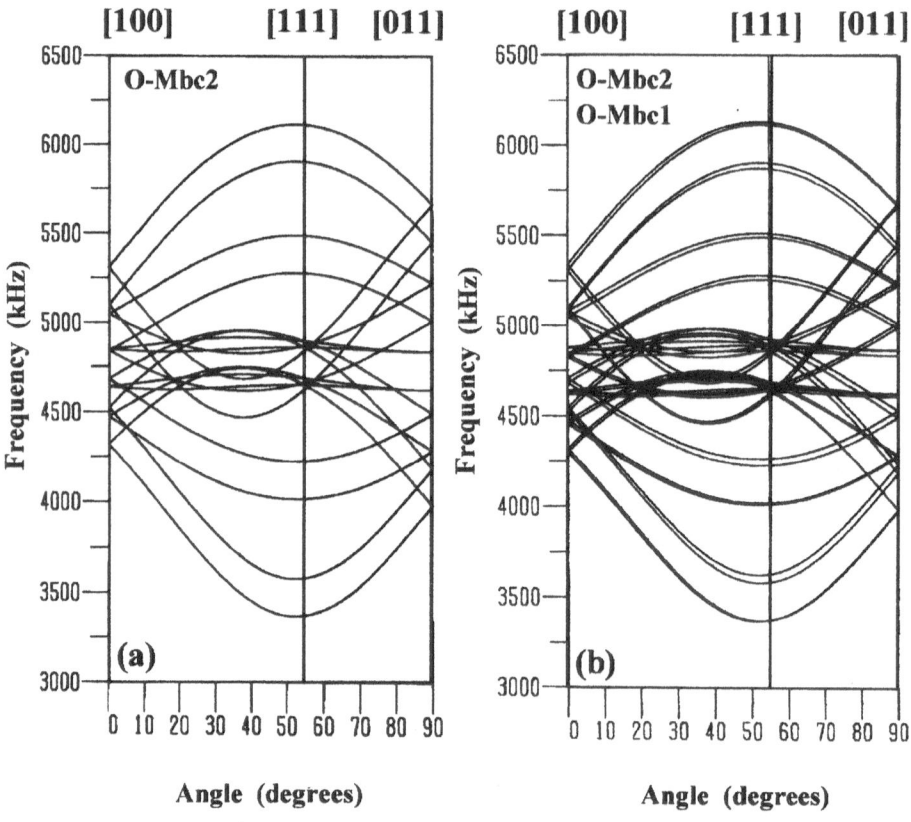

Figure 6. Computer simulation of the angular ENDOR pattern corresponding to a single EPR orientation (*bc*) for (**a**) one tensor (O-Mbc2) and (**b**) two similar tensors (O-Mbc1, O-Mbc2) corresponding to two different Si-NL10 species.

In order to fully analyze the oxygen ENDOR data not only the hyperfine but also the quadrupole interaction had to be taken into consideration. The hyperfine interaction tensor A was found to be nearly isotropic and of small magnitude; in terms of the LCAO analysis is corresponded to a minute spin localization of less than 0.01%. This results shows again an extremely delocalized character of the Si-NL10 center but bears very little information about the particular role of oxygen in the structure of this center. In this aspect the analysis of the quadrupole interaction was more fruitful. The experimentally measured quadrupole tensor could be interpreted as resulting from a superposition of two axial components, corresponding to two identical bonds within the mirror plane. The angle between these bonds was estimated as 155°. Taken together the ^{17}O ENDOR data resulted in a rather unexpected conclusion that the oxygen atom in the Si-NL10 center occupies a position very similar its usual bond site. No evidence for an "oxygen core" of the center, with a prominent spin localization, has been found.

The [17]O ENDOR experiment has also been performed for the Si-NL8 center [12]. Also in this case the a series of similar interaction patterns related to different species has been detected. The observed hyperfine interactions were again nearly isotropic but a total spin localization was considerably higher. The quadrupole interaction was found to be identical to that discussed earlier for the Si-NL10. In view of the preceding discussion we conclude that the local structure around oxygen must be identical for both centers. This observation is crucial for their mutual identification. Also for the Si-NL8 center no evidence of an electrically active oxygen core of TD has been found.

We finally note that, regretfully, both studies failed to produce any evidence that the individual species of the Si-NL8 and Si-NL10 centers differ by the actual number of oxygen atoms participating in the cluster. Such a role of oxygen is assumed in a vast majority of the so-far proposed TD models. Consequently, the particular role of oxygen in the formation and transformation of the heat-treatment centers remains unclear.

3.3 ALUMINUM ENDOR

In ENDOR experiments on the Si-NL8 center only hyperfine interactions with silicon and oxygen nuclei were observed. This is in line with the identification of this center as the TD^+ state and the general view that TD's are impurity independent, at least as far as their structure is concerned. In contrast to that, for the other TD-related EPR center Si-NL10, the hyperfine interactions with aluminum and hydrogen were also detected. Fig.7 shows aluminum ENDOR data as obtained for the Si-NL10 center generated in an oxygen-rich Al-doped sample. The observation of Al ENDOR was unexpected and consequently similar investigations were performed for differently doped samples. It has been established that the interactions with ^{27}Al could exclusively be observed in material where aluminum was introduced intentionally. In this way the possibility that the Si-NL10 center could be related to a residual presence of Al contamination has been excluded. Since, as discussed in the preceding section, the Si-NL10 center can be generated in any oxygen-doped silicon, such a conclusion meant that aluminum could be incorporated in the Si-NL10 structure without seriously changing its EPR spectrum. One should note here that the observation of well-defined ENDOR pattern ("characteristic" as opposed to "distant" ENDOR) means that the aluminum atom takes a characteristic position within the structure of the center, and the interaction cannot be viewed as that with Al nuclei randomly dispersed in the silicon crystal. In the latter case only one broad resonance at the nuclear Zeeman frequency of aluminum would have been seen. Following its first observation aluminum ENDOR of the Si-NL10 center has been studied in detail [14].

Similarly to the earlier described oxygen ENDOR also here a series of rather similar tensors has been observed; by the FSE technique each individual tensor was related to a different species of the Si-NL10 center. Fig.7a shows a computer simulation of the prominent ENDOR pattern as observed for a single EPR orientation. The pattern is clearly symmetric with respect to the nuclear Zeeman frequency of ^{27}Al and the symmetry type is orthorhombic-I. This means that this particular Si-NL10 species contains in its structure an aluminum nucleus located on the 2-fold <100> axis of the defect.

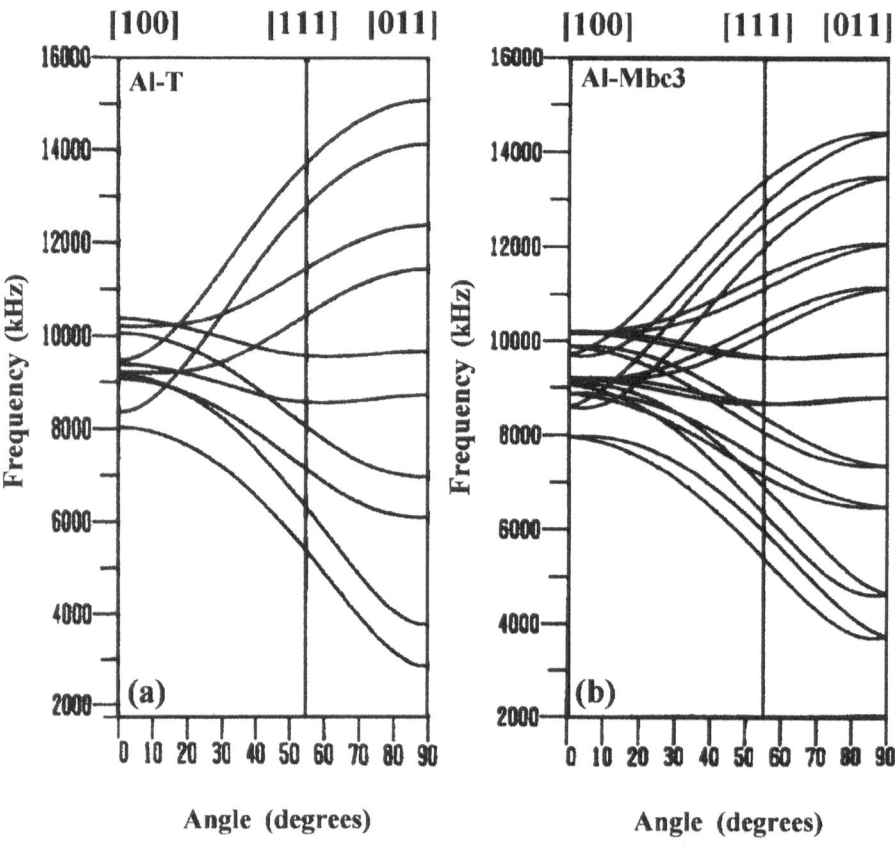

Figure 7. Computer simulation of the angular ENDOR pattern corresponding to a single EPR orientation (*bc*) for two similar tensors (**a**) orthorhombic Al-T and (**b**) monoclinic Al-Mbc3, related to two different Si-NL10 species.

In Fig.7b two Al ENDOR patters corresponding to different Si-NL10 species are shown; it is clear that the second one - Al-Mbc3 - is of the lower monoclinic-I symmetry type. Following the presented reasoning about different ENDOR shells, this means that the Si-NL10 species characterized by the Al-Mbc3 tensor contains two Al atoms located at one of the mirror planes. Such a conclusion stems from the assumed orthorhombic symmetry of the defect as a whole. Indeed the orthorhombic symmetry is concluded from the EPR spectrum. However, it might also be possible that the actual symmetry of the center is lower than that disclosed by EPR, but that the splitting due to the lowering of symmetry is small and therefore hidden within the experimental width of the resonance line. This possibility has been checked experimentally. The idea of the experiment and its outcome are depicted in Fig.8.

73

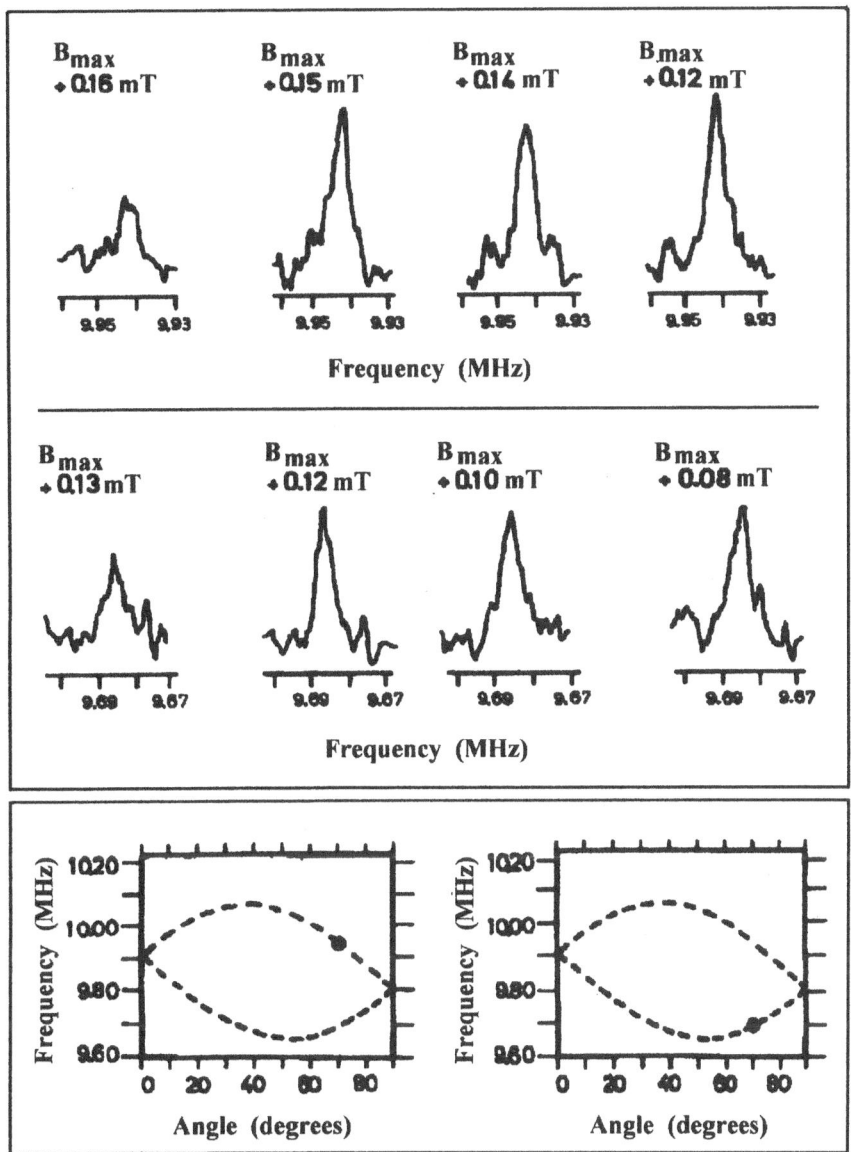

Figure 8. The application of the FSE technique to reveal the true symmetry of a Si-NL10 species related to a particular Al ENDOR tensor of monoclinic symmetry. The intensity of the ENDOR signal for the two resonances of one loop of the Al-Mbc3 tensor corresponding to the same EPR orientation, as indicated in the lower part of the figure, was recorded as a function of the magnetic field, with respect to the center of the EPR line B_{max}.

In the experiment, for one angle, two ENDOR signals belonging to the same loop of the Al-Mbc tensor were chosen, and their intensities were measured as a function of the field. For a truly orthorhombic center both branches of the loop are generated by the same EPR orientation and should attain maximum intensity for the same field value. If, however, the EPR orientation itself is split due to symmetry lowering then the intensity of ENDOR lines will follow the actual position of EPR components and will, in this way, reveal the splitting. In the figure the intensity of ENDOR lines is shown for several different magnetic field values measured with respect to the center of the EPR line B_{max}. As can be concluded from the figure, the ENDOR signals corresponding to different branches reach the maximum for different magnetic field values; this result proves that the real symmetry of the Si-NL10 species related to the Al-Mbc3 is monoclinic. Consequently, this ENDOR shell contains only a single Al atom located in a mirror plane of a monoclinic Si-NL10 defect.

Summarizing the results of the ^{27}Al ENDOR experiment, we conclude that the Si-NL10 center generated in an aluminum-doped material contains in its structure one Al atom. Out of several Si-NL10 species which were analyzed, only one has true orthorhombic symmetry and in this case the aluminum atom is found on the intersection of the two mirror planes; all the other species were monoclinic, with the Al atom in the remaining mirror plane. Spin localizations as found for the Al nuclei were also very small - $\eta^2 \approx 0.1\%$. One should note that this small localization value is nevertheless bigger than that determined earlier for oxygen atom sites.

3.4 HYDROGEN ENDOR

The evidence of the influence of hydrogen on the TD formation process has been accumulating for some time [15]. The effect was being linked to the hydrogen-enhanced oxygen diffusion and no direct participation of hydrogen in the structure of TD was expected. More recently it has been found that hydrogen can easily be introduced into silicon as contamination in a variety of ways including plasma exposure, high-temperature annealing in hydrogen gas or water vapor, and also some common technological operations such as etching and polishing. At the same time several reports appeared on generation of donor centers related to hydrogen. Shallow donor centers have been detected following hydrogen-plasma treatment of neutron irradiated silicon [16]. Also a series of hydrogen-related donors was found in Cz-Si after hydrogenation and heat treatment at 350°C [17].

Being aware of these findings we decided to investigate the possible involvement of hydrogen in the structure of the Si-NL10 center. In a sample with a strong Si-NL10 EPR signal we have indeed detected a new ENDOR spectrum. Due to the fact that the nuclear g-value of a proton differs from that of the other magnetic nuclei, the new ENDOR lines appear in a different frequency region and require very different experimental conditions than the ones used in our previous studies [14]. The observed ENDOR spectrum (see Fig.9a) is symmetrical with respect to the Zeeman frequency of a free proton. By recording ENDOR spectra for different resonance field values and monitoring the frequency shift - as depicted in Fig.9b - we unambiguously identify hydrogen as being responsible for the detected hyperfine interaction.

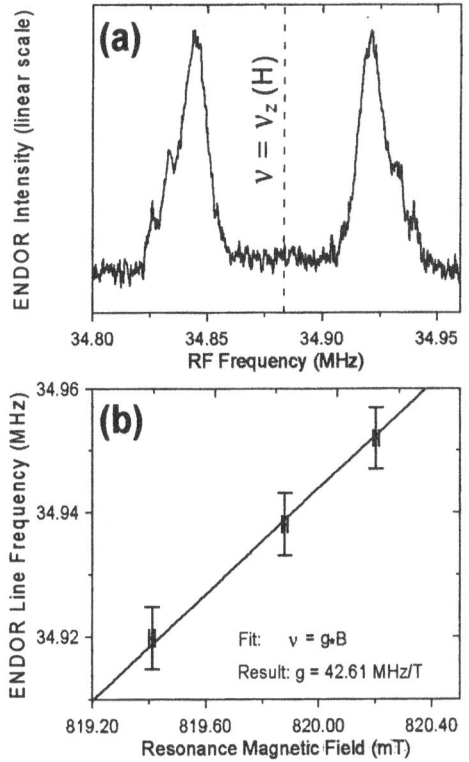

Figure 9. Hydrogen ENDOR spectrum observed in a Cz-Si:Al sample after 470°C/55 hrs heat treatment. (a) – spectrum, recorded with **B**‖<110> direction, B=819.324 mT; (b) – displacement of the ENDOR line as a function of the magnetic field shift, fitted with a linear function. The value of the proportionality coefficient is equal within the experimental accuracy to the nuclear Zeeman frequency of a proton $\nu(MHz)=42.5758\cdot B(T)$.

It should be emphasized that hydrogen has *not* been intentionally introduced into this material. In order to further investigate the influence of hydrogen on the formation of Si-NL10 centers, we have prepared a series of samples, intentionally doped with hydrogen. The samples were sealed in quartz ampoules, containing argon atmosphere and a few milligrams of water. Also an identical set of samples without any water in the ampoules has been prepared. All the samples, with or without water, were subjected to a standard procedure of heat treatment at 1250°C, followed by a quench to room temperature. This procedure served to disperse interstitial oxygen to ensure uniform starting conditions and/or to diffuse hydrogen. Subsequent to the quench the samples were annealed at 470°C for various periods of time. In all the samples diffused with hydrogen, an ENDOR spectrum, similar to that depicted in Fig.9a, has been observed. Its traces were also observed in almost all the samples that were not intentionally diffused with hydrogen, confirming the hydrogen contamination of commercially available high-grade silicon [18].

The angular dependence of the hydrogen ENDOR has been studied in hydrogen-doped Cz-Si:P and is shown in Fig.10. While the detailed analysis is difficult in view of a rather broad line width, the data are best fitted with two similar hyperfine tensors of

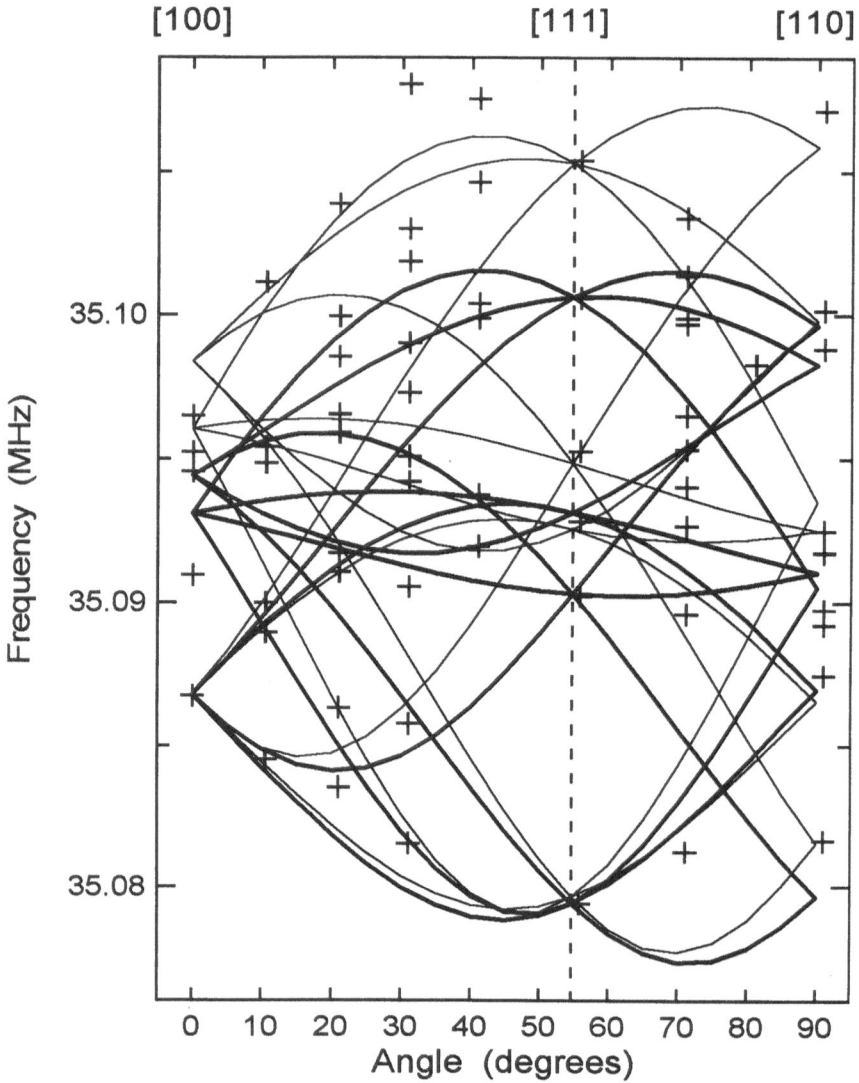

Figure 10. Angular dependence of hydrogen ENDOR as observed for Cz-Si:P sample. Experimental points and a computer fit of two tensors of triclinic symmetry are shown.

a triclinic symmetry. In view of the small value of the hyperfine interaction it is understandable that it cannot be observed in EPR. The isotropic part of the hyperfine tensor is proportional to the localization of the paramagnetic electron on hydrogen. Using the parameters of the hydrogen $1s$ wave function, we arrive at the localization $\eta^2 \approx 6 \times 10^{-3}\%$ for both species. Extremely small as it may appear, the value of localization is, nevertheless, comparable to that found for oxygen [14]. That means that hydrogen plays an important role in the Si-NL10 defect structure.

In order to propose a microscopic defect model for the H-containing Si-NL10 center it was essential to establish how many hydrogen atoms are actually involved in its structure. The results of an experiment relevant to this question are shown in Fig.11.

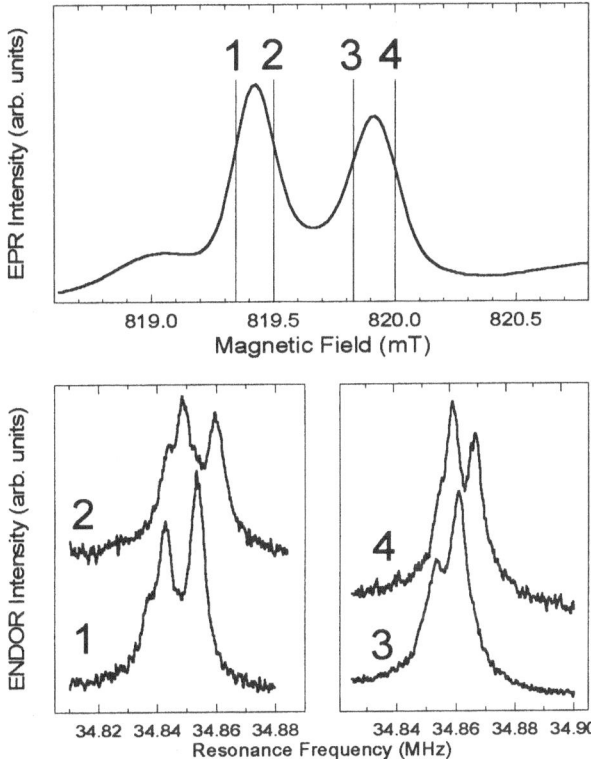

Figure 11. Determination of the true symmetry of the Si-NL10 defect. Top panel – EPR spectrum with **B**$\|<111>$ direction. Vertical lines, numbered 1 to 4 indicate the magnetic field values for which ENDOR spectra were recorded. Bottom left and bottom right panels present ENDOR spectra, recorded on low-field and high-field EPR lines, respectively.

From ENDOR data the triclinic symmetry of the hydrogen hyperfine interaction was found, whereas from EPR the g-tensor was concluded to be orthorhombic. Consequently, with the symmetry of the hyperfine interaction with the hydrogen nucleus, forming an essential constituent of the center, lower than the overall symmetry of the defect, this implies that there are more, in this case four, equivalent nuclei in the shell in question. However, in analogy to the earlier discussed case of ^{27}Al ENDOR, in EPR a small lower symmetry distortion of an otherwise almost orthorhombic tensor can be missed due to insufficient resolution of this technique. Following the detection of hydrogen ENDOR we have used field-stepped ENDOR to determine the actual symmetry of the Si-NL10(H) center. Suppose, two ENDOR peaks correspond to two unresolved EPR lines with slightly different positions of maximum intensity. In this case if the resonance magnetic field is shifted towards the maximum of one of these lines, the relative intensity of the ENDOR peak associated with it will increase. If, on the other hand, both ENDOR peaks are due to the same EPR line, their relative intensities are independent from the position of the magnetic field and reach their maximum simultaneously, i.e., at the position of the EPR line maximum intensity. The FSE measurements were carried out for magnetic field $B\|<111>$. The orthorhombic EPR spectrum consists in this direction of two lines, each of them giving rise to two ENDOR peaks. Changes of the relative intensities of the peaks in the ENDOR spectra upon shifting of the magnetic field were observed for both EPR lines as illustrated by Fig.11. We conclude that the symmetry of the Si-NL10(H) center is, in fact, triclinic and each defect contains only one hydrogen atom.

The triclinic symmetry of the Si-NL10(H) center appears to be different from that determined for the Si-NL10(Al) defect in the preceding subsection. In order to investigate that we have performed FSE studies of Si-NL10 in a hydrogen-diffused Cz-Si:Al sample. The results of this experiment are shown in Fig.12. In this sample ENDOR lines due to hydrogen (Fig.12a) and aluminum (Fig.12b) were detected – see the upper-right corner inserts in both figures. By the FSE technique we could then find the EPR "images" of centers responsible for these lines of different origin. As can be concluded from Fig.12a the hydrogen-containing EPR species have a higher anisotropy of the g-tensor than that of the total EPR spectrum. This means that not all the Si-NL10 defects in aluminum-doped material are hydrogen-containing. In the same sample [27]Al FSE revealed the species with lower anisotropy (see Fig.12b).

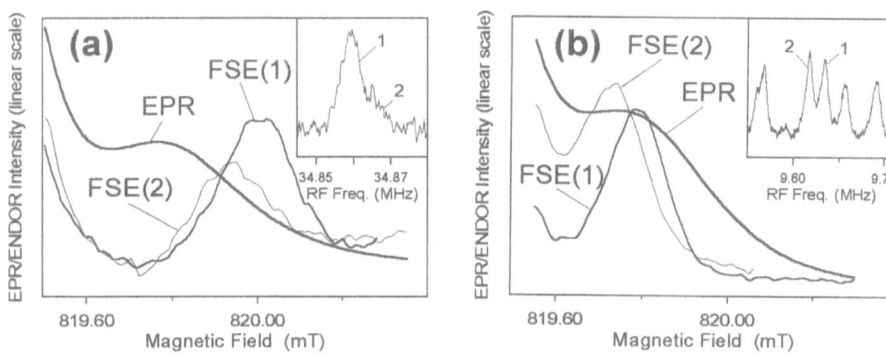

Figure 12. Hydrogen (**a**) and aluminum (**b**) FSE spectra in a Cz-Si:Al sample after hydrogen diffusion and 470°C/55 hrs heat treatment, compared to the EPR spectrum. On the inserts fragments of the respective ENDOR spectra are shown, lines indicated 1 and 2 were used to make FSE scans. T=7 K, ν=22.923 GHz, $B\|<110>$.

The detection of the hydrogen ENDOR brought again forward the puzzling question of the microscopic identification of the Si-NL10 center and its particular relation to TD's.

As mentioned in the introduction the identification of this center is an issue of a considerable dispute. The detection of the hydrogen ENDOR added here an important clue to the earlier revealed features of Si-NL10 which included: i) the center is closely related to TD's, ii) it is created subsequently to Si-NL8, iii) its oxygen structure is identical to that of Si-NL8, iv) the electron spin value is $S = 1/2$, and v) the wave function of the Si-NL10 paramagnetic electron has a more shallow character than that of Si-NL8.

Consistent with the above-listed features and in the light of the new finding the identification of the Si-NL10 center as a neutral thermal double donor with one of its two electrons passivated by hydrogen has been put forward. Following this model Si-NL8 and Si-NL10 would have an identical oxygen core, determining their symmetry and hyperfine interactions with the ^{17}O nucleus. However, the fact that one of the TD electrons is passivated, and not lost to the conduction band, would mean that the remaining electron is weaker bound than in the case of the $(TD)^+$ state that gives rise to the Si-NL8 spectrum.

To follow on our hypothetical identification we have studied the influence of hydrogen on the formation of both Si-NL8 and Si-NL10 spectra. In this case we used B-doped material, where the process of Si-NL10 formation is known to be rather slow [4]. The intensities of the Si-NL8 and Si-NL10 spectra in hydrogen-diffused and hydrogen-free Cz-Si:B samples were studied for various annealing times. In the samples diffused with hydrogen, the production of the Si-NL10 spectrum was significantly enhanced, while the formation of Si-NL8 was suppressed. We also could not detect any hydrogen ENDOR on the Si-NL8 spectrum. This brings us to the conclusion that hydrogen is not involved in the Si-NL8 defect formation, consistent with its assignment to $(TD)^+$ and in line with the infrared absorption measurements.

Being a dominant defect in Cz-Si after prolonged heat treatment, a singly passivated TD should manifest itself in a series of infrared-absorption lines due to transitions to the excited states of a shallow single donor. In line with this reasoning experimental evidence has recently been obtained relating the Si-NL10 EPR centers with the IR absorption series of *so-called* shallow thermal donors (STD's) [19].

Finally, we come back to the question of aluminum incorporation in the Si-NL10 center. So far aluminum was the only nucleus having a characteristic hyperfine interaction with the paramagnetic electron of the Si-NL10 defect, apart from silicon and oxygen. Following the current result the roles of aluminum and hydrogen in the defect structure seem, somehow, similar. An aluminum atom can be removed from its substitutional position by a self-interstitial, created during the process of oxygen clustering (Watkins kick-out mechanism [20]). Once made interstitial, aluminum, similar to hydrogen, is a fast diffusant. It would be able to diffuse towards the core of a TD, where it might compensate one of the two TD's electrons. According to the model the Si-NL10 defect must always contain either hydrogen or aluminum, as the presence of a passivating element in the structure is required. Future experiments should confirm this hypothesis. Although the roles played by hydrogen and aluminum in the Si-NL10, i.e., converntion from a double to a single donor, are similar, because of the different chemical nature of these atoms the microscopic structure of Si-NL10(H) and Si-NL10(Al) should also be different. This manifests itself by the small shift of the g-value, as evidenced by the FSE experiments illustrated in Fig.12. Regardless of this difference, the overall EPR symmetry for both centers is determined by their (oxygen) core structure and remains nearly orthorhombic.

While the hyperfine interaction carries information about the spin density within the center, the quadrupole interaction is determined by the electric field gradient. Since for Si-NL10 the spin localization is very low the observed quadrupole interaction is primarily indicative of local arrangements around the magnetic nucleus. To study these the samples were doped with deuterium which has a nuclear spin I=1 and, consequently, a quadrupole moment. If such a nucleus is placed in an inhomogeneous electric field additional energy shifting, detectable in an ENDOR experiment, can occur. In the sample treated in a heavy-water-vapor atmosphere an ENDOR spectrum, symmetrical with respect to the nuclear Zeeman frequency of a deuteron, was observed. This, together with the notion that the spectrum was only found in the samples treated in a D_2O atmosphere, allows us to identify deuterium as being responsible for the observed interaction. For the sake of comparison we prepared a sample containing both hydrogen and deuterium and measured D- and H-ENDOR in this sample. Fig.13 presents the angular dependence of deuterium ENDOR in this sample.

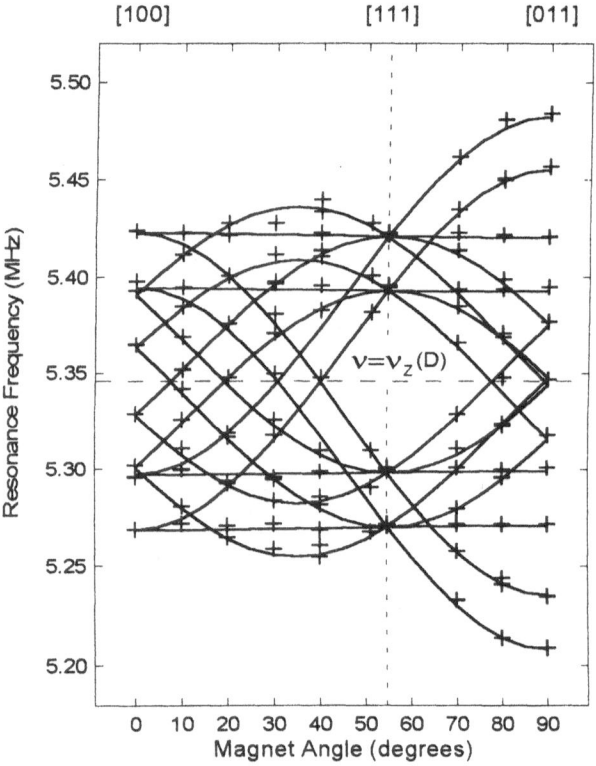

Figure 13. Angular dependence of ENDOR line positions of Si-NL10 in Cz-Si:In sample, treated 1/2 h at 1250°C in D_2O atmosphere and annealed 42 hrs at 470°C. Experimental points and computer fit to the data are shown.

It was fitted with an isotropic hyperfine and an orthorhombic quadrupole term. It should be noted here that since the symmetry of the TD-H center was established to be triclinic, the symmetry of deuterium hyperfine and quadrupole tensors must not be higher than triclinic. However, the departures from the cubic and orthorhombic symmetry for these tensors, respectively, were below the resolution of our experiment. Similar to the hydrogen ENDOR experiment different deuterium-containing Si-NL10 species were found. These differed in the magnitude of the hyperfine interaction, but the quadrupole tensor was identical for all the observed TD-D species. This indicated a unique bonding arrangement of deuterium (hydrogen) in the structure of a passivated TD.

Since the magnitude of the observed quadrupole interaction by far exceeds that expected for the field of distant charges, it was interpreted in a usual manner [14] as arising from the unbalanced charge density on the deuteron which, in turn, is determined by the bond(s) that deuterium is forming with neighboring atom(s). The quadrupole interaction tensor appears to be almost axial, consistent with a single bond usually formed by hydrogen. However, the direction of the tensor axis (and therefore of the bond) is along the $<011>$ crystalline direction, perpendicular to the oxygen-containing symmetry plane, and therefore at variance with the normally observed $<111>$ direction for bond-centered or antibonding configuration of hydrogen. This particular observation is expected to play an important role for the future theoretical work on the modeling of the TD structure.

4. References

1. Michel, J. and Kimerling, L.C. (1994) Electrical properties of oxygen in silicon, *Semicond. and Semimetals* **42**, 251-287.

2. Elliot G. (1957) Oxygen as a donor element in germanium, *Nature* **180**, 1350-1351.

3. Muller S.H., Sprenger M., Sieverts E.G., and Ammerlaan C.A.J. (1978) EPR spectra of heat-treatment centers in oxygen-rich silicon, *Solid State Commun.* **25**, 987-990.

4. Gregorkiewicz T., Wezep van D.A., Bekman H.H.P.Th., and Ammerlaan C.A.J. (1987) EPR studies of heat-treatment centers in *p*-type silicon, *Phys. Rev. B* **35**, 3810-3817.

5. Bekman H.H.P.Th, Gregorkiewicz T., Wezep van D.A., and Ammerlaan C.A.J. (1987) Electron-paramagnetic-resonance study of heat-treatment centers in *n*-type silicon, *J. Appl. Phys* **62**, 4404-4405.

6. Sieverts E.G. (1983) Classification of defects in silicon after their *g*-values, *Phys. Stat. Sol. (b)* **120**, 11-29.

7. Lee K.M., Trombetta J.M., and Watkins G.D. (1985) Identity of the NL8 EPR spectrum with thermal donors in silicon, in Johnson N.M., Bishop S.G., and Watkins G.D. (eds.), MRS Symposia Proceedings Vol.46, *Microscopic Identification of Electronic Defects in Semiconductors*, Materials Research Society, Pittsburgh, Pennsylvania, pp.263-268.

8. Gregorkiewicz T., Bekman H.H.P.Th., and Ammerlaan C.A.J. (1990) Comparative study of Si-NL8 and Si-NL10 thermal-donor-related EPR centers, *Phys. Rev. B* **41**, 12628-12636.

9. Michel J., Niklas J.R., Spaeth J.-M., and Weinert C. (1986) Thermal donors in silicon; A study with ENDOR, *Phys. Rev. Lett.* **57**, 611-614.

10. Bekman H.H.P.Th., Gregorkiewicz T., and Ammerlaan C.A.J. (1989) Silicon electron-nuclear double-resonance study of the NL10 heat-treatment center, *Phys. Rev. B* **39**, 1648-1658.

11. Gregorkiewicz T., Wezep van D.A., Bekman H.H.P.Th., and Ammerlaan C.A.J. (1987) Oxygen incorporation in thermal-donor centers in silicon, *Phys. Rev. Lett.* **59**, 1702-1705.

12. Michel J., Niklas J.R., and Spaeth J.-M. (1988) The structure of thermal donors (NL8) in silicon: A model derived from ^{17}O and ^{29}Si ENDOR experiments, in Stavola M., Pearton S.J., and Davies G. (eds.), MRS Symposia Proceedings Vol.104, *Defects in Electronic Materials*, Materials Research Society, Pittsburgh, Pennsylvania, pp.185-188.

13. Sprenger M., Muller S.H., Sieverts E.G., and Ammerlaan C.A.J. (1987) Vacancy in silicon: Hyperfine interactions from electron-nuclear double resonance measurements, *Phys. Rev. B* **35**, 1566-1581.

14. Gregorkiewicz T., Bekman H.H.P.Th., and Ammerlaan C.A.J. (1988) Microscopic structure of the NL10 heat-treatment center in silicon: Study by electron-nuclear double resonance, *Phys. Rev. B* **38**, 3998-4015.

15. Newman R.C. and Jones R. (1994) Diffusion of oxygen in silicon, *Semicond. and Semimetals* **42**, 289-352.

16. Hartung J. and Weber J. (1993) Shallow hydrogen-related donors in silicon, *Phys. Rev. B* **48**, 14161-14166.

17. McQuaid S.A., Newman R.C., and Lightowlers E.C. (1994) Hydrogen-related shallow thermal donors in Czochralski silicon, *Semicond. Sci. Technol.* **9**, 1-4.

18. Martynov Yu.V., Gregorkiewicz T., and Ammerlaan C.A.J. (1995) Role of hydrogen in formation and structure of the Si-NL10 thermal donor, *Phys. Rev. Lett.* **74**, 2030-2033.

19. Newman R.C., Tucker J.H., Semaltianos N.G., Lightowlers E.C., Gregorkiewicz T., Zevenbergen I.S., and Ammerlaan C.A.J. (1996) Infrared absorption in silicon from shallow thermal donors incorporating hydrogen and a link to the NL10 paramagnetic resonance spectrum, *to be published*.

20. Watkins G.D. (1965) A review of E.S.R. studies in irradiated silicon, in *Radiation Damage in Semiconductors*, Dunod, Paris pp.97-114.

MAGNETIC RESONANCE OF HEAT TREATMENT CENTRES IN SILICON

J.-M. SPAETH
Fachbereich Physik, Universität-GH Paderborn
33095 Paderborn, Germany

1. Introduction

Silicon crystals grown in quartz crucibles with the Czochralsky (Cz)-method contain a large concentration of interstitial oxygen (10^{18}cm^{-3}). Upon annealing at a temperature of 450-460 °C "shallow" electrically active defects are formed which are often called "thermal donors". Although their existence has been known for a long time [1,2], there is not yet a clear picture about their microscopic structures. These "heat treatment centres" have been investigated with numerous experimental methods, amongst those also with magnetic resonance techniques being the most important structure sensitive tools.

Upon annealing at 450 °C, first a shallow double donor is formed consisting of a series of defects which are formed one after another upon increasing the annealing time and which have been identified by optical infrared (IR) absorption measurements [3,4]. For both, the neutral and singly ionized charge states, there are IR optical transitions. Up to now, 16 species of this heat treatment centre [5] have been measured.

We will show below with magnetic resonance that these heat treatment centres (within the limits of magnetic resonance measurements) do not incorporate any impurity which is contained in the crystal, except for oxygen. They are thus heat treatment centres, which must always have the same microscopic structures whenever they are formed. Their electron paramagnetic resonance (EPR) spectrum measured in the singly ionized charge state in p-type silicon has been called "NL8" previously [6]. It was shown that these centres have energy levels which are effective-mass-like, differing by about 2meV from species to species. In this contribution, the name "thermal double donor" (TDD) shall be restricted to this kind of heat treatment centres. It was also shown that the first two species have metastable properties [3,7].

Besides the NL8 EPR spectrum, another EPR spectrum associated with heat treatment is observed, which appears after annealing for a longer time than needed to generate TDDs, but less than about 100 hours. This spectrum was called "NL10"; the corresponding defects are often called NL10 defects [6,8]. So far, no IR absorption spectra could be observed for these defects, nor do they appear as a clear series of species such

R. Jones (ed.), Early Stages of Oxygen Precipitation in Silicon, 83–101.
© *1996 Kluwer Academic Publishers.*

as the TDDs. In spite of this, it was previously suggested that there is a correlation between the formation of NL10 centres and TDDs [9].

The question of the microscopic structure of TDDs and other heat treatment centres, which from early on were supposed to be oxygen aggregate centres, is both of fundamental and technological interest. Unfortunately, EPR spectra of TDDs and NL10-type defects are structureless and do not allow to draw conclusions on the microscopic defect structure. Most experimental structural information was provided by electron nuclear double resonance (ENDOR) investigations. With ENDOR, the superhyperfine (shf) interactions with ^{29}Si neighbour nuclei could be resolved as well as shf and quadrupole interactions with ^{17}O, the magnetic oxygen isotope, which was diffused into (oxygen-free) float zone silicon. In some cases the shf and quadrupole interactions with impurity atoms involved in some of the heat treatment centres of the NL10-type could also be measured. It is from these experimental data that some conclusions could be drawn on the microscopic structures of heat treatment centres.

Two groups have intensely studied heat treatment centres with ENDOR. The Paderborn group studied in great detail the TDDs (NL8) and, to a lesser degree, various NL10-type defects, while the Amsterdam group dealt in more detail with various NL10 defects, in particular in Al- and H-doped silicon. In this contribution, the work of the Paderborn group will be reviewed and discussed as to the possible structure models of heat treatment centres.

2. Experimental details

2.1 SAMPLE PREPARATION

For the study of TDDs, mainly B-doped Cz-silicon was used. The B-concentration was in the range of $(1-5) \cdot 10^{15}B/cm^3$, the ^{16}O content was about 10^{18} cm^{-3}. Some samples had different acceptor dopings (Al, Ga, In). In order to have the same starting conditions for the formation of TDDs, the samples were pre-annealed at 770 °C for 15 minutes under N_2-gas atmosphere with a few vol% O_2 to avoid oxygen out-diffusion.

The magnetic oxygen isotope ^{17}O was diffused into FZ-Si doped with about $5 \cdot 10^{15}B/cm^3$ in an IR heat chamber under extremely clean conditions in order to avoid contamination with transition metals. In the IR heat chamber, the sample was placed into an O_2-gas atmosphere enriched to 70% with ^{17}O in a cooled quartz ampoule. The sample was held for about 14 days at an oxygen pressure of 3 bars at 1400 °C. After the diffusion, the sample typically contained a total oxygen concentration of $1.1 \cdot 10^{18}$ cm^{-3}, of which 54% were ^{17}O. No contamination was discovered. For further details see ref. [10,11].

For the study of NL10-type heat treatment centres in addition Cz and Fz-Si doped with P (conductivity $\sigma = 40$ Ω cm at room temperature) was used. The concentration of oxygen was $9.7 \cdot 10^{17}cm^{-3}$ and of carbon $< 5 \cdot 10^{15}cm^{-3}$. Some samples were γ-irradiated at room temperature in a ^{60}Co facility at a flux rate $1 \cdot 10^{13}cm^{-2}$ sec^{-1} and an integrated dose $> 10^{19}cm^{-2}$ in order to produce deep acceptors such as the A centre [12].

2.2 EPR AND ENDOR MEASUREMENTS

EPR and ENDOR measurements were performed in an X-band custom-built computer-controlled spectrometer using the stationary ENDOR method, in which the nuclear magnetic resonance transitions are detected through a change of the partially saturated EPR transitions, the latter being measured as a microwave absorption in the sample which is held at the resonance magnetic field [13]. The sample temperature could be varied between 3.5 K and room temperature. Due to the low centre concentration of about $3 \cdot 10^{15}$ cm^{-3}, low abundance of the magnetic isotope ^{29}Si (4.7%), the ENDOR

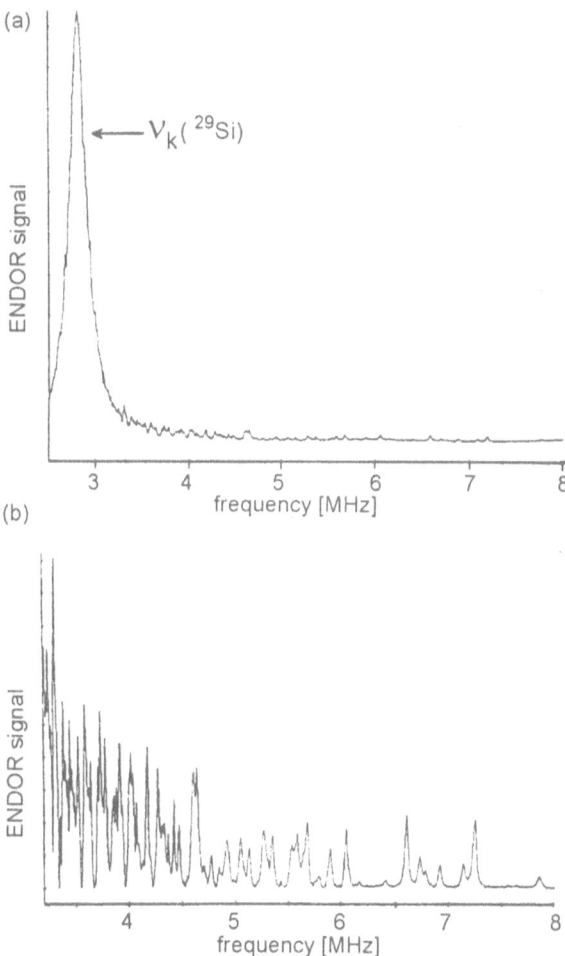

Fig. 1: Section of the TDD$^+$ ENDOR spectrum for Cz-Si after 4 hours of annealing. $B_0 \parallel$ [111], T = 42 K (a) as measured, (b) with the strong line at ν_k (^{29}Si) omitted [11].

signals were very weak and multiple scanning and averaging was necessary. In contrast to the usual stationary ENDOR, the magnetic field was also modulated with a frequency of the order of 100 kHz [11]. This resulted in a gain in signal-to-noise by about a factor of 10. This observation is in contrast to systems with a high abundance of magnetic nuclei, where such a modulation has a detrimental effect [13]. The frequency positions of the measured ENDOR lines were determined using numeric mathematic procedures described in detail in [11] and [13]. The sample could also be illuminated in situ.

Measurements of defect concentration with EPR were performed relative to a ruby standard, whereby the transition probability of the ruby EPR was taken into account. The concentrations could only be determined with an accuracy of ± 30%.

Some measurements were performed using electrical detection of EPR (EDEPR), a method particularly sensitive for low concentrations of defects [14]. In this method, the change of photoconductivity is monitored, which occurs as a result of an EPR transition within weakly exchange-coupled donor-acceptor pairs.

EPR and ENDOR measurements of TDD$^+$s (NL8) were mostly performed at about 30 K at which temperature NL10-type centres do not appear due to a short spin lattice relaxation time. At 6-8 K, the EPR signals of TDD$^+$s are completely saturated, and no ENDOR effect of TDD$^+$ can be seen any more at this temperature, while the EPR and ENDOR of NL10-type centres can be measured. Thus, in the simultaneous presence of both TDDs and NL10-type centres, the centres can be measured separately using either 30 K or 6-8 K as measurement temperature [11,15,16].

3. Thermal Double Donors (TDDs)

3.1 EPR AND ^{29}Si-ENDOR SPECTRA

The EPR spectrum of singly ionised TDDs in a B-doped Cz-Si sample consists of a rather narrow structured line for each centre orientation. If the concentration of acceptors exceeds twice that of TDDs, no EPR is observable, since all TDDs are doubly ionised. If the concentration is of the same order, EPR is observable. The signals can be enhanced by illumination with above band gap light for 0.5 N_B < N_{TDD} > N_B [11]. From the angular dependence follows that there are 6 centre orientations in orthorhombic symmetry (C$_{2v}$) with the principal g-values of g_1 = 1.99323 (axis ∥ [001]), g_2 = 2.00091 (axis ∥ [110]) and g_3 = 1.99991 (axis ∥ [001]). The symmetry of the TDD$^+$ defect is characterised by two (110) mirror planes perpendicular to each other which intersect in the [001] axis [6]. Fig. 1 shows a stationary ENDOR spectrum of a ^{16}O containing sample measured after 4 hours of annealing. The ^{29}Si-ENDOR lines are very weak, while the so-called "distant" ENDOR line at the Larmor frequency of ^{29}Si is intense (fig. 1a) (the sample contained about 3· 10^{15} TDD$^+$s). The spectrum of fig. 1a is the result of a 10 hours' accumulation [11]. Fig. 1b shows the ^{29}Si lines omitting the intense line at 3 MHz. Many ^{29}Si-ENDOR lines of lattice nuclei having shf interactions are clearly seen. It was verified that all the ENDOR lines are due to ^{29}Si, no signal of ^{10}B or ^{11}B or any other impurity was detected (for details see [11]). Measurement of the ^{29}Si-ENDOR spectra of TDD$^+$s generated in Cz-Si doped with different acceptors such as Al, In and Ga, after annealing for about the same time, show exactly the same ^{29}Si-ENDOR lines [17]. Thus, in all these samples, the TDD$^+$ defect must have the same microscopic structure. No acceptor is incorporated in the core of the defect, nor can an acceptor be near enough to experience the wave function of the unpaired electron. This is in contrast to what was found for NL10-type centres (see section 4).

The observation made in the IR and DLTS spectra, that there is a series of TDDs generated subsequently one after another upon annealing, is also reflected in the ^{29}Si-ENDOR spectra. In fig. 2, which represents a section of the ENDOR spectrum, the influence of annealing on the spectrum is shown: Fig. 2 (lower trace) is the spectrum after 2 hours annealing at 460 °C, fig. 2 (middle trace) after 4 hours, fig. 2 (upper trace) after 8 hours (for B ∥ [111]). The same ENDOR frequencies are observed in all samples. There is, unfortunately, no reliable information on the centre concentration from the ENDOR line intensities. However, relative intensities of lines reflect the relative concentrations of different defect species corresponding to these lines. The relative ENDOR line intensities vary upon annealing time. For example, the two lines at 7.07 and 7.14 MHz change their relative intensities quite markedly with the line at 7.07 MHz decreasing upon increasing annealing time. Such changes of intensity are also found for the other lines indicating that several species of TDD$^+$ centres are present simultaneously and that their relative concentrations change with annealing time. This is qualitatively quite similar to what was observed in IR spectroscopy [4]. The five species thus identified are labelled A-E in fig. 2. Also with illumination with 1.17 eV light the ENDOR line intensities vary, e.g. the lines of species A decrease, while those of species B increase. The assignment of lines belonging to a centre species agrees with that made on the basis of different annealing times. By correlation with IR measurements it was shown that the earlier species A identified by ENDOR is TDD$^+$3 [11]. Furthermore, the general trends observed in IR were also found in ENDOR: The ENDOR lines at high frequencies, i.e. for ^{29}Si nuclei with high shf interactions [13], which means a relatively high electron localisation, do anneal out first and the later species have smaller interactions. This correlates with the IR observations that the earlier species have deeper energy levels compared to the later ones. The more shallow a state, the more delocalised the electron distribution.

The analysis of the ^{29}Si-ENDOR spectra is described in detail in ref. [11] and is not repeated here. For the analysis, the angular dependence of the ^{29}Si-ENDOR lines had to be measured in small angular steps because of the large number of ENDOR lines and the fact that they often overlapped. Using the appropriate spin Hamiltonian and the symmetry of the g-tensor, the angular dependencies could be described assuming that the defect possesses C_{2v} symmetry, i.e. with respect to the ^{29}Si-ENDOR lines no deviation of that symmetry has been noted. A lower symmetry would have been seen in an additional splitting of the ENDOR lines (their width is typically 40-50 kHz). This C_{2v} symmetry was observed for all the species measured: i.e. the transformation from

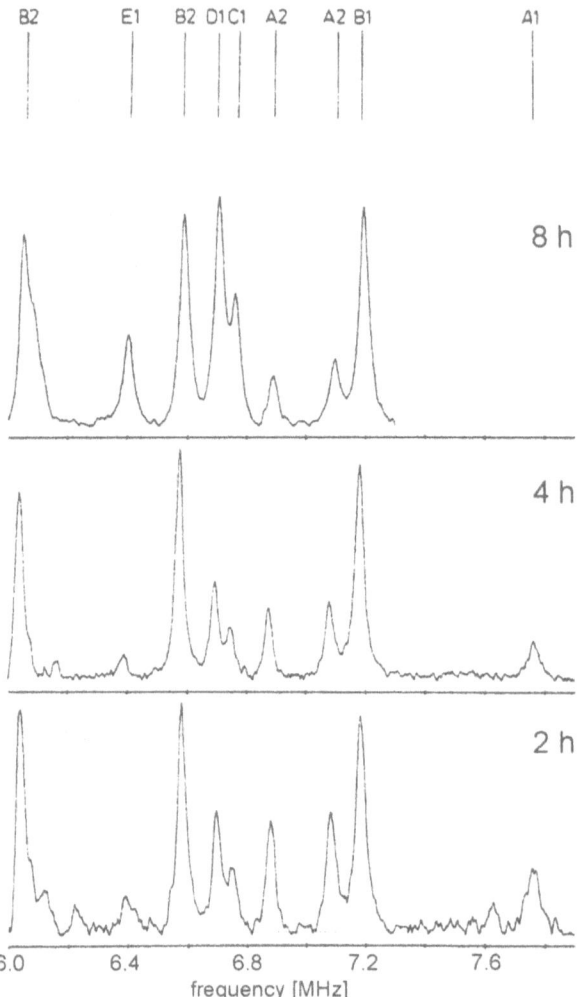

Fig. 2: Influence of annealing (2,4 and 8 hours) at 460 °C on the TDD$^+$ ^{29}Si ENDOR lines. A-E denote the TDD$^+$ species, the numbers denote the symmetry type of ^{29}Si neighbours (see text) [11].

the deeper to the shallower species by further annealing (and further addition of oxygen, see sec. 4) has not broken the C_{2v} symmetry within experimental error.

In table 1, the shf interactions are given for the species A as an example. 7 shells of ^{29}Si neighbour nuclei could be identified. The interactions are given in terms of the isotropic shf constant a and the anisotropic shf constants b and b' which are related to the principal values of the usual shf tensor A_{xx}. A_{yy}. A_{zz} by:

$$a = (A_{xx} + A_{yy} + A_{zz}) / 3$$

$$b = (A_{zz} - a)/2 \tag{1}$$
$$b' = (A_{xx} - A_{yy})/2$$

The orientation of the shf tensors is described by the angle ϑ between the z-axis and [001].

The many ^{29}Si nuclei of lattice neighbours of the "core" of the TDD$^+$ defects can be classified because of the centre symmetry by only a few types of shells. These are (see fig. 3):

Type 1 shell: The neighbour shell consists of just one nucleus. The principal axis system of its shf tensor is parallel to that of the g tensor. The ^{29}Si nucleus must be on the [001] axis. Its location on the axis is not determined by symmetry constraints.

Type 2 shell: The location of the neighbour nuclei is somewhere between the two {110} mirror planes present because of C_{2v} symmetry. The shell contains 4 neighbours. All principal axes of the shf tensors of this shell were found to be parallel to <100> directions of the crystal (see fig. 3). This suggests that they are on {100} planes.

Type 3 shell: All neighbours of this shell are in one (110) mirror plane. There are two of these neighbours. The z-axes (largest interaction) and their x-axes are in that plane (the angle ϑ between z and [001] is not fixed by symmetry).

Type 4 shell: This shell has the same symmetry property as the type 3 shell, but in the (110) plane perpendicular to that of shell 3.

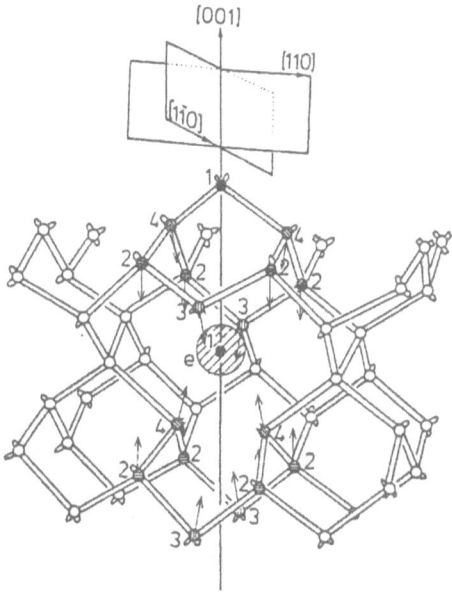

Fig. 3: Possible positions of Si neighbour nuclei in the lattice corresponding to different neighbour-shell symmetries denoted 1-4. The principal axes of the shf tensors (with largest interaction) are indicated by arrows [11].

In fig. 3 it was assumed that the different neighbour shells are as close to the core (centre) of the defect as their symmetry properties allow (there is no information on the distance from the symmetry arguments). Shells of type 2,3 and 4 are present in the upper and lower part of the lattice. Indeed, two shells of types 2,3 and 4 were identified experimentally (see table 1). In table 2 the isotropic shf constants of species A-E are compared.

Table 1: Shf interaction constants of ^{29}Si neighbour shells of TDD$^+$ A. a/h, b/h, b'/h in MHz, ϑ (in degrees) is the angle between the z-axis of the tensor and [001].

shell type	a/h	b/h	b'/h	ϑ
1	9.89	0.07	0.03	0
2a	8.53	0.06	0.03	0
2b	8.12	0.06	0.03	0
3a	6.79	0.45	0.10	3.5
3b	6.36	0.49	0.19	9.2
4a	6.03	0.17	0.09	11.5
4b	5.87	0.15	0.08	12.5

Table 2: Isotropic shf interaction constants (in MHz) of type 1 ^{29}Si neighbour shells of TDD$^+$s species A-E.

TDD$^+$	A	B	C	D	E
a/h	9.89	8.72	7.86	7.75	7.15

3.2 ^{17}O-ENDOR SPECTRA

Fig. 4a shows the ENDOR spectrum of TDD$^+$ defects measured in ^{17}O-diffused FZ-Si for the low frequency range (upper trace) in comparison to a sample containing TDD$^+$ defects in Cz-Si ([B] = $4.5 \cdot 10^{15}$ cm^{-3}) after 3 hours of annealing (lower trace). Both

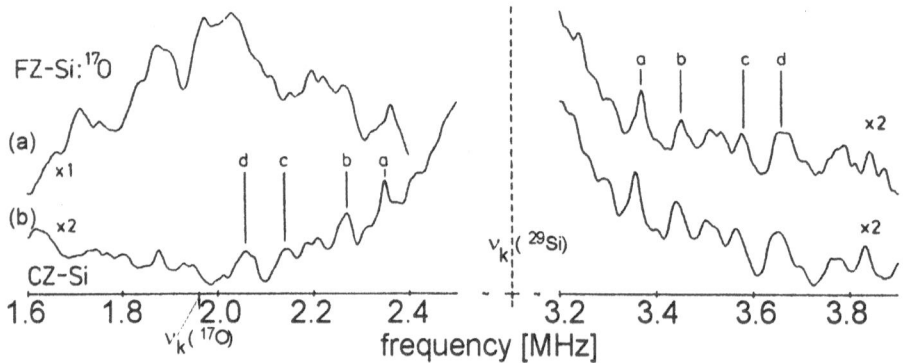

Fig. 4: TDD$^+$ ENDOR spectra in FZ-Si with diffused ^{17}O (upper trace) and in Cz-Si (lower trace). The lines belong to defects A$^+$ and B$^+$. a-d refer to ^{29}Si ENDOR lines for ms = ± 1/2 [11].

samples contained species A and B. The spectra of the two samples differ significantly between 1.6 and 2.5 MHz, that is around the Larmor frequency ν_k of ^{17}O, while they are identical in the frequency range between 3.2 and 3.9 MHz. The lines denoted by $a^{\pm}$ - $d^{\pm}$ are due to ^{29}Si, the + and - signs referring to $m_s = \pm 1/2$.

The lines about ν_k (^{17}O) are due to ^{17}O. This was confirmed by measuring the shift of the frequency position as a function of the magnetic field (For details of the shift methods see [13]). The shift $\Delta\nu/B$ is clearly different for ^{17}O and ^{29}Si lines and corresponds to the nuclear g factors of ^{17}O and ^{29}Si, respectively. The frequency position of the ^{17}O lines is largely determined by a quadrupole interaction. ^{17}O has the nuclear spin $I = 5/2$ and a quadrupole moment, while ^{29}Si has $I = 1/2$ and no quadrupole interaction. In first order, the ENDOR frequencies are given by

$$\nu = 1/h \mid m_s W_{shf} - g_I \mu_N B_0 + W_q \mid \qquad (2)$$

where W_{shf} is the shf interaction and W_q is the quadrupole interaction energy [13]. Since neither W_{shf} nor W_q depend on B, ν will depend linearly on B (see above). The oxygen lines in fig. 4a belong to the two species A and B. The analysis showed that for each species there are two shells of oxygen nuclei which differ mainly in their isotropic shf constants, while the other shf constants and the quadrupole interaction constants are almost identical. As was found for the ^{29}Si interaction, the isotropic constants for species B are smaller than those for species A reflecting again that the unpaired defect electron is less localised in the B species compared to the A species. The shf interaction parameters and the quadrupole interaction parameters q are collected in table 3, [11] whereby

$$q = 1/2 \, Q_{zz}$$
$$q' = 1/2 \, (Q_{xx} - Q_{yy}), \qquad (3)$$

Q being the quadrupole interaction tensor.

Table 3: Shf and quadrupole interaction constants (in MHz) of ^{17}O nuclei of the TDDs measured after 200 h heat treatment and of ^{17}O nuclei of type 1 and 2 of TDD$^+$ A and B [11, 15]. The data of NL10 - Al for comparison are from [8].

Species	shell	a/h	b/h	b'/h	ϑb [deg]	q/h	q'/h	ϑq [deg]
TDD$^+{}_{200h}$		<0.04	<0.07	<0.03	~50	0.11	0.03	90
TDD$^+$ A	1	-0.52	0.08	-0.07	58	0.11	0.09	
TDD$^+$ B	1	-0.47	0.09	-0.08	58	0.11	0.09	
TDD$^+$ A	2	-0.18	0.08	-0.07	62	0.12	0.05	
TDD$^+$ B	2	-0.11	0.08	-0.06	62	0.11	0.04	
NL10-Al	mbc1	0.2543	-0.003	-0.001	43	0.1056	0.0829	38

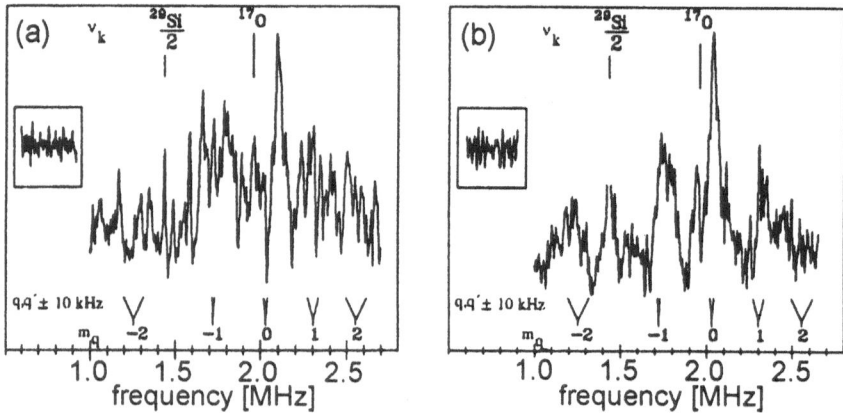

Fig. 5: Development of the oxygen ENDOR lines in ^{17}O diffused B-doped FZ-Si as a function of annealing time for $B_0 \parallel [110]$ for one centre orientation. The intensity is scaled to the distant ENDOR line. The inset gives a criterion for the noise. a) 4 hours annealed at 470 °C, b) 90 hours annealed at 470 °C. The quadrupole transitions are indicated [17].

At this stage of the investigation, no further oxygen interactions, except the two shells for species A and B, could be identified. It was thought that the two different oxygen atoms are contained in the core of the defect. Since it is generally believed that the symmetry of the defect is reflected in the symmetry of the ENDOR angular dependence [13] and since the latter showed a clear C_{2v} symmetry, it was assumed that the core of the TDDs contains 4 oxygen atoms [11]. It was, of course, realised that this would require a substantial lattice relaxation in order to accommodate four oxygen atoms in the core (a silicon vacancy). However, it remained open whether or not the transformation of one species to a subsequent one by longer annealing was accompanied by further addition of oxygens or whether silicon interstitials are incorporated as

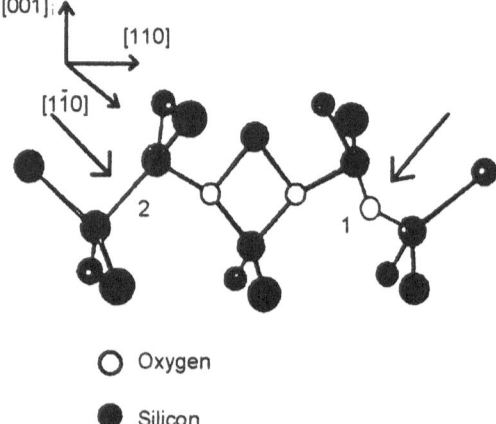

○ Oxygen

● Silicon

Fig. 6: Proposed defect model of the TDDs. The arrows mark the possible positions for oxygen atoms at interstitial sites added to the defect core. Such a defect can form a double donor as was suggested by Déak et al [21].

was suggested previously [18]. Therefore, attempts were made to study the ENDOR lines of ^{17}O in ^{17}O diffused FZ-Si after prolonged annealing. In fig. 5, the ^{17}O ENDOR lines are shown after a relatively short annealing of 4 hours (fig. 5a) and after a prolonged annealing of 90 hours at 470 °C in a new series of experiments with new ^{17}O-diffused samples. After the short annealing, the ENDOR lines are exactly the same as those in fig. 4a measured by Michel et al [11]. This was checked for various orientations. The ^{17}O lines are dominant in fig. 5a [17]. Fig. 5b shows the ENDOR spectrum of the ^{17}O lines for the same crystal orientation obtained after annealing the sample for 90 hours. According to IR measurements, this sample contains species 1-8, predominantly species 4-7, very few early ones. Compared to 4 hours annealing (fig. 5a), it is qualitatively seen that the ^{17}O lines become more numerous and group together into broader lines, which is seen more clearly around 1.2-1.3 MHz, 1.7 MHz and 2.3 MHz, for example. (The noise level is shown in the insets of fig. 5a and 5b). After a heat treatment of 200 hours, the ^{17}O ENDOR spectra look similar, i.e. broad lines are also seen as a result of the grouping together of the single sharp ENDOR lines. An angular dependence was measured by determining the central position of each of the discernible broad lines. The most important qualitative result is the fact that upon growth of the TDD$^+$ from "earlier" to "later" species, the number of ^{17}O ENDOR lines increases. Qualitatively it is seen that the integral of the ^{17}O ENDOR lines increases upon the formation of more and later TDD$^+$ species.

Upon longer annealing, the measurement of the ^{29}Si and ^{17}O ENDOR lines becomes difficult. The reason is that upon growing more and more TDDs, the fixed compensation by B is exhausted and the sample becomes increasingly n-type with the result of a deterioration of the spectrometer sensitivity. Therefore, a quantitative comparison of the growth of the integral over the ^{17}O ENDOR lines and the growth of the ^{29}Si ENDOR lines could be made only for an annealing time up to about 50 hours. All the line intensities were measured under the same experimental conditions. In accordance with the experience made comparing ^{29}Si ENDOR lines (of type 1 nuclei) with IR measurements [11], it was assumed that the ENDOR signal intensities are proportional to the number of nuclei measured, whereby the proportionality factor can be different for ^{17}O and ^{29}Si. When comparing the integral of the ^{29}Si ENDOR lines of type 1 nuclei for the same sample subjected to different annealing times, we obtained an increase by about a factor of 2 between 4 and 48 hours of annealing, respectively (each TDD$^+$ has one type 1 nucleus). For the ^{17}O ENDOR lines we obtained an increase by a factor of 4 (see table 4). From this, it can be qualitatively concluded that the number of oxygen lines increases more strongly than that of the type 1 ^{29}Si lines indicating that not only more centres were measured but that more oxygen per centre must be present. When comparing the so-called "distant" ENDOR line at the Larmor frequency of ^{29}Si, the increase is less (table 4) again showing the overproportional growth of the ^{17}O lines (note that the error margin for the ^{29}Si lines is larger because of the very weak ^{29}Si lines). From the data it was concluded that oxygen ENDOR lines increase overproportionally compared with the ^{29}Si ENDOR lines upon annealing. However, the exact value of the increase could not be determined precisely enough to allow a conclusion on the number of oxygen atoms added. When comparing the development of the integral of ^{29}Si lines at shorter annealing times, a good proportionality between their growth and the IR line growth was found [15] giving confidence that the result on the overproportional growth of oxygen is a correct result and not an artefact caused by changes of relaxation times [17].

Table 4: Intensities of the ^{17}O ENDOR lines and the ^{29}Si ENDOR lines in ^{17}O-diffused FZ silicon (AU stands for 'arbitrary unit', which is the same unit for all three columns.) After [17].

Annealing time [h]	Integral of the ^{17}O lines[a] [Hz x AU]	Integral of the ^{29}Si neighbour shells[b] [Hz x AU]	Integral of the ^{29}Si line[c] [Hz x AU]
4	0.013	0.0009	0.14
48	0.052	0.0019	0.18
Error	±15%	±30%	±5%

[a] Integral of the ^{17}O lines with $B_0\|[110]$

[b] Integral of the ^{29}Si lines of shell 1 of the TDD$^+$s (A-E) at $B_0\|[111]$

[c] Integral of the 'distant' ^{29}Si ENDOR line

The analysis of the angular dependence of the ^{17}O ENDOR lines after 200 hours of annealing yielded that there is almost no shf interaction. The angular dependence can be described with a pure quadrupole interaction with q/h = 0.11 MHz and q'/h = 0.03 MHz, whereby the orientation of the largest interaction is approx. along a <110> direction [17]. An approximate value of the shf interaction was roughly determined from the line width of the quadrupole ENDOR lines. Their upper bounds are listed in table 3. The ^{17}O ENDOR lines for species A and B are not measurable any more after 200 hours of annealing.

No new interactions could be observed for the ^{29}Si ENDOR lines beyond those already known. ENDOR lines of the ^{29}Si nuclei with very small shf interactions could not be analysed separately because of their low signal-to-noise ratio (see above).

Thus, after prolonged annealing when the later species are formed, more oxygen is added and the ^{17}O shf interactions are very small, if not zero, while the quadrupole interaction is the dominant one in the spectra. Surprisingly, the quadrupole interaction of the ^{17}O nuclei analysed in the early species is practically the same as that found in the later species. This latter result will be important when discussing the possible structure models in the next section.

3.3. DISCUSSION OF POSSIBLE STRUCTURE MODELS FOR TDDs

From the experimental results it is clear that oxygen is incorporated in the TDDs and that more oxygen is incorporated in the TDDs of later species. It was also found that no other impurities are part of the defect. No centre model can be derived directly from the experimental ENDOR data. Without theoretical interpretation of the shf and quadrupole interaction constants, no suggestion for a defect model can be made. The models proposed by Ourmazd et al [19], the so-called OBS model, as well as the Y-Lid by

Stavola et al [20] require the presence of one oxygen atom with an shf tensor having [001] symmetry. Since no such oxygen was found, but oxygens with other symmetries, these models cannot be the correct ones.

One key, which may lead to the correct model, is the quadrupole interaction of ^{17}O. It is practically the same for oxygen in early stages (TDD3 and TDD4) as in the later ones. The same magnitude is also found in one NL10 centre in Al-doped silicon [8]. The quadrupole interaction constant is caused by the electrical local field gradient at the site of the ^{17}O nucleus:

$$q = \frac{eQ_m}{4I(2I-1)} \left. \frac{\partial^2 V}{\partial z^2} \right|_{v=0} \qquad (4)$$

The electrical field gradient cannot be explained by unpaired spin density leading to the small anisotropic shf constant of ^{17}O nuclei, nor by a point charge in the distance of an Si-Si bond (2.35 Å) which can be easily estimated using the relation:

$$\frac{\partial^2 V}{\partial z^2} = \frac{e}{2\pi \varepsilon_0 r^3} (1-\gamma_\infty) \qquad (5)$$

where γ_∞ is the Sternheimer antishielding factor for ^{17}O (γ_∞ = -4.098). Q_m is the quadrupole moment of the ^{17}O nucleus [13].

For the point charge, we would obtain q/h = 18 kHz, an order of magnitude too small. Even for a ^{17}O near the core, where a point charge might be, one cannot explain the measured quadrupole interaction, let alone for the oxygens added later being probably much further away from the core. Since for the later species, the quadrupole interaction is practically the same as for the earlier ones, its explanation cannot depend on the wave function of the unpaired electron. Vice versa, the oxygen added later to a species cannot influence the symmetry of the defect as measured, for example in the shf tensors, i.e. the ENDOR spectra of the ^{29}Si nuclei, which are exclusively determined by the unpaired spin density. Therefore, the apparent C_{2v} symmetry as seen in the ^{29}Si interactions does not necessarily reflect defect symmetry, and, therefore, the conclusion mentioned earlier about 4 oxygens in the core is not stringent any more, nor the idea that oxygen is added pairwise upon growth. The charge density of a single 2p electron in a ^{17}O atom would cause a value of q/h = 0.71 MHz [17]. This value is too large, but of the right order of magnitude. Oxygen is normally incorporated interstitially with a bond angle of 164°. Form O^{2-} with 6 spin-compensated 2p electrons at right angles, q/h = 0, but through the bonding to 2 Si atoms and the bonding angle a q/h of the right order of magnitude can occur. This was shown in detail by a model calculation by Meilwes et al [15,25]. The conclusion is that all oxygens incorporated in the TDD^+ defect are interstitial oxygens. A similar suggestion was made for an NL10-Al defect [8] for the oxygen identified there which has about the same quadrupole interaction (see table 3).

The model proposed by Déak et al [21], which has a double donor in the core, seems to be one which is compatible with all the ENDOR data (see fig. 6). Two oxygen

96

atoms and one silicon interstitial form the defect core. If we identify the ^{29}Si of type 1 [11], which has the main axis of the shf tensor along [001] with the ^{29}Si interstitial and the type 1 ^{17}O nucleus with the core oxygens, then there would be two type 1 oxygens with the {110} plane mirror symmetry in the core. The next oxygen would be bound interstitially outside the core and maybe a type 2 oxygen. It does not seem necessary that at position 2 (see fig. 6) another oxygen of type 2 is added simultaneously. Further growth may occur by adding interstitial oxygen atoms one by one along a <110> chain. This model is also in agreement with the recent result that the thermal activation energy for the destruction of the TDDs is the same as the activation energy of the diffusion of interstitial oxygen (2.31 ± 0.19eV) [22,23]. Assuming this model, there are (2+14x1) oxygen atoms necessary to form TDD16 (the first two species are metastable and have not been seen in EPR or ENDOR). The number is not inconsistent with the recent result [24] that a loss of dissolved oxygen of an average of (10 ± 2) atoms per TDD is needed. The cluster calculation by Déak et al [21] yields shf interactions far too large by one order of magnitude, probably because the cluster was too small to adequately describe such an extended effective-mass-like defect.

4. EPR and ENDOR of other Heat Treatment Centres

NL10-type centres are heat treatment centres generated after prolonged annealing at about 450 °C. In an early stage, they have monoclinic, later C_{2v} orthorhombic symme-

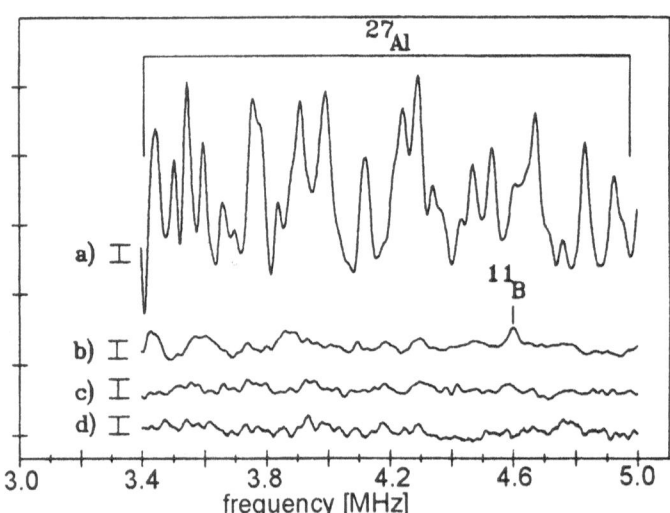

Fig. 7: ENDOR spectra of the NL10-like defects in differently doped Cz-Si, which was pre-annealed for 15 min. at 770 °C and afterwards for 100 hours at 460 °C. The bars give a criterion for the noise.

a) 5·10^{15}cm^{-3} Al doped, NL10-Al5   b) 4.5·10^{15}cm^{-3} B doped, NL10 (B)

c) 4.5·10^{15}cm^{-3} Ga doped, NL10 (Ga)   d) 4.5·10^{15}cm^{-3} In doped, NL10 (In)

try with g-factors of g_1 = 1.99747. g_2 = 1.99957, g_3 = 1.99959.The EPR line is of similar line width as that of TDD⁺s and has no resolved hf structure. The prolonged heat treatment with the generation of NL10-type defects does not appear to result in a unique defect structure as in the case of the TDD⁺s. The statement is the conclusion from the fact that different ENDOR spectra were measured depending on the chemical nature of acceptor dopings or other impurities and the dependence on the pre-annealing conditions. In particular, Al-doping influences the ENDOR spectra very much. Fig. 7a shows ENDOR lines of one ²⁷Al nucleus. in fig. 7b with B doping a distant ¹¹B ENDOR line is seen, while for Ga and In doping no corresponding ENDOR lines were observed (fig. 7b and c). In fig. 7. the distant ENDOR line at the Larmor frequency of ²⁹Si is measured in all cases, but not shown.

In particular, the Al-doping results in various NL10-like defects which have ENDOR spectra due to one Al nucleus. Altogether, 5 different centres have been identified. The appearance of each centre depends on the pre-annealing temperature and the measurement temperature, the latter reflecting different spin lattice relaxation times. It seems that the pre-annealing is decisive for the type of O-Al complex formed. At lower pre-annealing temperatures (e.g. 770 °C) only one type of NL10-Al defect is formed predominantly (NL10-Al5. see fig. 7a), while at higher pre-annealing temperatures several defects are produced with comparable concentrations [8, 25, 26].

Interestingly. the shf interactions of the Al nuclei vary only somewhat, but are, of course, clearly distinguishable by ENDOR. One finds a/h ≈ 0.95-1.5 MHz. b/h ≈ 0.04-0.08 MHz, b'/h ≈ 0.006-0.02 MHz. q/h ≈ 0.5 MHz and q'/h ≈ 0.5 MHz [8, 26]. Thus, probably Al occupies always the same type of lattice site, while the immediate envi-

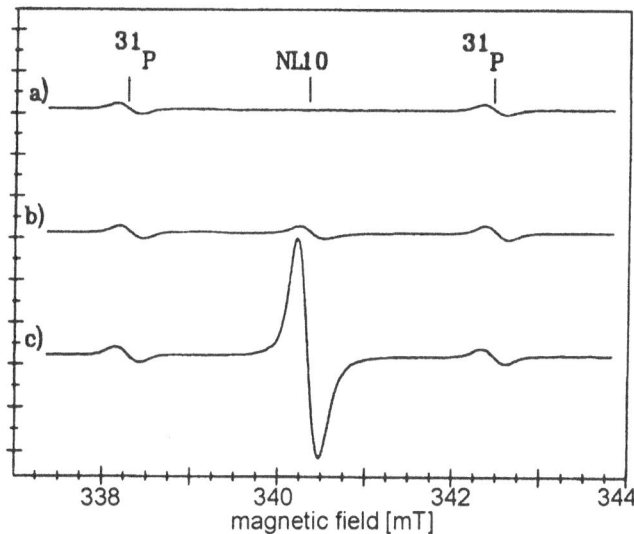

Fig. 8: EPR measurements in P-doped Cz-Si as a function of annealing time. The EPR intensity was scaled to a ruby standard a) P-doped starting material b) 24 hours annealed at 460 °C c) 90 hours annealed at 460 °C.

ronment varies, probably because of the oxygen atoms. The participation of oxygen in NL10-Al defects was demonstrated by the Amsterdam group using FZ material diffused with ^{17}O. From the analysis of the ^{17}O quadrupole tensor, an interstitial site for oxygen between two silicon atoms was postulated [8].

The quadrupole tensor of Al is extremely non-axial, with the principal value in a <001> direction being very small, the other two principal values, which point along <110> directions, are almost equal (about 1 MHz). These principal values are not explainable by the unpaired spin density at the Al nucleus nor by a point charge at a distance of 2.35 Å. Using equation (5), the latter would yield q/h = 0.054 MHz ($(1-\gamma_\infty)$ = 2.683), an order of magnitude too small. For an Al 3p function, one would obtain q/h = 1.05 MHz, which is in the right order of magnitude. A regular substitutional site for Al with an sp^3 hybrid for Al$^-$ would yield a small, if not negligible quadrupole interaction. However, a distorted substitutional site cannot be excluded and can, in principle, explain the quadrupole interaction. Likewise, an interstitial site between two Si atoms analogous to the interstitial oxygen is possible.

In ref. [8] it was argued that in P-doped Si those NL10 centres without Al could be identical to TDD$^+$s (NL8) in their singly ionised charge state, NL10 being an acceptor.

Fig. 8 shows the experimental proof that this is not the case [16]. In fig. 8, upper trace, the EPR spectrum of the shallow P donors is shown before annealing, in fig. 8, lower trace, after the formation of NL10 defects after 90 hours of annealing. The sample was cooled in the dark. The P-EPR signals did not change at all (± 20%) after NL10 defects have been formed to a concentration which exceeded that of the P donors by about a factor of 7. Therefore, the unpaired electron of NL10 cannot come from the P donors. This result is consistent with that of ref. [27]. Unfortunately, no ENDOR measurements have been possible for the NL10 centres in P-doped silicon.

For the determination of the paramagnetic level of NL10 defects in P-doped Si, one sample was annealed for 24 hours at 460 °C and then additionally γ-irradiated to form

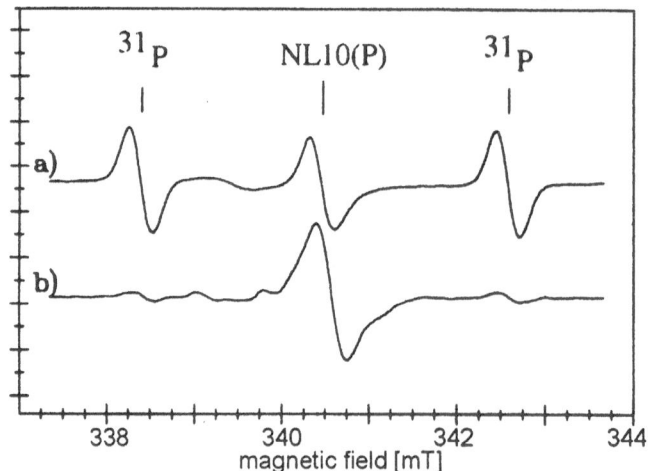

Fig. 9: EPR of a P-doped Cz-Si annealed for 24 hours at 410 °C
a) without irradiation, as annealed   b) with additional γ-irradiation of $1 \cdot 10^{19} cm^{-2}$

oxygen-vacancy pairs (A-centres [28]). The A centres act as acceptors. Fig. 9 shows that the P signal decreases as a result of the γ-irradiation, while the NL10 signal varies little (it is superimposed on the A centre). After γ-irradiation, the activation energy of 44meV due to P changes to 65meV. The P changes from the paramagnetic neutral state to the diamagnetic P^+ state, while the paramagnetic NL10 signal changes from a concentration of $8 \cdot 10^{13} cm^{-3}$ to $1.1 \cdot 10^{14} cm^{-3}$. This change in apparent concentration can be explained by an overlapping of the EPR line with that of the A centre. It was concluded from the experiment [25] that there is no electrical level of the NL10 in P-doped silicon in this region of the band gap. The NL10 (P) has to be a donor with its electrical level below 65meV below the conduction band.

A broad ^{17}O ENDOR line was formed in B-doped FZ-Si after ^{17}O diffusion when measured in the NL10 EPR spectrum. However, the ^{17}O interactions were much smaller compared to the one found in the NL10-Al defects. The maximum quadrupole interaction constant can be $q/h \leq 0.013$ MHz, an order of magnitude smaller than found for the TDD^+s and the NL10-Al centres (for this estimate a vanishing shf interaction was assumed) [25]. Oxygen seems to be aggregating differently here compared to the TDD^+s and the NL10-Al defects.

Finally, a ^{13}C-doped Cz-Si sample showed for NL10 besides the distant ^{29}Si ENDOR line only a small line of distant ^{13}C, analogously as the one of ^{11}B seen in fig. 7.

Only for NL10-Al defects, there is some information from ENDOR to discuss centre models, not so for the other impurity contents. However, it is clear that the "NL10 defects" denote a group of defects which show the same EPR resonance but have different microscopic structures.

5. Electrical Detection of early Species of Oxygen Aggregates

As mentioned in section 3.1, the earliest TDD identified by EPR and ENDOR is TDD3. In order to measure an EPR spectrum, one has to anneal for about at least 60 minutes, to measure ENDOR 4 hours are needed, since ENDOR is less sensitive compared to EPR [13]. Electrical detection of EPR (EDEPR) was recently shown to be several orders of magnitude more sensitive than conventional EPR, provided it operates. EDEPR is detected as a change of photoconductivity induced by an EPR transition [14]. As far as it is understood up to now, for EDEPR to operate it is necessary to have a donor-acceptor pair (DAP) recombination [14]. Fig. 10 shows such measurements in P-doped Cz-Si in comparison to conventional EPR as a function of annealing time (upper traces). Below in fig. 10, the growth of the various TDD^+ species upon annealing time is measured by IR absorption spectroscopy. Surprisingly, at very low annealing times, when conventional EPR fails to detect the TDD^+ spectrum, EDEPR detects it with high signal-to-noise ratio. From the spectral shape we cannot say which species were measured. According to IR experiments in the samples annealed for only 10 minutes, TDD^+s 1-3 were present [29]. The EPR spectra for the species identified by ENDOR are not distinguishable [11]. Whether the EPR spectra of the metastable species 1 and 2 are the same as those for later species is not known. Surprisingly, upon a

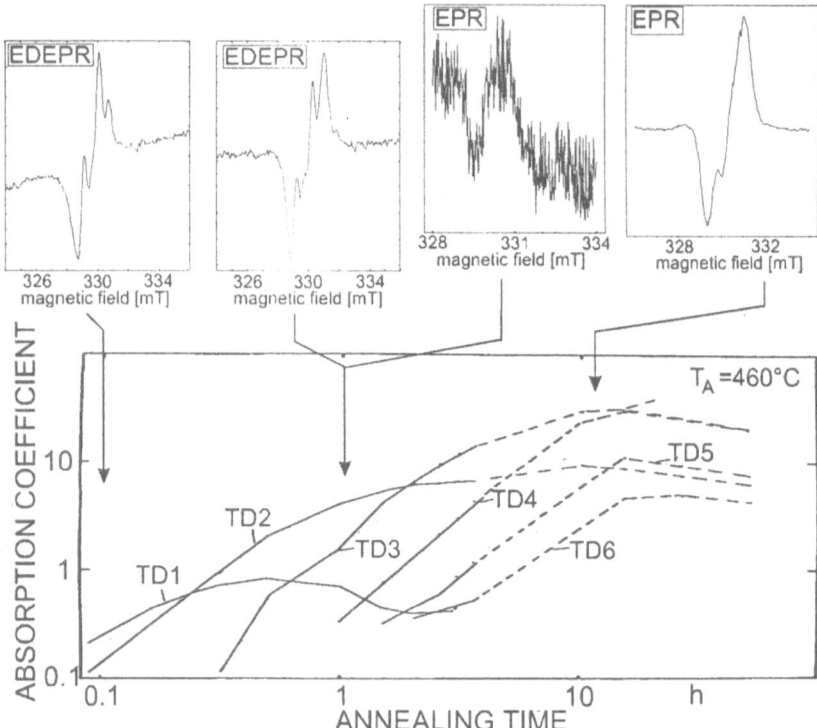

Fig. 10: Conventional EPR and EDEPR of TDD$^+$ centres in B-doped Cz-Si as a function of annealing time at 460 °C. B$_o$ ‖ [110]

mere quenching from 770 °C or 1200 °C, when in conventional EPR nor in IR absorption any TDD$^+$s are measurable. EDEPR spectra with good signal-to-noise ratio were detected. No systematic studies have yet been performed with EDEPR. However, this new measurement technique will allow a detailed study of the early species of oxygen precipitation. Since it was shown for phosphor in silicon that with this technique also ENDOR can be measured [30], there may be a new approach to study the microscopic structures of the first two species.

References

1. Fuller, C.S., Ditzenberger, J.A., Hannay, N.B., and Buehler, E. (1954) Phys. Rev. **96**, 833
2. Kaiser, W., Frisch, H.L., and Reiss. H. (1958) Phys. Rev. **B38**, 1546
3. Wagner, P. and Hage, J. (1989) Appl. Phys. **A49**, 123
4. Pajot, B., Compain, H., Lerouille, J. and Clerjaud, B. (1983) Physica B+C **117-118** B, 119
5. Götz, W., Pensl, G. and Zulehner, W. (1992) Phys. Rev. **B46**, 4312
6. Muller, H.S., Sprenger, M., Sieverts, E.G., and Ammerlaan, C.A.J. (1978) Solid State Comm. **25**, 987
7. Lathusko, A.I., Makarenko, L.F., Markevich, V.P., and Murin, L.I. (1986) Phys. Status Solidi **K181**, 93
8. Gregorkiewicz, T., Bekmann, H.H.B.Th., and Ammerlaan, C.A.J. (1988) Phys. Rev. **B38**, 3998
9. Bekmann, H.H.B.Th., Gregorkiewicz, T. and Ammerlaan, C.A.J. (1988) Phys. Rev. Lett. **61**, 227
10. Michel, J. (1987) *Die Struktur der thermischen Donatoren in Silizium*, Dissertation, Paderborn
11. Michel, J., Niklas, J.R. and Spaeth, J.-M. (1989) Phys. Rev. **B40**, 1732
12. Meilwes, N., Spaeth, J.-M., Emtsev, V.V. and Oganesyan, G.A. (1994) Semicond. Sci. and Technol. 9, 1346
13. Spaeth, J.-M., Niklas, J.R. and Bartram, R.H. (1992) *Structural Analysis of Point Defects in Solids*, Chapt. 5 and 8 (Springer, Heidelberg, New York)
14. Stich, B., Greulich-Weber, S. and Spaeth, J.-M. (1995) J. Appl. Phys. **77**, 1546
15. Meilwes, N. (1993) *Strukturuntersuchungen an thermisch gebildeten Sauerstoffaggregatzentren in Silizium*, Dissertation Paderborn
16. Michel, J., Meilwes N., Spaeth, J.-M. (1989) Phys. Rev. **B39**, 7978
17. Meilwes, N., Spaeth, J.-M., Götz, W. and Pensl, G. (1994) Semicond. Sci. and Technol. 9, 1623
18. Mathiot, D. (1987) Appl. Phys. Lett. **51**, 904
19. Ourmazd, A., Schröter, W. and Bourret, A. (1984) J. Appl. Phys. **56**, 1670
20. Stavola, M. and Snyder, L.C. (1983) in *Defects in Silicon*, ed. by W.M. Bullis and C.C. Kimmerling (Electrochemical Soc., Pennington, N.J.)
21. Déak, P., Snyder, L.C. and Corbett, J.W. (1992) Phys. Rev. **B20**, 11612
22. Götz, W., Pensl, G., Zulehner, W., Addinall, R. and Newman, R.C. (1994) Solid State Comm. **93**, 5, 454 1995
23. Götz, W., Pensl, G., Zulehner, W. (1993) Mater. Sci. Forum **117-118**, 213
24. McQuaid, S.A., Louders, C.A., Binns, M.J., Newman, R.C. and Tacker, J.H. (1994) Mater. Sci. Forum **143-147**, 963
25. Meilwes, N., Spaeth, J.-M., Emtsev, V.V., Oganesyan, G.A., Götz, W. and Pensl, G. (1994) Mater. Sci. Forum **143-147**, 141
26. Meilwes, N., Niklas, J.R. and Spaeth, J.-M. (1990) Mater. Sci. Forum **65-66**, 247
27. Suezawa, M., Sumino, K. and Iwaizumi, M. (1983) J. Appl. Phys. **54**, 6594
28. Watkins, G.D. and Corbett, J.W. (1961) Phys. Rev. **121**, 1001
29. Pensl, G. Private communication
30. Stich, B., Greulich-Weber, S. and Spaeth, J.-M. (1996) Appl. Phys. Lett. **68**

EFFECT OF HYDROGEN ON OXYGEN-RELATED DEFECT REACTIONS IN SILICON AT ELEVATED TEMPERATURES

V.P. MARKEVICH, I.F. MEDVEDEVA, and L.I. MURIN
Institute of Solid State and Semiconductor Physics
P. Brovki str. 17, Minsk 220072, Belarus

ABSTRACT

Peculiarities of oxygen-related defect reactions at elevated temperatures ($\leq$ 500°C) have been studied in hydrogen-rich Czochralski-grown silicon crystals. Hydrogen was introduced into the crystals by high-temperature (900-1200°C) annealing in hydrogen gas ambient or by exposure in hydrogen plasma at temperatures 350-450°C.

The presence of hydrogen is found to influence significantly the behaviour of thermally- and radiation-induced defects and to result in formation of several new electrically and optically active centres. The formation kinetics of these centres as well as the thermal double donors in the crystals with different content of oxygen, hydrogen and radiation-induced defects have been investigated.

The states of hydrogen in Si:O,H crystals and possible reactions between the hydrogen- and oxygen-related centres are discussed.

1. Introduction

Oxygen is known to be one of the main residual impurities in crystalline silicon and to play an important role in various aspects of semiconductor science and technology. Being incorporated into Czochralski-grown (Cz) silicon crystals up to amount of 10^{18} cm^{-3} oxygen atoms have significant effect on mechanical and electronic properties of the material. In particular, oxygen is an aid to high temperature processing since it strengthens the material. At temperatures higher than 350°C oxygen atoms are able to diffuse and form a large body of various

103

R. Jones (ed.), Early Stages of Oxygen Precipitation in Silicon, 103–122.
© 1996 *Kluwer Academic Publishers.*

aggregates which often are electrically active and contribute to gettering of some undesirable contaminants.

In contrast to the rather good understanding of the oxygen-related processes at temperatures higher than 600°C, oxygen behaviour at T < 600°C and effects associated with it remain active areas of investigation [1]. It is found that oxygen-related processes at these lower temperatures are clearly dependent upon the Si wafer history and other impurity content. But the origins of the many features observed continue to be unclear.

Recently it has been shown that atomic hydrogen incorporated in Cz-Si crystals influences strongly the oxygen behaviour at elevated temperatures. In particular, significant enhancements in the rate of the oxygen loss from the solution and in the rate of the oxygen-related thermal donor (TD) formation were observed under heat-treatments at T < 500°C for hydrogen-rich Cz-Si crystals [1-4]. Besides, a number of new hydrogen-oxygen-related defects was revealed recently in irradiated and heat-treated Si:O,H crystals [5-8]. So it is evident that hydrogen-oxygen interaction is a factor which plays a significant role in the processes of formation of electrically and optically active defects in Cz-Si crystals. However, our understanding of the phenomena associated with hydrogen-oxygen interaction in silicon is still rather poor.

The present paper reports our recent experimental results on the study of effect of hydrogen on oxygen-related defect reactions in the temperature range 50 to 500°C.

2. Experiment

Phosphorus doped ($N_P = 5 \times 10^{13}$-5×10^{15} cm^{-3}) Czochralski-grown silicon crystals with different initial concentrations of interstitial oxygen ($N_O = 5 \times 10^{17}$-1.2×10^{18} cm^{-3}) and substitutional carbon ($N_C \leq 2 \times 10^{16}$ cm^{-3} or $N_C = 3.0$-3.5×10^{17} cm^{-3}) were investigated. The concentrations of oxygen and carbon were deduced from the magnitudes of the 9 and 16.5 μm infrared absorption bands at room temperature using calibration coefficients of 3.14×10^{17} cm^{-2} for oxygen and 1.1×10^{17} cm^{-2} for carbon. Floating-zone-grown n-type Si crystals with oxygen content lower than 5×10^{15} cm^{-3} were used as control ones in some experiments.

Hydrogen (deuterium) was introduced into the samples of 2.0-2.5 mm thick by annealing in $H_2(D_2)$ gas ambient at different temperatures in the range 900 to 1200°C. Heat-treatments (usually for 2 hours) were followed by the fast cooling of the sealed quartz ampoules with the samples. After heat-treatments the surface layer (about 150 μm) of each sample was removed by etching in a

1HF + 5HNO$_3$ mixture. Exposure of the samples in a dc hydrogen plasma was carried out in a reactor for the reactive ion etching with a plate voltage of 300-600 V (ion current was $\leq$ 50 μA/cm^2) in the temperature range 350 to 450°C. After exposure the samples were cooled down to 230°C for 30 min and then were taken out of the reactor.

Irradiation with fast electrons (3.5 MeV in energy, 2x10^{12} cm^{-2}s^{-1} in electron beam intensity) or γ-rays from ^{60}Co source was performed at room temperature. A series of isochronal and isothermal heat-treatments were carried out in the temperature range 50 to 600°C in argon ambient or in air.

Concentrations of the electrically active defects and location of their energy levels were determined by means of Hall effect and DLTS measurements which were carried out in the temperature range 30 to 400 K. Electrical measurements were combined with grinding away the thin surface layers for determination of the defect concentration depth profiles in the samples. Optical absorption in the wavenumber range 200 to 4000 cm^{-1} was measured at 6 K and at room temperature by means of dispersive Specord 75 IR and Fourier transform JEOL J-100 infrared spectrometers.

3. Experimental results and discussion

3.1. ELECTRICAL AND OPTICAL PROPERTIES OF H$_2$(D$_2$)-SOAKED n-TYPE Cz-Si CRYSTALS

3.1.1. *Oxygen-hydrogen pairs and some their features*
In accordance with the theoretical investigations [9, 10] hydrogen and oxygen atoms can form stable complexes in silicon lattice. According to [9] energetically favoured configuration of oxygen-hydrogen pair in crystalline Si is one where H and O atoms are located at the bond-centred sites and are bound to the same silicon atom. Jones *et al.* [10] found that in the most stable configuration hydrogen and oxygen atoms are bound to the same host atom but H atom is located at the anti-bonding site opposite O atom. As both the oxygen and hydrogen atoms being at the above-mentioned positions in Si lattice give rise to the local vibrational modes (LVMs) [11, 12] it would be reasonable to expect of the appearance of the LVM lines due to the vibrations of these atoms when they are included in O-H complex.

A new absorption band at 1075.1 cm^{-1} was observed recently in H$_2$-soaked Cz-Si crystals [8]. Replace of hydrogen by deuterium resulted in the shift of the band to 1076.4 cm^{-1}. These lines were identified as ones related to the local vibrational modes of the complex containing one oxygen and one hydrogen (deuterium) atoms [8].

Figure 1a shows the optical absorption in the wavenumber range from 1070 to 1090 cm^{-1} measured at 6 K in the n-type Cz-Si sample annealed at 1200°C for 2 hours in hydrogen gas ambient. Beside the well-known LVM line at 1084.8 cm^{-1} related to the stretching mode of interstitial ^{18}O atoms the line at 1075.1 cm^{-1} is clearly seen. By analogy with the situation in Cz-Si samples doped with other light impurities, carbon and lithium [13], it is reasonable to suggest that the 1075.1 cm^{-1} LVM line is related to the stretching mode of interstitial ^{16}O atom slightly modified by the presence of a hydrogen atom in the nearest vicinity. If so the oscillator strengths should be similar for 1075.1 cm^{-1} and 1084.8 cm^{-1} lines. It allows one to estimate the concentration of the oxygen-hydrogen complexes by comparing the intensity of the 1075.1 cm^{-1} line with that of 1084.8 cm^{-1} line. The concentration of the ^{18}O atoms was 2.1×10^{15} cm^{-3} in the sample spectrum of which is shown in Fig. 1a. The intensity of the 1075.1 cm^{-1} line is about half of that of the 1084.8 cm^{-1} line (see Fig. 1a), so the concentration of O-H complexes is about 1.1×10^{15} cm^{-3} in the sample. This is comparable with the solubility of atomic hydrogen at 1200°C ($\approx 6 \times 10^{15}$ cm^{-3}) [14], i.e. a significant part of the hydrogen atoms introduced into Cz-Si crystals during high-temperature heat-treatments can interact with oxygen atoms and result in the formation of the O-H pairs.

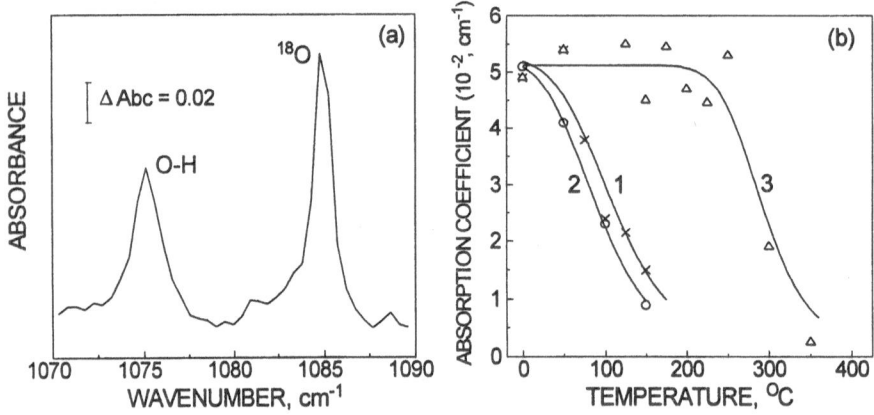

Figure 1. (a) Absorption spectrum at 6 K of a Cz-Si sample heat-treated at 1200°C for 2 h in H$_2$ gas ambient and (b) variations in the absorption coefficient at 6 K in the maximum of 1075.1 cm^{-1} line upon 30-min isochronal annealing (IA) for the samples heat-treated at 1200°C for 2 h in H$_2$ gas ambient. The sample 2 (according to the numbers on the curves) was irradiated with fast electrons (F = 1.6×10^{16} cm^{-2}) after heat-treatment at 1200°C. 30-min IA was followed by the annealing at 50°C for 1 h for the samples 2 and 3.

A preliminary study of the thermal stability of O-H complexes was carried out in [8]. It was shown from the isochronal annealing experiments that the complexes are rather unstable: heat-treatments of the H_2-soaked Cz-Si samples at temperatures higher than 100-125°C resulted in disappearance of the appropriate LVM band. However, further experiments showed that keeping the annealed samples at room temperature or heat-treatment at 50°C [8] led to the regrowth of the band. After a long time at room temperature the intensity of 1075.1 cm^{-1} line returned to its original value. The results of the two step annealing (30-min isochronal annealing up to 400°C followed by the heat-treatment at 50°C for 1 h) for a H_2-soaked Cz-Si sample are shown by triangles (curve 3) in Figure 1b. Complete recovery of the 1075.1 cm^{-1} line intensity was observed up to the 250°C isochronal annealing step. The 1075.1 cm^{-1} line disappeared without recovery only after annealing at temperatures higher than 350°C.

The results obtained can be explained either by a) dissociation of the pairs at temperatures higher than 100°C and their subsequent regeneration at lower temperatures (assuming that hydrogen atoms are mobile at room temperature and oxygen atoms are the most effective traps for them), or by b) reversible transformation of the O-H complex configuration with the appearance/disappearance of the local vibrational modes.

To check the validity of above-mentioned possibilities we studied the annealing behaviour of the O-H-related absorption band in the crystals irradiated with fast electrons, i.e. in the crystals containing additional traps for hydrogen atoms. It was found that irradiation (up to fluence of 2×10^{16} cm^{-2}) did not influence the intensity of the 1075.1 cm^{-1} line, but in irradiated samples the line disappeared completely (without recovery) after annealing at 125-150°C (see Fig. 1b, curve 2). It was inferred from these results that annealing behaviour of the line is consistent with the first suggestion, i.e., O-H(D) pairs are stable only at temperatures lower than 100-125°C and dissociate at higher temperatures that results in the appearance of mobile hydrogen atoms in Si lattice. In irradiated crystals these atoms are captured by radiation-induced defects and then there is no recovery of the O-H-related LVM band. Complete recovery of the 1075.1 cm^{-1} line intensity in non-irradiated H_2-soaked crystals after heat-treatments (HTs) in the temperature range 100 to 300°C means that there are no other effective traps except oxygen atoms for mobile hydrogen atoms under given conditions.

A further question is how to explain the absence of the regrowth of O-H pairs after anneals at temperatures higher than 300°C. As it will be seen from the results presented in Sections 3.2 and 3.3 in spite of the undetectability of

O-H complexes after HTs in this temperature range hydrogen continues to influence significantly the oxygen-related defect reactions. It can be inferred from those results that the absence of the regeneration of O-H complexes after heat-treatments at $T \geq 300°C$ is associated with incorporation of the main part of hydrogen atoms into complexes relatively stable at these temperatures and not with hydrogen out-diffusion from the samples.

3.1.2. *Effect of $H_2(D_2)$-soaking treatments on the electrical properties of n-type Cz-Si crystals*

It was found by means of DLTS and Hall effect measurements that high-temperature heat-treatments of n-type Cz-Si crystals in hydrogen (deuterium) gas ambient led neither to the appearance of defects with the energy levels in the upper part of the band gap nor to the changes in the free electron concentration. No changes in the electron concentration were observed either after a two-step annealing.

In accordance with the estimation given in the section 3.1.1 the concentration of the oxygen-hydrogen pairs reached the value of 1×10^{15} cm^{-3} in H_2-soaked crystals. Since the appearance/disappearance of this complex did not effect the electrical properties of the crystals investigated it implies that oxygen-hydrogen pairs as well as the single H atoms are neutral in n-type silicon crystals.

3.2. EFFECT OF HYDROGEN ON OXYGEN-RELATED THERMAL DOUBLE DONOR FORMATION

3.2.1 *Thermal double donor formation in H_2-soaked Cz-Si crystals*

To obtain further information on the oxygen-hydrogen interaction in silicon we studied peculiarities in the formation of oxygen-related thermal double donors (TDDs) in H_2-soaked Cz-Si crystals.

Figure 2 shows the dependencies of a) total concentrations of all the TDD species (N_{TDD}) and b) TDD formation rates ($\Delta N_{TDD}/\Delta t$) upon annealing duration at $T = 427°C$ for as-grown and pre-heat-treated samples cut from the same carbon-lean n-type Cz-Si ingot. Samples 2 and 3 (according to the numbers on the curves in Fig 2) were subjected to the preliminary heat-treatments at 1000°C for 2 hours in air and in H_2 gas ambient, respectively, before annealing at 427°C. It is clearly seen that preliminary heat- treatments, especially in the case of H_2 gas ambient, led to the significant changes in the TDD formation process. In the as-grown sample the TDD generation rate was a maximum at the initial stages of annealing and decreased monotonically with

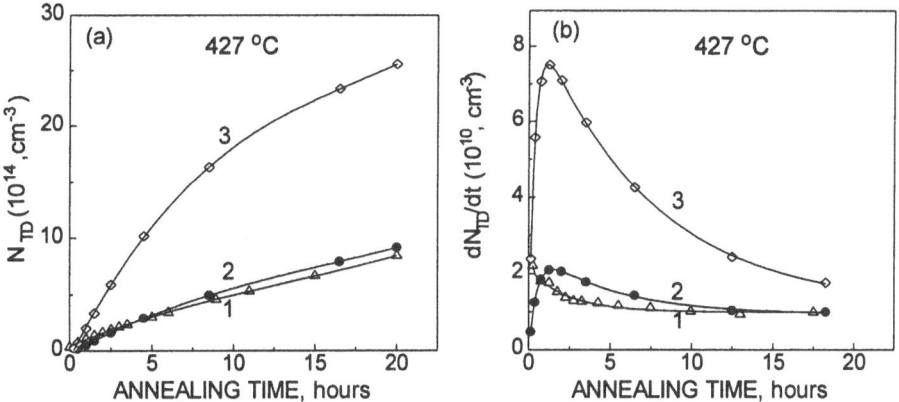

Figure 2. Dependence of (a) TDD concentration and (b) their generation rate on annealing time at 427°C in an as-grown (1) and pre-heat-treated for 2 hours in air (2) and H_2 gas (3) ambient Cz-Si samples; $(N_O)_{in} \approx 9.5 \times 10^{17}$ cm^{-3}, $N_P \approx 2.0 \times 10^{14}$ cm^{-3}.

the annealing duration to its steady-state value. Such a behaviour was found to be typical for many of as-grown Cz-Si crystals and could be associated with the transformation of electrically inactive TDD precursors or nuclei, which were formed during the post-growth cooling the ingots, into the TDDs [15, 16].

High-temperature heat-treatments resulted in the disappearance of TDDs as well as of their nuclei, and is fully consistent with our earlier observations [15, 16]. But the subsequent anneals of the pre-heat-treated samples at temperatures in the range 350 to 450°C led to the very rapid rise in the TDD formation rate (see Fig. 2b). This effect is especially pronounced for the H_2-soaked samples. In these samples $\Delta N_{TDD}/\Delta t$ rapidly reached their maximum values which were significantly higher than those in as-grown samples. Thereafter the TDD formation rates decreased monotonically going to the same values as in as-grown and control samples. In control samples, similar features in the TDD formation rate were observed as in H_2-soaked ones but the maximum values of $\Delta N_{TDD}/\Delta t$ were much lower and like to those for the as-grown samples.

Figure 3a shows the formation kinetics of the first species belonging to the TDD family (TDD-1) in the same samples the data for which are presented in Fig. 2. It is seen that the concentration of TDD-1 (N_{TDD-1}) in H_2-soaked sample was significantly higher than N_{TDD-1} in control and as-grown samples for all range of annealing time except the earliest stages.

So the presented data demonstrate that H_2-soaking treatments lead to the significant enhancement in the generation rate of oxygen-related thermal double donors in silicon. The data are consistent with the previous observations of enhanced rates of TDD formation in Cz-Si crystals subjected to treatments in

110

hydrogen plasma at T = 350-450°C [2-4]. The latter results were reasonably interpreted by assuming that the enhanced rates of the TDD formation were caused by the hydrogen-catalysed enhanced oxygen diffusion [3, 4]. However, as it was shown earlier [17] the process of successive transformation of thermal double donor centres cannot be described by only the attachment of mobile oxygen atoms to the TDD species. It is commonly supposed that some oxygen-related species whose diffusivity is much higher than the diffusivity of interstitial oxygen atoms are responsible for the process of successive transformation of TDD complexes [1]. But the origin of these fast diffusing species (FDSs) is still unknown.

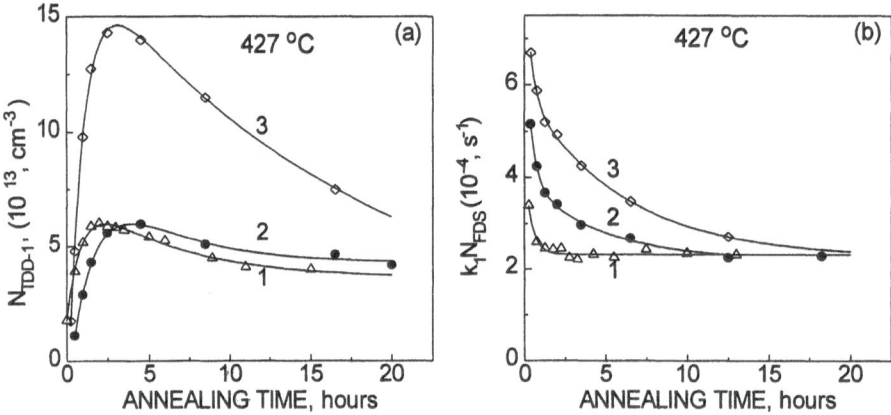

Figure 3. Dependence of a) TDD-1 concentration and b) the value of $k_1 N_{FDS}$ on annealing time at 427°C in as-grown (1) and pre-heat-treated at 1000°C for 2 hours in air (2) and H_2 gas (3) ambient Cz-Si samples.

Some important information on the behaviour of FDSs can be deduced from the study of TDD formation process. For the initial stages of heat-treatments at temperatures lower than 430°C the process of TDD species transformation can be described by the following simple equation [16]:

$$\sum_{m>n} \frac{d N_{TDD-m}}{dt} = k_n N_{FDS} N_{TDD-n} , \qquad (1)$$

where N_{TDD-m} and N_{TDD-n} are the concentrations of thermal donors of m and n types, respectively, N_{FDS} is the concentration of fast diffusing species and $k_n = 4\pi r_n D_{FDS}$ is the appropriate reaction rate constant (D_{FDS} is the diffusion coefficient of FDSs and r_n is the capture radius). On the basis of Eq. (1) $k_n N_{FDS}$ values for the first TDD species can be directly determined from the kinetics of their formation.

Figure 3b shows the variations in the product of $k_1 N_{FDS}$ with annealing duration at 427°C for H_2-soaked sample in comparison with those for control and as-grown ones. It is clearly seen that the values of $k_1 N_{FDS}$ are higher in H_2-soaked sample, i.e. presence of hydrogen leads to the enhancement in the reaction rate of TDD-1 transformation into TDD-2.

Further analysis of the TDD formation process in H_2-soaked samples allowed us to show that the hydrogen-related enhancements in reaction rates for the TDD transformation are different for the first TDD species. This finding is not consistent with the assumption that the same fast diffusing species are responsible for all the process of TDD formation and confirms our suggestion that the first members of TDD family are mobile and under their motion can interact with interstitial oxygen atoms [18]. Only on the basis of this suggestion can significant variations in the reaction rate constants for the transformation of the first TDD species (TDD-1 - TDD-3) [16] and changes of these constants in H_2-soaked samples be explained self-consistently.

Further questions relate to the origin of the decays of the TDD formation rate, the concentration of TDD species and the reaction rate constants in H_2-soaked samples (see Fig. 2b, Fig. 3). In the frame of the hydrogen-catalysed oxygen diffusion model these decays can be explained by the escape of hydrogen atoms either by binding in the stable complexes or by out-diffusion from the samples. If the decays are caused mainly by out-diffusion then one should anticipate a non-uniform distribution of generated TDDs. Indeed, by means of Hall effect measurements combined with etching back the surface of H_2-soaked samples subjected to heat-treatments at temperatures from 350 to 450°C, it was found that significant spatial gradients in TDD concentration exist in such samples. The TDD concentration was minimum at the surface of the samples and reached the maximum values in the central parts. It was inferred from these results that the enhancement effect of hydrogen decreases with annealing time because of the hydrogen out-diffusion.

The study of the spatial distribution of thermal donors generated in hydrogen-enhanced way can provide important information on the processes of oxygen-hydrogen interaction as well as the hydrogen diffusion in the silicon crystals. So we studied depth profiles of thermal donors generated in the samples subjected to hydrogen plasma treatments of various duration in the temperature range 350 to 450°C.

3.2.2. Hydrogen plasma-assisted thermal double donor formation in Cz-Si crystals

Plotted in Figure 4a are the depth profiles of the added electrons due to donors generated in Cz-Si samples subjected to treatments in hydrogen plasma at

400°C for different duration. Dashed lines in the figure show the added electron concentration on the back side of the same samples. Clearly enhanced generation of donor centres occurs near the surface subjected to H-plasma exposure. Analysis of the temperature dependencies of the free carrier concentration in the subsurface layers has shown that the donor centres with the energy levels at about E_c-0.07 eV and E_c-0.15 eV are induced by hydrogen plasma treatments. The results obtained confirm the conclusion made earlier on the basis of IR absorption measurements that the donors generated in hydrogen-plasma-treated Cz-Si samples are mainly well-known oxygen-related thermal double donors [2, 3]. Hall effect measurements have shown also that the concentration of the first species of TDD family, bistable TDD-1 and TDD-2 [19], is very high in the subsurface layers, that is also fully consistent with the previous data [2, 3].

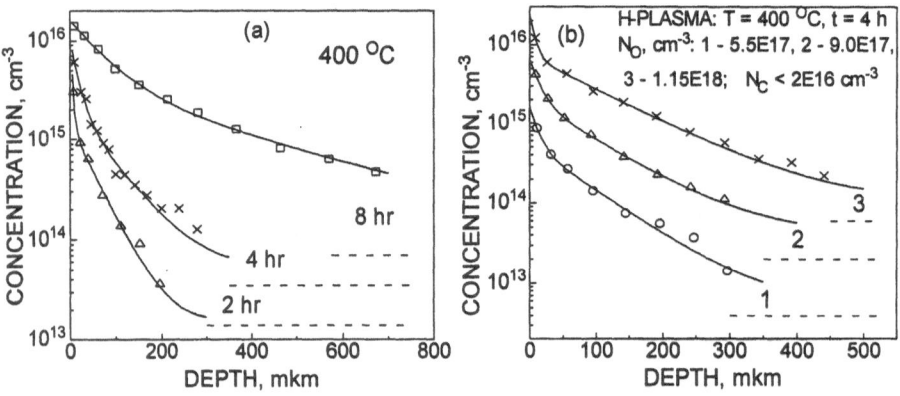

Figure 4. Depth distribution for the electron concentration from thermal donors after hydrogen plasma exposure at 400°C (a) for 2, 4 and 8 hours of the samples with initial oxygen content of about 9.2×10^{17} cm^{-3} and (b) for 4 hours of the samples with different oxygen content.

It has been found that the insertion and diffusion of hydrogen atoms and their effect on oxygen diffusivity is the origin of H-plasma-assisted TDD generation [3, 4]. So the TDD concentration depth profiles in H-plasma-treated Cz-Si samples can be considered as the manifestation of atomic hydrogen distribution in the samples.

Fig. 4b shows the depth profiles of the added electrons due to donors generated in Cz-Si samples with various oxygen content under hydrogen plasma exposure at 400°C for 4 hours. The dependence of the concentration of generated donors on oxygen content is apparent, but the shape and depths of the profiles are very similar in the samples with different N_O that implies an insignificant influence of the oxygen content on hydrogen penetration depth.

We have studied also the effects of annealing temperature, of carbon content, of thermal pre-history and preliminary irradiation on the TDD concentration depth profiles in H-plasma-treated samples. The results of this study will be published in detail elsewhere [20], but the following important findings should be noted.

i) TDD concentration depth profiles in H-plasma-treated Cz-Si samples can be described assuming that TDD formation rate is expressed by the following equation

$$\frac{dN_{TDD}}{dt}(x,t) = \left(\frac{dN_{TDD}}{dt}\right)_b \left[1 + A\, N_H(x,t)\right] \quad , \qquad (2)$$

where $(dN_{TDD}/dt)_b$ is the TDD generation rate in the bulk of the samples (non-enhanced by hydrogen), $N_H(x,t)$ presents the depth distribution of atomic hydrogen and A is an hydrogen-related enhancement coefficient. It has been found that the factors which effect the $(dN_{TDD}/dt)_b$ value, e.g., oxygen and carbon content, thermal pre-history of the samples, etc., do not influence significantly the process of hydrogen incorporation, i.e., $N_H(x,t)$ is independent on N_O, N_C, thermal pre-history, etc. Only radiation-induced defects were found to effect the penetration depth of atomic hydrogen.

ii) Diffusion profiles of hydrogen in the temperature range 350 to 450°C can be described assuming that hydrogen molecule formation is the dominant process which influences the hydrogen diffusivity.

3.3. HYDROGEN INTERACTION WITH RADIATION-INDUCED DEFECTS IN Cz-Si CRYSTALS

It was found in our previous work [5] from the results obtained by electrical measurements that H_2-soaking treatments of n-type Cz-Si crystals did not influence the type and introduction rates of the main defects induced by room-temperature irradiation with fast electrons. Further photoluminescence and DLTS measurements justified this conclusion. Vacancy-oxygen (V-O or A centre) and C_i-O_i complexes as well as divacancies (W) were the dominant defects induced by irradiation in both the as-grown and H_2-soaked crystals. This result is not unexpectable because oxygen and carbon due to their high concentration in Cz-Si crystals act as the most effective traps for mobile vacancies and self- and carbon-interstitials generated by irradiation.

As it was shown in section 3.1 heat-treatments of the H_2-soaked samples at temperatures higher than 100°C led to the release of hydrogen atoms from the weak-bound complexes. Interaction of this mobile H atoms with radiation-induced defects (RDs) can result in both the elimination of electrical activity of RDs and formation of new hydrogen-related electrically active defects. Indeed, as it was found recently by means of photoluminescence [21], electrical [5] and optical [6] measurements the hydrogen incorporation into the Cz-Si crystals led to the significant changes in annealing behaviour of radiation-induced defects and to the effective formation of new hydrogen-related centres.

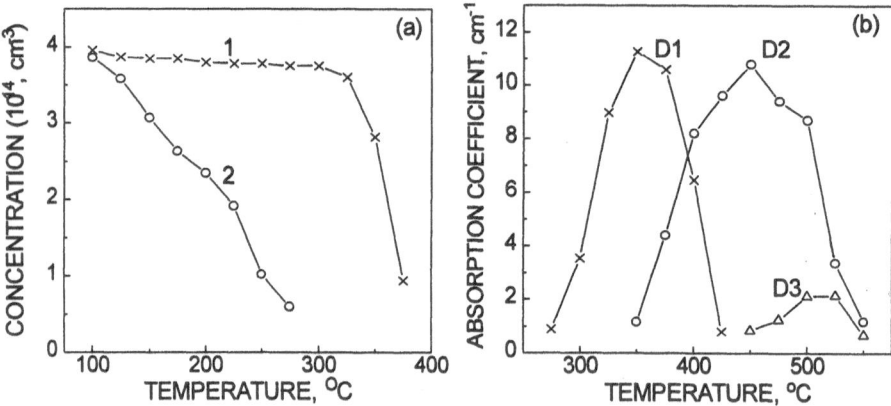

Figure 5. Variations upon 30-min isochronal annealing (a) in the A centre concentration for (1) as-grown and (2) H_2-soaked (1000°C, 2 h) Cz-Si samples and (b) in the absorption coefficients in the maximum of bands due to $1s \rightarrow 2p_{\pm}$ electronic transitions for D1-D3 centres in H_2-soaked sample. The samples were irradiated with fast electrons, fluence of irradiation was 3×10^{15} cm^{-2}.

For instance, curves 1 and 2 in Figure 5a show the annealing behaviour of A centres in the as-grown and H_2-soaked silicon samples irradiated with fast electrons. It is seen that A centres disappeared in the H_2-soaked sample at much lower temperatures than in as-grown one. However, it should be noted that similar decreases in annealing temperatures for the main RDs were observed also in control samples subjected to high-temperature treatments in air or in inert gas ambient with following fast cooling [22]. It was proposed in [22] that such effects in the control samples are related to the interaction of RDs with the mobile quenched-in defects preliminary identified as fast diffusing metallic impurities (e.g. Cu or Fe). As both the process of RDs passivation by hydrogen and elimination of their activity by interaction with quenched-in defects take place simultaneously in the same temperature region it was difficult to distinguish clearly these two processes at annealing temperatures lower than 250°C. But the heat-treatments of the samples at T > 250°C resulted in big difference in the processes of defect transformation for the H_2-soaked and control crystals. The most prominent feature of the former crystals was the effective formation of some new electrically active centres. It was found by means of IR absorption measurements [6, 7] that three shallow donor centres (termed D1-D3 centres in [6, 7]) were successively formed and eliminated when the H_2-soaked crystals were annealed in the temperature range 270-600°C. The formation kinetics of D1-D3 centres are shown by curves 1-3 in Figure 5b.

The electronic properties of all the D1-D3 centres were described well in the frame of the effective mass approximation, but further analysis showed that D1 is not a simple hydrogen-like shallow donor. It was found from the analysis of the data obtained by optical and electrical measurements that D1 centre-related features observed in IR absorption spectra are associated with the neutral charge state of the amphoteric negative-U centre having the occupancy level $E(-/+) = E_c - 0.076$ eV. The results of the study of the electronic structure of this defect were presented in [23].

Some preliminary results on the formation mechanism and composition of the D1 centre were given in [5, 6]. It was proposed that a few hydrogen atoms and an oxygen-related radiation-induced defect (most probably, the A centre) are included in the D1 centre. Here further experimental results devoted to elucidating the mechanism of D1 centre formation are presented.

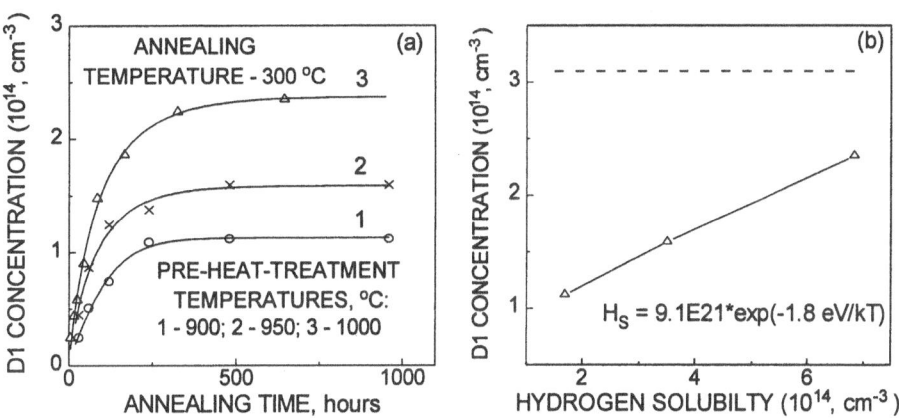

Figure 6. a) Variations in the D1 centre concentrations with annealing time at 300°C in Cz-Si samples pre-heat-treated in hydrogen gas ambient at various temperatures for 2 hours and irradiated with fast electrons ($F = 3 \times 10^{15}$ cm^{-2}) and b) maximum D1 centre concentrations versus possible amount of hydrogen in the samples (dashed line displays the average A-centre concentration after irradiation).

Figure 6a shows the generation kinetics of the D1 centres at 300°C in the samples pre-annealed in hydrogen gas ambient at different temperatures. The effect of different annealing temperatures is apparent: higher annealing

116

temperatures result in more D1 centres. This can be considered as undirect evidence for incorporation of hydrogen atoms into the D1 centre, because higher annealing temperatures result in higher solubility and consequently in greater amounts of hydrogen in the samples [14]. Figure 6b shows maximum concentrations of D1 centres versus expected amount of hydrogen [14]. The average concentration of A centres after irradiation is given by the dashed line in this Figure as well. It is seen that at low concentrations of hydrogen the maximum concentration of D1 centres is significantly lower than the concentration of A centres and is roughly equal to half of the hydrogen concentration. This may suggest that two hydrogen atoms are included into D1 centre. When the hydrogen concentration is high the D1 centre concentration is close to the concentration of A centres but never exceeds this latter one.

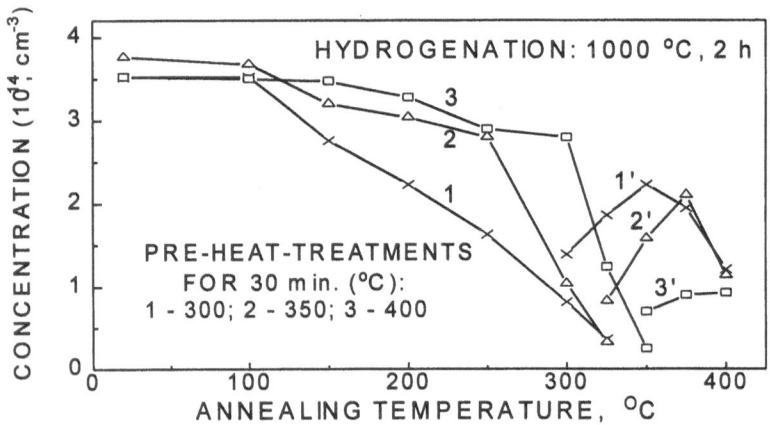

Fig. 7. Variations in the (1-3) A centre and (1'-3') D1 centre concentrations with temperature of 30 min isochronal annealing in H_2-soaked Cz-Si samples subjected to preliminary heat-treatments for 30 min at 300 (1, 1'), 350 (2, 2') and 400°C (3, 3') before irradiation with fast electrons.

It was shown in Section 3.1 that heat-treatments of H_2-soaked samples at temperatures higher than 350°C led to the immobilisation of the main part of the mobile hydrogen-related species. If transformations of radiation-induced defects in H_2-soaked samples are caused by the interaction of RDs with mobile hydrogen-related species then the heat-treatments of such samples at T > 350°C before irradiation would result in some changes in RDs transformation kinetics.

Figure 7 shows the annealing behaviour of A centres and formation kinetics of D1 centres in the H_2-soaked samples pre-annealed at 300, 350 and 400°C before irradiation with fast electrons. Clearly the heat-treatments before irradiation led to the significant changes in the processes of A centre elimination and D1 centre formation. It is evident that the number of hydrogen atoms that were able to interact with RDs was significantly reduced by pre-heat-treatments, especially at 400°C, before irradiation.

3.4. ON THE HYDROGEN-ENHANCED OXYGEN DIFFUSION

It was found earlier [4] that hydrogen influences significantly the oxygen diffusivity in the temperature range 300 to 500°C. The study of the properties of the recently discovered oxygen-hydrogen pairs can provide some useful information in understanding the mechanism for hydrogen-enhanced oxygen diffusion.

If hydrogen-oxygen interaction is the origin of enhanced oxygen diffusion then the effective diffusion coefficient of oxygen can be presented as [17]:

$$\left(D_O\right)_{eff} = \frac{D_O N_O + D_{OH} N_{OH}}{N_O + N_{OH}} \quad , \tag{3}$$

where N_O and N_{OH} are the concentrations of single oxygen atoms and O-H complexes, D_O and D_{OH} are the appropriate diffusion coefficients. The amount of atomic oxygen in the bulk of Cz-Si crystals usually exceed significantly the amount of single atomic hydrogen and, consequently, the concentration of oxygen-hydrogen pairs. So $N_O + N_{OH} \approx N_O$ and we can rewrite (3) as

$$\left(D_O\right)_{eff} = D_O + D_{OH}\frac{N_{OH}}{N_O} \quad . \tag{4}$$

It is thought that $D_{OH} \gg D_O$ and hydrogen-enhanced oxygen diffusion occurs if the concentrations of O-H pairs exceed some critical value $(N_{OH})_{crit}$ at which the first and the second terms in the right part of Eq. (4) are equal.

Analysis of the experimental results presented in Section 3.1 allowed us to obtain some new information regarding the oxygen-hydrogen pairs behaviour under thermal equilibrium conditions. In accordance with the study of the annealing behaviour of O-H pairs there are no other effective traps for mobile hydrogen atoms except oxygen atoms at temperatures below 300°C in moderately doped n-type Cz-Si crystals. Since no changes in the free carrier density were detected under the appearance and disappearance of O-H pairs these complexes as well as the single H atoms are thought to be neutral in n-type silicon. So the formation and dissociation of oxygen-hydrogen pairs at T < 300°C can be described by the following reaction and kinetic equations [24]:

$$O + H \Leftrightarrow OH, \tag{5}$$

$$\frac{dN_{OH}}{dt} = k_{OH} N_O N_H - \frac{N_{OH}}{\tau_{OH}}, \tag{6}$$

where k_{OH} and τ_{OH} are the formation rate constant and lifetime of the O-H complexes. k_{OH} and τ_{OH} can be presented as

$$k_{OH} = 4\pi R_{OH} D_H, \tag{7}$$

$$(\tau_{OH})^{-1} = k_{OH} N_{SOH} \exp\left(-\frac{E_{OH}}{kT}\right), \tag{8}$$

where R_{OH} is the capture radius of H by O, D_H is the diffusivity of hydrogen, N_{SOH} is the effective density of sites for the O-H complex in silicon lattice (see [24] for detail) and E_{OH} is the binding energy of O-H complex.

If only the reaction (5) occurs then the sum of the H and O-H concentrations is constant in the considered temperature range and is equal to the total amount of atomic hydrogen ($N_H + N_{OH} = N_{Ht}$). Taking into account that $N_{OH} \ll N_O$, the ratio of the concentration of oxygen-hydrogen pairs to the total atomic hydrogen concentration in thermal equilibrium will be

$$\frac{(N_{OH})_{eq}}{N_{Ht}} = \frac{k_{OH} N_O}{k_{OH} N_O + (\tau_{OH})^{-1}} = \frac{1}{1 + \frac{N_{SOH}}{N_O} \exp\left(-\frac{E_{OH}}{kT}\right)}. \tag{9}$$

As the configuration and electronic structure of the oxygen-hydrogen pairs is not defined exactly the precise value of N_{SOH} parameter is unknown. Roughly it can be taken as the density of lattice sites for silicon (5×10^{22} cm^{-3}). Under this assumption we calculated the temperature dependencies of the ratio of N_{OH}/N_{Ht} for the different values of E_{OH} in silicon crystal with the oxygen content of 10^{18} cm^{-3}. The results are plotted in Figure 8. The normalised steady-state values of the absorption coefficient in the maximum of the 1075.1 cm^{-1} IR absorption line after heat-treatments in the temperature range 50 to 150°C are presented in the Figure 8 as well. By comparing the data of experiment and calculations one can estimate the value of binding energy for O-H pairs. It is equal to 0.35±0.05 eV.

Based on Eq. (9) it would be possible now to estimate the equilibrium concentration of O-H pairs at a given temperature if the concentration of atomic hydrogen is known. Our consideration shows that at

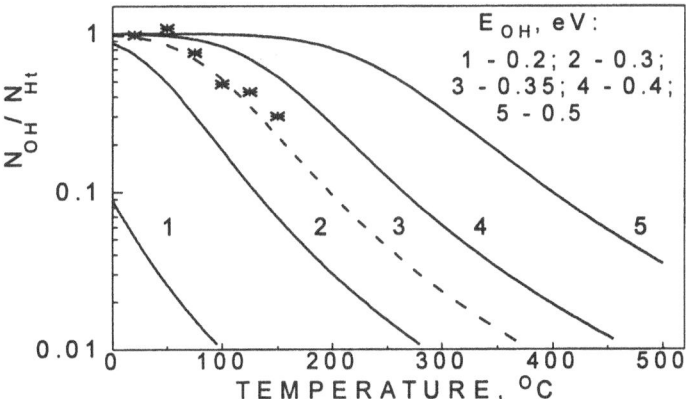

Figure 8. Temperature dependencies of the ratio of oxygen-hydrogen pair concentration to the concentration of single hydrogen atoms for different values of O-H complex binding energy. The normalised steady-state values of the absorption coefficient in the maximum of the 1075.1 cm^{-1} IR absorption line are shown by stars.

temperatures higher than 300°C the main part of hydrogen atoms are included in the immobile hydrogen-related complexes, most probably hydrogen molecules (H_2), but dynamical equilibrium exists between the single and bound forms of hydrogen. However, the information available at present on the properties of hydrogen in Si crystals is not sufficient for the description of the process of interchange between the single and bound forms of hydrogen and for the correct determination of the atomic hydrogen concentration.

4. Concluding remarks

The experimental data presented here and previous investigations [1-8] have shown that hydrogen influences significantly the oxygen-related defect reactions in silicon at elevated temperatures. In the non-irradiated H_2-soaked Cz-Si crystals the most prominent effect of hydrogen is its effect on oxygen diffusivity. The process of hydrogen-enhanced oxygen diffusion is still not well understood and we think that some properties of the recently revealed oxygen-hydrogen pairs could help to solve this problem.

Along with the O_i diffusion the mobility of other oxygen-related species (e. g., oxygen dimers) is necessary for the explanation of oxygen clustering at T = 300-500°C. The preliminary results of the study of TDD formation in H_2-soaked Si crystals show that hydrogen can influence not only O_i diffusion but the mobility of small oxygen clusters.

It has been inferred from the analysis of the TDD concentration depth profiles in the samples subjected to hydrogen plasma exposure that hydrogen molecule formation is the dominant process which influences the hydrogen diffusion in Si crystals at T = 350-450°C. This result and other experimental data indicate that in this temperature range most hydrogen atoms exist in the form of H_2 molecules but equilibrium between the free and bound forms of hydrogen exists. Hydrogen molecules can be considered as a reservoir for supply of atomic hydrogen into the volume of the crystal at T = 300-500°C and since the H_2 molecules are thought to be rather immobile [24] the effect of hydrogen on defect reactions can exists rather long in H_2-soaked crystals.

In summary, it should be noted that the information available now on the oxygen-hydrogen-related phenomena in Si crystals at elevated temperatures is inadequate for the good description of the appropriate processes and further experimental and theoretical studies are necessary. Further investigations of the properties of oxygen-hydrogen pairs are thought to be very useful for the better understanding the oxygen-hydrogen interaction.

5. Acknowledgements

The authors are grateful to Prof. M. Suezawa and Prof. K. Sumino for the help in carrying out IR absorption measurements and helpful discussions. We thank Dr. Yu.A. Bumai and Dr. A.G. Ulyashin for the hydrogen-plasma-treatments of the samples. The work was supported in part by grant No RWNOOO from the International Science Foundation.

6. References

1. McQuaid, S.A., Binns, M.J., Londos, C.A., Tucker, J.H., Brown, A.R., and Newman R.C. (1995) Oxygen loss during thermal donor formation in Czochralski silicon: New insights into oxygen diffusion mechanisms, *J. Appl. Phys.* **77**, 1427-1442.
2. Brown, A.R., Clayborn, M., Murray, R., Nandhra, P.S., Newman, R.C., and Tucker, J.H. (1988) Enhanced thermal donor formation in silicon exposed to a hydrogen plasma, *Semicond. Sci. Technol.* **3**, 591-593.

3. Stein, H.J. and Hahn, S (1990) Hydrogen-assisted thermal donor formation in silicon, in K. Sumino (ed.), *Defect Control in Semiconductors*, Elsevier Science Publishers B.V., North-Holland, pp. 211-220.

4. Newman, R.C., Tucker, J.H., Brown, A.R., and McQuaid, S.A. (1991) Hydrogen diffusion and the catalysis of enhanced oxygen diffusion in silicon at temperatures below 500 °C, *J. Appl. Phys.* **70**, 3061-3070.

5. Korshunov, F.P., Markevich, V.P., Medvedeva, I.F., and Murin, L.I. (1994) Electrically active hydrogen-related defects in irradiated n-type silicon, *Doklady Akad. Nauk Belarus* **38** 35-39.

6. Markevich, V.P., Suezawa, M., Sumino, K., and Murin, L.I. (1994) Radiation-induced shallow donors in Czochralski-grown silicon crystals saturated with hydrogen, *J. Appl. Phys.* **76**, 7347-7350.

7. Hatakeyama, H., Suezawa, M., Markevich, V.P., and and Sumino, K. (1995) Formation of hydrogen-oxygen-vacancy complexes in silicon, *Mater. Sci. Forum* **196-201**, 939-944.

8. Markevich, V.P., Suezawa, M., and Sumino, K. (1995) Optical absorption due to vibration of hydrogen-oxygen pairs in silicon, *Mater. Sci. Forum* **196-201**, 915-919.

9. Estreicher, S.K. (1990) Interstitial O in Si and its interactions with H, *Phys. Rev. B* **41**, 9886-9891.

10. Jones, R., Oberg, S., and Umerski, A. (1992) Interaction of hydrogen with impurities in semiconductors, *Mater. Sci. Forum* **83-87**, 551-562.

11. Bosomworth, D.R., Hayes, W., Spray, A.R.L., and Watkins, G.D. (1970) Absorption of oxygen in silicon in the near and the far infrared, *Proc. Roy. Soc. Lond. A.* **317**, 133-152.

12. Stavola, M. and Pearton, S.J. (1991) Vibrational spectroscopy of hydrogen-related defects in silicon, in J.I. Pankove and N.M. Johnson (ed.), *Hydrogen in Semiconductors*, Academic Press, San Diego, pp. 139-183.

13. Newman, R.C. and Smith, R.S. (1969) Vibrational absorption of carbon and carbon-oxygen complexes in silicon, *J. Phys. Chem. Solids* **30**, 1493-1505.

14. Binns, M.J., Newman, R.C., McQuaid, S.A., and Lightowlers, E.C. (1994) Hydrogen solubility and defects in silicon, *Mater. Sci. Forum* **143-147**, 861-866.

15. Markevich, V.P. and Murin, L.I. (1989) Thermal donor formation in pre-heat-treated n-Si:O crystals, *Phys. Stat. Sol.* **A111**, K149-K154.

16. Murin, L.I. and Markevich, V.P. (1990) Effects of various pre-treatments and impurity content on thermal donor formation in silicon, in K. Sumino (ed.), *Defect Control in Semiconductors*, Elsevier Science Publishers B.V., North-Holland, pp. 199-210.

17. Ourmazd, A., Schroter, W., and Bourret, A. (1984) Oxygen-related thermal donors in silicon: a new structural and kinetic model, *J. Appl. Phys.* **56**, 1670-1681.

18. Murin, L.I. and Markevich, V.P.(1996) Thermal double donors in silicon: a new insight into the problem, in *Proc. of This Conf.*

19. Latushko, Ya.I., Makarenko, L.F., Markevich, V.P., and Murin, L.I. (1986) Electrical and optical characterization of thermal donors in silicon, *Phys. Stat. Sol.* **A93**, K181-K184.

20. Markevich, V.P., Murin, L.I., Bumai, Yu.A., and Ulyashin, A.G., unpublished.

21. Safonov, A.N. and Lightowlers, E.C. (1994) Hydrogen related optical centers in radiation damaged silicon, *Mater. Sci. Forum* **143-147**, 903-907.

22. Korshunov, F.P., Makarenko, L.F., Markevich, V.P., Medvedeva, I.F., and Murin, L.I. (1990) Enhanced annealing of radiation defects in pre-heat-treated Si crystals, in K. Sumino (ed.), *Defect Control in Semiconductors*, Elsevier Science Publishers B.V., North-Holland, pp. 541-545.

23. Markevich, V.P., Medvedeva, I.F., Murin, L.I., Sekiguchi, T., Suezawa, M., and Sumino, K. (1995) Metastability and negative-U properties for hydrogen-related radiation-induced defect in silicon, *Mater. Sci. Forum* **196-201**, 945-950.
24. Herring, C. and Johnson, N.M. (1991) Hydrogen migration and solubility in silicon, in J.I. Pankove and N.M. Johnson (ed.), *Hydrogen in Semiconductors,* Academic Press, San Diego, pp. 225-350.

PASSIVATION OF THERMAL DONORS
BY ATOMIC HYDROGEN

JÖRG WEBER
Max-Planck-Institut für Festkörperforschung
Postfach 80 06 65, D-70506 Stuttgart, Germany

AND

DIRK I. BOHNE
Present address: Phoenix Contact GmbH & Co,
Flachsmarktstr. 8-28, D-32825 Blomberg, Germany

1. Abstract

Thermal Donors (TDs) form electrically inactive complexes with atomic hydrogen. Only one hydrogen atom is necessary to passivate the double donor levels. The TDs are completely reactivated at temperatures $(T \leq 200\,^{\circ}\mathrm{C})$ well below their formation temperature $(T \approx 450\,^{\circ}\mathrm{C})$. The dissociation of the hydrogen complexes is studied by means of infrared absorption, capacitance voltage profiling and deep level transient spectroscopy. Characteristic differences in the dissociation process for the individual TD complexes are detected: the dissociation enthalpies differ for different TD species and the attempt frequencies are at least three orders of magnitude larger than expected for an atomic jump process. Our results require different core structures for TD1 and TDn with $n \geq 2$ and large lattice relaxations, which accompany the dissociation of the TD - hydrogen complexes.

2. Introduction

Thermal donors (TDs) are generated in oxygen-rich silicon at temperatures around $450\,^{\circ}\mathrm{C}$ [1]. At least seventeen species, all exhibiting double donor characteristics, have been reported (TDn with $n = 0, 1, \ldots, 16$) [2]. The different TDs appear sequentially with continued heat treatment and

123

R. Jones (ed.), Early Stages of Oxygen Precipitation in Silicon, 123–140.
© *1996 Kluwer Academic Publishers.*

exhibit a decrease in ionization energy. Deep level transient spectroscopy (DLTS) is not able to resolve the individual species, only two signals at $E_C - 0.07$ eV and $E_C - 0.15$ eV are observed, which correspond to the ionization of the neutral and singly ionized centres. Although the atomic structure of the TDs is still in dispute, it is generally assumed that all TDs contain the same electrically active core. The slight differences in ionization energies arise from the successive incorporation of additional oxygen atoms in chains along a ⟨110⟩ crystal direction [3].

Thermal Donors TD0, TD1 and TD2 exhibit bistable properties with negative U ordering of the energy levels [4]. A schematic configuration coordinate diagram for TD1/2 was proposed by Latushko *et al.* and is shown in figure 1(left) [5].

There are two different configurations of TD1/2: the stable configuration S is electrically inactive, whereas the metastable configuration M reveals the two electronic levels of the double donor. The configurational state is controlled by the charge state of the defects. Either configuration can be 'frozen-in' by cooling the defects in different charge states to temperatures low enough that the structural transformations cannot be exceeded [6, 7, 8]. Cooling the samples from room temperature in the dark or without applied reverse bias 'freezes-in' TD1/2 in the stable configuration, due to the lower energy for the neutral charge state. The detection of the shallow donors by infrared (IR) absorption or deep level transient spectroscopy (DLTS) is not possible. The metastable configuration (M) is accessible only from the ionized charge states of the TDs. This configuration is 'frozen-

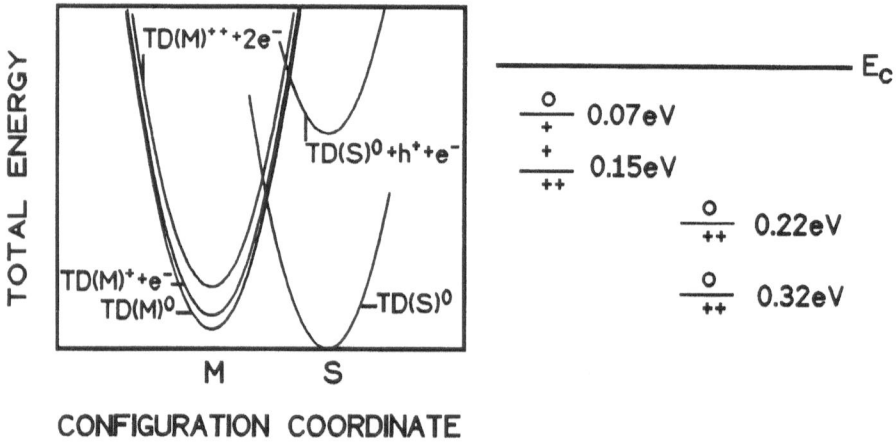

Figure 1. left: Schematic configuration coordinate diagram for TD1/2 [5]. right: Electronic levels for TD1/2 and occupancy levels for TD1 at $E_C - 0.32$ eV and TD2 at $E_C - 0.22$ eV.

in' by cooling the sample from room temperature under illumination or by applying a reverse bias to the Schottky diode. Now the IR absorption lines or the DLTS signal of the second TD state at $E_C - 0.15$ eV can be detected. Figure 1(right) gives the energy level scheme for the $(0/+)$ and the $(+/++)$ levels of TD1/2. All TDs exhibit a deep stable single donor level X with an energy which corresponds to the energy difference between the minima of the $TD(M)^+ + e^-$ and the $TD(S)^0$ total energy curves. This level is, however, not observable due to the negative U ordering. Instead the occupancy levels — which are half way between the single deep donor level and the second ionization level of the TD double donor — determine the population of the different states. The occupancy levels at $E_C - 0.32$ eV for TD1 $(0/++)$ and $E_C - 0.22$ eV for TD2 $(0/++)$ are given in figure 1(right) and correspond to an emission of two electrons simultaneously. At the same time, a structural transformation from the stable to the metastable configuration of the defects occurs.

Thermal Donors are passivated by atomic hydrogen leading to the formation of electrically inactive TD-hydrogen (TD-H) complexes [9]. The dissociation kinetics of TD-H were studied by several authors [10, 11, 12]. The experimental methods used in these studies (capacitance voltage profiling (CV), deep level transient spectroscopy (DLTS), admittance spectroscopy) are unable to separate the small differences in binding energies of individual TDs. In DLTS and admittance spectroscopy, the signals of individual TDs superimpose to an unique broadened line. The dissociation of TD-H was found to occur in a single stage, and an identical dissociation enthalpy of 1.1 eV was derived for the different TDn-H species and for both charge states of the double donors [10]. These results support the oxygen chain model with a common core as mentioned above. In addition, the degree of TD passivation was reported to vary monotonically with TD size [13]. Recently, in contrast to the earlier studies the existence of a single donor state for a TD-H complex was postulated from electron-nuclear-double-resonance (ENDOR) measurements [14]. The new hydrogen-TD complex is, however, formed at much higher temperatures ($T \approx 450\,°C$) and reactivation of the TDs was not determined.

In this paper we summarize our study of the formation and dissociation kinetics of the TD-H complexes [15, 16, 17]. We first describe the sample preparation and the experimental techniques to distinguish between individual TD complexes and then present the experimental results on the dissociation kinetics. Our results give new information about the core structure of the different TDs.

3. Experimental

3.1. SAMPLES AND THEIR PREPARATION

The samples used in this study are cut from dislocation free Czochralski (Cz) grown silicon with a phosporus concentration $[P] = 8.0 \times 10^{14} - 1.0 \times 10^{15}$ cm^{-3} corresponding to a Fermi level at 300 K of $E_C - E_F \approx 0.27 - 0.28$ eV. Oxygen and carbon concentrations were determined by IR absorption ($[O_i] = 1.0 \times 10^{18}$ cm^{-3}, $[C_s] = 1.6 \times 10^{16}$ cm^{-3}). The samples are chemically cleaned and preannealed at 770 °C for 15 min. (argon ambient) in order to destroy any TDs present initially, and to obtain uniform starting conditions [7]. The subsequent heat treatment at 460 °C (argon ambient) for the first set of samples is performed for 15 min. or for 20 min. Such short annealing times generate predominantly TD1 and TD2 ($N_{TD1/2} \approx 10^{13}$ cm^{-3} from DLTS) and only low concentrations of other TDs. We have no evidence for the bistable Thermal Donor TD0 reported by Makarenko *et al.* in any of our samples [6]. The samples annealed for 15 min. contain a larger contribution of TD1 and are used to investigate TD1-H, whereas the 20 min. anneal generates a larger contribution of TD2. These samples are used to study TD2-H. A second set of samples was annealed for 46 h at 460 °C (argon ambient). In this case the dominant TD observed by IR are with decreasing absorption coefficient ordering: TD4, TD5, TD3, TD8, TD6, TD7, TD2. Taking into account the double donor character, the total concentration of TD calculated from CV measurements at room temperature is $N_{\Sigma TD} \approx 1 - 2 \times 10^{16}$ cm^{-3}.

The passivation is performed in the downstream region of an rf-driven (13.56 MHz) hydrogen plasma at 100 °C for 10 h. The samples were etched in CP6 (15–45 s) to remove the damaged surface layer and palladium was evaporated through a metal mask to form Schottky contacts. The Ohmic contacts consist of an In/Ga alloy scratched onto the back side of the samples.

Isothermal anneals of the diodes up to 200 °C are carried out in the dark under reverse bias in order to dissociate the TD-H complexes. The anneals are terminated by quenching the samples to room temperature with the help of liquid nitrogen.

3.2. EXPERIMENTAL TECHNIQUES

Infrared absorption in the range from 100 cm^{-1} to 650 cm^{-1} is carried out using a Fourier transform spectrometer (Bruker IFS 113v). The samples were mounted in a cold finger cryostat. Sample thickness was 2 mm with a wedge of 2° and all surfaces were polished with Syton.

The capacitance DLTS measurements were performed using a Lock-In technique or by recording the whole capacitance transients. The sensitivity of our setup is $\delta C/C \approx 10^{-5}$. The temperature of the sample holder could be varied between 20 K and 470 K with an accuracy of ± 1 K. In order to minimize the influence of the Poole Frenkel effect, the DLTS spectra are recorded only in the low field region of the space charge layer [18].

For the CV measurements we used a CV profiler (HP 4280A). The passivation and reactivation of the shallow dopants and of the TDs was monitored by CV and DLTS profiling. Current-voltage characteristics of the diodes are measured regularly, yielding ideality factors within a range of $n = 1.0 - 1.3$.

4. Results

4.1. PROPERTIES OF TD1/2 BEFORE PLASMA TREATMENT

We use the bistability of TD1 to distinguish between TD1 and TD2 in DLTS. It was therefore necessary to prepare the samples according to the following criteria:
1) The shallow dopant concentration of the sample is chosen so that the Fermi level at room temperature is located between the occupancy levels of TD1 and TD2.

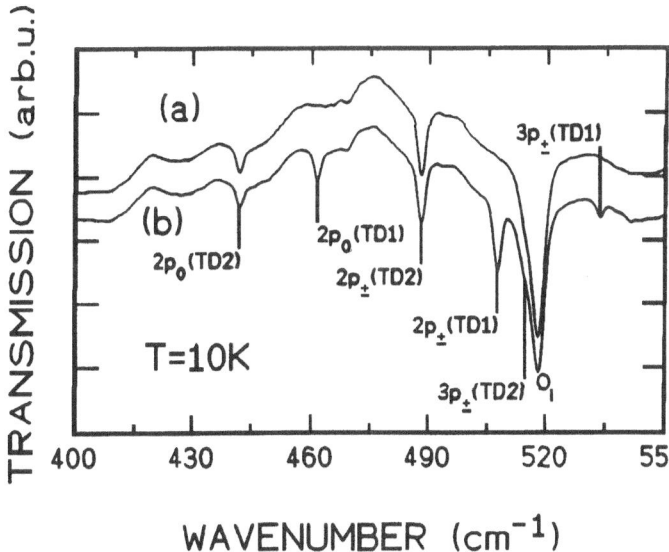

Figure 2. Infrared transmission spectra of Cz Si (770 °C 15 min. + 460 °C 20 min.),
(a) after cooling from room temperature in the dark,
(b) after cooling from room temperature under illumination with band gap light.

2) The annealing time to generate the TDs is chosen sufficiently short to create only TD1 and TD2.
Under these conditions TD2 behaves as a stable centre whereas TD1 is still bistable.

Infrared absorption measurements are performed to verify that only TD1 and TD2 are generated during the 460 °C anneal, and that TD2 behaves as a stable centre. The IR spectra in figure 2 are recorded after cooling the sample from room temperature in the dark (a) and under illumination with band-gap light (b). Apart from TD1 and TD2 no other TDs are observed in measurable concentrations. By cooling the sample from room temperature in the dark, the absorption lines characteristic of TD1 vanish. Therefore in these samples, TD2 behaves as a stable centre whereas TD1 is bistable.

The two configurations of the bistable TD1 can also 'frozen-in' by cooling down the sample with or without applied reverse bias from room temperature. Using both cooling methods allows us to determine by DLTS the concentrations of TD1 and TD2 separately. An example of the determination of the TD1 and TD2 concentration is given in figure 3.

Our standard procedure for reverse bias cooling starts with an anneal at 340 K for 20 min. of the Schottky diodes in the dark at an applied reverse bias of 8 V. Then the diodes are cooled to 30 K at a rate of ≈ 80 K/min. with

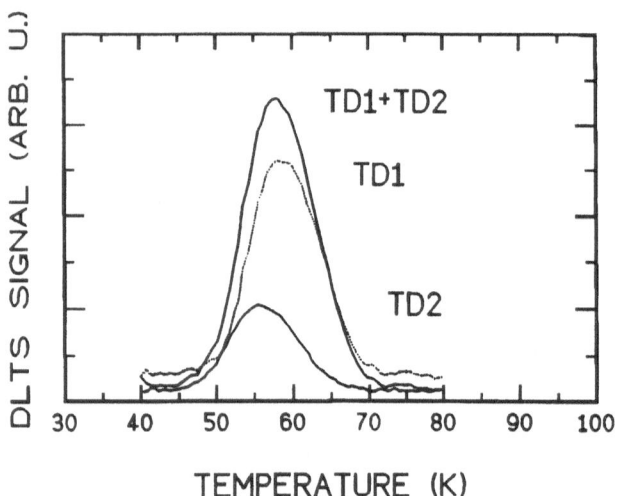

Figure 3. DLTS spectra of Cz-silicon (770 °C 15 min. + 460 °C 20 min):
TD2: zero bias cooling in the dark
TD1 + TD2: reverse bias cooling $U_B = -8$ V
TD1: numerically determined difference.

reverse bias still applied. DLTS from 30 K to 100 K monitors TD1 and TD2. Several DLTS scans with successively decreasing reverse bias, are performed in order to calculate concentration profiles. Structural transformations from the metastable configuration to the stable configuration take place only for temperatures higher than 200 K [8].

Standard zero bias cooling is achieved by first storing the samples at 300 K for 30 min. in the dark without applied reverse bias. The subsequent cooling was performed with the identical cooling rates as before but without applied reverse bias. Only the DLTS signal due to TD2 remains and yields the concentration of TD2. No change in the measured DLTS signals due to Fermi level effects are observed by decreasing the cooling rate to 2 K/min. or increasing the storage time to 24 h.

The determination of the TD1 concentration in our samples allows us to study the structural changes of the bistable TD1 [19]. The energy barriers and attempt frequencies which we derive from our measurements are in agreement with the data published in reference [8].

4.2. NEW DLTS LEVELS AFTER HYDROGEN PLASMA TREATMENT

The samples which contain only TD1/2 are treated in the hydrogen plasma (100 °C). At least five DLTS levels in comparable concentrations appear in these samples in the temperature range from 30–300 K. The energies ΔH and the prefactors σ are derived from the Arrhenius plots and summarized in table 1.

TABLE 1. energies ΔH and the prefactors σ of the levels found in the hydrogen passivated samples.

level	ΔH (eV)	σ (cm^2)
E1	0.07	3.6×10^{-16}
E2	0.11	8.5×10^{-17}
E3	0.14	1.3×10^{-15}
E4	0.19	6.3×10^{-16}
TD1/2	0.13	2.4×10^{-16}

Only the E3 level can be correlated with the carbon-hydrogen complex [20, 21]. All other defects are tentatively assigned to different C-O-H complexes. Figure 4 gives the concentration profiles of the levels. The TD concentration is strongly reduced in the near surface region due to the formation of neutral TD-H complexes. The concentration of TDs

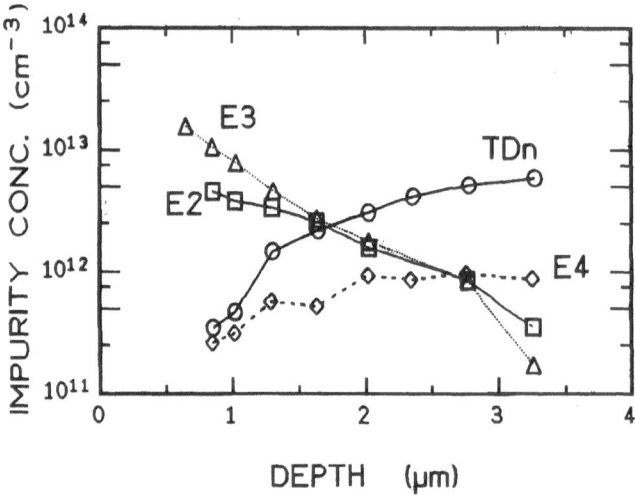

Figure 4. Concentration profiles of the levels TD*n*, E2, E3 and E4 in Cz-Silicon (770 °C 15 min. + 460 °C 20 min. + hydrogen plasma 100 °C 2 h).

increases with depth due to the small penetration depth of hydrogen, all other defects (except E4) show an increased concentration at the surface. The stability of the new defects is very weak. Annealing the diode at 120 °C for 2 h destroys the defects and only the TDs remain detectable.

4.3. DISSOCIATION OF THE TD1-H AND TD2-H COMPLEXES

The hydrogen plasma treatment at 100 °C leaves the shallow dopant concentration unaffected. The CV profiles show no passivation of the phosphorus donors. The Fermi level is therefore not changed and the bistability of TD1 remains in the hydrogenated samples. Figure 5a shows DLTS spectra recorded after reverse bias cooling (solid line) and after zero bias cooling (dotted line) of a sample, which was treated in the hydrogen plasma.

The solid line corresponds to TD1 and TD2 and the dotted line is due to TD2. The spectra are recorded at a depth where the new levels discussed above are not present. The increase in amplitude of both signals after a 150 min. anneal at 140 °C under a reverse bias of 8 V in the dark, accounts for the dissociation of TD-H and therefore a recovery of the electrical activity of the TDs. The electrical field during the reverse bias annealing reduces the retrapping of the hydrogen [22, 23, 21]. Since the difference of the reverse-bias/zero-bias signals remains constant, only reactivation of TD2 occurs in this temperature range (figure 5b). The reactivation of TD1 takes place at elevated temperatures. A subsequent anneal at 200 °C leads to a dissociation of TD1-H (figure 5c).

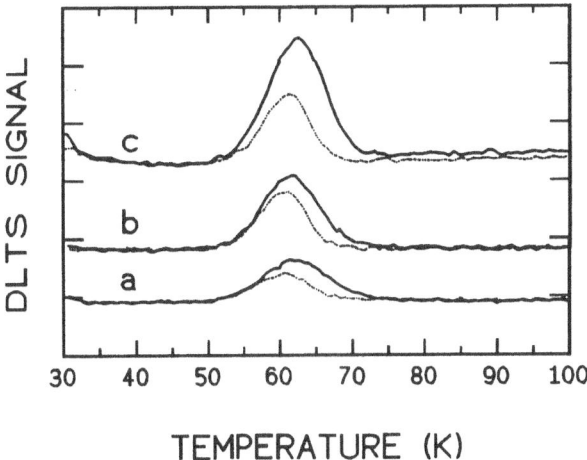

Figure 5. DLTS spectra of Cz-silicon (770 °C 15 min. + 460 °C 20 min.) recorded after reverse bias cooling (solid line) and after zero bias cooling (dotted line):
(a) after hydrogenation,
(b) after a subsequent 150 min. anneal at 140 °C under a reverse bias of 8 V in the dark,
(c) after a subsequent 20 min. anneal at 200 °C under a reverse bias of 8 V in the dark.

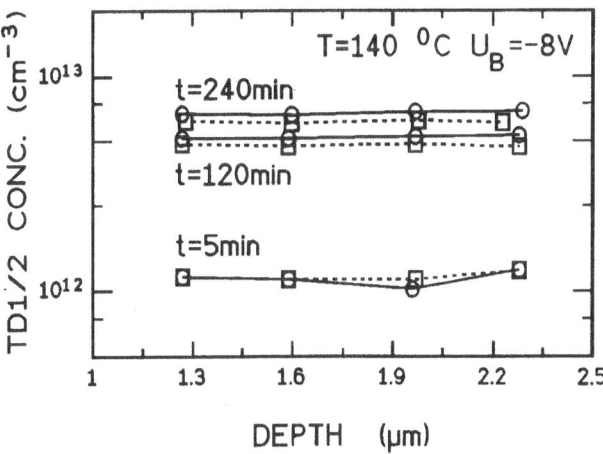

Figure 6. Concentration profiles of TD1/2 after isothermal anneal at 140 °C under a reverse bias of 8 V in the dark. Circles correspond to concentrations determined after cooling down with reverse bias (TD1+TD2) and squares correspond to concentrations determined after cooling down without bias (TD2).

We measure the TD1/2 profiles between 1.1 μm and 2.3 μm below the sample surface after different annealing times at various temperatures. Figure 6 gives typical depth profiles for TD1/2 after isothermal annealing at 140 °C under a reverse bias of 8 V in the dark for different annealing times. An average concentration $N_{TD1/2}(t)$ was determined by numerically integrating over this region. The concentration of the TD1/2-H, $N_{TD1/2-H}(t)$, is given by the equation $N_{TD1/2-H}(t) = N^0_{TD1/2} - N_{TD1/2}(t)$. The first term is the total concentration of TD1/2 and refers to the concentration measured after very long annealing times. This concentration was found to be identical to the original TD1/2 concentration before hydrogen passivation, indicating a fully reversible passivation process.

The exponential time dependence of $N_{TD1/2-H}(t)$ shown in figure 7 and figure 8 confirms that the reactivation of TD1/2 follows first order kinetics and satisfies the equation $N_{TD1/2-H}(t) = N^0_{TD1/2-H} \times \exp\left(-\nu_{TD1/2} \times t\right)$.

Directly after the hydrogen passivation ($t = 0$) the concentration of TD1/2 is below the detection limit of our DLTS system (10^{11} cm^{-3}). The degree of passivation depends on the duration of the plasma treatment and varied between 65% and 100% in our experiments. We also observe a strong passivation of TD1/2 in reference samples, which are not plasma treated, due to the wet chemical etching before the metal evaporation of the Schottky contacts.

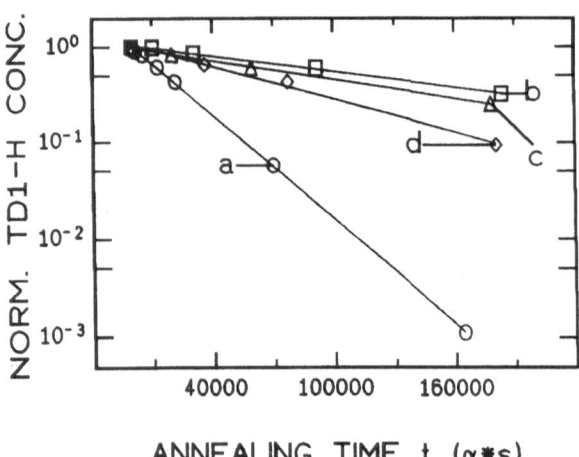

Figure 7. Normalized TD1-H concentration vs. annealing time at different temperatures. The time axis is scaled by a factor; (a) $\alpha = 1.0$, $T = 160\,°C$; (b) $\alpha = 17.0$, $T = 169\,°C$; (c) $\alpha = 13.0$, $T = 180\,°C$; (d) $\alpha = 86.0$, $T = 190\,°C$.

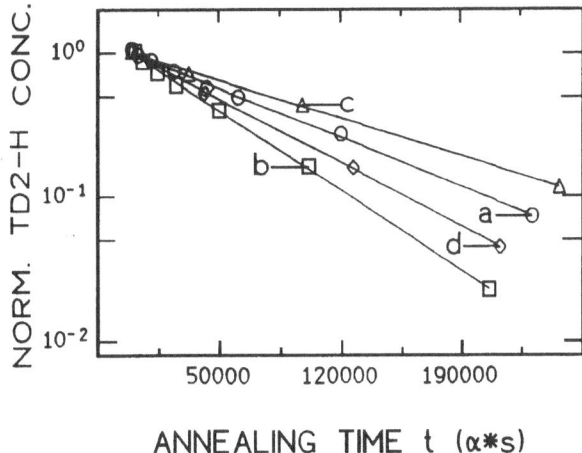

Figure 8. Normalized TD2-H concentration vs annealing time at different temperatures. The time axis is scaled by a factor; (a) $\alpha = 1.0$, $T = 119\,°C$; (b) $\alpha = 7.1$, $T = 139\,°C$; (c) $\alpha = 13.0$, $T = 150\,°C$; (d) $\alpha = 86.0$, $T = 159\,°C$.

4.4. DISSOCIATION OF THE OTHER TD-H COMPLEXES

For the second set of samples, which contain a mixture of different TDs, the total TD concentration was more than one order of magnitude higher compared to the concentration of the phosphorus dopant. According to the IR absorption, TD2 was bistable in this sample. If we consider the double donor nature of TD, we have equal concentrations of shallow levels ($TD^{0/+}$) and deep levels ($TD^{+/++}$) and standard DLTS can no longer be applied. Therefore, we determined the total TD concentration by CV profiling at 60 K, after applying a reverse bias at 340 K for 20 min. to the diode and subsequent reverse bias cooling. The CV profiles are measured by successively decreasing the applied voltage in order to avoid an influence due to the emission of electrons from the $TD^{+/++}$ level. The concentration profile of the TDs in the plasma treated sample are given in figure 9 along with the profiles after isothermal annealing for different times under reverse bias.

The reactivation of the donor concentration is clearly visible. The upper profile was produced after a complete reactivation of the TDs by an anneal at 200 °C for 20 min. The decrease of the profile towards the surface is due to outdiffusion of oxygen during the generation of TDs at 460 °C.

Figure 10 shows again a first order reactivation kinetics for the dissociation of the ΣTD-H complexes. This indicates that the dissociation of TDn-H for $n > 2$ occurs in a single stage.

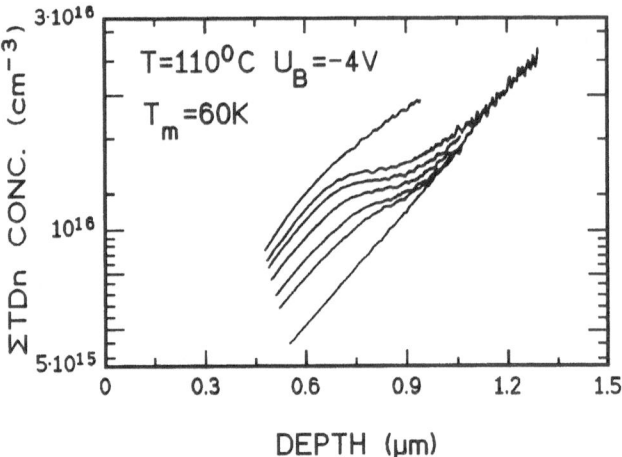

Figure 9. CV profile of Cz-silicon (770 °C 15 min. + 460 °C 46 h + H-plasma 100 °C 10 h) after isothermal anneals at 110 °C in the dark under reverse bias ($U_B = -4$ V). The reactivation increases with increasing annealing time. The top profile corresponds to an anneal at 200 °C for 20 min.

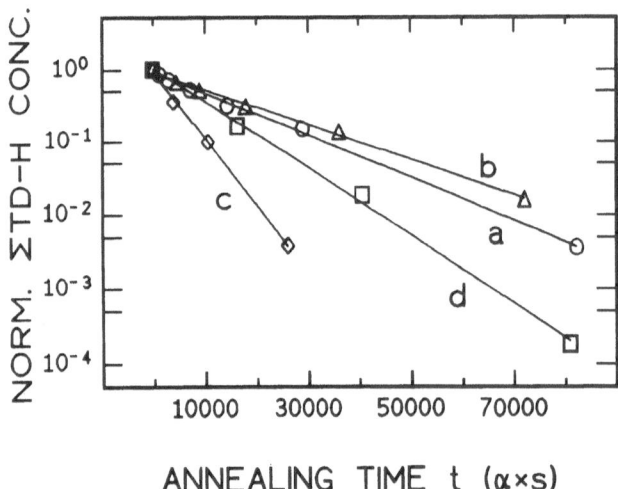

Figure 10. Normalized ΣTD-H concentration versus annealing time at various temperatures. The time axis is scaled by a factor; (a) $\alpha = 1.0$, $T = 100$ °C; (b) $\alpha = 5.0$, $T = 110$ °C; (c) $\alpha = 13.3$, $T = 120$ °C; (d) $\alpha = 27.0$, $T = 130$ °C.

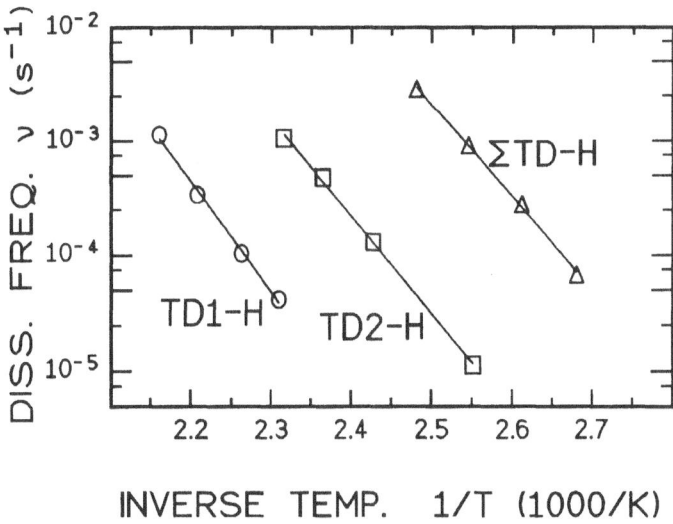

Figure 11. Arrhenius plot of the dissociation frequencies for the various TD-H complexes.

4.5. DISSOCIATION FREQUENCIES

Figure 11 shows the Arrhenius plot of the dissociation frequencies. It is clear that dissociation of TD1-H, TD2-H and ΣTD-H occurs in different temperature regimes. The dissociation enthalpies calculated from the slopes of these curves and the attempt frequencies are given in Table 2.

TABLE 2. Dissociation enthalpies and attempt frequencies ν for the different Thermal-Donor hydrogen complexes.

TD-H species	ΔH_D [eV]	ν [Hz]
TD1-H	1.90 ± 0.07	$5.0 \times 10^{17\pm1}$
TD2-H	1.67 ± 0.05	$4.0 \times 10^{16\pm1}$
ΣTD-H	1.61 ± 0.05	$4.0 \times 10^{17\pm1}$

5. Discussion

5.1. DISSOCIATION ENTHALPIES

The dissociation of atomic complexes in a solid is a thermally activated process hindered by an energy barrier ΔU. This barrier is called the dissociation energy and is an important quantity for describing the dissociation process. In addition, the so-called frequency factor is the second quantity necessary to determine the kinetics of the process. Conventionally, in dissociation processes the frequency factor is called attempt frequency. If ΔU is much larger than $k_B T$ (k_B Boltzmann constant, T temperature), and quantum mechanical tunneling processes can be neglected, the theoretical approaches of Kramers [24], Vineyard [25], and Rice [26] lead to similar expressions for the temperature dependent dissociation rate or dissociation frequency

$$\nu(T) = \nu_0 \exp\left(-\Delta F / k_B T\right), \tag{1}$$

with ΔF beeing the free energy needed to carry the mobile part of the complex from an equilibrium position to a saddle point during the dissociation. The change in the free energy ΔF and the energy barrier ΔU are connected by the relation

$$\Delta F = \Delta U - T \Delta S, \tag{2}$$

where ΔS is the change in entropy of dissociation. In the experiments described above, isothermal and isobaric conditions prevail, and the internal energy ΔU has to be replaced by the enthalpy ΔH and the free energy ΔF by the Gibbs free energy

$$\Delta G = \Delta H - T \Delta S. \tag{3}$$

The attempt frequencies can be calculated using the rate theories mentioned above. Following the simple approach of the transition state theory, in which quantum mechanical tunneling processes and coupling to a heat source are neglected, the attempt frequency is given by the vibrational frequency of the mobile particle of the complex at the minimum of the effective energy potential [25].

$$\nu_0 = \left(\frac{K^*}{m^*}\right)^{1/2}, \tag{4}$$

where m^* equals the mass of the jumping atom. K^* is a factor, given by a combination of the force constants between neighboring atoms at the equilibrium position and at a saddle point of the effective potential, respectively.

The dissociation frequency is given by

$$\nu(T) = (K^*/m^*)^{1/2} \exp\left(+\Delta S / k_B\right) \exp\left(-\Delta H / k_B T\right). \tag{5}$$

The attempt frequency could be a localized vibrational mode or a gap mode depending on the mass of the jumping atom and is in general of the order of a typical phonon frequency of the lattice. In silicon, the frequency of optical phonons (at 300 K) is approximately 1.5×10^{13} Hz.

The temperature dependent dissociation frequency is determined experimentally from isothermal anneals [27]. According to equation 5, an Arrhenius plot of the dissociation frequency yields both the dissociation enthalpy and the prefactor. The prefactor equals the attempt frequency only if the dissociation entropy can be neglected, otherwise the prefactor is changed and can be much larger or smaller than a typical phonon frequency. In most investigations of the dissociation of atomic complexes in crystalline semiconductors, the dissociation entropy was neglected or was negligible.

We have generated the Thermal Donors TD1 and TD2 after a short anneal of Cz-silicon at 460 °C. The Fermi level at room temperature is located between the occupancy levels of TD1 and TD2. Therefore, only TD1 is in its bistable configuration whereas TD2 is stable. This fact allows us to distinguish between TD1 and TD2 in DLTS. We separately observed the dissociation kinetics of TD1-H and TD2-H. The reactivation of TD1/2 follows first order kinetics. The dissociation of TD1-H occurs at considerably higher temperatures as is observed for TD2-H. The different dissociation enthalpies clearly indicate that the dissociation of TD-H is in contradiction to the data reported in the literature [10, 11, 12].

Regarding the known binding energies of hydrogen to silicon and oxygen in the silicon lattice [28], the dissociation energies seem to be consistent with the hydrogen atom bonded to a silicon self interstitial. In this respect our results support the oxygen interstitial-self interstitial model of TDs proposed in reference [29]. The significantly higher dissociation enthalpy for TD1-H compared to TD2-H and ΣTD-H implies that TD1 and TDn ($n \geq 2$) have different microscopic structures. This is in accordance with the suggestion given for TD1 (self interstitial bonded to one oxygen) and TD2 (self interstitial bonded to two oxygens) in reference [28].

We attribute the higher effectivity of hydrogen passivation of early TD [9, 13] to the higher thermal stability of the TD1-H complex compared to the ΣTD-H complexes.

5.2. ATTEMPT FREQUENCIES

In contrast to the enthalpies, the attempt frequencies are determined by extrapolating onto a logarithmic scale leading to large error bars. Nevertheless, the attempt frequencies are at least three orders of magnitude larger than those expected for an atomic jump process. According to equation 5, dissociation entropies of $\Delta S \approx 8 - 10\,k_B$ are required to account for the measured attempt frequencies.

A mechanism for the hydrogen passivation of thermal donors was proposed by Deák *et al.* [3]: Adding a hydrogen atom to the core structure of TDs results in a stabilization of the electrically inactive S configuration, and passivation of the electrical activity. After complexing with hydrogen the TD-H complex gains energy while transforming to the stable configuration (S-H). This reaction involves a large lattice relaxation, which is reversed in the dissociation process. We suggest that this lattice relaxation during the dissociation process accounts for the large attempt frequencies.

The similar, large attempt frequencies for all TD-H complexes — even for those which do not exhibit bistable properties $(n \geq 3)$ — require a similar S configuration for all TDn. The same suggestion was made in reference [30] to explain the bistability of only TD0, TD1, and TD2. Apparently, hydrogen leads to a stabilization of the S configuration even for $n \geq 3$.

We estimate the dissociation entropy for the dissociation of the TD-H complexes according to the procedure given in reference [31]. Four different entropy contributions are involved in the dissociation process:

(1) a vibrational contribution of the complex in its ground state configuration,

(2) the entropy of the saddle point configuration,

(3) the entropy of the migrating hydrogen atom, and

(4) the entropy of ionization due to the emission or capture of electrons or holes during the dissociation process.

We determined the different contributions for the thermal donor model proposed by Deák *et al.* [3], with TD1 being the self-interstitial-oxygen complex (IO) and TD2 the IO_2 complex. Under the assumption that no ionization entropy term is involved in the dissociation, a total dissociation entropy of $\Delta S \approx 8k_B$ is derived in good agreement with the experimental data.

6. Summary

We have performed a systematic study of the hydrogen passivation of TDs. Our results support those microscopic models of the TDs, which suggest different core structures for TD1 and TDn with $n \geq 2$. Hydrogen passivates the TDs by stabilizing their neutral stable configuration. Only one hydrogen atom is necessary for this passivation process.

Acknowledgement

It is a pleasure to thank Prof. H. J. Queisser for his interest and support throughout this work. The authors are grateful to W. König for performing the IR measurements and P. Deák for useful discussions. We appreciate the technical support of W. Heinz and W. Krause. We are indebted to P. Wagner and W. Zulehner (Wacker-Siltronic) for supplying the silicon material. This work was supported in part by the BMBF under contract number NT 2786.

References

1. Fuller, C. S., Ditzenberger, J. A., Hannay, N. B., Buehler, E. (1954) Resistivity changes in silicon induced by heat treatment, *Phys. Rev.* **96**, 833.
2. Götz, W., Pensl G., and Zulehner, W. (1992) Observation of 5 additional thermal donor species TD12 to TD16 and of regrowth of thermal donors at initial-stages of the new oxygen donor formation in czochralski-grown silicon, *Phys. Rev. B* **46**, 4312–4315.
3. Deák, P., Snyder L. C., and Corbett, J. W. (1992) Theoretical-studies on the core structure of the 450°C oxygen thermal donors in silicon, *Phys. Rev. B* **45**, 11612–11626.
4. Anderson, P. W. (1975) *Phys. Rev. Lett.* **34**, 953.
5. Latushko, Ya. I., Makarenko, L. F., Markevich V. P., and Murin, L. I. (1986) Electrical and optical characterization of thermal donors in silicon, *Phys. Stat. Sol. (a)* **93**, K181–K184.
6. Makarenko, L. F., and Murin, L. I. (1986) Nature of alpha-traps in heat-treated Si-O crystals, *Fizika i Tekhnika Poluprovodnikov* **20**, 1530 [*Sov. Phys. Semicond.* **20**, 961–962].
7. Wagner, P., and Hage, J. (1989) Thermal double donors in silicon, *Appl. Phys. A* **49**, 123–138.
8. Chantre, A. (1989) Introduction to defect bistability, *Appl. Phys. A* **48**, 3–9.
9. Johnson, N. M., and Hahn, S. K. (1986) Hydrogen passivation of the oxygen-related thermal-donor defect in silicon, *Appl. Phys. Lett.* **48**, 709–711.
10. Pearton, S. J. (1991) in *Hydrogen in Semiconductors, Semiconductors and Semimetals* Vol. **34**, 84, Academic Press Inc., London.
11. Pearton, S. J., Chantre, A., Kimerling, L. C., Cummings, K. D., and Dautremont-Smith, W. C. (1986) Hydrogen passivation of oxygen donors in Si, *Mat. Res. Soc. Symp. Proc.* **59**, 475–480.
12. Chantre, A., Pearton, S. J., Kimerling, L. C., Cummings, K. D., and Dautremont-Smith, W. C. (1987) Interaction of hydrogen and thermal donor defects in silicon, *Appl. Phys. Lett.* **50**, 513–515.
13. Johnson, N. M., Hahn, S. K., and Stein, H. J. (1986) in J. von Bardeleben (ed.), *Defects in Semiconductors Materials Science Forum* **10–12**, 585, Trans. Tech. Publications Ltd., Switzerland.
14. Martynov, Yu. A., Gregorkiewicz, T., and Ammerlaan, C. A. J. (1995) Role of hydrogen in the formation and structure of the Si-NL10 thermal donor, *Phys. Rev. Lett.* **74**, 2030–2033.
15. Bohne, D. I., Deák, P., and Weber, J. (1992) Hydrogen passivation and reactivation of bistable thermal donors in silicon, *Mat. Res. Soc. Symp. Proc.* **262**, 395–400.
16. Bohne D. I., and Weber, J. (1993) Hydrogen passivation and reactivation of thermal donors in silicon, *Phys. Rev. B* **47**, 4037–4040.
17. Bohne D. I., and Weber, J. (1994) *Materials Science Forum* **143–147**, 879.
18. Komarov, B. A., Korshunov, F. P., and Murin, L. I., (1994) Role of field effects in a determination of the concentration of thermal donors in silicon by DLTS method, *Fizika i Tekhnika Poluprovodnikov* **28**, 499, [*Sov. Phys. Semicond.* **28**, 305–309].

19. Bohne, D. I., PhD Thesis, Stuttgart 1992, unpublished.
20. Endrös, A. (1989) Charge-state-dependent hydrogen-carbon-related deep donor in crystalline silicon, *Phys. Rev. Lett.* **63**, 70–73.
21. Endrös, A. L., Krühler, W., and Grabmaier, J. (1991) Hydrogen in phosphorus-doped and carbon-doped crystalline silicon, *Physica B* **170**, 365–370.
22. Tavendale, A. J., Pearton, S. J., and Williams, A. A. (1990) Evidence for the existence of a negatively charged hydrogen species in plasma-treated n-type Si *Appl. Phys. Lett.* **56**, 949–951.
23. Zhu, J., Johnson, N. M., and Herring, C. (1990) Negative-charge state of hydrogen in silicon, *Phys. Rev. B* **41**, 12354–12357.
24. Kramers, H. A. (1940) *Physica* **7**, 284.
25. Vineyard, G. H. (1957) *J. Phys. Chem. Solids* **3**, 121.
26. Rice, S. A. (1958) *Phys. Rev.* **112**, 804.
27. Bourgoin, J., and Lannoo, M. (1983) *Point Defects in Semiconductors II*, in Springer Series in Solid-State Sciences **35** Springer.
28. Deák, P., Snyder, L. C., Heinrich, M., Ortiz, C. R., and Corbett, J. W. (1991) Hydrogen complexes and their vibrations in undoped crystalline silicon, *Physica B* **170**, 253–258.
29. Deák, P., Snyder, L. C., and Corbett, J. W. (1991) Silicon-interstitial oxygen-interstitial complex as a model of the 450 °C oxygen thermal donor in silicon, *Phys. Rev. Lett.* **66**, 747–749.
30. Watkins, G. D. (1991) Metastable defects in silicon – hints for DX and EL2, *Semicond. Sci. Technol.* **6**, B111–B120.
31. Dobson, T. W., Wager, J. F., and Van Vechten, J. A. (1989) Entropy of migration for atomic hopping, *Phys. Rev. B* **40**, 2962–2967.

OXYGEN-CARBON, OXYGEN-NITROGEN AND OXYGEN-DIMER DEFECTS IN SILICON

C. P. EWELS AND R. JONES

Department of Physics, University of Exeter, Exeter, EX4 4QL, UK

AND

S. ÖBERG

Department of Mathematics, University of Luleå, Luleå, S95187, Sweden

1. Abstract

An *ab initio* local density functional cluster method, AIMPRO, is used to examine a variety of oxygen related point defects in silicon. In particular results are given for X-O_n complexes where X is interstitial C, N or O. For $n = 2$, the first defect, C–O_2 has been assigned to the P-centre giving a PL line at 0.767 eV and seen in Cz-Si annealed around 450°C. The second, N-O_2, has properties consistent with a nitrogen related shallow thermal donor. We have also found that a (C-H)O_{2i} defect has very similar electronic properties, and this implies that shallow thermal donors do not have a unique composition. The structure and migration energy of the oxygen dimer is considered and the dimer is found to migrate very much faster than a single oxygen atom.

2. Introduction

Oxygen precipitation in silicon is strongly affected by the presence of impurities, and light impurities such as C and N are especially important. Carbon is a common impurity in Si and usually takes the form of a substitutional defect [1]. This can be displaced into an interstitial site, becoming C_i by irradiation or by trapping an interstitial silicon atom, Si_i, released

141

R. Jones (ed.), Early Stages of Oxygen Precipitation in Silicon, 141–162.
© *1996 Kluwer Academic Publishers.*

by oxygen aggregation. A recent review of carbon related defects in Si has been given in ref. [2].

Nitrogen can be introduced into Si by adding Si_3N_4 to the melt, annealing in an atmosphere of N_{2i}, or by implantation. N is known to enhance O precipitation and inhibit C enhanced O precipitation at 750°C [3]. Photoluminescence (PL) experiments showed a decrease in oxygen precipitate strain-induced defects in samples containing a higher N content.

The principal N defect in silicon is a nitrogen pair [4] composed of two adjacent [100] oriented nitrogen interstitials, N_i. Substitutional N is a rare defect although it has been detected by EPR [5] and local vibrational mode spectroscopy [6].

Isolated N_i defects also exist in implanted material, but lie below the detection limit in doped as-grown material. They possess a local vibrational mode (LVM) around 691 cm^{-1} [7], which is in close agreement with a calculated N stretch mode of the [100] oriented nitrogen-silicon split interstitial at 700 cm^{-1} [4]. In this defect N and Si share a lattice site. Berg Rasmussen *et. al.* [8] observed N_i in N-implanted Si singly annealed at 600°C. They concluded that N_i was mobile during annealing, and rapid cooling was necessary if N was to be trapped as N_i rather than forming other defect complexes.

The N_i defect is structurally similar to the carbon interstitial, C_i. EPR and uniaxial stress experiments show the latter exists in the [100] oriented split-interstitial form [9, 10, 11, 12]. The C and Si radicals have dangling bonds along orthogonal $\langle 011 \rangle$ directions notionally occupied by one electron. But as the electronegativity of C exceeds Si, the lower C-related level fills at the expense of the upper Si-related level. The resulting donor and acceptor levels lie at $E_v + 0.28$ and $E_c - 0.10$ eV [10, 13] respectively. The defect is associated with a photoluminescence (PL) line at 0.856 eV. First-principles calculations [14] confirm the character of the states. An important point is that the small atomic radius of C causes a tensile strain along [011] leading to dilated Si-Si bonds close to the defect.

Let us consider next how C_i complexes with O_i. EPR investigations confirm that an O atom resides at the centre of one of the two dilated Si-Si bonds nearest the defect [15] as expected from a strain release mechanism. But because the Si radical is positively charged and has an empty dangling bond, we expect it to try to pull the negatively polarised O atom towards it. This displacement must be substantial because the highest O related LVM due to O_i at 1132 cm^{-1} is reduced to 742 cm^{-1} [16] in C_iO_i, implying weak Si-O bonding [17]. This defect has a filled state at $E_v + 0.34$ eV and gives a PL line at 0.79 eV. It is however only stable to about 350°C and so is not important in silicon annealed at 450°C where oxygen atoms become mobile.

Larger oxygen complexes with interstitial carbon are believed to exist [1, 2]. There is the possibility of further O attack at the second dilated Si–Si bond along [011]. This would yield a defect with C_{2v} symmetry. Alternatively, the second oxygen atom might lie adjacent to the first one and be separated from C_i. Such a complex has C_{1h} symmetry consistent with the defect giving the 0.767 eV PL [16], and known as the P- line. This line is observed to increase in intensity around 450°C. An early assignment to C_i complexed with VO_{2i} [16] seems to us unlikely.

The 0.767 eV PL suggests a hole trap at $E_v + 0.39$ eV implying that the additional O atom displaces the levels of C_i upward. Indeed this is seen in the series C_i, C_i-O_i, C_i-O_{2i} where the donor levels are $E_v + 0.28$, $E_v + 0.34$, and $E_v + 0.39$ eV respectively. The defects are increasingly stable against thermal dissociation with the P-centre stable to 600°C.

This work on C_i provides a useful guide to the electronically active N_i-O_n complexes. N is also more electronegative than Si and the additional electron in N_i over C_i must occupy the upper state originating from the Si radical whose wave-function is small in the vicinity of the N atom. Like C, the small size of the N atom suggests that the Si-Si bonds along [011] are again dilated and are sites where O atoms would preferentially precipitate. We would then expect the partially occupied level to be pushed upwards through the proximity of the Si radical with polarised oxygen atoms. This would lead to more shallow donor behaviour. We shall show that N_i-O_{2i} has an extremely shallow donor level and it is tempting to assign the shallow donor defect (STD) observed in N-doped material to this defect.

A complication with the unravelling of the processes by which oxygen atoms aggregate in the presence of these impurities is the existence of more than one type of oxygen complex. For example in N-doped material the dominant nitrogen defect is the pair which itself forms electrically inactive NNO defects [18, 19]. These have to be distinguished from the shallow donor defects originating from N_i. In the same way substitutional carbon forms defects with oxygen which are stable to high temperature and are electrically inactive [20]. An interesting question is whether C_i can form STDs. We shall show that this is indeed possible if, in addition to oxygen, a H atom complexes with C.

A crucial aspect of the kinetics of growth of oxygen complexes is the motion of oxygen dimers. We shall discuss this below but first review experimental studies of shallow thermal donors.

3. Shallow Thermal Donors

The shallow thermal donors (STDs) are a family of defects which can be formed by annealing N rich Cz-Si around 650°C [21, 22, 23, 24, 25]. As single donors they have ionisation energies from 0.0347 to 0.0374 eV [26]; this is comparable with P, and is much shallower than the thermal double donor (TDD) (0.0530 to 0.0693 eV). N is known to suppress TD formation [27].

EPR experiments of Hara $et.$ $al.$ [28, 29] found STD defects to have $S = 1/2$ with a g-tensor extremely close to that of the NL10 defect. This led them to suggest that they were the same defect. The g-tensor shows that the STD possesses an apparent C_{2v} symmetry. However EPR and ENDOR studies have failed to resolve any hyperfine splitting in the defect core [28] and so the presence of N in the defect is only inferred from its presence in the material and remains controversial. Although several studies claim to have shown that both O and N must be present in the Si for STDs to form [27, 28, 25], other workers [32] find STDs in nominally N free Cz-Si. In addition, large N concentrations can actually *suppress* STD formation [58]. We believe that the explanation of these conflicting observations lies in a) that other defects, not containing N, can be STDs, and b) there are other electrically inactive nitrogen-oxygen defects which form in competition with the N-related STDs.

The IR electronic absorption lines observed by Suezawa $et.$ $al.$ [25] originated from several different families of STDs. In addition, calculations using effective mass theory by Griffin $et.$ $al.$ [26] suggest that the different families of STDs are due to different numbers of oxygen atoms surrounding a common core, with larger oxygen complexes leading to increasingly shallow donor behaviour.

Recently, Yang $et.$ $al.$ annealed N-doped Si containing STDs. As their concentration decreased, the concentrations of O_i and N-pairs, as monitored by IR spectroscopy, increased. Conversely when STDs formed, the concentration of both O_i and N-pairs decreased. This unfortunately cannot be taken as direct evidence that these elements are part of the STD, because it is known that electrically inactive NNO defects [18] can be formed around the same temperature as STDs and this confuses the issue. It seems to us that while N pairs would complex with O_i creating inactive NNO defects [19], N_i defects could form active centres.

Liesert $et.$ $al.$ [32] observed PL from STDs in samples that were nominally N-free. It may be that there was an undetected level of N in these materials but it is worth considering whether STDs can be formed with different core centres. One such possibility to be considered below is an oxygen complex with C_i-H. The latter defect is to be distinguished from the marginally stable substitutional carbon – hydrogen defect [33, 34]. A C_i-H defect complexed with a substitutional carbon atom is believed to form the T-centre [35] which is observed when Cz-Si containing carbon is annealed between 450°C and 600°C [36]. There is no deliberate H doping of these materials showing that H is a common and unintentional impurity in Cz-Si. The chemistry of C_i-H seems very similar to that of N_i and hence it is plausible that $(CH)_iO_n$ defects could form in C, O and H rich Si.

4. Rapid Diffusion of Oxygen Dimers

Stress induced dichroism experiments show that interstitial oxygen, O_i, in Si hops between bond centre (BC) sites at temperatures around 380°C with an activation energy of 2.54 eV [37, 38, 39]. Secondary ion mass spectrometry (SIMS) measurements [40, 41] also lead to the same activation energy for the oxygen diffusivity at temperatures above 700°C, while small angle scattering measurements on the growth of small oxygen precipitates show that the precipitation is controlled by oxygen diffusion at temperatures above 650°C [42].

Below this temperature, the process is sufficiently slow to become unusable. An important question is whether the *long range* diffusion of oxygen and the growth of oxygen precipitates at temperatures below 400°C is controlled by the same process: namely O_i hopping between BC sites. Now, it is known that some impurities, eg H, can increase the low temperature hop rate [43, 7] and reduce its activation energy to about 1.6 eV [44, 45]. This then increases the rate of oxygen precipitation as the thermal donor production rate also increases. There are 17 thermal donors which are likely to involve increasing numbers of oxygen atoms. The growth rates of these donors have been monitored and it is found that they grow with an activation energy around 1.6 eV – much lower than the energy for a single O_i hop. There have been suggestions that VO_{2i} complexes could diffuse rapidly [46] but the evidence is not conclusive. An early suggestion [47] was that an oxygen dimer could diffuse rapidly, and this has been supported by SIMS measurements of the out-diffusion of oxygen [48] between 500 and 700°C, and recent experiments [49, 50] studying carefully the loss of O_i from solution at temperature as low as 350°C. These show that the process is second order below 400°C but of higher orders, up to 8, as T increases to 500°C. The low temperature process must be controlled by the formation of oxygen dimers. The change in order of the process is then explained by an instability of the dimer above about 450°C and thus larger numbers of oxygen atoms are required to diffuse together to form a stable precipitate.

Previously, semi-empirical CNDO/S techniques have shown that the dimer has a binding energy of 0.1 eV and the saddle point for its migration is 1.36 eV [51]. The saddle point for dimer diffusion was found to lie close to an over-coordinated oxygen square defect – rather similar to the known structure of the nitrogen pair [4]. A plane wave supercell calculation [52] gave a binding energy for the oxygen dimer of 1.0 eV.

Here we describe preliminary results on the structure of the oxygen dimer as well as its dissociation and migration energies.

5. Method

We present here results from *ab initio* local density functional calculations carried out on large H-terminated clusters, using both the serial and parallel versions of the AIMPRO code (described in detail elsewhere [53]). The electronic wave-functions are expanded in $s-$ and $p-$ Gaussian orbitals centred at nuclei as well as at bond centres. Norm-conserving pseudo-potentials [54] are used in order to exclude core electrons. The self-consistent energy E and the force on each atom are calculated and the atoms moved by a conjugate gradient algorithm to equilibrium. All calculations are spin averaged, and a Fermi finite temperature term is used to simulate temperature dependent 'smearing' of the electrons between levels.

All the N, C and O defects considered here have been constrained to have C_{2v} symmetry unless stated otherwise, however we also breached the symmetry of the defects and allowed them to relax; in every case they returned to the C_{2v} symmetrical structure. A variety of different cluster sizes have been used. Initial results were obtained using 113 atom clusters, and the final results shown here come from 148-152 atom clusters. The results are qualitatively identical, showing them to be independent of cluster size.

The NO_{2i} defect described below was optimised using a 150 atom cluster, $Si_{79}H_{68}NO_2$, centred on the N-Si bond centre lying on the C_2 axis; this ensured that all atoms up to second shell neighbours of the defect atoms were bulk Si atoms.

The O dimer calculations were also performed on a variety of clusters, and the results presented here are from a large 113 atom cluster containing up to three oxygen atoms, similar to those used to explore the structure and vibrational modes of O_i, VO, VO_{2i} and VO_{3i} defects [55].

The Kohn-Sham energy levels arising from the calculations give only an approximate estimate of donor and acceptor levels. This being so, it is difficult to conclude that an occupied level lying close to the conduction band of unoccupied levels is a shallow donor as opposed to a donor whose ionisation energy is a few tenths of an eV. It certainly appears in this work that certain levels are very close to the conduction band and will be described as shallow to distinguish them from deep mid-gap states possessed by other defects.

In order to investigate diffusion barriers and determine the saddle point configuration of O_i and O_{2i} it is necessary to introduce constraints on the positions of atoms during the relaxation. Unless this is done, the cluster

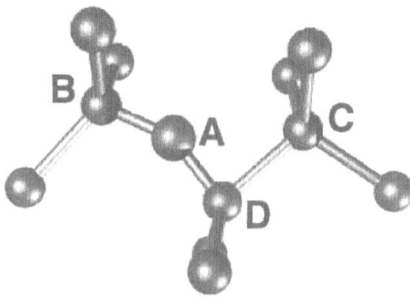

Figure 1.   Interstitial oxygen in silicon, x-direction is $\langle 110 \rangle$, y-direction is $\langle 001 \rangle$. For explanation of the labelling, see text.

would relax away from the saddle to the nearest stable configuration. Suppose that an atom A is hopping from one site to another during which one bond $A - B$ is broken and the bond $A - C$ is created (see Figure 1). Then a constraint is introduced on the relative bond lengths, r_{A-B} and r_{A-C}, and the cluster relaxed maintaining this constraint. The actual constraint used is

$$c_1 = r^2_{A-B} - r^2_{A-C} \tag{1}$$

Now c_1 is clearly negative in the configuration where the bond $A - B$ is short and positive in the configuration where the bond $A - C$ has been formed.

The saddle point usually, but not always, corresponds to a value of c_1 around zero where the $A - B$ and $A - C$ bond lengths are equal. For interstitial O motion, a constraint is selected so that A, B and C correspond to the O atom and the two Si atoms which swap bonds with O. However, the imposition of one constraint is insufficient [45]. During the hop, the O atom initially bonded to the Si atoms D and B, becomes bonded to D and C, then the Si atom D also breaks a bond with C and makes a bond with B. To deal with this a second constraint is required. Here,

$$c_2 = r^2_{D-B} - r^2_{D-C}. \tag{2}$$

In the initial configuration c_1, c_2 are both negative but become positive after the hop. The cluster is relaxed using a conjugate gradient algorithm subject to these constraints for a range of values of c_i and the saddle point can then be directly found by interpolation on the energy surface.

a.

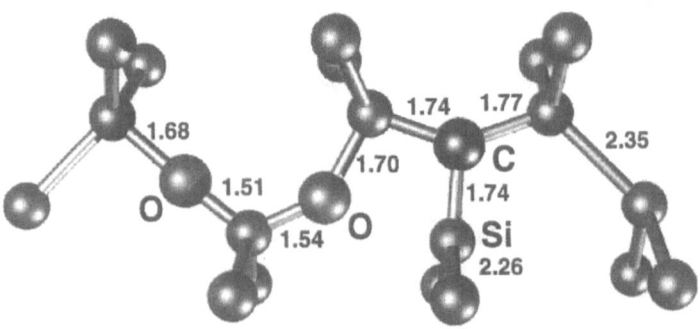

b.

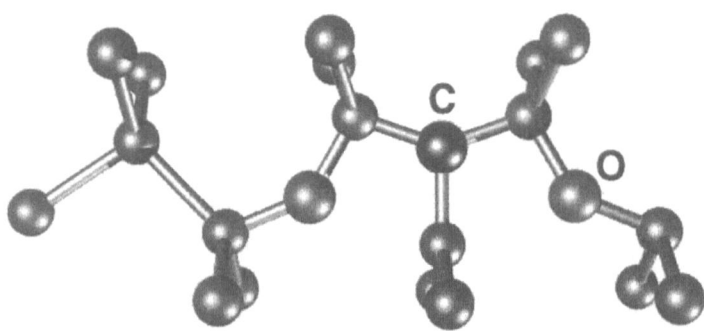

Figure 2. Two possible C_i-O_{2i} structures, with (a) C_{1h} and (b) C_{2v} symmetry. (a) has the lower energy by 0.27 eV.

6. C_i-O_{2i} Defects

There are two likely structures of C_i-O_{2i} which possess C_{1h} or C_{2v} symmetries. These were investigated using 150 atom $Si_{79}H_{68}CO_2$ clusters.

The resultant structures are shown in Figures 2a and 2b. The C_{1h} structure is 0.27 eV lower in energy than the C_{2v} structure. This is consistent with the symmetry of the P-centre assigned to C_i-O_{2i} [16].

The Kohn-Sham energy levels consist of a filled level near E_v and an empty one deep in the gap (see Figure 3). For the C_{2v} structure the un-

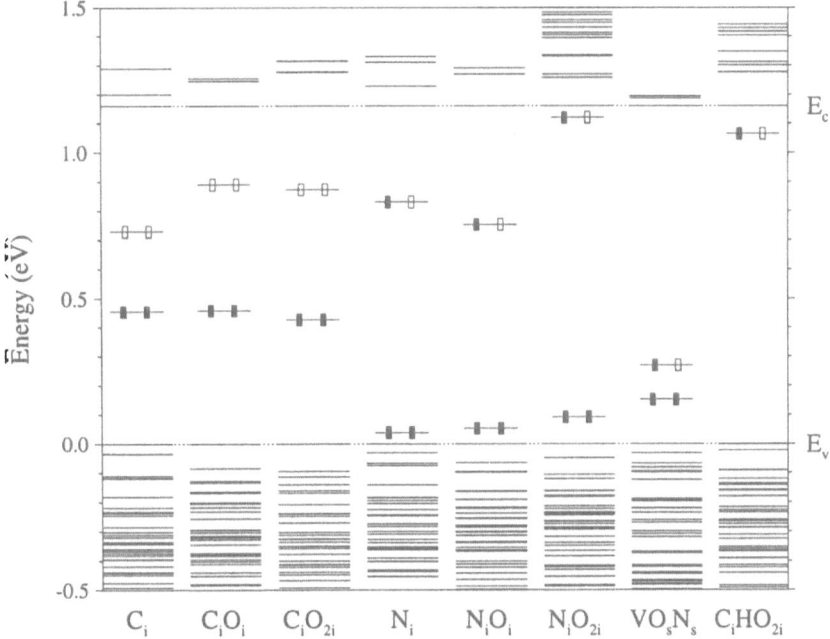

Figure 3. Kohn-Sham eigenvalues in the vicinity of the band gap for various interstitial defects in silicon. The band gap has been arbitrarily scaled to the experimental Si band gap of 1.16eV. Filled (empty) boxes on states near the band edges denote 1 (0) occupancy.

occupied state is much higher in the gap because of the proximity of the Si_i to oxygen. An estimate of a PL transition involving the deep filled level and a conduction band state is found from these levels; scaled by the band gap, it is 0.75 eV. Thus we consider the defect to be a candidate for the P-line at 0.767 eV.

7. Interstitial Nitrogen-Oxygen Complexes

7.1. THE N_I DEFECT

A single interstitial N atom was inserted into a 148 atom cluster, $Si_{79}H_{68}N$ and all atoms were allowed to relax. The Kohn-Sham eigenvalues are given in Figure 3, and it can be seen that the single donor state lies quite low in the gap. The top N–Si bonds oriented along [011] are 1.83 Å and the [100] N–Si bond is 1.75 Å . These Si atoms are pulled towards the core leading to a dilation of the Si-Si backbonds. This makes these bonds attractive sites for oxygen precipiation. Vibrational modes of the defect have been given earlier [4] and are in good agreement with an observed 691 cm^{-1} line.

7.2. N_I–O_I – THE INTERSTITIAL NO PAIR

We examined an N_i–O_i defect using the 149 atom cluster, $Si_{79}H_{68}NO$. The defect has C_{1h} symmetry with a deep donor level close to that of the isolated N_i (see Figure 3). The O atom sits in a stable bond-centred location with Si-O lengths of 1.64 and 1.68 Å, and a Si-O-Si bond angle of 137°.

The central Si atom moves away slightly from the O. The asymmetry of the defect allows the wavefunction to concentrate on the lobe of the $p-$ orbital furthest from the oxygen, thus minimising the Coulombic repulsion with O. This explains why the donor level remains deep in the gap.

Although N_iO_i cannot be a candidate for a STD, it seems reasonable that it would form as an intermediate structure between N_i and N_iO_{2i}.

7.3. N_I–O_{2I}

There are two likely structures for the N_iO_{2i} defect which have C_{2v} and C_{1h} symmetries respectively. Unlike the case of C_iO_{2i}, the first defect has the lower energy and hence the N_iO_{2i} has the same symmetry as that found for nitrogen related STDs by Hara *et. al.*[28].

The 150 atom cluster, $Si_{79}H_{68}NO_2$, was used for investigating this defect. Its relaxed C_{2v} defect structure is shown in Figure 4. The oxygen atoms have an Si–O–Si bond angle of 140°, considerably lower than the standard interstitial angle of $\approx 170°$, and closer to that of substitutional oxygen ($\approx 120°$). In addition the Si–O bond lengths are 1.69 Å and 1.64 Å for the outer and bonds respectively, again lying somewhere between normal interstitial and substitutional bond lengths (1.61 Å and 1.75 Å [55]). The distance between the O atoms and the central Si atom is 2.54 Å: considerably longer than would be expected if there was a Si–O bond forming.

The N atom has two short 1.65 Å top Si–N bonds and a longer 1.72 Å [100] Si–N bond, so the presence of the O_i causes a 9.5% compression of the top bonds, slightly compensated by a 4.1% extension of the lower N–Si bond.

A plot of the wavefunction of the partially filled shallow level is given in Figure 5. The level is diffusively localised on the $p-$ type dangling bond of the central Si radical. There is negligible localisation on the interstitial nitrogen, consistent with the lack of any hyperfine interaction with N observed by EPR/ENDOR experiments. There is also some anti-bonding character of this level on the neighbouring oxygen atom lone pairs. This is due to the lone-pair repulsion of this dangling bond, and confirms the

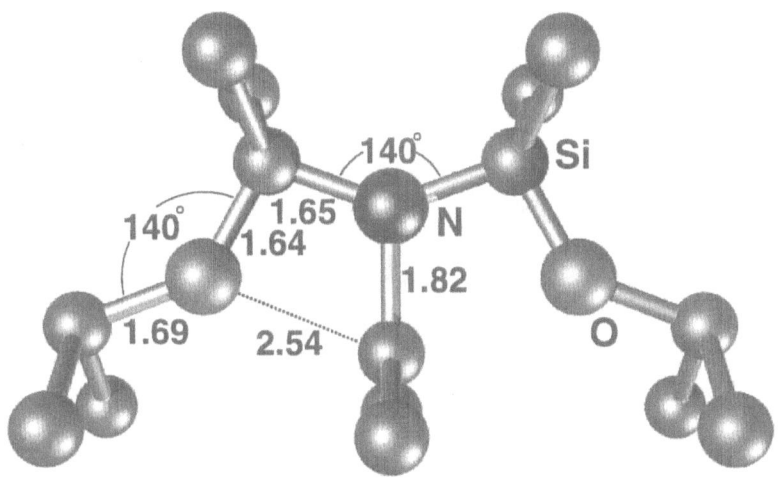

Figure 4. The core structure of the neutral N_i-O_{2i} defect. The symmetry axis lies along [100] and the O atoms are aligned along [011]. All bond lengths are in Å.

mechanism whereby this repulsion pushes up the energy of the Si dangling bond state, making it a shallow level.

The local vibrational modes for the defect are given in Table 1.

TABLE 1. Local vibrational modes of NO_{2i}

$^{14}N^{16}O^{16}O$	$^{14}N^{18}O^{16}O$	$^{14}N^{18}O^{18}O$	$^{15}N^{16}O^{16}O$
1051.398	1050.989	1050.573	1022.763
893.550	883.316	857.225	892.129
857.923	829.617	817.026	857.085
720.268	719.198	718.112	707.957
673.382	671.403	669.570	661.252

As would be expected, the longer Si–O bonds lead to oxygen related LVMs that are lower than those of the standard interstitial.

Preliminary calculations show that the defect with two more interstitial oxygen atoms in the ⟨110⟩ plane is stable. These atoms sit in standard interstitial sites but are able to relieve some of their compressive strain through a compression of the weaker central Si-O-Si bonds. These O atoms

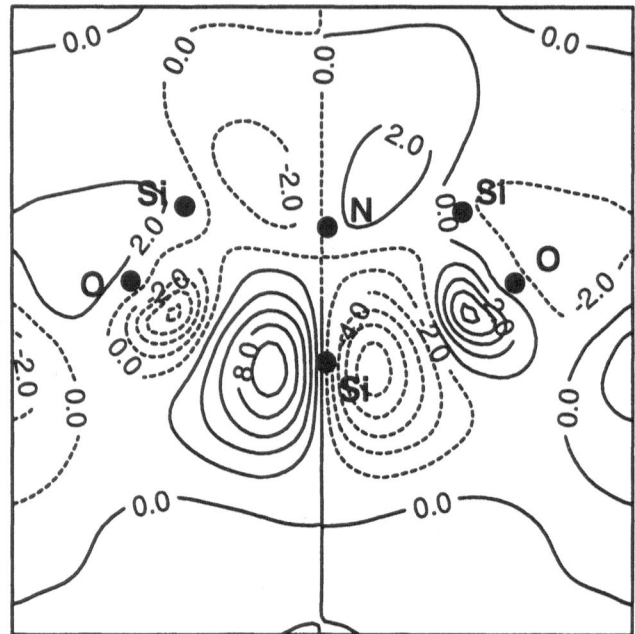

Figure 5.   The pseudo-wavefunction (× 100 a. u.) of the highest occupied orbital of N_i-O_{2i}. Note that it has little amplitude on N and is localized on the Si radical. There are nodal surfaces lying between this atom and the O atoms demonstrating anti-bonding behaviour.

move closer to the Si radical. This should result in pushing upwards the shallow donor level but the limitations of the cluster method prevent a quantitative estimate of the change.

This mechanism whereby larger oxygen aggregates push the nearest oxygen atoms towards the Si radical demonstrates how the family of shallow donor states develops. It also explains Suezawa *et. al.*'s observation that the absorption lines they observed for different shallow donor defects on first sight appeared to be single absorption line splitting due to small strains at the defect core [25].

8. Other models

We also considered a variety of other defects as candidates for the nitrogen related shallow thermal donor.

8.1. $(N_{2I})_N$–O_I – SUEZAWA'S MODEL

A kinetic study by Suezawa *et. al.* [25] linking the rate of loss of N-pairs to the increase in STDs (using IR electronic absorption intensities), suggested that STDs consisted of an unknown O-containing core, to which N-pairs attached themselves. However this modelling neglected dissociation of N_n-O_m complexes, which will be crucial in the kinetics of these defects. In addition it is known that the NNO defect consisting of an N-pair neighbouring a single O_i is electrically inactive [18, 19].

It was proposed that the STD could consist of a core containing O_i, surrounded by N_{2i} pairs in the $\langle 110 \rangle$ direction [25]. However in order to maintain C_{2v} symmetry with only one O_i, the O_i must lie along the central C_2 axis. This structure will therefore be of a [100] split interstitial type, similar to N_i. Since it contains an even number of N atoms it is EPR inactive in the neutral state.

We investigated the cluster $Si_{57}H_{56}N_4O$, containing this defect and constrained to possess C_{2v} symmetry. No shallow donor level was found and we therefore conclude that this cannot be a valid model for the STD. However, N_{2i} pairs could still co-exist with the N_iO_{2i} shallow thermal donor core proposed above. Although O_i atoms will continue to agglomerate at the defect core due to the polar bond formation, the outer O_i atoms will not benefit from the tensile strain field of the defect core. They will themselves still exert tensile stress on the surrounding lattice, and therefore there will be a driving force for N_{2i} pairs to also collect along the same $\langle 110 \rangle$ plane. This formation mechanism is identical to that of the NNO defect [18, 19].

8.2. SUBSTITUTIONAL NITROGEN

Earlier we had proposed that *substitutional* N_sO_{2i} pair could form the core of an STD [57]. The problem here is that the substitutional defect seems to be a rare defect. However, in analogy with the VO_{2i} centre, we examined a substitutional split N_s-O_s pair. This, however, was found to possess a deep level, and the addition of extra O_i in neighbouring bond centres did not provide sufficient strain to push this up to give a shallow donor level.

8.3. THE $(C-H)_IO_{2I}$ DEFECT

An 151 atom cluster, $CSi_{79}H_{69}O_2$, containing a C_i atom passivated with H was relaxed. The C-H bond is parallel to $[0\bar{1}1]$ and lies perpendicular to the plane containing the three C-Si bonds. Thus the defect has C_{1h} symmetry rather than C_{2v}. This might seem to rule it out as a candidate for a STD whose symmetry, according to Hara *et. al.* [28] is C_{2v}. However, if the spin-polarised density is negligible on the H atom, then the symmetry might be misinterpreted. This had been the case for the H-related NL10 EPR centre [56]. In this defect, the H atom also appears to lie along $[0\bar{1}1]$.

The Kohn-Sham eigenvalues are shown in Figure 3. They show that this defect does possess a single shallow donor level. The C-H unit moves

154

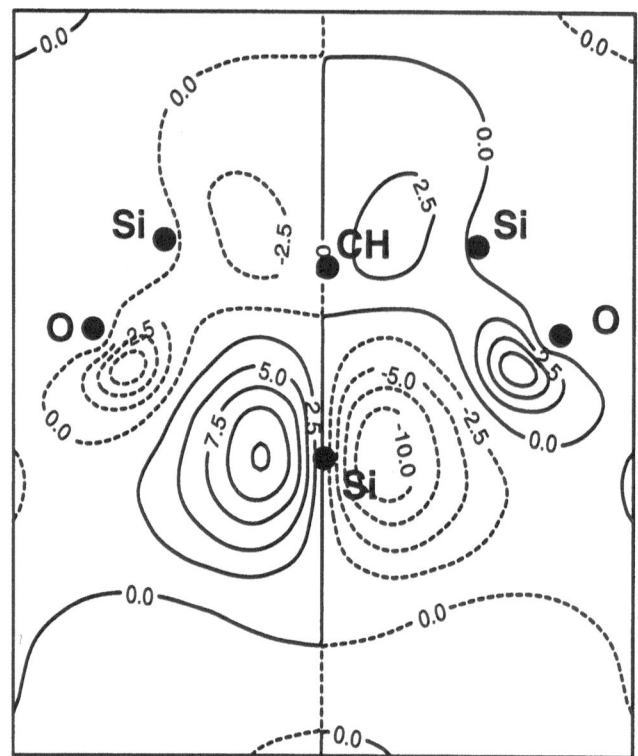

Figure 6. The pseudo-wavefunction (× 100 a. u.) of the highest occupied orbital of $(CH)_i$-O_{2i}. Note the similarity to N_i-O_{2i}, shown in Figure 5.

slightly out of plane until the Si-C-H bond angles are all 100°, with Si–C bond lengths of 1.89, 1.76 and 1.76 Å. A plot of the wavefunction of the donor level is given in Figure 6, and reveals that it has very little overlap with H or C. There is a remarkable similarity between the wavefunction for this donor level, and that of the $N_i O_{2i}$ STD.

In conclusion, it appears that the $C_i HO_{2i}$ defect has a shallow donor level, possesses C_{1h} symmetry and contains a C-H bond lying along $[0\bar{1}1]$. There is very little spin-polarised density residing on C or H.

8.4. DISCUSSION

It has been shown previously [4] that the dominant N defect in Si is a N_{2i} pair, consisting of two neighbouring N-Si split [001] interstitials which together form a N_{2i} square. When this complexes with O, electrically inactive NNO is formed. On the other hand, we have shown here that if N_i complexes with a pair of O atoms, then an STD is created. The question then arises as to which defect is actually formed.

This will be largely governed by kinetics and the relative binding energies of the different complexes. If N_i atoms are sufficiently densely packed in the lattice and given sufficient time to diffuse, then they will preferably bond together to form N_{2i} pairs.

A kinetic model explains many experimental observations which at first sight appear contradictory. Sun *et al* [3] showed that for N to have an effect on O precipitation it is essential to pre-anneal at 1100°C; this step will be required to break up the N_{2i} pairs and homogenously distribute the N_i.

The formation kinetics of the NO_{2i} defects will be dependent on the quantities of each reaction component, and the rate constants for many different reactions:

$$N_i + N_i \rightleftharpoons N_{2i}$$
$$N_i + O_i \rightleftharpoons N_iO_i$$
$$O_i + O_i \rightleftharpoons O_{2i}$$
$$N_{2i} + O_i \rightleftharpoons N_{2i}O$$
$$N_iO_i + N_i \rightleftharpoons N_{2i}O$$
$$N_iO_i + O_i \rightleftharpoons N_iO_{2i}$$
$$N_i + O_{2i} \rightleftharpoons N_iO_{2i}$$

Here N_{2i} is the nitrogen pair defect, $N_{2i}O$ is the electrically inactive NN–O defect, and N_iO_{2i} is the defect proposed here as a STD. In theory there could be other reaction components such as $N_{2i}O_{2i}$, as well as reaction paths involving other point defects such as C.

High temperature annealing is normally used to break up any defect complexes present in the material, and this will include N_{2i} pairs. On cooling, the diffusing N_i defects are trapped either by O or N_i. A high concentration of N_i implies that N_{2i} pairs will be preferably formed, whereas a high oxygen concentration results in N_iO defects. Both defects subsequently complex with oxygen to create electrically inactive NNO or STD defects in the form of NO_{2i}. It is likely that the N-pair is more stable than N_iO and hence slow cooling probably results in high concentrations of NNO.

Hara *et. al.* [28] examined the effect of cooling rate on N-rich Cz-Si, and found that in quenched Si they obtained high concentrations of STDs, whereas in slow cooled Cz-Si the STD concentration was much lower. This is consistent with the above explanation. Slow cooling will also lead to out-gassing of nitrogen, leaving less N_{2i} pairs to mop up O_i forming NNO. Similarly, evidence that high N concentrations do not necessarily result in

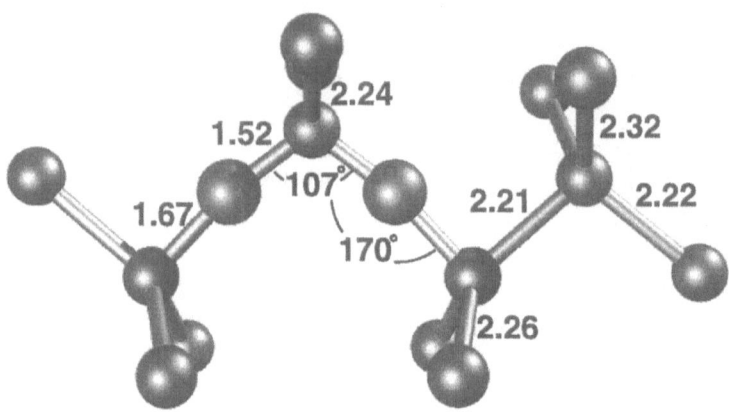

Figure 7. The oxygen dimer in silicon. Symmetry axis lies along [100] and the O atoms are aligned along [011]. All lengths in Å.

high concentrations of STDs has been shown by Griffin *et. al.* [58] using PTIS on N doped samples annealed at 450°C. They found that N suppressed TD growth — behaviour similar to that of carbon doped Si where C–O complexes are formed. In addition, for low doping levels of N (1.4×10^{15} cm^{-3}), they observed a large number of STDs, whereas for a higher N doping level (9×10^{15} cm^{-3}), the STD concentration was strongly reduced. This result is consistent with N_{2i} pairs forming preferentially at higher concentrations, since N_i atoms are able to combine before encountering any O_i.

9. Structure and Motion of Oxygen atoms and Dimers

In the absence of impurities, especially those with tensile strain fields which promote oxygen aggregation, the first oxygen complex to be formed is likely to be a dimer. In this section we explore its stability and kinetics of motion.

A single oxygen atom was placed at a central BC site of an 113 atom cluster and all the atoms relaxed. The Si-O lengths are 1.63 Å. The next Si-Si bonds along the (0$\bar{1}$1) plane are compressed by $\approx 1\%$.

The saddle point for hopping to an adjacent bond centre was found using the above constraints, and corresponds [45] to $c_1 = c_2 = 0$. The saddle configuration then possesses C_{2v} symmetry with O equidistant from the Si atoms B and C, and the Si atom D is equidistant from B and C (see Figure 1). The configuration in which an O atom is equidistant from two second neighbour Si atoms, such as B and C is called the Y-lid as there is some

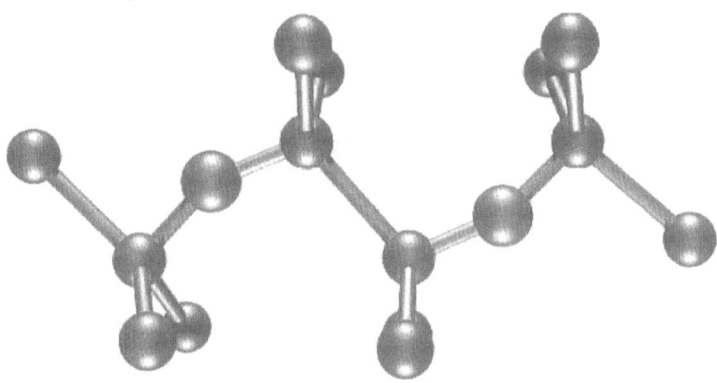

Figure 8. Partially dissociated dimer, 0.8–1.1eV higher in energy than the ground state. Symmetry axis lies along [100] and the O atoms are aligned along [011].

bonding with the third Si atom D. The energy barrier is 2.6 eV in excellent agreement with the experimental value of 2.55 eV. This is an improvement on our previous value [45] of 2.8 eV. The saddle point structure consists of B-O and D-O bonds of length 2.5 and 1.55 Å respectively. These long bonds reflect the inability of the O atom at the Y-lid position to strain the lattice and pull in the Si atoms along [011]. There is then a net tensile strain at the saddle point pulling the Si atoms B and C inwards along [011]. This contrasts with the compressive strain at the stable BC site where the backbonds to D and B are compressed.

An oxygen dimer consisting of two neighbouring BC sited O_i was placed at the centre of the cluster and subsequently relaxed. The dimer defect was stable with O-Si-O bonds of 1.52 Å which are shorter and stronger than the other two Si-O bonds of 1.67 Å. The O-Si-O angle is 107° (see Figure 7). There are no gap levels. The influence of cluster size on the energy of the dimer was investigated by displacing it so that it is centred on a nearby Si atom, *ie* it is displaced by $a/4$ (011). This energy changed by less than 0.08 eV, showing the cluster surface has negligible effect.

We then investigated dimers which were partially dissociated. In the first configuration (Figure 8) the O atoms were pulled apart along [011] and are then separated by $a/4$ (022). A Si-Si bond lies between them. The energy of this partially dissociated configuration is 0.8 to 1.1 eV above that of the dimer. There is some uncertainty in this value because the number of bond centres used for this partially dissociated dimer is different from the number used in the stable dimer. The larger basis yields the lower value and the other value is obtained by removing bond centres to achieve the

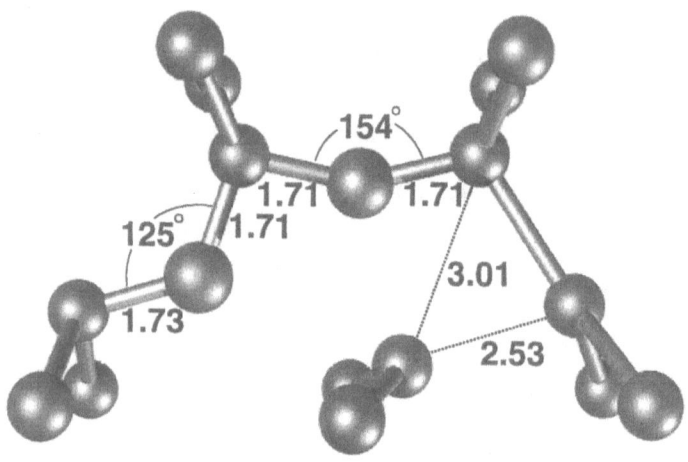

Figure 9. The 'y-lid' constrained oxygen dimer structure. Symmetry axis lies along [100] and the O atoms are aligned along [011]. All lengths in Å.

same sized Hamiltonian. A second configuration was also investigated where the two O_i are again separated by a Si-Si bond but the Si-O bonds lie in different planes. This roughly corresponds to cis- and trans- isomers. The separation of the O atoms is now $a/8$ (233) and has an energy between 0.68 and 0.82 eV above the dimer. This second dissociated pair isomer is more stable than the [011] oriented pair because the compressive strain fields of each O atom are now perpendicular to each other. Since the stress field are do not reinforce each other, these energies are closer to the dissociation energy of the dimer which is then around 0.7 eV. The stability of the dimer is attributed to the creation of a [011] chain of aligned polar bonds as in $Si^{\delta+}-O^{\delta-}-Si^{\delta+}-O^{\delta-}-Si^{\delta+}$.

9.1. MIGRATION OF THE DIMER

We turn now to investigating the movement of the dimer. There are several ways in which a diffusive jump could be executed. We have investigated two mechanisms.

The first involves cooperative dimer diffusion where one of the O atoms drags the other into a Y-lid configuration. This process was considered in detail by Snyder *et. al.* [51]. So far, we have failed to find a low energy route by this mechanism. If the constraints are chosen so that each O atom lies in a Y-lid configuration, this means that the O atoms bond to the Si atoms that are second neighbours to each other, and these two Si-O bonds have equal length. The energy of this configuration is 2.4 eV above that of the stable dimer. One reason for the high energy is that there is a large tensile stress acting on the dimer although this is offset by the absence of any broken Si bonds. It may be possible that there are low energy migration paths by this mechanism but they must involve other configurations.

An alternative diffusive mechanism occurs when the dimer partially dissociates in the $(0\bar{1}1)$ plane. This involves a diffusive jump where one O atom of the dimer moves to its Y-lid configuration and the other O atom is free to remain at its BC site. This configuration was found by imposing a pair of constraints so that the two arms of the Y-lid are equal and the Si atom at the base of the Y-lid is equidistant from the arms (see Figure 9). Remarkably, the energy of this structure is only 1.1–1.5 eV above that of the stable dimer. Again the uncertainty in energy is due to the different numbers of bond centres employed. The crucial result is that the dimer energy is now lowered by the presence of the other O atom. The structure of the Y-lid is very revealing. The Si-O bonds across the arms of the Y-lid are now strong having length of 1.69 Å while the third Si-O bond is 2.62 Å and essentially broken. This structure of two strong Si-O bonds has to be contrasted with that of the Y-lid obtained for single O diffusion where only one strong Si-O bond was found. The extra strong Si-O bond has been made because of the compressive force acting on the arms of the Y-lid from the other O atom. It is interesting that a shallow double donor level also occurs in this structure reinforcing the claim that overcoordinated oxygen atoms are the source of donor activity.

This configuration where the two Si-O bonds have equal lengths is not necessarily the saddle point because of the asymmetry imposed by the second O atom. Calculations are underway to determine the migration barrier.

9.2. DISCUSSION

In summary, we have indicated how a low migration energy for the oxygen pair involving partial dissociation can arise. It suggests that diffusion is fast along one of the [011] directions depending on the orientation of the dimer. Fast diffusing dimers will rapidly encounter relatively slow moving single O_i, and this means that oxygen trimers will be rapidly created below the dimer dissociation temperature.

At some point during the oxygen aggregation process, a Si interstitial must be ejected. Initial calculations suggest that this may occur when the linear trimer is created, as in this case the energy of the central Y-lid configuration is lower than the linear case. In the Y-lid configuration, the central Si atom is displaced downwards away from the O atom and forms a [011] Si self-interstitial. If this migrates away, then a rearrangement can

160

occur leading to a VO_3 defect. Further details of this will be given in a later publication.

10. Conclusions

The theoretical investigations have shown that interstitial defects like C_i and N_i are expected to form several types of complexes with oxygen. Whereas the C related ones have empty levels in the upper band gap, the N_i-O_{2i} complex possesses a shallow donor level. It is possible for an interstitial carbon di-oxygen defect to possess a shallow donor level when complexed with H. However, it is likely to be ionised if it is formed along with C_iO_{2i}. In the absence of impurities, oxygen dimers can be formed which have marginal stability, but appear to migrate with a low activation energy.

11. Acknowledgements

S. Öberg thanks NFR and TFR in Sweden for financial support. He also thanks PDC at KTH in Sweden, for computer time on the SP2. We also thank the HPCI committee of the EPSRC for computer time on the T3D where some of these results were derived. We would also like to thank the British Council for financial support and P. Deák, J. Miro, J. Goss, P. Leary, C. L. Latham, and F. Berg Rasmussen for useful discussions.

References

1. Bean, A. R. and Newman, R. C. (1970), *Solid State Commun.* **8**, 175.
2. Davies G. and Newman R. C. (1994), in *Handbook on Semiconductors* **3**, ed. by S. Mahajan, Elsevier, p. 1557.
3. Sun Q., Yao K. H., Gatos H. C., and Lagowski J. (1992), Effects of nitrogen on oxygen precipitation in silicon, *J. Appl. Phys.* **71**.
4. Jones R., Öberg S., Berg Rasmussen F., and Bech Neilsen B. (1994), Identification of the Dominant Nitrogen Defect in Silicon, *Phys. Rev. Lett.* **72**, 1882.
5. Brower K. L. (1982), Deep-level nitrogen centres in laser-annealed ion-implanted silicon, *Phys. Rev. B* **26**, 6040.
6. Stein H. J. (1988), Implanted nitrogen in germanium, *Appl. Phys. Lett.* **52**, 153.
7. Stein H. J. and Hahn S. K. (1990), Hydrogen-accelerated thermal donor formation in Czochralski silicon, *Appl. Phys. Lett.* **56**, 63.
8. Berg Rasmussen F., Jones R. and Öberg S. (1994), Nitrogen in Germanium - identification of the pair defect, *Phys Rev B* **50** 7, 4378.
9. Watkins G. D. and Brower K. L. (1976), *Phys. Rev. Lett.* **36**, 1329.
10. Song L. W. and Watkins G. D. (1990), EPR identification of the single-acceptor state of interstitial carbon in silicon, *Phys. Rev. B* **42**, 5759.
11. Wooley R., Lightowlers E. C., Tipping A. K., Claybourn M. and Newman R. C. (1986), Electronic and vibrational absorption of interstitial carbon in silicon, *Mat. Sci. Forum* **10-12**, 929.
12. Zheng J. F., Stavola M. and Watkins G. D. (1994), Structure of the neutral charge state of interstitial carbon in silicon, *The Physics of Semiconductors*, ed. D. J.

Lockwood, World Scientific, Singapore, 2363.
13. Kimerling L. C., Blood P. and Gibson W. M. (1978), *Inst. Phys. Conf. Ser.* **46**, 273.
14. Jones R., Leary P., Öberg S. and Torres V. J. T. (1995), Peculiarities of Interstitial Carbon and Di-Carbon Defects in Si, *Mater. Sci. Forum* **196-201**, 785.
15. Trombetta J. M. and Watkins G. D. (1988), Identification of an interstitial carbon interstitial oxygen complex in silicon, *Mat. Res. Symp.* **101** 93.
16. Kürner W., Sauer R., Dörnen A. and Thonke K. (1989), Structure of the 0.767-eV oxygen-carbon luminescence defect in 450°C thermally annealed Czochralski-grown silicon, *Phys Rev B* **39** 18, 13327.
17. Jones R. and Öberg S. (1992), Oxygen frustration and the interstitial carbon-oxygen complex in Si, *Phys Rev Lett* **68** 1, 86.
18. Jones R., Ewels C., Goss J., Miro J., Deák P., Öberg S. and Berg Rasmussen F. (1994), Theoretical and isotopic infrared-absorption investigations of nitrogen-oxygen defects in silicon, *Semicond. Sci. Tech.* **9** 2145-48.
19. Berg Rasmussen F., Öberg S., Jones R., Ewels C., Goss J., Miro J., Deák P. (1996), The nitrogen-pair oxygen defect in silicon, these proceedings.
20. Kaneta C., Sasaki T., Katayama-Yoshida H. (1992), Atomic configuration, stabilizing mechanism, and impurity vibrations of carbon-oxygen complexes in crystalline silicon, *Phys. Rev. B* **46** 13179.
21. Abe T., Masui T., Harada H. and Chikawa J. (1985), in *VLSI Science and Technology*, edited by Bullis W. M. and Broydo S. (Electrochemical Society, Pennington, NJ), p.543.
22. Stein H. J. (1986), Nitrogen in crystalline Si, in *Oxygen, Carbon, Hydrogen and Nitrogen in Crystalline Silicon, Mat. Res. Soc. Symp.*, Pittsburgh PA, ed Mikkelsen Jr. J. C., Pearton S. J., Corbett J. W. and Pennycook S. J., **67**, 523.
23. Navarro H., Griffin J., Weber J. and Gentzelm L. (1986), New Oxygen Related Shallow Thermal Donor Centres in Czochralski-grown Silicon, *Solid St. Commun.* **58**, 151.
24. Suezawa M., Sumino K., Harada H. and Abe T. (1986), Nitrogen-Oxygen complexes as shallow donors in silicon-crystals, *Jpn. J. Appl. Phys.* **25** L859.
25. Suezawa M., Sumino K., Harada H. and Abe T. (1988), The Nature of Nitrogen-Oxygen Complexes in Silicon, *Jpn. J. Appl. Phys* **27** 62.
26. Griffin J. A., Navarro H., Weber J., Genzel L., Borenstein J. T., Corbett J. W. and Snyder L. C. (1986), The new shallow thermal donor series in silicon, *J. Phys. C: Solid State Phys.* **19** L579-L584.
27. Yang D., Que D. and Sumino K. (1995), Nitrogen effects on thermal donor and shallow thermal donor in silicon, *J. Appl. Phys.* **77** (2) 943.
28. Hara A., Fukuda T., Miyabo T., Hirai I. (1989), Electron Spin Resonance of Oxygen-Nitrogen Complex in Silicon, *Jpn. J. Appl. Phys.* **28** 1, p.142-143.
29. Hara A., Hirai L. and Ohsawa A. (1990), NL10 defects formed in Czochralski silicon-crystals, *J. Appl. Phys.* **67**, 2462.
30. Hara A., Aoki M., Koizuka M. and Fukuda T. (1994), Model for NL10 thermal donors formed in annealed oxygen-rich silicon-crystals, *J. Appl. Phys.* **75** 6.
31. Chen C. S., Li C. F., Ye H. J., Shen S. C. and Yang D. R. (1994), Formation of nitrogen-oxygen donors in N-doped Czochralski-silicon crystal, *J. Appl. Phys.* **76** 6.
32. Heijmink Liesert B. J., Gregorkiewicz T. and Ammerlaan C. A. J. (1993), Photo-luminescence of silicon thermal donors, *Phys. Rev. B* **47** (12) 7005.
33. Endrös A. (1989), Charge-state-dependent hydrogen-carbon-related deep donor in crystalline silicon, *Phys Rev Lett* **63**, 70.
34. Endrös A. L., Krüher W. and Grabmaier J. (1991), Hydrogen in phosphorus-doped and carbon-doped crystalline silicon, *Physica B* **170**, 365.
35. Safonov A. N., Lightowlers E. C., Davies G., Leary P., Jones R. and Öberg S. (1996), unpublished.
36. Minaev N. S. and Mudryi A. V. (1981), *Phys. Stat. Solidi A* **68**, 561.

162

37. Bosomworth D. R., Hayes W., Spray A. R. L. and Watkins G. D. (1970), Absorption of oxygen in silicon in the near and the far infrared, *Proc. Roy. Soc.* **A 317** 133.
38. Stavola M., Patel J. R., Kimerling L. C. and Freeland P. E. (1983), Diffusivity of oxygen in silicon at the donor formation temperature, *Appl. Phys. Lett.* **42**, 73.
39. Newman R. C., Tucker J. H. and Livingston F. M. (1983), Radiation-enhanced diffusion of oxygen in silicon at room-temperature, *J. Phys. C: Solid State Phys.* **16**, L151.
40. Mikkelsen J. C. (1982), Excess solubility of oxygen in silicon during steam oxidation, *Appl. Phys. Lett.* **41** 9, 871.
41. Lee S.-T. and Nichols D. (1985), Outdiffusion and diffusion mechanism of oxygen in silicon, *Appl. Phys. Lett.* **47** 1001.
42. Newman R. C., Jones R. (1994), in *Oxygen in Silicon*, edited by Shimura F., *Semiconductors and Semimetals*, edited by Willardson R. K. and Beer A. C., **42**, p. 289, Academic Press.
43. McQuaid S. A., Newman R. C., Tucker J. H., Lightowlers E. C., Kubiak A. and Goulding M. (1991), Concentration of atomic-hydrogen diffused into silicon in the temperature range 900-1300 °C, *Appl. Phys. Lett.* **58**, 2933.
44. Estreicher S. K. (1990), Interstitial-O in Si and its interactions with H, *Phys Rev.* **41**, 9886.
45. Jones R., Öberg S. and Umerski A. (1992), Interaction of hydrogen with impurities in semiconductors, *Mater. Sci. Forum* **241** 551.
46. Corbett J. W., Watkins G. D. and McDonald R. S. (1964), New oxygen infrared bands in annealed irradiated silicon, *Phys. Rev. B* **135**, A1381.
47. Gösele U. and Tan T. Y. (1982), Oxygen diffusion and thermal donor formation in silicon, *Appl. Phys. A* **28**, 79.
48. Lee S.-T., Fellinger P. and Chen S. (1988), Enhanced and wafer-dependent oxygen diffusion in Cz-Si at 500-700 °C, *J. Appl. Phys.* **63** 1924.
49. McQuaid S. A., Binns M. J., Londos C. A., Tucker J. H., Brown A. R. and Newman R. C. (1995), Oxygen loss during thermal donor formation in Czochralski silicon: New insights into oxygen diffusion mechanisms, *J. Appl. Phys.* **77**, 1427.
50. Yamanaka H. (1994), Formation kinetics and infrared-absorption of carbon-oxygen complexes in Czochralski silicon, *Jpn. J. Appl. Phys.* **33**, 3319.
51. Snyder L. C., Corbett J. W., Deák P. and Wu R.(1988), On the Diffusion of Oxygen Dimre in a Silicon Crystal, *Mat. Res. Symp. Proc.* **104** 179.
52. Needels M., Joannopoulos J. D., Bar-Yam Y. and Pantelides S. T. (1991), Oxygen complexes in silicon, *Phys. Rev B* **43**, 4208.
53. Jones R. (1995), *Phil. Trans. Roy. Soc. Lond. A* **350** 189.
54. Bachelet G. B., Hamann D. R. and Schlüter M. (1982), Pseudopotentials that work - from H to Pu, *Phys. Rev. B* **26** 4199.
55. Ewels C., Jones R. and Öberg S. (1995), First Principles investigation of vacancy oxygen defects in Si, *Mater. Sci. Forum* **196-201**, 1297.
56. Martinov Yu. V., Gregorkiewicz T. and Ammerlaan C. A. J. (1995), Role of Hydrogen in the formation and structure of the Si-NL10 thermal donor, *Phys. Rev. Lett.* **74** 2030.
57. Jones R., Öberg S. and Umerski A. (1991), Ab initio calculations on interstitial O clusters in Si, *Mater. Sci. Forum* **72** 287, Trans Tech Publications, Zürich.
58. Griffin J. A., Hartung J., Weber J., Navarro H. and Genzel L. (1989), Photothermal Ionisation Spectroscopy of Oxygen-Related Shallow Defects in Crystalline Silicon, *Appl. Phys. A.* **48** 41-47.
59. Gregorkiewicz T., Th Bekman H. H. P. and Ammerlaan C. A. J. (1988), Microscopic Structure of the NL10 heat-treatment center in silicon - study by electron-nuclear double-resonance, *Phys. Rev. B* **38**, 3998.

THE ROLE OF TRIVALENT OXYGEN
IN ELECTRICALLY ACTIVE COMPLEXES

PETER DEÁK
Department of Atomic Physics, Technical University Budapest
Budafoki út 8., Budapest, H-1111, HUNGARY
e-mail: deak@bigmac.eik.bme.hu

Abstract: Oxygen forms two bonds in silicon or silica but, generally, it can also be trivalent. *Ab initio* calculations on the molecules H_2O and $(SiH_3)_2O$ on the one hand, and H_3O^+ and $(SiH_3)_3O^+$ on the other, are used to deduce properties of trivalent oxygen. It is explained why a trivalent oxygen is a donor. During diffusion in c-Si, oxygen may form a third bond. The possibility of Bourgoin-Corbett diffusion is discussed for a single oxygen and for a dioxygen complex. A third oxygen bond can also be formed in defect complexes with other impurities. Based on *semi-empirical* calculations, stable complexes of trivalent oxygen with a Si self-interstitial (w/o the presence of hydrogen) or with nitrogen are predicted. The relation of these defects to thermal double donors and shallow thermal donors are discussed.

1. Introduction

An oxygen atom, as isolated interstitial in silicon or as constituent of silica, forms two (polarized) covalent bonds. This fact often makes one forget that this is not the only possibility for oxygen in semiconductors. Of course, one knows that oxygen in metal oxides can have higher coordination numbers but, of course, these compounds are mostly very ionic. However, this is not always the case. The relative partial charge on oxygen is -0.19 in semiconducting ß-Ga_2O_3, (less than in SiO_2 where it is -0.23) [1], even though, two out of every three oxygens are 3-fold coordinated and one is 4-fold coordinated. Therefore, these oxygen atoms can be regarded as trivalent and tetravalent, respectively, with (polarized) covalent bonds, just as those of divalent oxygen in silica. The aim of this paper is to show that trivalent oxygen atoms may play a very important role in defect complexes of silicon. First, the properties of trivalent oxygen will be deduced from calculations on molecules, then specific examples will be shown, how it may influence the properties of a silicon crystal.

The considerations presented here are based on theoretical computations. *Ab initio* Hartree-Fock-Roothaan (HFR) calculations with second order Moeller-Plesset (MP2) perturbation for correlation correction, as well as *first principles* Local Density Approximation (LDA) of the density functional theory has been used together with *semi-empirical* quantum chemical methods. To describe these methods is beyond the scope of this paper. (A nice and short comparison of the former two can be found in ref. [2].) Suffice it to say that there is no *exact* method for solving the many-body Schrödinger equation. *Ab initio* or *first principles* mean non-empirical approximations, i.e., *a priori* mathemat-

R. Jones (ed.), Early Stages of Oxygen Precipitation in Silicon, 163–177.

ical simplifications but no adjustable parameters or direct use of experimental data. Accordingly, the performance of these methods is, in principle, independent of the system they are applied to. In fact, since the one-electron wave functions have to be expanded in terms of known functions, a lot of freedom is left for finding the best possible basis set to solve a given problem. Semi-empirical methods of quantum chemistry are intuitive approximations based on the effective valence shell Hamiltonian theory [3]. They utilize experimental data on small systems directly (substituting one-center integrals in the calculation with data measured on atoms) and/or indirectly (by using parametrized empirical functions with adjustable parameters for two-center integrals). As a consequence, the performance depends very much on the parametrization and may fluctuate in different systems. Nevertheless, the vast amount of information on small molecules allow to evaluate these methods for practically every situation, and with sufficient care, they can successfully be applied to predict properties of large molecules or defect complexes in solids. In conclusion, it can be stated that proper caution is needed in use of any kind of theoretical method, ab initio or semi-empirical, alike (no black-box mode!). I have found it very effective to use semi-empirical methods to run over a multitude of complex possibilities, then use ab initio methods for accurate results on the most promising ones. All the calculations presented here utilize standard basis sets to expand the one-electron wave functions in terms of localized orbitals : $6\text{-}31G^{*}$ for HFR-MP2 in the GAMESS package [4], Slater-orbitals and standard parametrization in the semi-empirical methods MINDO/3 [6] and PM3 [7]), 8 Gaussians for each s and p orbitals on oxygen, 5 for the orbitals of silicon, 2 for hydrogen atoms, and 3 additional Gaussians for every bond center in the pseudopotential-based LCGTO-LDA code, AMPRO [5]. For defect calculations, LDA has been used within the molecular cluster model, while semi-empirical methods were applied within the framework of the cyclic cluster model. The latter provides a better description of the crystalline background.

2. Trivalent Oxygen in Molecules

If trivalency of oxygen is a real possibility, it should occur also in molecules. Indeed, three methyl groups bonded to a positively charged oxygen can be found, e.g., in tetra-fluoro-borate salts. Compounds are also known in which tri-tert-butyl silicon is bonded to the oxygen of a water molecule in a positive ion, or other ones with two ethyl groups and a silyl group bonded to O^{+}. The most obvious example is the molecular ion, H_3O^{+}. This ion can mostly be found in molecular crystals with water content (e.g., $HClO_4{:}H_2O$, $HNO_3{:}H_2O$, $H_2SO_4{:}H_2O$, etc.). It has a planar structure with the hydrogen atoms equidistant from oxygen and from each other. The O-H bond length, 0.99 Å, is 3 % longer than in water (0.96 Å). A simple chemical picture featuring the localized electronic orbitals shows the difference in the bonding of trivalent oxygen in H_3O^{+} and divalent oxygen in H_2O (Figure 1).

The situation is somewhat similar to the case of diamond and graphite. In the water molecule, 4 of the 8 valence electrons of the system occupy two O-H bonds. The other four occupy two so-called *lone pair* orbitals. The two bonds and the two lone pairs form a distorted tetrahedron (the bond angle is 104.5°) similar to carbon in diamond. In case of H_3O^{+} three bonds are formed in trigonal configuration similar to graphite. The three bonds and the one lone pair (p_z of oxygen) accommodates all together 8 of the 9 valence electrons. The system is positively ionized because there is simply no place for the ninth valence electron to go! This simple picture explains that a trivalent oxygen atom must be a donor in a covalent crystal (where all other bonds are saturated): the sixth electron of oxygen will be given up to the conduction band.

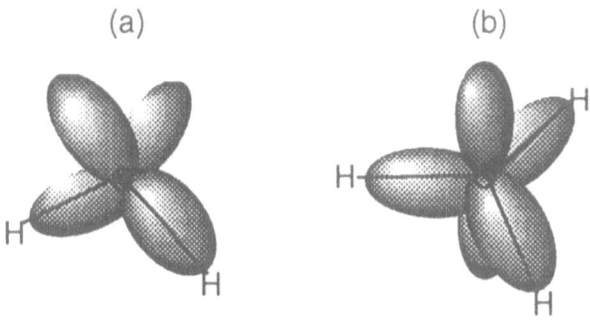

(a)   (b)

Figure 1. The bonding in H_2O (a) and H_3O^+ (b).

Table 1. compares calculated and experimental bond dissociation energies relevant to these hydrogen oxygen systems. The experimental data have been derived from heats of formation at 298 K. The parametrization of PM3 has also been fitted to yield formation energies at room temperature. The spin polarized LDA calculations refer to 0 K. The reactions (1-2) in Table 1 serve as test for the accuracy of the calculations, reaction (6) proves that the "basis set error" in calculating the dissociation energy is negligible.

TABLE 1. H-O bond dissociation energies (enthalpies)

	reaction	ΔH [eV]		
		LDA	PM3	exptl.
1	$H_2 \rightarrow H + H$	4.08	5.10	4.52
2	$O_2 \rightarrow {}^3O + {}^3O$	5.30	5.35	5.16
3	$H_3O^+ \rightarrow H^+ + H_2O$	6.55	6.12	
4	$H_2O \rightarrow H + OH$	5.82	4.70	5.18
5	$OH \rightarrow H + {}^3O$	4.78	4.71	4.43
6	$H_3O^+ \rightarrow H^+ + H + H + {}^3O$	17.17	15.54	

As both type of calculations show, the third O-H bond is even harder to break than any of the normal bonds in water or the OH radical. Nevertheless, to create H_3O^+ and OH⁻ from two water molecules is a very endothermic reaction (9.87 eV according to LDA and 10.77 eV according to PM3). It is worth noting that for these small systems the ab initio LDA and the semi-empirical PM3 methods perform on about an equal level.

Turning to silicon, the molecule with Si-O bonds, in analogy of H_2O, is $(SiH_3)_2O$ (disiloxane). The bonding picture in the Si-O-Si core is similar to that of water in Figure 1(a), except that the experimental bond angle is 144° (the Si-O bond length is 1.634 Å). The ion $(SiH_3)_3O^+$, with bonding similar to that of H_3O^+ in Figure 1(b), has not been observed. This is no surprise: even disiloxane is a rather volatile compound. Nevertheless, calculations find the ion $(SiH_3)_3O^+$ stable. Table 2. compares bond dissociation energies calculated by LDA, PM3, and HFR-MP2 calculations.

TABLE 2. Si-O bond dissociation energies (enthalpies)

	reaction	ΔH [eV]		
		LDA	PM3	HFR-MP2
7	$Si_2H_6 \rightarrow SiH_3 + SiH_3$	2.72	2.95	
8	$(SiH_3)O^+ \rightarrow (SiH_3)^+ + (SiH_3)_2O$	1.17	3.94	2.52
9	$(SiH_3)_2O \rightarrow SiH_3 + O(SiH_3)$	5.11	5.05	4.14
10	$O(SiH_3) \rightarrow SiH_3 + {}^3O$	4.82	4.71	
11	$(SiH_3)O^+ \rightarrow (SiH_3)^+ + SiH_3 + SiH_3 + {}^3O$	10.96	13.70	

While LDA has reproduced the H-O bond lengths in H_2O (0.96 Å) and in H_3O^+ (0.98 Å), the geometry of the Si-O compounds is not se well predicted, as it is shown in Table 3. Increasing the basis on the hydrogens improves the Si-O bond lengths but the Si-O-Si bond angle could probably be improved only by d orbitals on silicon.

TABLE 3. Calculated geometry of Si-O systems

	LDA	PM3	HFR
$(SiH_3)_2O$			
d(Si-O) [Å]	1.513	1.682	1.639
ϑ(Si-O-Si) [°]	180	118	150
$(SiH_3)O^+$			
d(Si-O) [Å]	1.691	1.725	1.799

Despite the difference in geometries, the only major discrepancy among the calculations appear in the dissociation energy for the third oxygen bond (see Table 2). Apparently, LDA underestimates, while PM3 overestimates the tendency for oxygen to become trivalent. Based on the HFR-MP2 results, it can be stated, however, that the third oxygen bond has still about half the energy of a normal Si-O bond. The Si-O bond length for trivalent oxygen is again longer than for the divalent one by about 10 %.

The difference between the bonds of divalent and trivalent oxygen can also be seen in the vibration frequencies. Calculated frequencies for the asymmetric stretch modes of the HOH and SiOSi units are shown in Table 4. (It should be noted that even HFR-MP3 is not good enough for predicting accurate vibration frequencies, while, surprisingly, LDA delivers consistently good values. Results of semi-empirical calculations cannot be used without scaling [8].) The bonds of trivalent O are softer. The difference is especially distinct in case of Si-O bonds.

TABLE 4. Asymmetric stretch modes of the HOH and SiOSi units [cm^{-1}]

	LDA	exptl. (298 K)	HFR-MP3
H_2O	3646	3652	
H_3O^+	3576		
$(SiH_3)_2O$	1132	1107	1234
$(SiH_3)_3O^+$	846		886

3. Trivalency of Oxygen Diffusing in Crystalline Silicon

Apparently a (polarized) covalent third bond of oxygen to silicon is possible in a free radical. The question is whether it can occur in silicon crystal as well. In 1983, Stavola and Snyder [9] called attention to the fact that an oxygen interstitial passes through a so-called YLID position (Figure 2) on its diffusion path, where it can be trivalent.

A simple examination of the geometrical situation of oxygen diffusion helps to decide whether oxygen can, indeed, become trivalent. As it is well known, oxygen is a puckered bond center interstitial in silicon [10]. Since the vibration frequency of interstitial oxygen (O_i) is very close to that in

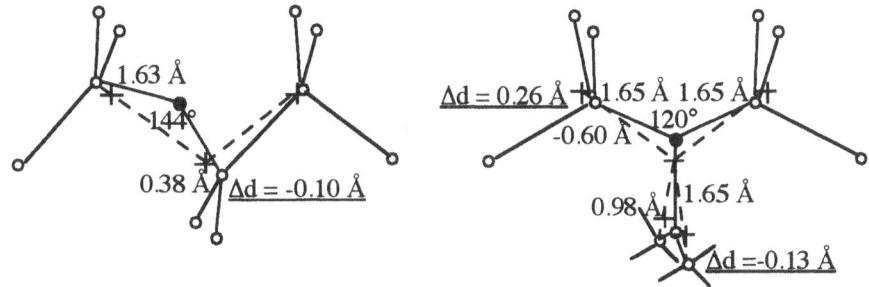

Figure 2. The geometry of oxygen diffusion in silicon

disiloxane, one may assume in a first approximation that the ideal geometry of the Si-O-Si unit in that molecule can be transferred to the crystal. The silicon atoms will then be away from their lattice position by 0.38 Å, along [111] directions. This decreases the lengths of their backbonds by 0.10 Å. Using the stretching force constant of silicon bonds in the crystal, 9.5 eV/Å^2, the strain energy necessary to insert this ideal Si-O-Si unit can be calculated to be:

$$\Delta E(O_i) = 6 \cdot (9.5 \cdot 0.10^2)/2 = 0.29 \text{ eV} \qquad /1/$$

Compared to the energy gained by substituting a Si-Si bond by two Si-O bonds, this is almost negligible. With experimental bond energies in silicon (2.3 eV) and silica (4.7 eV), the energy gain is

$$E(Si:O_i) = E(Si-Si) - 2E(Si-O) + \Delta E(O_i) = -6.8 \text{ eV} \qquad /2/$$

Now let's consider a single diffusion jump from one bond center site to the next. Let's assume that halfway in the motion (in the YLID position) the oxygen (O_Y) forms three equivalent bonds to the silicon atoms at the vertex of the Y. If we assume bond lengths equal with those in silica, 1.65 Å, the two upper Si atoms have to move inward along [111] directions by 0.60 Å, while the lower one has to move down in [001] direction by 0.98 Å. Consequently, the backbonds of the upper ones will be elongated by 0.26 Å, and those of the second neighbors of the lower one[*] compressed by 0.13 Å. This gives rise to a strain energy of

$$\Delta E(O_Y) = 6 \cdot (9.5 \cdot 0.26^2)/2 + 6 \cdot (9.5 \cdot 0.13^2)/2 = 2.40 \text{ eV} \qquad /3/$$

Now, in this case, two ideal Si-Si bonds are traded for the three Si-O bonds of the trivalent O_Y. We can estimate the strength of these three bonds using the fact that the total energy difference between O_Y and O_i is the activation energy for diffusion, which is known experimentally: $E_d = 2.5$ eV [11].

$$E(Si:O_Y) = E(Si:O_i) + E_d \qquad /4/$$

The energy of O_Y is

$$E(Si:O_Y) = 2E(Si\text{-}Si) - 3E(Si\text{-}O_Y) + \Delta E(O_Y) + Eg \qquad /5/$$

The appearance of the energy of the silicon gap, E_g, needs some explanation. The sixth valence electron — originally in one of the lone pair orbitals of the divalent oxygen, O_i, with energy below the valence band (VB) edge — cannot be accommodated by the trivalent oxygen, O_Y. Since one of the silicon neighbors (the lower one in Figure 2) is trivalent, this electron can find place in the singly occupied dangling bond of this silicon atom. However, the repulsion of the electrons will push this level up, possibly close or above the conduction band (CB) edge. Either way, E_g is a good estimate for the energy needed to promote one electron of O_i (below the VB edge) to a level close to the CB edge. From equations /4-5/ the energy of the 3 Si-O bonds of O_Y can be expressed, and using the experimental energies $E(Si\text{-}Si) = 2.3$ eV, $E_d = 2.5$ eV, and $E_g = 1.1$ eV, as well as the values obtained in equations /2-3/:

$$3E(Si\text{-}O_Y) = -12.4 \text{ eV} \qquad /6/$$

This is comparable to the values obtained for reaction (11) in Table 2, so indeed, we may except trivalency to happen, the same way as in the case of $(SiH_3)O^+$ (possibly with two "normal" and one weaker Si-O bond). Consequently, if the YLID state could be stabilized, it would be a donor (as suggested in [9] to explain the core of thermal double donors) and would have a characteristic vibration frequency below (or, taking into account the compression, around) 900 cm^{-1}.

Theoretical calculations have shown [8] that a single YLID cannot be stabilized during the initial stage of oxygen aggregation. However, it is interesting to pause at the thought that oxygen could be a "transient" shallow donor during its diffusion. If the donor level related to the YLID position is shallow enough to be ionized, the diffusion could be of the Bourgoin-Corbett type [12]. This means that the motion switches back and forth between the total energy surfaces of the neutral and the positively charged oxygen atom. The energy minima in the former correspond to the bond center sites (O_i), while in the latter to the YLID positions (O_Y). The diffusion barrier should so be lower than for diffusion on the neutral potential surface (see Figure 3).

---

[*] To avoid having to deal with the bond strength of trivalent Si, the length of its backbonds are kept fixed.

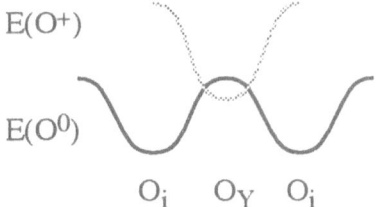

Figure 3. Bourgoin Corbett diffusion of oxygen

Based on theoretical calculations, it is difficult to decide whether this is really the case. A good answer to that question would require the accurate prediction of level positions relative to the gap, which is a weak point of every theory (ab initio or semi-empirical alike), aggravated by the deficiencies of the model for simulating the crystalline background. It is telling, that calculated results for the diffusion barrier scatter from 1.2 to 4.1 eV [8,13-18]. Since typically, calculational results are better than that, it is suspected that the Bourgoin-Corbett mechanism is really at work here. In that case very high calculated values are due either to the "small-cluster" effect or to the HFR-type approach (much to big a "gap"), while too small values may come from LDA or semi-empirical methods, if the model system is, in principle, sufficiently big.

Oxygen aggregation kinetics, especially in the low temperature (< 500 °C) range cannot be understood based on the 2.5 eV migration barrier of isolated interstitial oxygen. Earlier proposals [19] of fast diffusing dioxygen complexes are being confirmed recently [20]. Different calculations [8,21,22] agree that an oxygen molecule is not stable in the silicon crystal, and the most likely dioxygen complex, O_{2i}, is formed in the (110) plane from two O_i atoms with a common silicon

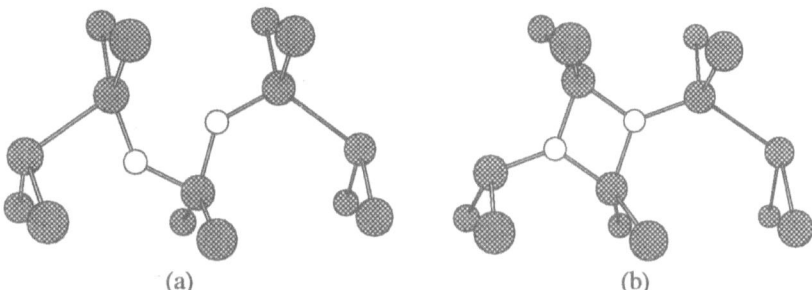

(a)   (b)

Figure 4. Diffusion of a di-oxygen complex: the O_{2i} (a) and the O_{2r} (b) configurations. The oxygens move from right to left between (a) and (b). The saddle point configuration is between (a) and (b) [13].

neighbor (Figure 4). Along the diffusion path of this complex the oxygen atoms will pass through trivalent configurations as it was shown above. It was predicted [8,13] that the correlated motion of the two oxygen will lead to a four-member ring structure, O_{2r}, also shown in Figure 4. While in case of a single oxygen interstitial, the possibility of Bourgoin-Corbett diffusion is an open question, it is hardly questionable for the dioxygen complex. In the O_{2r} configuration there are two excess electrons which cannot be accommodated in bonds or lone pairs of the trivalent oxygens, and there are no dangling bonds around. Consequently, those electrons will occupy EMT type orbitals (weakly localized shallow orbitals formed from the bottom states of the CB, due to the electric and stress field of the complex). Unless n-type doping pushes the Fermi level above them, these orbitals are necessarily ionized. From the point of view of calculations, good results for the barrier can only be achieved if these EMT orbitals are well reproduced. The Bourgoin-Corbett nature of the diffusion is confirmed by the fact that the calculated diffusion barrier — which is significantly lower than the one obtained for isolated oxygen — lies between O_{2i} and O_{2r} [13].

The existence of the O_{2r} complex seems likely but at present no direct experimental evidence supports it. The same is true for a number of other oxygen complexes which have been suggested to contain trivalent oxygen [21,23,24]. There are, however, defects where the case of oxygen trivalency can be better defended.

4. Trivalent Oxygen in Defect Complexes

4.1. OXYGEN - SELF-INTERSTITIAL COMPLEXES AND THE THERMAL DOUBLE DONORS

A third oxygen bond can also be formed in complexes with other impurities. An obvious candidate is the complex of an interstitial oxygen, O_i, and a silicon self-interstitial (I). The isolated self-interstitial is a negative-U defect: in the neutral charge state it is in a puckered, asymmetric bond center site, in the doubly positive charge state it is at the tetrahedral interstitial site, and it does not exist in the singly positive charge state [8,25,26].

Encounter of a double-positive I^{2+} and an O_i may trigger the trivalency in the latter, and the I^{2+} will be captured. This is depicted in Figure 5 (a), where the electron configuration is also marked: two electrons of the oxygen participate in normal Si-O bonds, two are in a lone pair orbital, and two are forming the third bond to I^{2+}, whose two electrons occupy its (non-bonding) 3s orbital. This $(IO)^{2+}$ defect induces EMT levels at the bottom of the conduction band, into which two conduction electrons can be trapped, thereby neutralizing the system [8]. In the neutral charge state, the puckered interstitial I^0 and O_i may form a complex to the analogy of (a) or (b) in Figure 4. Semi-empirical MINDO/3 calculations [8] predict that the latter is realized. The electron configuration is shown in Figure 5(b): the excess electron of the trivalent oxygen is accommodated by an sp^3-like lone pair orbital on the self-interstitial. The calculations predicted the corresponding level close to the VB edge. Seemingly, the complexes in Figure 5 (a) and (b) can be transformed into each other by a simple interstitialcy motion (I switching role with the nearest lattice Si). According to the calculations, IO^{2+} is stable in the electrically active (a) configuration, and IO^0 in the inactive (b) configuration. The singly positive charge state is unstable, so this is also a negative-U defect.

The bistability with negative-U, the nature of the calculated double donor EMT orbital in the (a) configuration resembles the behavior of the family of oxygen thermal double donors (TDD), discovered long ago [27]. It has been postulated [8] that this system is a precursor (TDD1) to the core of higher TDDs. Indeed, another oxygen can be added to

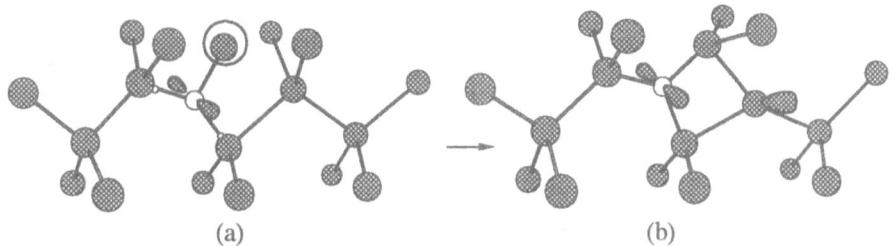

Figure 5. The bistable IO defect: (a) doubly positive, (b) neutral charge state.

these configurations (or a self-interstitial can be captured by dioxygen complexes), as shown in Figure 6. The C_{2v} complex in Figure 6(a) is the most stable defect involving a self-interstitial and two oxygen atoms. The oxygens are trivalent: both form two normal Si-O bonds with two of their electrons, two electrons of each are in lone pair orbitals and another two of each oxygen form bonds with I^{2+}. The latter has its two electrons in an sp^2-like non bonding hybrid pointed in the [001] direction. The energy level of the perpendicular p-orbital is above the CB edge. The electric and stress field of the complex again induces EMT levels at the bottom of the conduction band, into which two conduction electrons can be trapped. This configuration and the calculated spin distribution fitted the ENDOR data, therefore, the IO_2 looked as an excellent candidate for the core of higher TDDs, and was assigned to TDD2 [8].

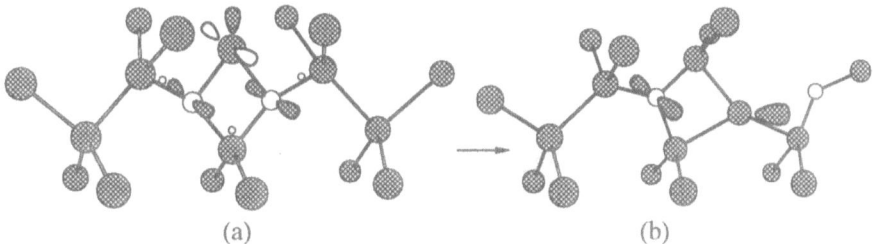

Figure 6. The double positive (a) and neutral (b) configuration of the IO_2 complex.

Apart from fitting existing data about TDDs, this model allowed two important predictions [8]. The first one was related to the hydrogen passivation of TDDs. The binding energy of hydrogen to the self-interstitial in the electrically inactive (b) configurations in Figure 5 and 6, were predicted to be higher than in the electrically active (a) configurations. That means that the passivation occurs by stabilizing the (b) configurations. Due to the reconstruction involved, a substantial change (8-10 k_B) in the configurational entropy, and correspondingly high attempt frequencies in the reactivation rate are expected. Since in this model higher TDDs were assumed to have a core identical with TDD2, the reactivation energy was expected to be similar for all of them but TDD1. These predictions have been confirmed experimentally later on [28,29].

While talking about hydrogen "passivation", it should be emphasized that bonding a hydrogen to the self-interstitial in the (b) configurations of IO and IO_2 may remove the TDD related electronic transitions, but does not passivate the complex completely in the electrical sense. Recounting the electrons in the (IO)H and (IO_2)H complexes leads to the conclusion that these systems are EMT-type single donors. Namely, with the dangling hybrid of the self-interstitial saturated by the bond to hydrogen, the excess electron of the trivalent oxygen cannot be accommodated locally. Recent calculations [30] confirmed that the structure in Figure 7 is the energetically most stable configuration for hydrogen in the vicinity of IO or IO_2 and it gives rise to a shallow single donor level. The symmetry of the system and the direction of the IH bond supports the assignment of the low symmetry NL10 components to the shallow thermal donors arising from hydrogen "passivation" of the TDDs [30,31].

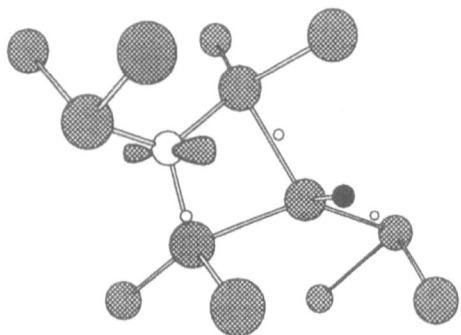

Figure 7. The shallow EMT donor arising from passivation of the TDD core.

The other prediction of the I-O complex model for the core of TDD is concerned with the localized vibrational mode of oxygen in the TDDs, which has never been observed before. As outlined in Section 1, the characteristic vibrations of a trivalent oxygen are around 900 cm^{-1}. The MINDO/3 calculations predicted (after approximate scaling) a pair of frequencies: 873 and 622 cm^{-1} for TDD1, 966 and 868 cm^{-1} for TDD2, in their double positive charge state [8]. Recently, pairs of frequencies in the 975 to 1006 cm^{-1} and in the 724 to 744 cm^{-1} range, respectively, have been shown to correlate with TDDs [32,33]. Taking into account the known problems in calculating frequencies and the fact that divalent oxygen always produces stretching frequencies between 1050 and 1280 cm^{-1}, this finding is a very strong support for the involvement of trivalent oxygen in TDDs. Regarding the uncertainties in the results of semi-empirical calculations, we have carried out cluster calculations on the $(IO_2)^{2+}$ complex using LDA. As can be seen from Table 2., LDA underestimates the tendency for oxygen trivalency. Still, the structure shown in Figure 6 (a) was found to be stable. Tests of basis set effects and calculation of the vibration frequencies are under way .

4.2. OXYGEN - NITROGEN COMPLEXES AND THE SHALLOW THERMAL DONORS

The hydrogen passivated TDDs constitute [34,31] one of the families termed shallow thermal donors (STD) [35,36]. However, some authors believe that nitrogen is involved with another family of STDs [36-42]. The low symmetry NL10 signal was attributed to the former [30,31], the C_{2v} center to these latter ones [41]. Nitrogen, inherently trivalent, is a very good candidate to replace one of the oxygens in Figure 4(b) or the self-interstitial in Figure 5(b). In fact, Figure 4 (b) could depict just as well the structure of the NN pair in silicon [43]. In a recent work [44] we have investigated the stability and properties of complexes based on an (NO) ring, similar in structure to O_{2r} or the NN pair.

Using the semi-empirical PM3 method, it has been found that the (NO) ring is stable with respect to isolated nitrogen and oxygen interstitials. The binding energy of the (NO) complex is about 10 % smaller than the one calculated for the (NN) ring, but considerably bigger than that of di-oxygen complexes. It was also found that both (NN) and (NO) rings may capture further interstitial oxygens in the (110) plane. These findings can explain, why the presence of nitrogen diminishes the concentration of TDDs. On the other hand, based on the calculated stabilities, the concentration of (NO) is more than one order of magnitude smaller than that of (NN), after a heat treatment at 450 °C. Generally, in case of high nitrogen concentrations, the generation of (NN) pairs are more likely. If created, however, (NO) rings may survive further annealing, because the encounter with an (NN) pair leaves them intact. In the (NO) ring structure, the bonds and the lone pair orbital of nitrogen is saturated by its five electrons, so the excess electron of the trivalent oxygen will occupy an EMT orbital. Therefore, (NO) may serve as core for STDs.

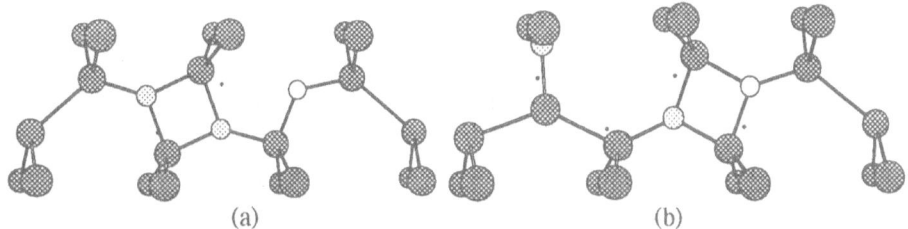

(a)   (b)

Figure 8. Equilibrium structures of the neutral (a) and singly positive (b) charge state of an NNO complex

The triatomic complex of an oxygen and two nitrogen atoms was found to be bistable. In the neutral charge state it has an (NN)O, in the positive charge state an N(NO) configuration (Figure 8). This also means that (NO) rings will not be created upon the encounter of an (NN) ring and an oxygen interstitial, both being electrically inactive and so inherently neutral. On the other hand, since (NO) will generally be positively charged, the encounter with a single nitrogen may preserve the (NO) ring. (Alas, in the N(NO) structure the electron donated by oxygen could be accommodated in the dangling bond of the trivalent silicon atom.)

The (NO) ring, attracting further oxygen atoms or (NN) pairs in the (110) plane, can form families of EMT donors. The stable systems in both cases are of C_{1s} symmetry. The donor wave function of these (NO) based systems is generally an asymmetric combination from the pair of states at the conduction band edge in the [001] direction. The spin distribution is very low on the N and O atoms of the core. One metastable (ONO) complex with C_{2v} symmetry has been found (Figure 9 (a)). This (ONO) double-ring might be expected to be stabilized by adjacent oxygens in the (110) plane. In this interesting structure, the central silicon is overcoordinated. The extra bond to the nitrogen accommodates the sixth electron of one of the oxygens (with energy in the valence band), so only the other is given up for an EMT orbital. Details of these results are given in a contributed paper to this proceedings [45].

The stability and EMT donor nature of the (NO) ring have been confirmed by LDA cluster calculations as well. In contrast to PM3, however, the (ONO) structure with C_{2v} symmetry is stable, but does not have trivalent oxygens (Figure 9 (b)) [46]. This may well be the consequence of the underestimated bond strength of the third Si-O bond by LDA (see Table 2). Still, the structure in Figure 9 (b) exhibits EMT donor properties, due to the strong electrostatic interaction of the oxygen lone pairs with the dangling bond of the central silicon which is eventually pushed above the CB edge.

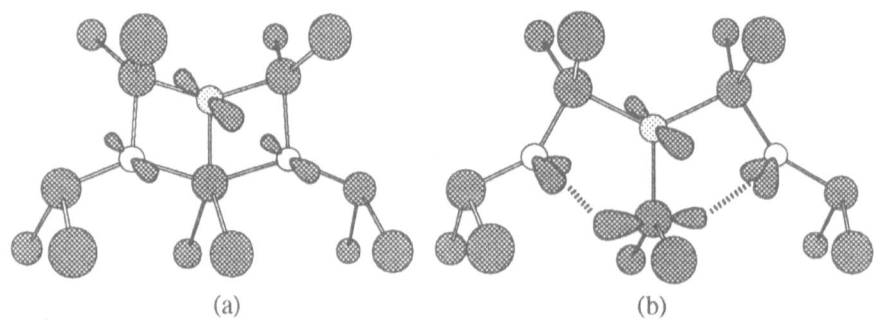

(a)   (b)

Figure 9. The structure of the (ONO) center as obtained from PM3 (a) and LDA (b) calculations

5. Conclusion and Outlook

Electron donating groups based on trivalent oxygen atoms are found in many chemical compounds, some of them containing Si-O bonds as well. Calculations on the positive trisiloxane ion show that such compound are possible with three Si-O bonds. Based on geometrical considerations it was shown that trivalency of oxygen can even occur during the diffusion of interstitial oxygen or dioxygen complexes in the silicon crystal. Since trivalent oxygen is inherently an electron donor, the possibility of Bourgoin-Corbett diffusion has to be invoked. Semi-empirical calculations predict stable oxygen - self-interstitial and oxygen - nitrogen complexes with EMT donor properties. These complexes can explain most of the properties of thermal double and shallow thermal donors. The prediction of the semi-empirical calculations have been partially verified by new experimental results and more accurate LDA calculations. Further work in that direction is in progress. In course of that, the stability of some of the complexes given here might not be confirmed, but I believe the examples shown convince the reader about the possibility and significance of oxygen trivalency in semiconductors. To emphasize the fact that this phenomenon can be expected in other materials as well, I would like to make two last remarks regarding oxygen in germanium.

It has been found that — while oxygen is electrically inactive in a-Si:H — it is an effective donor in a-Ge:H [47]. Calculations [48] have shown that unlike in silicon, the YLID position (O_Y) is not a saddle point configuration in germanium but a metastable structure (the potential between two stable bond center sites has a camel-back shape with a small dip). The reason for that are the bigger space for the relatively smaller oxygen and the smaller band gap. The oxygen can move easily out of the metastable YLID position, but if hydrogen is present it can stabilize the oxygen there (Figure 10).

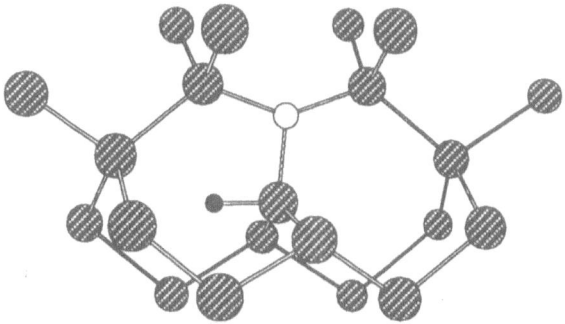

Figure 10. The OGeH donor complex in germanium.

The OGeH complex shown here is a single donor. The calculated oxygen vibration frequency, 640 cm^{-1}, is very close to the band observed in a-Ge:(H,O) systems, 670 cm^{-1}, which had been assigned to a Ge-O vibration at a Ge atom bonded to H [49].

The other remark concerns thermal double donors in Ge which have been found similar to the ones in silicon in many respect. For TDDs in Ge, however, the characteristic oxygen vibrations are known [50]. If one scales that frequency with the ratio of interstitial oxygen vibration modes in Si and Ge, one obtains the value of 1011 cm^{-1}, close to that predicted for IO_2 and found experimentally for TDDs in silicon. So, it is well possible that TDDs are also caused by trivalent oxygens in germanium.

Acknowledgment

This work has been inspired by the original thoughts of L. C. Snyder. The author thanks J. Weber and R. Jones for years of collaboration and fruitful discussions. The help of the graduate students C. P. Ewels, J. Miro, A. Gali , and K. Kadas, in the calculations is gratefully acknowledged. The support of the British-Hungarian Intergovernmental S&T Cooperation (Project Nr. 11) and the Hungarian OTKA (grant Nr. 2744) are appreciated.

The results presented here are dedicated to the memory of **J. W. Corbett**.

176

References

1. Bodor, E. (1971) *Inorganic Chemistry* (in Hungarian), Tankönyvkiadó, Budapest.
2. Takada, A., Catlow, C.R.A., Lin, J.S., Price, G.D., Lee, M.H., Milman, V. and Payne, M.C (1995) *Phys. Rev. B* **51**, 1447.
3. Thiel, W. (1988) *Tetrahedron* **44**, 7393.
4. Schmidt, M.V., Baldridge, K.K., Boatz, J.A., Elbert, S.T., Gordon, M.S., Jensen, J.H., Koseki, S., Matsunaga, N., Nguyen, K.A., Su, S.J., Windus, T.L., Dupuis, M. and Montgomery, J.A. (1993) *J. Comput . Chem.* **14**, 1347.
5. Jones, R. (1992) *Phil. Trans. R.Soc. Lond. A* **341**, 351.
6. Bingham, R.C., Dewar, M.J.S., and Lo, D.H. (1975) *J. Am. Chem. Soc.* **97**, 1285. Dewar, M.J.S., Lo, D.H., and Ramsden C.A. (1975) *J. Am. Chem. Soc.* **97**, 1311. Edwrads, A.H. and Fowler, W.B. (1985) *J. Phys. Chem. Solids* **46**, 841.
7. Stewart, J.J.P. (1989) *J. Comput. Chem.* **10**, 209.
8. Deák, P., Snyder, L.C., and Corbett, J.W. (1992) *Phys. Rev. B* **45**, 11612.
9. Stavola, M. and Snyder, L.C. (1983), in L.C. Kimerling and M.W. Bullis (eds.), *Defects in Silicon* Electrochemical Society, Pennington, NJ 1983) p. 61.
10. Bosomworth, D. R., MacDonald, R.S., and Watkins, G.D. (1964) *J. Phys. Chem. Solids* **25**, 873.
11. Benton, J.L., Kimerling, L.C., and Stavola M. (1983) *Physica B* **116**, 271.
12. Bourgoin, J. C. and Corbett, J. W. (1972) *Phys. Lett.* **38A**, 135.
13. Snyder, L.C., Corbett, J.W., Deák, P. and Wu, R.Z. (1988) *MRS Symp. Proc.* **104**, 179.
14. Kelly, P. J. (1989) *Mater. Sci. Forum* **38-41**, 269.
15. Saito, M. and Oshiyama, A. (1988) *Phys. Rev. B* **38**, 10711.
16. Estreicher, S. K. (1990) *Phys. Rev. B* **41**, 9886.
17. C. Ewels, private communication
18. Needels, M., Joannopoulos, J.D., Bar-Yam, Y., Pantelides, S.T., and Wolfe, R. H. (1991) *MRS Symp. Proc.* **209**, 103.
19. Gösele, U. and Tan, T.Y. (1982) *Appl. Phys. Lett.* **45**, 454.
20. McQuaid, S.A., Binns, M.J., Londos, C.A., Tucker, J.H., Brown, A.R., and Newman, R. C. (1995) *J. Appl. Phys.* **77**, 1427.
21. Chadi, D.J. (1990) *Phys. Rev. B* **41**, 10595.
22. Needels, M., Joannopoulos, J.D., Bar-Yam, Y., and Pantelides, S.T. (1991) *Phys. Rev. B* **43**, 4208.
23. Snyder, L.C., Deák, P., Wu, R.Z., and Corbett, J.W. (1989)*Mater. Sci. Forum* **38-41**, 329.
24. Jones, R. (1990)*Semicon. Sci. Technol.* **5**, 255-260.
25. Carr, R., Kelly, P.J., Oshiyama, A., and Pantelides, S.T. (1984) *Phys. Rev. Lett.* **52**, 1814.
26. Bar-Yam, Y., Joannopoulos, J. D. (1984) *Phys. Rev. B* **30**, 2216.
27. Fuller, C.S., Ditzenberger, J.A., Hannay, N.B., and Buehler, E. (1954) *Phys. Rev. B* **96**, 833 .
28. Bohne, D.I. and Weber, J. (1993) *Phys. Rev. B* **47**, 4037.
29. Bohne, D.I. and Weber, J. (1994) *Mater. Sci. Forum* **143-147**, 879.
30. Martynov, Yu. V. , Gregorkiewicz, T., Ammerlaan, C.A.J., Gali, A., and Deák, P. (1996) *Phys. Rev. B* submitted.
31. Martynov, Yu. V. , Gregorkiewicz, T., Ammerlaan, C.A.J. (1995) *Phys. Rev. Lett.* **74**, 2030.
32. Lindstöm, J.L. and Hallberg, T. (1994) *Phys.Rev.Lett.* **72**, 2729.
33. Hallberg, T. and Lindstöm, J.L. (1996) *Phys. Rev. B* . submitted.

34. McQuaid, S.A., Newman, R.C., and Lightowlers, E.C. (1994) *Semicond. Sci. Techn.* **9**, 1736 (1994).

35. Navarro, H., Griffin, J., Weber J., and Genzel, L. (1986), *Sol. State Commun.* **58**, 151.

36. Suezawa, M., Sumino, K., Harada, H. and Abe, T. (1986) *Jpn. J. Appl. Phys.* **25**, L859.

37. M. Suezawa, K. Sumino, H. Harada and T. Abe (1988), *Jpn. J. Appl. Phys.* **27**, 62.

38. Griffin, J.A., Hartung, J., Weber, J., Navarro, H., and Genzel, L. (1989) *Appl. Phys. A* **48**, 41.

39. Steele, A.G., Lenchyshyn, L.C., and Thewalt, M.L.W., (1990) *Appl. Phys. Lett.* **56**, 148.

40. Hara, A., Fukuda, T., Miyabo, T., and Hirai, I. (1989) *Jpn. J. Appl. Phys.* **28**, 142.

41. Hara, A., Masaki, A., Koizuka, M., and Fukuda,T. (1994) *J. Appl. Phys.* **75**, 2929.

42. Chen, C.S., Li, C.F., Ye, H.J. , Shen, S.C., and Yang, D.R. (1994) *J. Appl. Phys.* **76**, 3347.

43. Jones,R., Öberg, S., Berg Rasmussen, F., and Bech Nielsen, B. (1994) *Phys. Rev. Lett.* **72**, 1882.

44. Gali, A., Miro, J. Deák, P., Ewels, C.P., and Jones, R. (1996) *Phys. Rev. B* submitted.

45. Gali, A., Miro, J. Deák, P., Eweles, C.P., and Jones, R. (1996) in this *Proceedings*.

46. Ewels, C.P., Öberg, S., Jones, R., Miro, J., and Deák, P. (1996) to be published.

47. Schröder, B., Annen, A., Drüsedau, T, and Deák, P. (1993) *Appl. Phys. Lett.* **62**, 1961.

48. Deák, P., Schröder, B., Annen, A., and Scholz, A. (1993) *Phys. Rev.B* **48** 1924.

49. Lucovsky, G., Chao, S.S., Yang, J., Tyler, J.E., Ross, R.C., and Czubatyj, W. (1985) *Phys. Rev. B* **31**, 2190.

50. Fukuoka, N., Saito, H., and Kambe, Y. (1991) *Jpn. J. Appl. Phys.* **30**, 784.

HYDROGEN – OXYGEN INTERACTIONS IN SILICON

STEFAN K. ESTREICHER
Physics Dept., Texas Tech University, Lubbock, TX 79409
YOUNG K. PARK
Physics Dept., University of California, Irvine, CA 92717
AND
PETER A. FEDDERS
Physics Dept., Washington University, St. Louis, MO 63130

Hydrogen plays many roles in crystalline Si, the best known of which are probably the shallow-dopant passivation reactions. But hydrogen also activates electrically inactive impurities, passivates many deep-level defects and some deep-level impurities, and creates its own electrically and optically active centers. Hydrogen also interacts with the A-center, and passivates O-related thermal donors (TDs) at low temperatures. However, the most exotic of all the reactions involving hydrogen is its ability to enhances the formation rate of O-related TDs in the $300 - 450\,°C$ range. Under some conditions, H becomes a component of TDs. None of the microscopic interactions involving H and O are understood. However, various pieces of the puzzle are becoming available from experiment as well as theory. This paper summarizes the experimental and theoretical work on H–O interactions, with emphasis on the issue of H-enhanced diffusion of interstitial O. Preliminary results of molecular dynamics simulations are also presented.

R. Jones (ed.), Early Stages of Oxygen Precipitation in Silicon, 179–195.
© *1996 Kluwer Academic Publishers.*

1. HYDROGEN IN SILICON

1.1. INTERSTITIAL HYDROGEN

Hydrogen is an unavoidable impurity in Si and other semiconductors.[1, 2] It may be introduced in an uncontrolled manner during the growth or processing of the material. Examples includes hydrogen ambients, source gases containing H, wet and dry etching, H in chemicals used to grow oxides or nitrides, etc... Hydrogen can also be voluntarily introduced by a variety of means, such as exposure to a H plasma, proton implantation, or simply boiling in water. Deuterium can be substituted for H to guarantee that one is monitoring the effects of the voluntarily introduced species, to increases the sensitivity of secondary ion mass spectrometry (SIMS), and to check local vibrational modes (LVMs) for isotope shifts.

In addition to any surface reaction, H diffuses through the bulk of the semiconductor. It seeks out and binds to stretched or dangling bonds at impurities, localized defects, and complexes. This modifies the local geometrical configuration and alters the energy eigenvalues associated with the defect, leading to profound changes in the electrical and optical properties of the material. The electrical changes may be *passivation* (a level moves from the gap to a band), *activation* (a level moves from a band to the gap), or *'level shifting'* (a shallow level becomes deep or vice-versa). These changes affect both the concentration and lifetime of charge carriers. The optical changes are new infra-red (IR) or Raman active LVMs as well as new photoluminescence (PL) lines.

These reactions are reversible. Annealing at a few hundred degrees Celsius is enough to break many imperfect bonds involving H and restore the sample to its pre-hydrogenation condition. In some cases, the injection of minority charge carriers greatly enhances the reactions.[3] The importance of the latter issue may grow as industry relies more on rapid thermal anneals, a process which involves the creation of lots of electron-hole pairs. High-temperature anneals are required to expel H from the crystal.

Interstitial H exists in a variety of states depending on the material, the Fermi level, the temperature, and the impurity concentration. In silicon, neutral H is found in two states. Bond-centered hydrogen (H_{BC}^0) bridges a Si–Si bond. The odd electron is localized in a non-bonding orbital on the two nearest neighbors (NNs) of H. A second center, isotropic and atomic-like, is labeled H_T^0. It is known to be metastable in diamond[4] and silicon,[5] but it may be the stable state in germanium.[6]

The BC and T species have a gap level in Si. The donor level, associated with H_{BC}, is near $E_c - 0.18\,eV$ from deep-level transient spectroscopy (DLTS)[7] and muon spin rotation (μSR)[8] measurements. H_{BC}^+ is the stable species above $\sim 150\,K$ and in p-type material. The acceptor level, associated with H_T, has been predicted to be midgap,[9] but this prediction conflicts with the μSR measurement[8] of the ionization energy $H_T^- \rightarrow H_T^0 + e^-$, which is $0.66\,eV$.[10] The acceptor level is therefore $E_c - 0.66 + \delta E$, where δE is the energy difference between the metastable (H_T^0) and the stable (H_{BC}^0) states (an additional correction arises from the difference in zero point energies of a muon and proton). The experimental value of δE is not known, but theory predicts it to be a few tenths of an eV.[1] This would place the acceptor level quite close to the donor level, and H may or may not have negative-U properties in Si.[11] Recent experiments involving periodic exposure to zero-bias during the hydrogenation of biased Schottky barrier structures do not support the presence of H^- in n-type Si.[12]

However, μSR data[8] in Si unambiguously show that above $400\,K$ and on the microsecond time scale, three states coexist in p-type material (Mu_{BC}^+, Mu_{BC}^0, and Mu_T^0) while four states coexist in n-type material (Mu_{BC}^+, Mu_{BC}^0, Mu_T^0, and Mu_T^-). These experiments do not probe the equilibrium situation. However, they show that up to four states of interstitial H may coexist in non-equilibrium situations, such as thermal anneals, plasma exposure, etc. Each state of H has a different diffusion path and activation energy, making it difficult to talk about *the* diffusivity of H.

1.2. HYDROGEN INTERACTIONS WITH DEFECTS AND IMPURITIES

The most stable complexes formed by H result from the interaction with *dangling bonds*. This can occur at surfaces, vacancies, vacancy aggregates, grain boundaries, dislocations, etc. The H-host covalent atom bonds are generally strong, and the thermal stability of such complexes is high. This role of H has been known for over 30 years.

Hydrogen passivates *shallow acceptors* with a great efficiency. Since H^+ is generally abundant in p-type material and diffuses readily at room temperature, the Coulomb attraction by an ionized acceptor results in a huge capture radius. However, the thermal stability of H-acceptor pairs is low (about $150\,^\circ C$ in Si).

Hydrogen sometimes passivates *shallow donors*, but much less efficiently. This could be due to one or both of the following reasons. *(i)* H^- is a large, slowly diffusing ion, making it difficult for it to reach the donor. *(ii)* H^- is less abundant than H^0 in n-type material. Then, the capture radius is not Coulombic, and the efficiency of the passivation reaction is low. Note also that the thermal stability of H-donor pairs is only around room temperature in Si under band-gap light. Thus, donor passivation may have to be achieved at low temperatures, where H diffuses slowly, or in the dark.

Hydrogen interacts with *deep-level impurities*. It reacts with some transition metals but does not appear to affect others. The deep levels of (substitutional) Au in Si are passivated, but those of (interstitial) Ti are not, while substitutional Pt has more gap levels after hydrogenation than before.

H–H interactions may lead to the formation of *dimers* and larger complexes. The H_2 molecule is predicted to be localized near a T site, but is IR inactive and has never been detected. However, the 'bond-centered-antibonding' pair (H_2^*) has been observed by FTIR.[13]

H interacts with O and with defects containing O. An overview is given in Sec. 2. The interactions between interstitial O and atomic H, the main focus of this paper, are discussed in Sec. 3.

2. OVERVIEW OF HYDROGEN-OXYGEN INTERACTIONS

Interstitial O (O_i) was recognized very early as a very important impurity in Si, in particular in Czochralski (CZ) material. Oxygen comes from the quartz crucible containing the melt, and O_i is found in concentrations of the order of 10^{18} cm^{-3}. The concentration of O_i in typical float-zone (FZ) Si is two orders of magnitude smaller. Many properties of O_i, including the IR active vibrational modes, have been measured and calculated by a variety of groups. Upon annealing at moderate temperatures ($\sim 450\,°C$ or so), a series of double donors called thermal donors (TDs) is generated, and their appearance coincides with the disappearance of O_i. The reader is referred to a recent extensive review[14] of the topic and to the discussion in Ref. [15] for details and references. The key points relevant to the present paper are as follows.

O_i bridges a Si–Si bond in a puckered BC configuration. Its measured diffusivity is $D_{O_i} = 0.13\,e^{-2.53\,eV/kT}$. The diffusivity of O_i and the formation rate of TDs are moderately affected by a variety of factors, but atomic H has a huge impact.[16] There is increasing experimental evidence[16, 17] that oxygen *dimers* $(O_i)_2$ diffuse with a substantially lower activation energy than isolated O_i. The dimer could be some arrangement of adjacent O_i interstitials, stabilized by producing less lattice strain than the same interstitials far apart (see e.g., Ref. [18]). There is theoretical support for the $(O_i)_2$ pair diffusing much faster than isolated O_i.[18] Note that the idea of oxygen *pairs* diffusing faster than isolated interstitials was first proposed by Gösele and Tan,[19] who proposed a molecular state. Theory[20] shows that an O_2 molecule inserted into a Si crystal dissociates into two O_i's with a large gain in energy.

The (rapidly-diffusing) vacancy in Si seeks O_i, resulting in the formation of the vacancy-oxygen pair or 'A-center'.[14] One or two H interstitials interact with this defect, although it is not clear if a single H will bind to O or saturate a Si dangling bond.[21, 22]

Hydrogen passivates the electrical activity of the TDs.[15, 23, 24] The passivation mechanisms are speculative, and probably involve some geometrical relaxation of the complex following the formation of an H–O, H–Si, or H-I$_{Si}$ bond, where I$_{Si}$ is a self-interstitial. The presence of H within the TDs has been reported from ENDOR,[25] electrical and IR absorption measurements,[26] and photoluminescence.[27] A detailed understanding of these phenomena is lacking, mostly because there is only speculation as to the nature and structure of the TDs themselves. Since H is well-known to be attracted to and interact with virtually any imperfect region of the Si crystal, it is not surprising that it interacts with TDs as well.

What is clearly lacking at this point is a microscopic understanding of the early stages of O$_i$ aggregation, with or without the help of hydrogen.

3. HYDROGEN-ENHANCED DIFFUSION OF O

3.1. SUMMARY OF THE EXPERIMENTAL INFORMATION

The first (in 1957) published evidence that the presence of H facilitates the formation of O-related TDs is due to Fuller and Logan[28] who showed that annealing in water vapor substantially increases the formation rate and total concentration of TDs. They also noted that samples grown in a hydrogen vs. helium atmosphere differed in that subsequent $450\,^\circ C$ anneals produce almost one order of magnitude more TDs with H than with He.

In 1985, Qi *et al.*[29] observed that three IR active Si-H stretching vibrational modes (2191, 2123, and $1894\,cm^{-1}$ at $10\,K$, respectively) strongly correlate with the $1136\,cm^{-1}$ line of O$_i$ (correlation coefficient of 0.97). This suggests that these modes correspond to H atoms located near O$_i$, i.e. that O$_i$ *attracts* H interstitials which become trapped in its immediate vicinity.

Further microscopic evidence that hydrogen (or rather, muonium) behaves differently in FZ- and CZ-Si came from μSR and positron (from muon decay) channeling. A summary of the key results was published as a part of an extensive review article[30] on μSR (the positron blocking experiments were also discussed in Ref. [31]). The results can be summarized as follows. *(i)* The metastable species Mu$_T^0$ is abundant at low temperatures in both materials, but it diffuses much faster in CZ-Si than in FZ-Si. So fast in fact that the conventional μSR signature of Mu$_T^0$ in not seen (the muon looses its spin polarization too fast). However, the channeling data show the presence of "an oxygen-trapped muonium state at a tetrahedral interstitial site."[31] *(ii)* The diamagnetic (μ^+) signal is not observed in CZ-Si, while it is abundant in FZ-Si. This is true at low temperatures ($38\,K$) as well as at room temperature. In FZ material, the Mu$_{BC}^0$ species converts to Mu$_{BC}^+$ when the temperature exceeds $\sim 200\,K$. In CZ-Si, the results are consistent with a conversion to a rapidly diffusing diamagnetic state (no

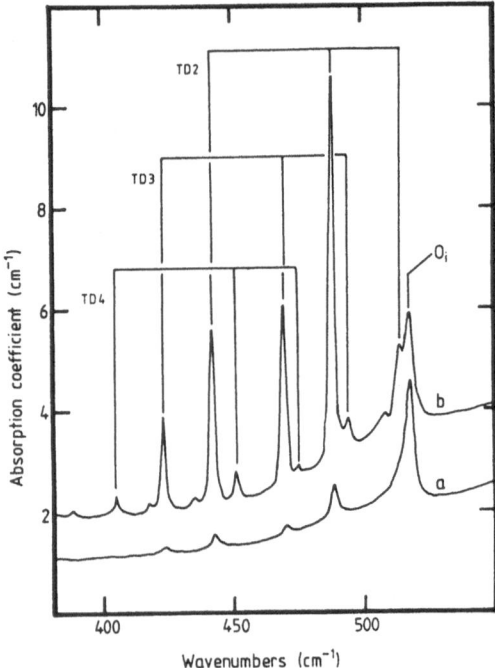

Figure 1. The absorption lines due to electronic transition of TDs for two samples heated at 400 °C for 2 hours (a) in an H_2 gas and (b) in a H plasma (from Ref. [32]).

signal is detected in CZ-Si by conventional μSR). *(iii)* The Mu_{BC}^0 signal is mostly unaffected by the concentration of O_i.

In 1988, Brown *et al.*[32] demonstrated dramatic differences in the formation rate of TDs in CZ-Si samples annealed in the range 350–450 °C in a H plasma vs. an H_2 gas (figure 1). The huge enhancement in the TD formation rate caused by the presence of H was confirmed independently by Stein and Hahn.[33] Qualitatively similar H-enhancements were reported to occur at much higher temperatures: H enhanced out-diffusion of O_i at 1000 – 1200 °C (Ref. [34]), and H aggregates related to enhanced O_i precipitation at 1270 °C (Ref. [35]).

Systematic studies by groups in the UK and the USA have revealed a number of puzzling features. The H-enhancement depends on how the hydrogenation is done (plasma vs. exposure to an H_2 gas), the hydrogenation temperature, and the cooling rate (for details, see the paper by R.C. Newman in the present proceedings and Ref. [16]). The H-enhanced diffusivity, $D_O^{enh.} = 2.3 \times 10^{-6}\, e^{-1.7\,eV/kT}$, is likely that of the *dimer*, $(O_i)_2$, discussed above.

Stein and Hahn[36, 37] hydrogenated their samples at low temperatures (275 K) in an ECR plasma for short and long times (up to 64 hours), then

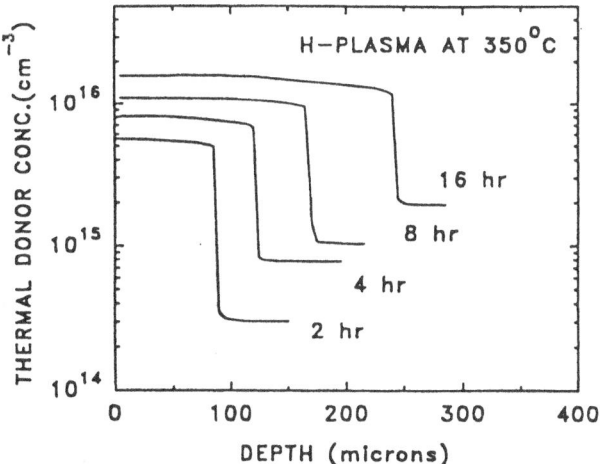

Figure 2. Depths profiles for TDs formed at times from 2 to 16 hours under hydrogen RF plasma exposure at 350 °C (from Ref. [37]).

annealed at temperatures ranging from 350 to 400 °C. Finally, they performed spreading resistance measurements (5 μm resolution) of the depth profiles for H-enhanced TD formation. One of their depth profiles is reproduced in figure 2.

A summary of their conclusions is as follows.[37] *(i)* The H-enhanced TD formation is a diffusion-limited process, as shown by a $\sqrt{t}$-dependence of the penetration depth. *(ii)* The $\sqrt{m}$-dependence when D is substituted for H proves that the diffusing species is hydrogen. *(iii)* The activation energy for the H-enhanced TD formation process is $\sim 1.5 \pm 0.2\,eV$, within error bars of the $1.7\,eV$ activation energy mentioned above. *(iv)* There is no easy explanation for the box-like shape of the depth profiles. Although the depth profile for any one exposure suggests saturation effects, the increase in TD concentration with exposure time rules out saturation as the cause of the box-like profiles. This shape may be explained if, under the conditions of the experiment, the diffusivities of H and O$_i$ become equal.[38]

3.2. STATIC MODELS

Two static $(0\,K)$ potential energy surface calculations have been performed to try to explain the large enhancement by H of the diffusivity of O$_i$.

The first study was performed by Estreicher[39] at the approximate ab-initio Hartree-Fock (HF), ab-initio HF, and post-HF levels, in molecular clusters. The following assumptions were made. *(i)* The hydrogen species responsible for the enhanced diffusion is the metastable H$_T^0$ state. The basis for this assumption are the μSR observations discussed above. *(ii)* H$_T^0$ is attracted to O_i. *(iii)* The diffusion of O_i from a puckered BC configuration to an adjacent one occurs in the $\{110\}$ plane.

The activation energy of O_i was calculated by moving O_i in steps from one equilibrium configuration to the adjacent one and, at every step, optimizing the coordinates of neighboring Si atoms, and of H when it was in the cluster. The goal was to compare the calculated activation energies of O_i, with and without H. The key results was that H lowers this activation energy by about a factor of two (from 2.5 to 1.3 eV) by forming a Si-H bond at the transition point configuration. Without H, this bond is considerably stretched. With H, the transition point configuration is stabilized, resulting in a substantially reduced activation energy for diffusion. In the final configuration, O_i had performed one jump, and H was in a near-BC configuration near O. The author did not speculate about what happens next.

The second study was performed by Jones *et al.*[40] They used the density functional (DF) approach, also in molecular clusters, with a Gaussian basis set of atomic-like orbitals. They obtained an activation energy of about 2.8 eV for the diffusion of isolated O_i. To obtain the effect of H, they also assumed that H is attracted to O_i, but optimized the configuration of H in the vicinity of O_i and found H to be at the AB site of one of the two Si atoms to which O_i is bound, opposite to O_i. As O_i was moved from one puckered BC configuration to the adjacent one, the Si-H bond strengthened. At the transition point, H was found to fully saturate the Si bond. This would be a dangling were H not present. The H-reduced activation energy for diffusion of O_i was 1.4 eV. As in the first study, no attempt was made to speculate about what happens next.

These two models have a number of very similar features. Both predict that the H-enhanced diffusivity of O_i is caused by H stabilizing the transition point configuration: A Si bond which would be a dangling bond when O_i diffuses in the absence of H, is tied up by H, and it is this binding energy which reduces the energy at the saddle point for diffusion. Further, both models start with the metastable state of H, namely H_T^0, rather than with H_{BC}^0. The only real difference between the two models is the orientation of this Si–H bond. Estreicher finds it near the BC site where O_i was prior to diffusion, Jones finds it at an AB site of the Si atom. Since the bending force constant of this H–Si bond is small, these two transition point configurations should be energetically very close. Indeed, both authors find a similar reduction by H of the activation energy of O_i, about a factor of two. In fact, almost any orientation of such a Si-H bond would likely result in a comparable reduction.

3.3. DYNAMIC STUDIES

All the static $(0\,K)$ calculations of the configuration(s) of H near O_i and of the potential energy surface for H-enhanced diffusion of O_i assume that hydrogen – which diffuses rapidly through the crystal at the temperatures of interest $(> 300°C)$ – not only is *attracted* to O_i but is *able to remain* near it long enough for at least one jump of O_i to take place. The processes leading to the 'capture' of H by O_i are not understood. Further, the nature of the H-assisted jump of O_i is not clear either. Finally, there is no information as to what happens to H after O_i jumped.

The problem at hand is a purely *dynamic* one, and the appropriate theoretical tool involves molecular dynamics (MD) simulations. The diffusion of H in Si has been successfully treated by MD simulations at various levels of theory.[41, 42, 43] The diffusion of vacancies and the formation of vacancy-hydrogen pairs have also been studied[44] at the MD level. Such simulations are possible because of the low activation energies for diffusion of H $(\sim 0.5\,eV)$ and of the vacancy (a few tenth of an eV, depending on its charge state[45]). These low values allow many diffusion jumps to be observed in a rather short computing time. Typical time steps for MD simulations involving H are of the order of $2 \times 10^{-16}\,s = 0.2\,fs$ (without H, five to ten times longer time steps may be used), and simulations involving 10^3 to 10^4 time steps are computationally tractable.

However, in the case of isolated O_i, the activation energy for diffusion is very high $(2.5eV)$, and observing even a single hop within a computationally reasonable time is a highly unlikely event. If H is present, the activation energy has been predicted to drop substantially, but does it drop enough to allow theoretical studies at the MD level (a few *ps*, may be a *ns* at most)? Although the simulation of unlikely events can in principle be addressed (see Ref. [46]), the problem is far from trivial. The key questions are the following.

(i) If interstitial H and O_i are far apart in the same supercell, will H spontaneously diffuse toward O_i?

(ii) If H is indeed attracted to O_i, will it remain trapped near it for a substantial amount of time, and if so, what is the equilibrium configuration?

(iii) How does O_i react to the presence of a nearby H?

(iv) How does H catalyze the diffusion of O_i?

(v) What happens after O_i makes one jump?

(vi) If H actually follows O_i as it hops, how does the resulting '$\{H, O\}$ pair' interact with O_i or another $\{H, O\}$ pair?

3.3.1. *Theoretical method*

We used the first-principles, pseudo-atomic orbital MD method developed by Sankey et al.[47] It is based on the DF theory and uses a basis set of pseudo-atomic orbitals. Its suitability to covalent systems such as Si and C is well documented.[48, 49, 50, 51]. The essential ingredients of the method are *(a)* a spin averaged (non self-consistent) version of DF theory using the Harris functional[52] and the local-density approximation; *(b)* non-local, norm-conserving pseudopotentials of the type developed by Hamann et al.[53] for the core states; *(c)* a minimal basis set with localized atomic-like wavefunctions for the valence orbitals (one s and 3 p functions per Si atom) with an orbital confinement radius for Si and H of 5.0 a_B. The exchange-correlation potential term assumes the Ceperley-Alder[54] form, as parameterized by Perdew and Zunger.[55] The classical equations of motion for the nuclei are integrated using Gear's algorithm[56] with time steps $\Delta t = 0.2 - 0.5\, fs$. The canonical ensemble was used and the temperature controlled by velocity scaling with the Drabold temperature controller.[49, 50]

Because of these approximations, our calculations are much faster than true ab-initio MD simulations of the Car-Parrinello type.[57] This allows us to use a large supercell and perform longer simulations than would be computationally tractable at the ab-initio level. A comparison of our results for vacancies and vacancy-hydrogen complexes to other calculations or experiment indicate that the loss of accuracy is not substantial.[44]

The host crystal is approximated by a 64-Si atom periodic supercell which contains interstitial H and O. The minimum distance between the center of the defects in adjacent supercells is about 11 Å. Since the orbital cutoff radius of the pseudo-atomic valence orbitals is 5.0 a_B, there is no direct overlap between defects in neighboring cells. Of course, there is indirect overlap caused by the host atoms displaced from their perfect crystal positions around the defect. The lattice constant of the perfect cell was optimized as discussed in Ref. [44] (optimized bond length Si−Si=2.381 Å).

3.3.2. *Preliminary results*

We have performed a number of simulations aimed at testing the nature of H-O interactions in the 64-Si atom supercell with the Fermi level at midgap. In order to test a variety of situations as rapidly as possible, we first picked a high temperature ($1600\,K$) and a long time step ($0.5\,fs$). Some key calculations have already been re-done at lower temperatures and with shorter time steps.

O_i was first placed near a centrally located BC site. The equilibrium configuration was obtained by the power quenching technique.[58] At 0 K, O_i forms a slightly puckered bond, but the angle increases with T, and O_i

vibrates easily in the plane perpendicular to the Si–Si bond it bridges. We first performed a series of simulations starting with H far away from O. Two initial configurations were tried: H_T and H_{BC}.

We should note here the following regarding the behavior of H^0 in a 64-atom supercell **which does not contain O_i**. When placed at a BC site, H diffuses from BC to BC site, without getting to the metastable state at (near) the T site. This result was also obtained by other authors[41, 42, 43] who performed MD simulations of H_{BC} in Si. The activation energy for BC-to-BC diffusion appears to be smaller than the energy that must be overcome to reach the metastable state. However, if one starts with the metastable state, H_T, it diffuses much faster than H_{BC} and along a T–H–T path, without reaching the BC site. The activation energy for T-to-T diffusion is lower than the energy needed for the T-to-BC jump. Thus, if started in the metastable state, H remains in that state, at least for the duration of our simulations. This behavior has not been mentioned before, and should be considered when discussing the wide range of diffusivities reported at low temperatures.[2]

If O_i and H are in the same supercell, H_{BC} first diffuses toward O_i, hopping from BC to BC site, but comes no closer to it than the third adjacent BC site. From there, it jumps away from O_i, then comes back, but does not come close enough to O_i to affect its diffusion. Since our simulations involved less than 2,000 time steps, it is possible that longer simulations could result in H_{BC} and O_i being at adjacent BC site, but we saw no evidence of this.

If we start with H_T far from O_i, hydrogen moves rapidly toward O_i. Its diffusion is faster than that of H_T in the same supercell, at the same temperature, and without O_i. However, when H_T reaches the region in the crystal strained by O_i, it encounters imperfect Si–Si bonds and self-traps at a BC site, second or third NN to that occupied by O_i (H–O=3.4 Å). This behavior is consistent with impurity- or defect-related strain lowering the T–BC barrier and allowing H to reach a BC site. Our simulations therefore suggest that several attempts ay be needed for H_T to reach a site very close to O_i, and/or that more than one H is involved. We intend to test this by including two or three H interstitials in the supercell to see what happens after the first (or the first few) self-trap near O_i.

Then, we tested the static models described in the previous subsection, and assumed that H reaches a site near O_i, with H in its metastable state. That is, we started a simulation with H either at the AB site of one of the Si atoms to which O_i is bound (Jones' model), or at a T site immediately adjacent to O_i (Estreicher's model). The former configuration is not a local minimum of the energy, or at least not deep enough to trap H at high temperatures. H_{AB} rapidly moves away from O: after 60 fs, H is near a

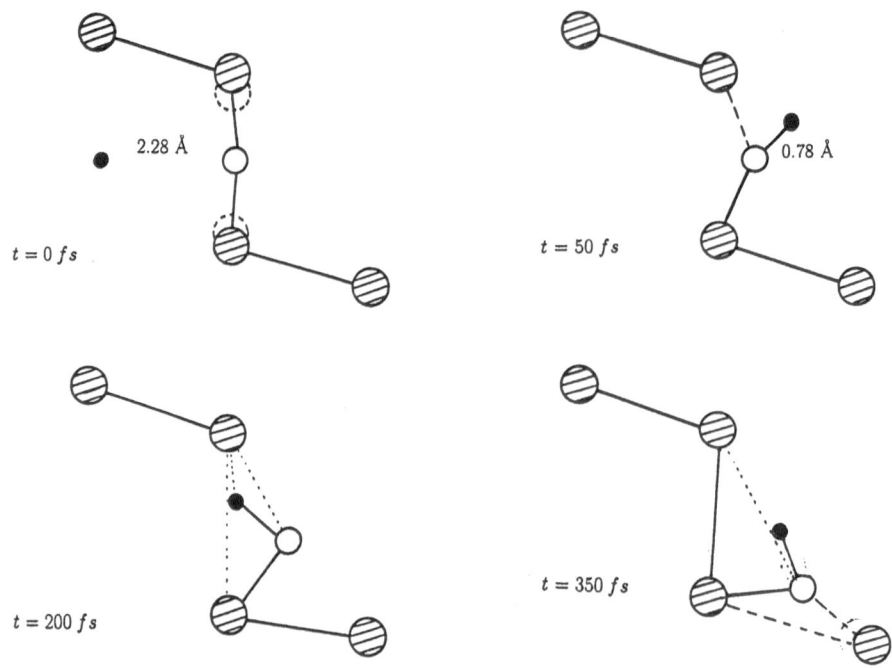

Figure 3. Schematic snapshots of the configurations calculated at the MD level at the initial time (0 *fs*) and after 50, 200, and 350 *fs*. The same behavior occurs at lower temperatures (1200 and 900 *K*) and with shorter time steps (0.2 *fs*). The shaded circles are Si atoms, the open circle is O, and the black circle is H.

T site, 4.8 Å away from O_i. However, the initial configuration with H at the T site nearest to O_i is very close to a local minimum of the potential energy. We quenched this initial configuration to make sure that it is not accidentally a high-energy state. It is not (actually, it is very close to a local minimum of the energy). Then, we raised the temperature to 1600 *K*, let the system evolve for 2,000 time steps, and something remarkable happened. We saw H-enhanced diffusion of O_i.

Figure 3 shows the situation at the times $t = 0, 50, 200$, and 350 *fs*. After 480 *fs*, O is at a site almost identical to that after 50 *fs*, but with O at the BC site adjacent to its initial one. The following is happening. First, H rapidly moves toward O_i, disrupts one of the two Si–O bonds, and attaches to O_i. Although these simulations do not provide a population analysis or bond indices, the average H–O bond length becomes and remains around 0.8 to 0.9 Å, a value very close to that in H_2O. Since O_i is now bound to one (fixed) Si atom and the mobile H, it can 'flip' fairly easily around that Si atom, while H remains attached to it. This perturbs the adjacent Si–Si bonds, and O has a good chance of hoping to an adjacent BC site.

We have already verified that this result is reproducible. MD simulations at three different temperatures (900, 1200, and 1600 K) and with two different time steps (0.5 and 0.2 fs) have been done. Every time, if we start with H at a T site neighboring O_i, the same happens. H binds to O very quickly (within 20 fs or so), and the $\{H, O\}$ pair begins to rotate almost freely around one host atom. After a few tenths of a picosecond, O is already some 4 Å (!) away from its original position. The game involves continuously strengthening or weakening the overlap between O, H, and one or two Si atoms.

The chemical reason for this can be guessed by comparing the various bond strengths involved: H–O is about 5.2 eV in H_2O, O–Si about 5.6 eV in HO-Si(CH$_3$)$_3$, and H–Si about 3.7 eV in H-Si$_2$H$_5$. Although the bond strengths in these molecules are not expected to be identical to those occurring in the present situation, they should be within may be 0.1 or 0.2 eV of the ones quoted. Thus, replacing one of the O–Si bonds by an H–O bond comes at a very low cost in energy ($\sim 0.4 eV$), but replaces a situation with O_i bound to **two fixed** atoms (the two Si NNs to O) by one with O_i bound to **one fixed and one mobile** atoms (one Si NN to O and the H atom). This allows O_i to pivot rather freely around the Si atom and probe the four adjacent BC sites.

Note that the static models discussed in a previous section both predict the formation of an H–Si bond, not of an H–O bond. In the former case, the two O–Si bonds are replaced by one H–Si bond and a O interstitial strongly bound to one Si atom and more weakly bound to one (or two) other Si atom(s). The activation energy is roughly given by the energy difference between the bond strengths (5.6 for O–Si and 3.7 eV for H–Si), i.e., about 1.9 eV. This is still high but provides a "H-enhancement" when compared to the 2.5 eV required for O alone to hop. In the latter case, MD simulations show the formation of an H–O bond, with energy much closer to that of O–Si (5.6 for O–Si, 5.2 eV for H–Si). As a result, O_i almost freely rotates at high temperatures, visiting several adjacent BC sites, and eventually hopping. This is illustrated in figure 4 which shows the motion of H and O for 2,000 time steps (the host atoms are not shown for clarity). The dotted line corresponds to the position of H, while the line that shows the motion of O appears continuous (since O moves slower, the 2,000 dots overlap). After H becomes trapped by O_i, the H–O internuclear distance stabilizes around 0.8 Å, and O hops to an adjacent BC site and begins to rotate rapidly around a host atom, with H buzzing around it.

At present, we think that the rate-limiting factor in the H-enhanced diffusion process is not the activation energy for diffusion of the $\{H, O\}$ pair, but how difficult it is for H to reach a site immediately adjacent to O_i

time $t = 0$

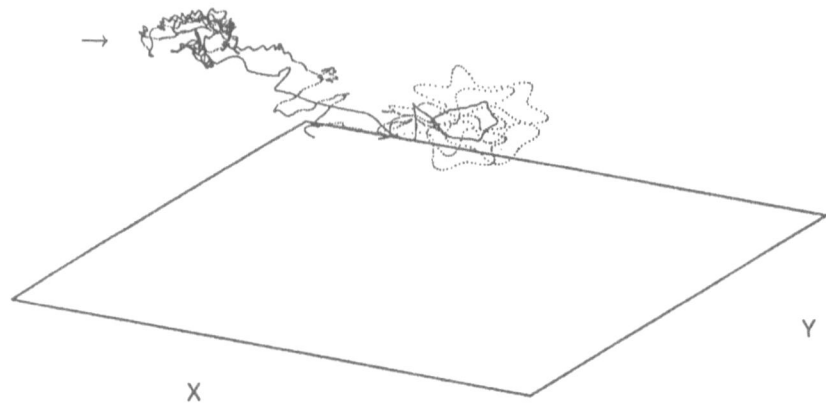

Figure 4. Trajectories of O_i (the successive dots overlap and its trajectory looks continuous) and H (dots) as a function of time for 2,000 time steps. The temperature is 1600 K and the time step 0.5 fs.

without self-trapping at some BC site away from it. We do not have this part worked out yet. It is possible that, as several H interstitials approach O_i, they occupy various BC sites in its vicinity, until one H finally reaches the desired initial configuration. Then, the {H,O} pair forms and becomes highly mobile. This picture is consistent with a number of previously unexplained observations.

(i) The interpretation of μSR data in CZ-Si suggest that Mu_T either traps at a BC site near O_i or rapidly diffuses with it. Further, the diamagnetic state is not seen by μSR, implying a rapidly diffusing state. This is very much the behavior predicted by the MD simulations.

(ii) The AA9 spectrum of H_{BC}^0 has been observed[59] in both FZ- and CZ-Si. However, the spectrum in CZ material appears[60] to be perturbed by O_i. This would be expected from our simulations.

(iii) The plots of the concentration of TDs vs. depth[37] in samples exposed to a plasma at 350 $°C$ for many hours could be understood if - under these conditions - the diffusivities of H and O_i become *equal.*[38] Again, this would be implied by our simulations which show that H and O move together.

We emphasize once again that we consider these results to be preliminary. Many more simulations are needed to explore all the possibilities. Further, we also want to examine the situation where several H interstitials approach the same O_i.

4. DISCUSSION

The theoretical results presented here provide a new picture for the inter-actions between H and O_i in Si. These are the first MD simulations for this problem. They reveal that the static models developed to explain H-enhanced diffusion of O_i are correct to assume that the metastable state of H plays the major role, but incorrect regarding the role of H. Indeed, MD simulations show that a H–O rather than a H–Si bond is involved.

While many details are still missing, our results suggest that the rate-limiting process in H-enhanced diffusion is getting H to the immediate vicinity of O_i in the desired state. Both H_{BC} and H_T may self-trap at a sec-ond or third NN BC site from O_i, in which case we saw no hint of enhanced diffusion of O_i. We speculate that either many attempts are required or that more than one H must be involved. More MD simulations are under way, and should provide answers to at least some of these questions.

Other important issues involve the interactions between a mobile $\{H,O\}$ pair and X, where X is O_i, another $\{H,O\}$ pair, a vacancy, a self-interstitial, etc. It is possible that a reaction such as $\{H,O\} + O_i \rightarrow (O_i)_2 + H$ occurs readily? This reaction would leave O-dimers which may or may not interact with H in a way that enhances their diffusivity. even if this reaction occurs, it is unlikely to be the only possible reaction. Mobile $\{H,O\}$ pairs could interact with all sorts of impurities and defects, including TDs. This may result in the incorporation of H into the TD itself, a situation consistent with experimental observations.

Clearly, a lot of work remains to be done. However, we have some con-fidence that we now understand the first step in a long series of reactions. We are also confident that we have one theoretical tool able to handle at least some of these issues. It is not unrealistic to hope that we will be able to simulate the formation of small O aggregates in Si under various (theoretically) controlled conditions.

Acknowledgements

The work of SKE is supported in part by the grant D-1126 from the R.A. Welch Foundation, and in part by the contract XAX-5-15230-01 from the National Renewable energy Laboratory. The work of PAF is supported by NSF grant DMR 93-05344.

194

References

1. Estreicher S.K. (1995), Mat. Sci. Engr. Reports **14**, 319.
2. Pearton S.J., Corbett J.W., and Stavola M. (1992), *'Hydrogen in Crystalline Semiconductors'*, (Springer-Verlag, New York).
3. Seager C.H. and Anderson R.A. (1990), Sol. St. Com. **76**, 285; Johnson N.M. and Herring C. (1992), Phy. Rev. B **45**, 11379.
4. Patterson B.D., Holzschuh E., Kündig W., Meier P.F., Odermatt W., Sellschop J.P.F., and Stemmet M.C. (1984), Hyp. Int. **17-19**, 605.
5. Westhauser E., Albert E., Hamma M., Recknagel E., Weidinger A., and Moser P. (1986), Hyp. Int. **32**, 589.
6. Estreicher S.K. and Maric D.M. (1993), Phys. Rev. Lett. **70**, 3963.
7. Holm B., Bonde Nielsen K., and Bech Nielsen B. (1991), Phys. Rev. Lett. **66**, 2360.
8. Kreitzman S.R., Hitti B., Lichti R.L., Estle T.L., and Chow K.H. (1995), Phys. Rev. B **51**, 13117.
9. Johnson N.M., Herring C., and Van de Walle C.G. (1994), Phys. Rev. Lett.**73**, 130.
10. Estle T.L. and Lichti R.L. (1995), private communication.
11. See the Comment on the above paper by Seager C.H., Anderson R.A., and Estreicher S.K. (1995), Phys. Rev. Lett. **74**, 4565.
12. Seager C.H. and Anderson R.A. (1996), submitted to Appl. Phys. Lett.
13. Holbech J.D., Bech Nielsen B., Jones R., Sitch P., and Öberg S. (1993), Phys. Rev. Lett. **71**, 875.
14. *Oxygen in Silicon*, ed. F. Shimura (1994), Semic. and Semimet. **42** (Academic, NY).
15. Deák P, Snyder L.C., and Corbett J.W. (1992), Phys. Rev. B **45**, 11612.
16. Newman R.C. and Jones R. (1994), in *Oxygen in Silicon*, Ref. [14].
17. McQuaid S.A., Binns M.J., Londos C.A., Tucker J.H., Brown A.R., and Newman R.C. (1995), J. Appl. Phys. **77**, 1427.
18. Corbett J.W., Deák P., Lindström J.L., Roth J.M., and Snyder L.C. (1989), Mat. Sci. Forum **38-42**, 579.
19. Gösele U. and Tan T.Y. (1982), Appl. Phys. A **28**, 79.
20. Roberson M.A., Estreicher S.K., and Chu C.H. (1993), J. Phys.: Condens. Matter **5**, 8943.
21. Artacho E. and Ynd. uráin F. (1989), Sol. St. Com. **72**, 393.
22. Gutsev G.L., Myakenkaya G.S., Frolov V.V., and Glazman V.B. (1989), Phys. Stat. Sol. (b) **153**, 659.
23. Johnson N.M. and Hahn S.K. (1986), Appl. Phys. Lett. **48**, 709.
24. Bohne D.I. and Weber J. (1993), Phys. Rev. B **47**, 4037 amd (1994), Mat. Sci. Forum **143-147**, 879.
25. Martynov Yu V., Gregorkiewicz T., and Ammerlaan C.A.J. (1995), Phys. Rev. Lett. **74**, 2030.
26. McQuaid S.A., Newman R.C., and Lightowlers E.C. (1994), Semic. Sci. Technol. **9**, 1736.
27. Lightowlers E.C., Newman R.C., and Tucker J.H. (1994), Semic. Sci. Technol. **9**, 1370.
28. Fuller C.S. and Logan R.A. (1957), J. Appl. Phys. **28**, 1427.
29. Qi M.W., Bai G.R., Shi T.S., and Xie L.M. (1985), Mater. Lett. **3**, 467.
30. Patterson B.D. (1988), Rev. Mod. Phys. **60**, 69 (see pages 110 and 131).
31. Patterson B.D., Bosshard A., Straumann U., Truöl P., Wüest A., and Wichert Th. (1984), Phys. Rev. Lett. **52**, 938.
32. Brown A.R., Claybourn M., Murray R., Nandhra P.S., Newman R.C., and Tucker J.H. (1988), Semic. Sci. Technol. **3**, 591.
33. Stein H.J. and Hahn S.K. (1990) Appl. Phys. Lett. **56**, 63.
34. Zhong L. and Shimura F. (1993), J. Appl. Phys. **73**, 707.
35. Hara A., Aoki M., Fukuda T., and Ohsawa A. (1993), J. Appl. Phys. **74**, 913.
36. Stein H.J. and Hahn S.K. (1994), Appl. Phys. Lett. **75**, 3477.

37. Stein H.J. and Hahn S.K. (1995), J. Electrochem. Soc. **142**, 1242.
38. Stein H.J. (1994), private communication.
39. Estreicher S.K. (1990), Phys. Rev. B **41**, 9886.
40. Jones R., Öberg S., and Umerski A. (1992), Mat. Sci. Forum **83-87**, 551. Jones R. (1995), Phil. Trans. R. Soc. Lond. A **350**, 189.
41. Buda F., Chiarotti G.L., Car R., and Parrinello M. (1989), Phys. Rev. Lett. **63**, 294.
42. Boucher D.E. and DeLeo G.G. (1994), Phys. Rev. B **50**, 5247.
43. Panzarini G. and Colombo L. (1994), Phys. Rev. Lett. **73**, 1636.
44. Park Y.K., Estreicher S.K., Myles C.W., and Fedders P.A. (1995), Phys. Rev. B **52**, 1718.
45. Watkins G.D. (1986), in *Deep Centers in Semiconductors*, ed. S.T. Pantelides (Gordon and Breach, New York), p. 147.
46. Bennett C.H., *Molecular Dynamics and Transition State Theory: the Simulation of Infrequent Events*, in "Algorithms for Chemical Computations", p. 63.
47. Sankey O.F. and Niklewski D.J. (1989), Phys. Rev. B **40**, 3979.
48. Sankey O.F., Niklewski D.J., Drabold D.A., and Dow J.D. (1990), Phys. Rev. B **41**, 12750.
49. Yang S.H., Drabold D.A., and Adams G.B. (1993), Phys. Rev. B **48**, 5261.
50. Drabold D.A., Fedders P.A., Klemm S., and Sankey O.F. (1991), Phys. Rev. Lett. **67**, 2179; Drabold D.A., Wang R., Klemm S., and Sankey O.F. (1991), Phys. Rev. B **43**, 5132.
51. Alfonso D.R., Yang S.H., and Drabold D.A. (1994), Phys. Rev. B **50**, 15369.
52. Harris F.J. (1978), Proc. IEEE **66**, 51.
53. Hamann D.R., Schlüter M., and Chiang C. (1979), Phys. Rev. Lett. **43**, 1494; Bachelet G.B., Hamann D.R., and Schlüter M. (1982), Phys. Rev. B **26**, 4199.
54. Ceperley D.M. and Adler B.J. (1980), Phys. Rev. Lett. **45**, 566.
55. Perdew S. and Zunger A. (1981), Phys. Rev. B **23**, 5048.
56. Gear C.W. (1966), ANL Report #7126, Argonne Nat. Lab.; C.W. Gear (1971), *Numerical Initial Value Problems in Ordinary Differential Equations* (Prentice-Hall, NJ), Chapt. 9.
57. Car R. and Parrinello M. (1985), Phys. Rev. Lett. **55**, 2471.
58. Park Y.K. and Myles C.W. (1995), Phys. Rev. B **51**, 1671.
59. Bech Nielsen B., Bonde Nielsen K., and Byberg J.P. (1994), Mater. Sci. Forum **143-17**, 909.
60. Bonde Nielsen K. and Bech Nielsen B. (1995), private communication.

OXYGEN DIFFUSION IN SILICON:
THE INFLUENCE OF HYDROGEN

M. RAMAMOORTHY AND S. T. PANTELIDES
Department of Physics and Astronomy, Vanderbilt University
Nashville, TN 37235

1. Introduction

Oxygen is incorporated at concentration levels of order 10^{18} cm^{-3} in single crystals of silicon grown using the Czochralski method. These silicon wafers are extensively used in device applications. During thermal processing, depending on the processing parameters, the oxygen atoms aggregate to form clusters of various sizes and also SiO$_2$ precipitates [1]. The precipitation of oxygen leads to the formation of extended defects in the silicon wafer, improving its mechanical properties and also "gettering" impurities from the the bulk. The rate-limiting step in oxygen aggregation is the diffusion of oxygen through the bulk of the material. Thus, there has been extensive investigation of the diffusion of oxygen in silicon, using a range of experimental techniques [2]. In this article, we will summarize recent theoretical results [3] which have allowed us to resolve many outstanding puzzles about oxygen diffusion and give a comprehensive account of the underlying atomistic processes. In addition, we report hitherto unpublished results on the effects of H on O diffusion.

2. Technique

The theoretical results in this article were obtained using density functional theory and the local-density approximation for exchange and correlation. The ultra-soft pseudopotentials of Vanderbilt [4] were used for Si, O and H. The calculations employed a plane wave basis set. Converged results were obtained with an energy cutoff of 25 Ry. A 32 site bcc supercell was used, with one k-point at $(0.5, 0.5, 0.5)$ in the irreducible Brillouin zone [5]. Each structure was relaxed until the force on each atom was less than 0.5 eV/Å.

R. Jones (ed.), Early Stages of Oxygen Precipitation in Silicon, 197–205.
© *1996 Kluwer Academic Publishers.*

3. The isolated oxygen impurity

From analysis of infrared absorption data, [6] it is well established that oxygen occupies an interstitial position between two neighbouring Si atoms, breaking the Si-Si bond, forming a puckered Si-O-Si unit. This is illustrated schematically in figure 1.

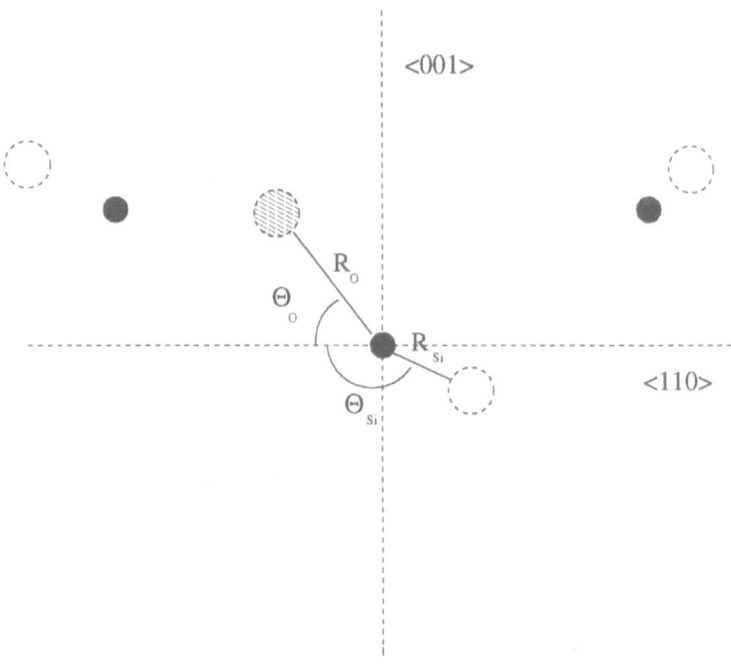

Figure 1. The equilibrium configuration of the isolated oxygen impurity, shown as a shaded circle. The relaxed positions of Si neighbours are shown using empty circles, while the filled black dots give their respective positions in the perfect solid.

The experimental data on the diffusion of oxygen in silicon are well described by an Arrhenius-like expression, over a wide range of temperatures (300–1200 °C):

$$D_{\mathrm{oxy}} = 0.13 \exp\left(-2.5\mathrm{eV}/k_B T\right) \ \mathrm{cm}^2 \, \mathrm{sec}^{-1}. \tag{1}$$

The diffusion has been presumed to occur, by early investigators, by means of simple jumps from one bond-centre site to the next, with the saddle-point for the process being at the midpoint of the jump. However, theoretical efforts to calculate the activation energy for this process, based on this assumption, have yielded a very wide range of values (1.2–4.1) eV [2]. Recently, Jiang and Brown (JB) [7] calculated the energy of the system along the entire migration path of the oxygen atom and found that the activation energy for the migration process was 2.5 eV, but that the saddle-point was well beyond the midpoint. Such a result violates the intrinsic symmetry of the system. Thus, there was still no satisfactory account of the experimentally determined activation energy.

We have recently completed a systematic study of the migration process of oxygen is silicon, using first-principles total-energy calculations and found that a single oxygen jump is an extremely complex process, involving two atoms (the oxygen atom and the silicon neighbour which is bonded to the oxygen atom in both the initial and final equilibrium configurations, called the central Si atom) [3]. The lowest energy pathway for the diffusion process is found to be along the (110) plane. In the following discussion, we describe systematic investigations of the energy of the system as a function of the positions of the oxygen atom and the central Si atom, using the polar coordinates (θ_O, θ_{Si}, R_O, R_{Si}), as shown in figure 1. The energy of the system was computed as a function of θ_O and θ_{Si}, with the values of R_O and R_{Si} being varied so as to obtain the local minimum in the energy.

In figure 2, the energy of the system is plotted as a function of θ_{Si}, with the O atom at the midpoint of its path, $i.e.$ ($\theta_O = 90°$). It is seen that this curve is symmetric about $\theta_{Si} = 90°$, as one might expect, with a maximum value of 2.2 eV. In the symmetric configuration, ($\theta_O = \theta_{Si} = 90°$), the O atom is three-fold coordinated and the central Si atom five-fold coordinated. There are two minima in the energy as a function of θ_{Si}, with $\theta_O = 90°$. In each minimum the O atom is twofold coordinated and the central Si atom fourfold coordinated. This is energetically favourable compared to the over-coordinated symmetric configuration. In fact, the minimum on the left (right) corresponds to the central Si atom being bonded to the Si atom on the left (right) in figure 1, having broken its bond with the other Si atom on the right (left). When the oxygen atom moves from the bond-centre position on the left to the midpoint of its path, where $\theta_O = 90°$, the central Si atom is in the minimum on the right, as shown in figure 2. The extra energy required to make this silicon atom go from the minimum on the right to that on the left, with oxygen at $\theta_O = 90°$, is about 1.5 eV. This makes it clear that during the oxygen migration process both the O atom and the central Si atom face large barriers to their motion. The energy of the symmetric configuration is 2.2 eV, which is close to the experimental activation energy. One might assume that this symmetric configuration is the saddle-point of the oxygen migration process.

However, the situation is even more complex. There exist many pathways by which the O atom and the central Si atoms can cross their respective barriers and complete the migration process. As we can see from figure 2, when the oxygen atom is at the midpoint of its path, the Si atom is well to the right of the central symmetry axis. When the oxygen atom moves past its midpoint to the right of the symmetry plane, the central silicon atom is still trapped in a local minimum which evolves from the minimum shown in figure 2, but which is separated from the global minimum, for that position of the O atom, by a considerable barrier. No matter at what point during the oxygen motion the central Si atom clears its barrier, the total energy of the system rises to a value of at least 2.2–2.5 eV.

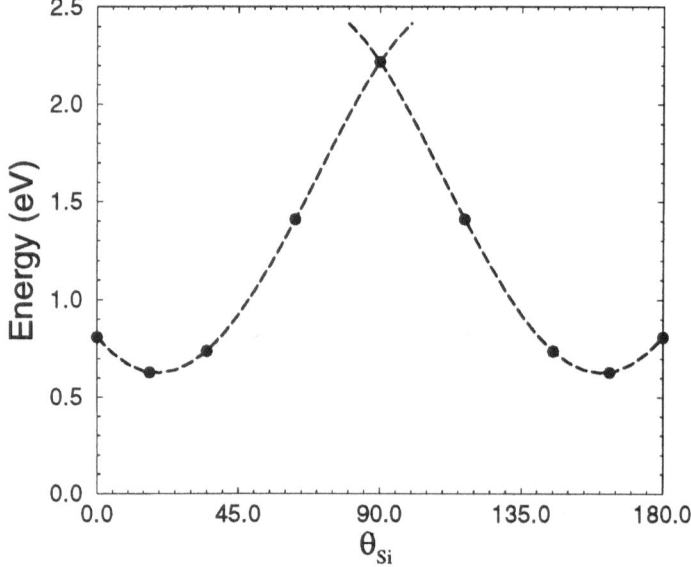

Figure 2. The total-energy variation of the system, as a function of θ_{Si} when the O atom is at $\theta_O = 90°$. The zero of the energy in this plot is taken as that of the equilibrium configuration.

A careful investigation of the energy as a function of the position of the O atom and the central Si atom in the (110) plane reveals that one needs all four coordinates (θ_O, θ_{Si}, R_O, R_{Si}) to give a proper description of the dynamics of oxygen migration in Si. A detailed description of one-dimensional and two-dimensional slices of this hypersurface has been published [3]. These reveal a "saddle ridge", of average height 2.5 eV. The migration process occurs along a multiplicity of paths over this ridge which has a height of ~ 2.5 eV over a considerable range. Finally, we find that the results of earlier authors correspond to different points or lines on the hypersurface.

4. The oxygen-hydrogen complex

A variety of experiments [2] have indicated that oxygen diffusion in silicon is considerably enhanced when a significant amount of hydrogen is present in the sample. The activation energy extracted in these experiments is in the range of 1.5–2.0 eV, which is considerably lower than the value of 2.5 eV for the isolated oxygen impurity. This enhanced diffusion was attributed to the formation of some kind of oxygen-hydrogen complex. Theoretical investigations to determine the nature of the oxygen-hydrogen complex have been carried out by Estreicher [8], and Jones *et al.* [9], both of whom used cluster calculations. While both describe migration pathways for the oxygen-hydrogen complex with reduced activation energies as compared to the isolated impurity, they disagree with each other in their identification of the equilibrium structure of the complex and consequently in the details of the migration pathway. Estreicher found the equilibrium structure of the oxygen-hydrogen complex to be such that the two impurity atoms occupy adjacent bond-centre positions. On the other hand, Jones *et al.* found that the H atom prefers an anti-bonding site adjacent to the Si-O-Si unit. Both authors restricted their calculations to the neutral complexes. This is a severe restriction, as theoretical investigations of the isolated H impurity in Si [10] have shown that the neutral state of the impurity is not stable for any position of the Fermi level with H^+ and H^- being the most stable charge states for p-type and n-type Si, respectively.

We have examined a number of possible configurations of the oxygen-hydrogen complex in the positive, neutral and negative charge states. We have found three stable configurations of the oxygen-hydrogen complex which are illustrated in figures 3, 4 and 5, which are similar to those considered by earlier investigators. In figure 3, the O atom and the H atom occupy neighbouring bond-centre sites, and we denote this configuration by BC. In figure 4, the H atom occupies an antibonding position relative to an Si atom bonded to an O atom, which we denote by AB. In figure 5, the H atom is directly bonded to the O atom, which we denote by OH. We find that the stability of each configuration and the nature of the equilibrium configuration is strongly dependent on the charge state. Our results for the energetics of these complexes are given in the following paragraphs. We estimate the thermodynamic stability of each complex in a given charge state with respect to the equilibrium configurations of the isolated impurities in the same charge state.

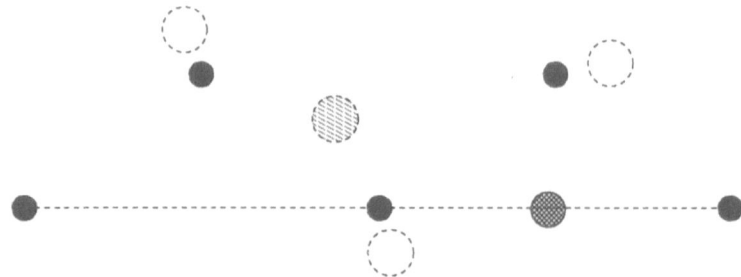

Figure 3. The oxygen and hydrogen atoms occupy neighbouring bond-centre positions in this configuration, denoted in the paper as BC. The oxygen atom is shown as a shaded circle and the H atom as a grey circle, with the relaxed Si neighbours shown as open circles.

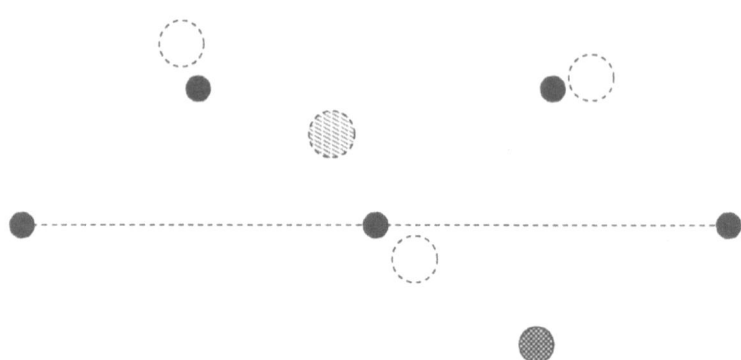

Figure 4. The hydrogen atom is bonded, in an antibonding direction, to one of the Si neighbours of the O atom, which occupies a bond-centre position in this configuration, which is denoted in the paper as AB. The oxygen atom is shown as a shaded circle and the H atom as a grey circle, with the relaxed Si neighbours shown as open circles.

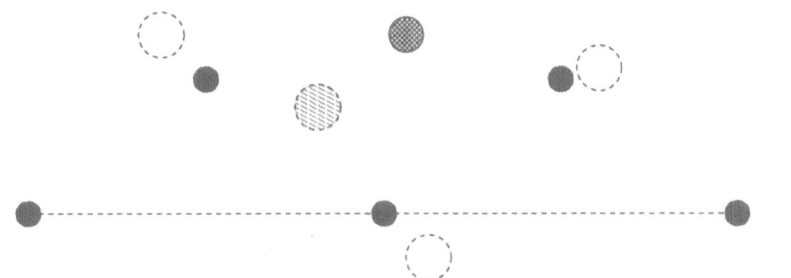

Figure 5. The hydrogen atom is bonded directly to the O atom, which occupies a bond-centre position in this configuration, which is denoted in the paper as OH. The oxygen atom is shown as a shaded circle and the H atom as a grey circle, with the relaxed Si neighbours shown as open circles.

- **positive charge state**: BC is the equilibrium structure, and is bound by 0.5 eV compared with isolated, bond-centred O and H^+ impurities. OH is about 0.1 eV higher in energy, and is also strongly bound compared to the same isolated impurities. AB is unstable with respect to spontaneously distorting into BC.

- **neutral charge state**: BC is the equilibrium structure, and is bound by 0.3 eV compared with isolated, bond-centred O and H^0 impurities. AB is 0.2 eV higher in energy and marginally bound, while OH is about 0.5 eV higher in energy and is unbound, in comparison with the isolated impurities.

- **negative charge state**: AB is the equilibrium structure, and is bound by 0.4 eV compared with isolated, bond-centred O and the intersititial H^- impurity. OH and AB are both much higher in energy (over 1 eV) and are thus unbound with respect to the isolated impurities.

We have carried out a systematic investigation of the migration of the oxygen-hydrogen complex in the positive and neutral charge states. We find a considerably lowered activation energy, of order 1.1 eV for the neutral charge state and 1.5 eV for the positive charge state, with an error bar of ~ 0.5 eV. We find that the oxygen atom follows the H-atom along the Si chains in the (110) plane, as shown in figure 6. The H atom breaks the Si-Si bond in advance of the oxygen atom's arrival. This facilitates the migration of the oxygen atom in comparison with that of the isolated impurity where a significant fraction of the energy barrier is spent in breaking and making Si-Si bonds. During the motion of the oxygen atom, at an intermediate stage, the H atom goes from a bond-centre position to an anti-bonding position adjacent to one Si atom. The H atom effectively saturates dangling bonds created during the migration of the oxygen atom, resulting in a substantial decrease in energy of the system at every step.

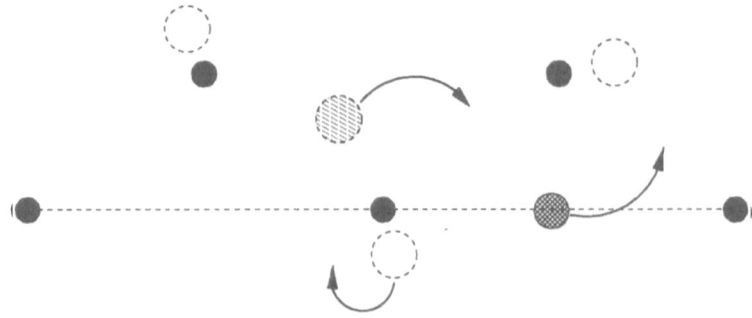

Figure 6. The diffusion of the oxygen-hydrogen complex in the positive and neutral charge states is shown. The O atom, the H atom and a Si atom move in the manner indicated by the arrows.

5. Conclusion

In conclusion, the motion of the isolated oxygen impurity is an extremely complex process, with both the O atom and an Si neighbour needing to overcome substantial energy barriers. The resulting energy surface describing the process is four-dimensional. In a two dimensional slice, there is a "saddle ridge", with an average height of 2.5 eV, in good agreement with experiment. The oxygen-hydrogen complex exhibits a variety of equilibrium and metastable configurations in its various charge states. There exist pathways for its migration with an activation energy of order 1.5 eV.

6. Acknowledgments

This work was supported in part by the Office of Naval Research Grant No. N00014-95-1-0906. Supercomputing time was provided by the Pittsburgh Supercomputing Center under Grant No. DMR950001P.

References

1. Oxygen in Silicon (1994) F. Shimura (ed.), *Semiconductors and Semimetals* **42**, Academic Press Inc., San Diego. See references therein.
2. Newman, R. C., and Jones R. (1994) Diffusion of oxygen in silicon, in F. Shimura (ed.), *Semiconductors and Semimetals* **42** Academic Press Inc., San Diego, pp. 289–352. See references therein.
3. Ramamoorthy, M., and Pantelides, S. T. (1996) Coupled-barrier diffusion - The case of oxygen in silicon, *Phys. Rev. Lett.* **76**, 267–270.
4. Vanderbilt, D. (1990) Soft self-consistent pseudopotentials in a generalized eigenvalue formalism, *Phys. Rev. B* **41**, 7892–7895.

5. Chadi, D. J., and Cohen, M. L. (1973) Special Points in the Brillouin Zone, *Phys. Rev. B* **8**, 5747–5753.
6. Hrostowski, H. J. and Kaiser, R. H. (1957) Infrared Absorption of Oxygen in Silicon, *Phys. Rev.* **107**, 966-972.
7. Jiang, Z., and Brown, R. A. (1995) Atomistic calculation of oxygen diffusivity in crystalline silicon, *Phys. Rev. Lett.* **74**, 2046–2049.
8. Estreicher, S. K. (1990) Interstitial-O in Si and its interactions with H, *Phys. Rev. B* **41**, 9886–9891.
9. Jones, R., Öberg, S., and Umerski, A. (1991) Interaction of hydrogen with impurities in semiconductors, *Mater. Sci. Forum* **83–87**, 551–561.
10. Van de Walle, C. G., Denteneer, P. J. H., Bar-Yam, Y., and Pantelides, S. T. (1989) Theory of hydrogen diffusion and reactions in crystalline silicon, *Phys. Rev. B* **39**, 10791–10808.

GENERATION OF THERMAL DONORS, NITROGEN-OXYGEN COMPLEXES AND HYDROGEN-OXYGEN COMPLEXES IN SILICON

M. SUEZAWA
Institute for Materials Research, Tohoku University
2-1-1 Katahira, Aoba-ku, Sendai 980-77, JAPAN

Abstract: We first discuss about the methods to investigate clusters of impurities and defects in semiconductors and point out the importance of investigating the generation process of clusters in order to clarify the kind and number of constituent impurities and defects included in clusters. We also point out the importance of spectroscopic measurement of concentrations of clusters and the effectiveness of analysis with the equation of chemical reaction. As examples of the above points, we show the results for thermal donors, nitrogen-oxygen complexes, and hydrogen-oxygen complexes. We studied the generation process of thermal donors, TD-1 through TD-6, with the measurement of their optical absorption after isothermal annealing. From the analysis of their generation process, we concluded that TD1 through TD-6 included 3 through 8 oxygen atoms. We found 6 kinds of nitrogen-oxygen complexes. Their thermal stablility was greater than that of thermal donors. We determin the compositions of two nitrogen-oxygen complexes, N-O-6 and N-O-3. They included one oxygen atom, and 2 and 3 pairs of nitrogen atoms, respectively. Hydrogen-oxygen complexes were generated due to the annealing around $50°C$ after doping of hydrogen in Czochralski-grown Si at high temperature. They were determined to be pairs of hydrogen and oxygen atoms from the measurement of the isotope effect of hydrogen.

R. Jones (ed.), Early Stages of Oxygen Precipitation in Silicon, 207–221.
© 1996 *Kluwer Academic Publishers.*

1. INTRODUCTION

Impurities and defects play essential roles in determining the electrical and optical properties of semiconductors. Hence quite a large number of studies have been conducted to clarify the electrical and optical properties related to impurities and defects in various semiconductors. In the course of these studies, it has been recognized that clusters and complexes, such as pairs and large aggregates, of impurities and/or defects have properties much different from those of isolated impurities and defects. The effect of impurities and defects on electrical and optical properties is usually determined by their energy levels. As for clusters, they have spatial extension. Such spatial extension may endow the clusters with other properties which cannot be derived from the energy levels alone. For example, the probability of radiative recombination at clusters may be much different from that of isolated impurities which have energy levels similar to those of clusters. If we can control the spatial distribution and density of clusters, we will be able to obtain new materials which cannot be realized by isolated impurities and defects.

Roughly speaking, there are two ways to study the properties of clusters. The first is the same as that used to study isolated impurities and defects, namely, to clarify their electrical and optical properties. Experimentally, we measure their optical absorption spectra, photoluminescence, DLTS, the Hall effect, electron spin resonance and so on, which are common methods for the study of isolated impurities and defects. We can determine energy levels, local symmetry and so on from these measurements. Sometimes we can determine the sort of impurity or defect which is responsible for such properties of clusters. We should be careful since one or a few impurity atoms or defects seem to contribute to such properties in many cases. Namely, other impurity atoms or defects are indispensable to form clusters but contribute to such properties only slightly. Therefore the first method of study is insufficient to clarify the nature of clusters. The second is the study to determine constituent impurities and defects, compositions, temperature ranges of generation and annihilation of clusters. The latter is characteristic of the study of clusters. Hence, this second way is important but cannot be done without the results of the former type of study.

We can determine composition of clusters, namely, the sort of impurities or defects, and their numbers, from the measurement and analysis of the generation process of clusters. It is neccesary to determine the concentrations of clusters as functions of annealing time and annealing temperature in such experiments. In the study of the generation process of clusters as a function of annealing time, we should measure the concentrations of clusters many times in the same specimen. Therefore, a non-destructive method is necessary for such experiments. In many cases, plural kinds of clusters are generated, namely, many sorts of clusters co-exist which have similar energy levels. They have different compositions. Hence, we should measure the concentrations of those clusters separately to determine the composition of each cluster. It is necessary to measure them by spectrocopic methods, such as optical absorption, to determine the concentrations of clusters separately. Regarding this, we need to know the results of the former type of study. There are several ways to analyze the experimental results of the generation process of clusters. Within the author's knowledge, the simplest and most fruitful way is analysis based on the chemical reaction equation. Kaiser et al. [1] first applied this method for the analysis of the generation process of thermal donors and proposed the well-known SiO_4 model of the thermal donor.

We present the results of our studies on the generation processes of thermal donors, nitogen-oxygen complexes and hydrogen-oxygen complexes in Si in this paper. We determined their concentrations from the measurement of their optical absorption and analyzed their generation processes with a slight extension of Kaiser et al.'s treatment of chemical reaction rate equations.

2. THERMAL DONORS

2.1. BACKGROUND

Generation of shallow donors in Si crystal supersaturated with oxygen due to annealing at temperatures around $450°C$ was found in 1954 [2]. Those donors are termed thermal donors (TD). They are annihilated due to annealing above $500°C$. The generation and annihilation of TD are thought to be related to generation and dissociation of some kind of clusters of oxygen atoms or self-interstitial atoms of Si generated by the generation of SiO_2.

Several models of TD have so far been proposed [3]. Kaiser et al. [1] investigated the generation process of thermal donors due to isothermal annealing from the measurement of electrical resistivity at room temperature for the determination of generated TD concentration and proposed the SiO_4 model on the basis of the following experimental results:

1) The initial generation rate of TD is proportional to the fourth power of the initial concentration $[O_I]$ of solute oxygen.

2) The maximum concentration of TD is proportional to the third power of $[O_I]$.

210

3) Many optical absorption lines are observed to be associated with TD. It suggests that there are many kinds of TD.

They analyzed the above results with the use of chemical reaction equations which described generation and dissociation of oxygen clusters. They concluded that the dominant TD was SiO_4, eventhough SiO_2 and SiO_3 might also be TD. They also concluded that oxygen clusters larger than four oxygen atoms were electrically inactive. There are many kinds of TD. Hence, we think that it seems necessary to trace the generation process of TD with the measurement of concentration of each TD separately.

2.2. EXPERIMENT [4]

Specimens were prepared from n-type Czochralski-grown Si (CZ.Si) crystals having different concentration of oxygen as shown in Table 1. The phosphorus concentration was about 1.4×10^{14} cm^{-3}. All specimens were annealed at 650°C for 30 min, so-called

TABLE 1. Characterization of the specimens

Crystal	Oxygen Conc. (atoms/cm^3)
CN - 1	5.65×10^{17}
CN - 2	6.39×10^{17}
CN - 3	7.85×10^{17}

donor-killer treatment, to annihilate TD generated during crystal growth. We annealed specimens which were sealed in evacuated quartz capsules between 430 and 490°C to generate TD. The concentration of TD was determined by measurement of optical absorption at about 6 K. In the next section, we show absorption intensity of TD instead of the concentration itself. The calibration factor from optical absorption intensity to the concentration is about 1×10^{13} cm^{-2} [5]. This factor is thought to be approximately independent of kinds of TD since they have similar energy levels.

2.3. EXPERIMENTAL RESULTS

First, we examine Kaiser et al.'s conclusion that the dominant TD is SiO_4 or that the the species of dominant TD is invariant during annealing of the specimen. Figure 1 shows the optical absorption spectra of specimen CN-3 after annealings of various durations at

471.3°C. TD-1 through TD-6 are the names of TD which are determined after classification of many absorption lines due to the effective-mass theory. Assigned absorption lines correspond to optical transition of electrons from 1s to 2p± states in the effective-mass theory. A broad absorption line at about 520 cm⁻¹ in the as received state (curve 1) is due to vibration of solute oxygen atoms. There is no peak due to TD in curve 1. Among many absorption lines in curves 2 through 4, the species of TD which gives the strongest absorption intensity changes successively from TD-3 to TD-4 and to TD-5 due to the change of annealing duration from 20 to 150 and to 1250 min, respectively. This implies that the species of dominant TD changes with the duration of annealing. The conclusion of Kaiser et al. is thus proved invalid. Figure 2 shows another examination of Kaiser et al.'s conclusion. In Fig. 2, the annealing duration is fixed at 150 min. The strongest absorption line among many lines depends on the specimen. That of TD-3 is the strongest in specimen CN-1, the smallest concentration of oxygen, while that of TD-4 is the strongest in specimen CN-3, the largest concentration of oxygen. These results imply that the species of dominant TD depends on the specimens.

Figure 3 shows the dependence of optical absorption coefficients of TD-3

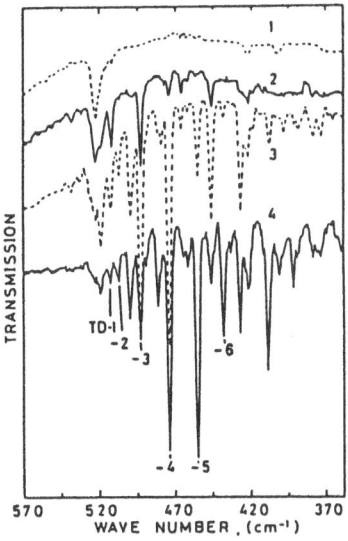

Figure 1. Optical absorption spectra of crystal CN-3 annealed at 471.3°C. The duration of annealing are as flows: 1. 0 min, 2. 20 min, 3. 150 min, 4. 1250 min. The species of TD responsible for each absorption line is shown in the figure.

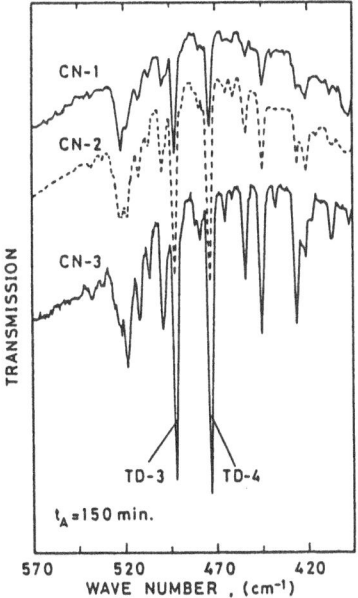

Figure 2. Optical absorption spectra of various crystals afeter annealing at 471.3°C for 150min.

212

through TD-6 on the annealing duration at 471.3°C. For purposes of simplicity, the data points of TD-3 are not plotted. Since the concentrations of TD-1 and TD-2 are small, their generation processes are not analyzed in the following. The solid lines and dashed line in Fig. 3 are the fitting curves determined in the next section. The fittings seem good. One of the characteristic features of

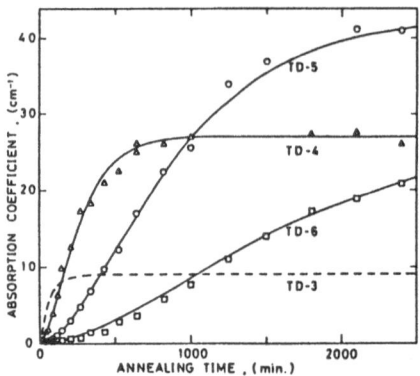

Figure 3. Generation behaviors of TD-3 through TD-6 at 471.3°C. Solid lines for TD-4 through TD-6 are the fitting curves due to Eq. (4).

TD-3 is that its absorption intensity (or concentration) is proportional to annealing duration t at small t. On the other hand, one of the characteristic features of TD-4 through TD-6 is that their absorption intensities are proportional to t^2 for small t. As already shown in Fig. 1, Fig. 3 also clearly shows that the species of TD of the maximum intensity changes with the duration of annealing.

2.4. ANALYSIS

We [4] slightly extend Kaiser et al.'s analysis to describe the generation processes of TDs in Fig. 3. We assume that clusters of oxygen atoms are the origin of TDs. Hence we describe the generation process of oxygen clusters with chemical reaction equations as follows.

$$A_1 + A_1 \underset{K_{-1}}{\overset{K_1}{\Leftrightarrow}} A_2 \tag{1a}$$

$- \quad - \quad - \quad -$

$$A_{n-1} + A_1 \underset{K_{-(n-1)}}{\overset{K_{n-1}}{\Leftrightarrow}} A_n \tag{1b}$$

$$A_n + A_1 \underset{K_{-n}}{\overset{K_n}{\Leftrightarrow}} A_{n+1} \tag{1c}$$

$$A_{n+1} + A_1 \underset{K_{-(n+1)}}{\overset{K_{n+1}}{\Leftrightarrow}} A_{n+2} \tag{1d}$$

$- \quad - \quad - \quad -$

Here A_n denotes the name of cluster consisting of n oxygen atoms and also its concentration; K_n and K_{-n} represent the forward and backward reaction constants, respectively. A_1 at t=0 is equal to the initial concentration of oxygen $[O_I]$. We assume that only A_1 is mobile. Equations (1a) through (1d) are solved under the following three assumptions :

1) The reactions (1a) and (1b) are balanced between the forward and backward reactions.

2) The backward reactions in Eqs. (1c), (1d)....are neglected.

3) A_1 is constant during annealing, i.e., $A_1=[O_I]$.

Simplified solutions can be obtained as follows under an additional assumption that all of the forward reaction constants are the same, namely, $K_1=K_2=....=K_n=K_{n+1}=...=K$;

$$A_n = B_n A_1^n , \quad B_n = \frac{K_{n-1}}{K_{-(n-1)}}...\frac{K_1}{K_{-1}} \tag{2}$$

$$A_{n+1} = A_n[1 - \exp(-KA_1 t)] \tag{3}$$

$$A_{n+2} = A_n[1 - (1 + KA_1 t)\exp(-KA_1 t)] \tag{4}$$

One characteristic feature of Eqs.(3) and (4) is that the concentrations of A_{n+1} and A_{n+2} increase from 0 at t=0 to saturation value A_n at a long duration of annealing. According to Eq. (4), duration of annealing enters the equation as a product with A_1. This means that the time for any fixed value of A_{n+2}, say half of the saturation value, is proportional to $1/A_1$. We define this time duration as $t_{1/2}$. Figure 4 shows that $t_{1/2}$ is proportional to $1/[O_I]$. This shows that the moving defect A_1 is a solute oxygen atom.

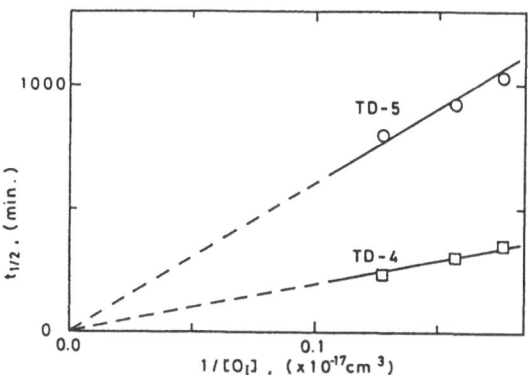

Figure 4. The relation between $t_{1/2}$, the time for the generation of a half of the saturation value, and the inverse of the initial concentration of oxygen $1/[O_I]$.

Equations (3) and (4) becomes as follows at small t :

$$A_{n+1} \propto A_1^{n+1} t, \quad A_{n+2} \propto A_1^{n+2} t^2 \tag{5}$$

Equation (5) shows that if the concentrations of generated TD are proportional to t or t^2 for small t, the number of oxygen atoms contained in individual TD is determined from the dependence of of proportional coefficient (Ke) on A_1 (=$[O_I]$). Figure 5 shows the

214

dependence of Ke on $[O_I]$. A numeral attached to each straight line shows its tangent. These numerals show the number of oxygen atoms contained in various kinds of TD according to Eq. (5). Hence, we can conclude that TD-3, TD-4, TD-5, and TD-6 contain 5, 6, 7, and 8 oxygen atoms, respectively. We have not determined the numbers of oxygen atoms contained in TD-1 and TD-2 experimentally, but we assume that they are 3 and 4 for TD-1 and TD-2, respectively. S in Fig. 5 shows the behavior of the sum of TD-1 through TD-6. Hence, the relation should coincide with Kaiser et al.'s result. However the size of clusters determined from line S is 5, which is larger by 1 than their result. This discrepancy may result from the difference of annealing temperature. Kaiser et al.

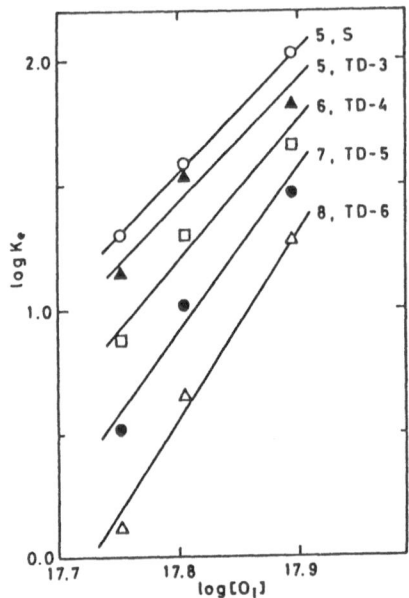

Figure 5. Logarithmic plots of the coefficients Ke of t and t^2 for small t and the initial concentration of oxygen $[O_I]$. Numerals attached to the lines are the values of the tangents.

annealed specimens at 450°C while we annealed at 471.3°C. Clusters of a larger size are thought to be generated at higher annealing temperature.

Equation (4) leads to

$$A_{n+2} (t \rightarrow \infty) \rightarrow A_n = A_1^n \qquad (6)$$

Figure 6 shows the dependence of saturation value on $[O_I]$. The tangents for TD-4 and TD-5 are 4 and 5 which are smaller by 2 than the number of oxygen atoms contained in those TDs.

As shown above, the generation of TD is well described as a clustering process of oxygen atoms in silicon.

Finally, we determined the activation energy necessary for the generation of each TD from the measurement of $t_{1/2}$ at various temperatures. As shown in Eqs. (3) and (4),

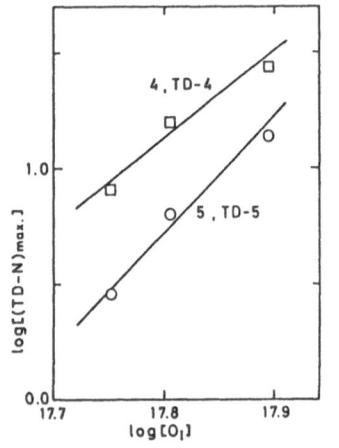

Figure 6. Plots of the saturation values of TD and $[O_I]$.

the duration of annealing in those equations is a product with reaction constant K. Hence, we can determine the activation energy of K from the above measurement. The activation energies for the generation of TD-4, TD-5, and TD-6 are 2.2, 2.2, and 1.7 eV, respectively. That of oxygen diffusion is about 2.5 eV. Hence those of TD-4 and TD-5 are near that of oxygen diffusion, but that of TD-6 is much smaller. We have not clarified the reason for this discrepancy.

3. NITROGEN-OXYGEN COMPLEXES IN SILICON

3.1. BACKGROUND [6]

We found nitrogen-oxygen complexes (N-O complexes) which act as shallow donors during the course of investigation of nitrogen behavior in Si. Optical absorption spectra due to these donors are almost identical to those reported by Navarro et al. [7], which they thought to be due to new kinds of TDs without any concrete reasons. As shown in this section, the origin of absorption is electronic transitions in N-O complexes. Figure 7 shows optical absorption spectra of as-grown CZ.Si doped with nitrogen. The shapes of lines of groups I and II are very similar. Actually they can be classified according to the effective-mass theory as shown in Table 2.

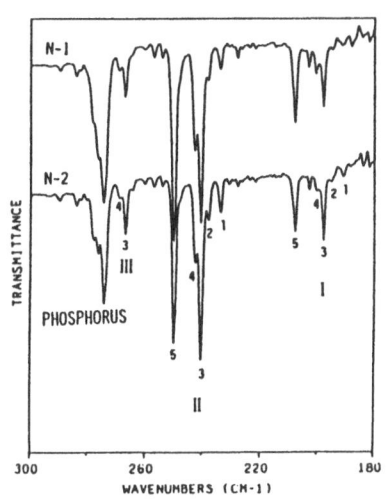

Figure 7. Optical absorption spectra of as-grown CZ.Si doped with nitrogen.

TABLE 2. Positions of absorption lines and their assignment s due to the effective-mass theory

Line		Wavenumber	Line		Wavenumber	Line		Wavenumber
I	1	190.8	II	1	233.7			
	2	195.0		2	237.9			
	3	197.7		3	240.5	III	3	266.7
	4	200.2		4	242.4		4	268.8
	5	207.4		5	249.9			
	6	204.3		6	247.0			

(Wavenumber; cm^{-1} unit. Group I, II, and III correspond to $2p_0$, $2p_\pm$, and $3p_\pm$, respectively.)

216

N-O complexes are more stable than TD as shown in Fig. 8. The nature of these complexes can be revealed most clearly using the approach adopted in this paper.

3.2. EXPERIMENT

Specimens used in this experiment were n-type CZ.Si doped with nitrogen during crystal growth.

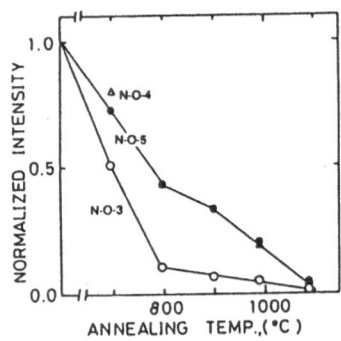

Figure 8. Isochronal annealing behavior of N-O complexes due to annealing for 30 min in specimen N-1 in the as-grown state.

TABLE 3. Characterization of the specimens

Crystal	Nitrogen Conc. (atoms/cm^3)	Oxygen Conc. (atoms/cm^3)
N - 1	1.0×10^{15}	8.6×10^{17}
N - 2	4.4×10^{14}	7.3×10^{17}
FZ(Ref.)	(high purity)	

Table 3 shows the characteristics of these specimens. All specimens were first annealed at 1200°C for 60 min to homogenize them.

3.3. EXPERIMENTAL RESULTS [8]

We investigated the generation process of N-O-3 and N-O-6 among 6 kinds of N-O complexes since their intensities are strong. Figure 9 shows the generation behaviors of N-O-3 and N-O-6 due to isothermal annealing at 479°C. One of the characteristic features of N-O-6 and N-O-3 is that their concentrations are proportional to t^2 and t^3, respectively, at short duration of annealing. Concentrations of N-O-6 and N-O-3

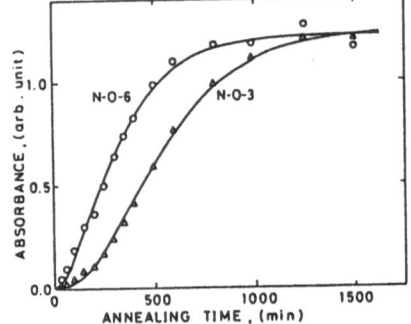

Figure 9. Generation of N-O-3 and N-O-6 due to annealing at 479°C in specimen N-1. Solid lines are fitting curves due to Eqs.(8) (N-O-6) and (9) (N-O-3).

saturate to a similar value at long duration of annealing. The solid lines in Fig. 9 are fitting curves of theoretical results which will be shown later. The fitting seems good.

3.4. ANALYSIS

To analyze the generation process of N-O complexes with the chemical rate equations, Eq.(1a) is modified as follows considering the reaction between the two components, while other equations are unchanged;

$$B + A_1 \underset{K_{-1}}{\overset{K_1}{\Longleftrightarrow}} A_2 \qquad (7)$$

We assume that A_1 is mobile in the above equation. The solutions of the chemical rate equations are obtained with the same assumption as adopted in the case of TD:

$$A_3 = B[1 - (1 + KA_1t)\exp(-KA_1t)] \qquad (8)$$

$$A_4 = B[1 - (1 - KA_1t + \frac{K^2A_1^2t^2}{2}) \exp(-KA_1t)] \qquad (9)$$

As in the case of TD, the time constant $t_{1/2}$ is inversely proportional to the concentration of moving impurity. Figure 10 shows that $t_{1/2}$ is proportional to the inverse of nitrogen concentration. Nitrogen atoms dissolved in Si are known to be paired. We therefore conclude that the moving impurity is a pair of nitrogen atoms. Impurity B is identified as solute oxygen atom since the saturation values of N-O-6 and N-O-3 are similar and proportional to the concentration of oxygen. Concentrations of N-O-6 and N-O-3 are known to be proportional to the square and the third power of nitrogen concentration. We therefore conclude that N-O-6 contains one oxygen atom and two pairs of nitrogen atoms and that N-O-3 contains one oxygen atom and three pairs of nitrogen atoms.

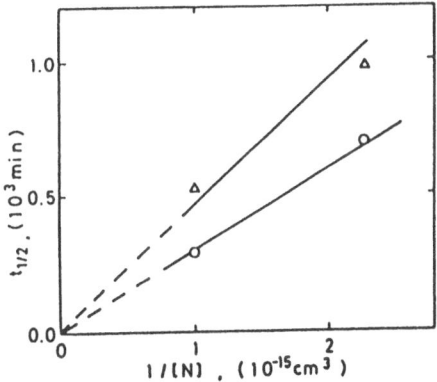

Figure 10. The relation between $t_{1/2}$ and the inverse of nitrogen concentration 1/[N].

4. HYDROGEN-OXYGEN COMPLEXES

4.1. BACKGROUND

Hydrogen in Si is known to remarkably enhance the diffusion rate of the oxygen atom [9]. The reason for such enhancement has not yet been clarified. Theoretical studies show that there is no significant bond between hydrogen and oxygen atoms [10,11]. We found a new optical absorption line in hydrogen-doped CZ.Si as shown below [12].

4.2. EXPERIMENT

Specimens were CZ.Si. Oxygen concentration was about 1×10^{18} atoms /cm^3. Hydrogen was doped by annealing specimens at 1200°C in hydrogen atmosphere followed by quenching in iced water.

4.3. EXPERIMENTAL RESULTS

Figure 11 shows the optical absorption spectra of specimens doped with hydrogen, hydrogen and deuterium, and deuterium. Specimens were annealed at 70°C for 50 min. The peak position of the hydrogen-doped specimen (spectrum a) is 1075.0 cm^{-1}. This peak was also observed in p-type CZ.Si. This means that this peak originates from the vibration of atoms and is not related to electronic transition. This peak was not observed in the case of floating-zone grown crystals. This means that this peak is related to oxygen. Hence, we conclude that this absorption is due to hydrogen-oxygen complexes.

We measured the isotope effect of hydrogen by doping deuterium to obtain more direct data about the participation of hydrogen in this optical absorption. The results are shown in spectrum (c) in Fig. 11. The peak position shifts to 1076.3

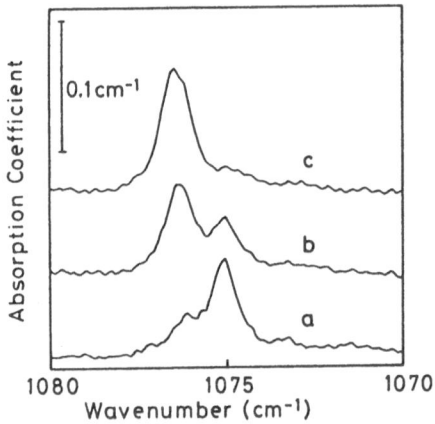

Figure 11. Optical absorption spectra of (a) hydrogen doped, (b) hydrogen and deuterium co-doped, and (c) deuterium doped specimens. Specimens were annealed 50 min at 70°C after quenching.

cm^{-1}. This shift is opposite that expected from ordinary vibration. In any case, there is an effect of isotopes. Therefore, it can be concluded that hydrogen atoms participate in the vibration responsible for the above absorption. We measured optical absorption of a specimen co-doped with hydrogen and deuterium. Spectrum (b) in Fig. 11 is the result. It has 2 peaks which are related to those of hydrogen and deuterium doped specimens. This suggests that one hydrogen atom is included in the hydrogen-oxygen complex. Most oxygen atoms are isolated at as-grown state. Hence, we conclude that this complex is a hydrogen-oxygen pair (H-O pair).

Figure 12 shows the optical absorption spectrum in a wide range of wavenumbers. We estimate the concentration of the H-O pair assuming the same oscillator strength for the H-O pair and ^{18}O. The natural abundance of ^{18}O is 0.2 %. Hence, the optical absorption due to ^{18}O in Fig. 12 corresponds to 2×10^{15} atoms/cm^3. The concentration of the H-O pair is estimated to be about 7×10^{14} cm^{-3} since the peak height of H-O pair is about 1/3 that of ^{18}O. The solubility of hydrogen at 1200°C under 1.5 atm is about 7×10^{15} cm^{-3}, and hence about 10 % of the hydrogen atoms make pairs with oxygen atoms.

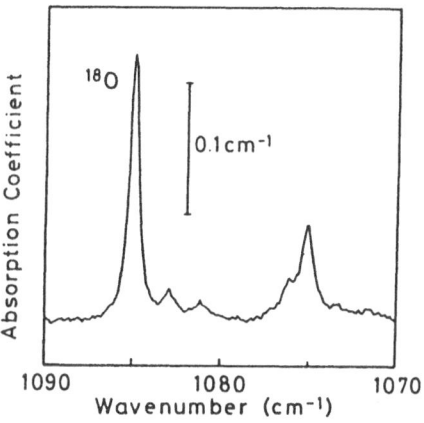

Figure 12. Optical absorption spectrum of hydrogen doped specimen.

Figure 13 shows the isochronal annealing behavior of the H-O pair. The annealing duration was 10 min at each temperature. There is no H-O pair in the as-quenched state. It is generated above 40°C, attained a maximum concentration at about 80°C, and then diminished. This behavior was the same in both n- and p-type specimens. Therefore, there is no effect of Fermi level on the generation and annihilation of the H-O pair. We have not yet determined the activation energy for

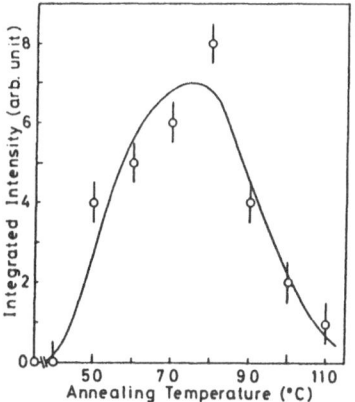

Figure 13. Isochronal annealing curve of H-O pair in a n-type specimen. Duration of annealing is 10 min at each temperature.

the generation of the H-O pair. Inferring from the isochronal annealing behavior, it is probably around 0.8 eV . This is much larger than the activation energy of hydrogen migration, 0.48 eV [13]. We therefore conclude that the generation process of the H-O pair is not determined by simple migration of the hydrogen atom to the oxygen atom. Probably, the rate determining process for the generation of the H-O pairs is the release of hydrogen atoms from hydrogen clusters which are generated during quenching after doping of hydrogen.

5. CONCLUSION

There are two ways to investigate properties of clusters. One is the way to clarify the electronic and optical properties of clusters. The other is the way to determine compositions of clusters and thermal stability. Many people are studying clusters with the former way. We emphasize the importance of the latter way since it gives a clear image of clusters and open a way to control clusters.

REFERENCES

1. Kaiser, W., Frisch, H., and Reiss, H. (1958) Mechanism of the formation of donor state in heat-treated silicon, *Physcal Review* **112**, 1546-1554

2. Fuller, C. S., Ditzenberger, J. A., Hannay, N. B., and Buehler, E. (1954) Resitivity changes in silicon induced by heat treatment, *Physical Review* **96**, 833

3. Patel, J. R. (1981) Current problems on oxygen in silicon, in H. R. Huff, R. J. Kriegler and Y. Takeishi (eds.), *Semiconductor Silicon 1981*, The Electrochemical Society, Inc., pp. 189-207

4. Suezawa, M., and Sumino, K. (1984) Nature of thermal donors in silicon crystals, *Physica Status Solidi* (**a**)**82**, 235-242

5. Kamiura, Y., Suezawa, M., Sumino, K., and Hashimoto, F. (1990) Optical properties of new kinds of thermal donors in silicon, *Japanese J. Applied Physics* **29**, L1937-L1940

6. Suezawa, M., Sumino, K., Harada, H., and Abe, T. (1986) Nitrogen-oxygen complexes as shallow donors in silicon crystals, *Japanese J. Applied Physics* **25**, L859 - L861

7. Navarro, H., Griffin, J., Weber, J., and Genzel, L. (1986) New oxygen related

shallow thermal donor centres in Czochralski-grown silicon, *Solid State Communications* **5 8**, 151-155

8. Suezawa, M., Sumino, K., Harada, H., and Abe, T. (1988) The nature of nitrogen-oxygen complexes in silicon, *Japanese J. Applied Physics* **2 7**, 62-67

9. Newman, R. C., Tucker, J. H., Brown A. R., and McQuaid, S. A. (1988) Hydrogen diffusion and the catalysis of enhanced oxygen diffusion in silicon at temperatures below 500°C, *J. Applied Physics* **7 0**, 3061-3070

10. Estreicher, S. K. (1990) Interstitial O in Si and its interactions with H, *Physical Review* **4 1**, 9886-9891

11. Jones, R., Oberg, S., and Umerski, A. (1992) Interaction of hydrogen with impurities in semiconductors, in G. Davies, G. G. DeLeo, and M. Stavola (eds.), *Materials Science Forum*, Trans Tech Publications, **83-87**, pp. 551-562

12. Markevich, V. P., Suezawa, M., and Sumino, K. (1995) Optical absorption due to vibration of hydrogen- oxygen pairs in silicon, in M. Suezawa and H. Katayama-Yoshida (eds.), *Materials Science Forum*, Trans Tech Publications, **196-201**, pp. 915-920

13. Van Wieringen, A., and Warmoltz, N. (1956) On the permeation of hydrogen and helium in single crystal silicon and germanium at elevated temperatures, *Physica* **XXII**, 849-865

THE ELECTRONIC STRUCTURE OF THE OXYGEN DONOR IN SILICON FROM PIEZOSPECTROSCOPY

MICHAEL STAVOLA
Department of Physics, Lehigh University
Bethlehem, PA 18015, USA

1. Introduction

It has been known for more than 40 years that annealing Czochralski-grown silicon (with [O] $\approx 10^{18}$cm^{-3}) at temperatures near 450°C produces shallow "oxygen donor" defects with concentrations up to 10^{16}cm^{-3} [1, 2]. Since this early discovery there have been numerous studies performed to determine the oxygen donor's structure and the mechanism for its formation [3,4]. In spite of a tremendous amount of effort, the atomic structure of the oxygen donor and an explanation of its formation kinetics remain controversial. This paper focusses on what has been learned about the *electronic structure* of the oxygen donor from experiments in which uniaxial stress was used in conjunction with infrared absorption [5-7], DLTS [8-11], and EPR [12] spectroscopies. These piezospectroscopic results have shown that the spectra measured by infrared absorption, DLTS and EPR are indeed due to the same oxygen donor defect and that the ground state of the oxygen donor is constructed from the effective-mass wave functions associated with a single pair of conduction band valleys. In this paper, a survey of the piezospectroscopic results is presented.

Historically, uniaxial stress results were first reported in an infrared absorption study [13] and two DLTS studies [8, 9]. An interpretation of these stress results in terms of a ground state constructed from the wave functions associated with a single pair of conduction band valleys was given in a comment by Watkins at the Thirteenth International Conference on Defects in Semiconductors held in 1984. Infrared absorption and EPR measurements made with stress were then undertaken at Bell Laboratories and Lehigh University, with each group aware of the results of the other. These infrared absorption and EPR studies provide complementary information that independently lead to similar conclusions. Here, the infrared absorption measurements made with stress [5-7] are described first, followed by a discussion of the DLTS [8-11] and EPR [12] results. Studies of the alignment of the oxygen donor by stresses

R. Jones (ed.), Early Stages of Oxygen Precipitation in Silicon, 223–242.
© *1996 Kluwer Academic Publishers.*

applied at elevated temperatures (~460°C) help to complete the picture by relating the electronic structure of the oxygen donor to the localized stresses that result from its atomic structure [14, 15]. These stress-alignment results will be described briefly here and are discussed in greater detail in the article by Watkins that appears elsewhere in this volume.

The oxygen donor is an example of a defect for which local strains perturb its effective-mass-like electronic states. The application of uniaxial stress can produce several effects for such a defect which have been exploited for the oxygen donors. Spectroscopic transition energies can be shifted or split, providing information about the symmetry of a defect. The defect's electronic levels can have their energies shifted by the stress which can lead to a modification of their populations. Finally, if the temperature is sufficiently high, the defect can reorient under the influence of stress, where the concentrations of defects with different orientations are given by Boltzmann statistics in equilibrium. Each of these effects manifests itself in different ways for different spectroscopic techniques. An important aspect of the interpretation of the stress data is distinguishing, from the sometimes subtle differences in the experimental results, the electronic degeneracy characteristic of a defect with high symmetry from the orientational degeneracy characteristic of a defect with low symmetry.

2. Spectroscopic Signatures of the Oxygen Donor

In this section, a few results from the effective mass theory (EMT) [16] for shallow donors in Si are summarized. EMT has been discussed in several reviews [17, 18]. The oxygen donor has shallow EMT-like levels but its low symmetry also leads to important departures from EMT as will be seen in the sections that follow. The spectroscopic signatures of the oxygen donor observed by infrared absorption, DLTS, and EPR spectroscopies are also summarized in this section. Each of these spectroscopic methods provides complementary information and each has its natural advantages. The infrared absorption spectra provide information about the different members of the family of oxygen donor complexes most easily. With EPR, the C_{2v} symmetry and orientational degeneracy of the oxygen donor are immediately apparent. And DLTS reveals the ionization of the electronic levels that is probed less directly with the other spectroscopies.

2.1. EFFECTIVE MASS THEORY FOR SHALLOW DONORS IN SILICON

Silicon has six equivalent conduction band minima that lie along the <100> directions in k-space. These conduction band valleys are characterized by an anisotropic effective mass with $m_{//} > m_{\perp}$. The EMT wave functions for a bound donor electron in silicon are a linear combination of the product wave functions associated with the conduction band valleys and are written as follows:

$$\psi(r) = \sum_j a_j \rho_j(r) F_j(r). \tag{1}$$

Here, $\rho_j(r)$ is a Bloch function and $F_j(r)$ is a hydrogenic envelope function associated with the jth conduction band valley. Linear combinations of the valley wave functions are chosen to be consistent with the site symmetry of the defect. For T_d site symmetry, the valley-orbit interaction of the bound electron with the central cell causes the 1s states to split into singlet A_1, doublet E, and triplet T_2 states. For substitutional donors in Si the A_1 state is the ground state. The np excited states are split by the anisotropy of the conduction band valleys into a singlet np_0 and doubly degenerate $np_{\pm}$ states. When the symmetry of a center is lower than T_d, as for the oxygen donor, the degeneracies of the states can be reduced further.

A hydrogenic series of strong infrared absorption lines due to optical transitions from the $1s(A_1)$ ground state to the np_0 and $np_{\pm}$ excited states are observed at low temperature for the substitutional donors in Si. (The spectroscopy of shallow impurities has been reviewed by Ramdas and Rodriguez [18].) At elevated temperatures, transitions originating from the thermally populated, higher lying 1s states are also observed. The irreducible representation (and the conduction-band-valley makeup) of the 1s states can be determined from infrared absorption spectra measured in the presence of uniaxial stress [18, 19]. The degeneracy of the conduction band is lifted by stress as is described by deformation potential theory. If EMT were exact then the bound states would shift rigidly with the conduction band valleys. This is the case for the np excited states. For the 1s states, a Hamiltonian matrix that includes the valley-orbit interaction and the uniaxial stress must be diagonalized. For T_d site symmetry, $1s(A_1) \rightarrow n$p transitions split into a doublet [because the $1s(A_1)$ shifts but does not split and the np excited states split like the conduction band], $1s(E)$ $\rightarrow n$p transitions split into a triplet, and $1s(T_2) \rightarrow n$p transitions do not show a splitting of the spectral features. This latter case is of particular interest here. Each of the three $1s(T_2)$ states is constructed from 1s wave functions associated with a pair of conduction band valleys that lie along the same direction in k-space. Thus both the $1s(T_2)$ states and np excited states are split by stress exactly like the conduction band valleys and, because optical transitions between states associated with different valleys are forbidden, the transition energies remain unchanged.

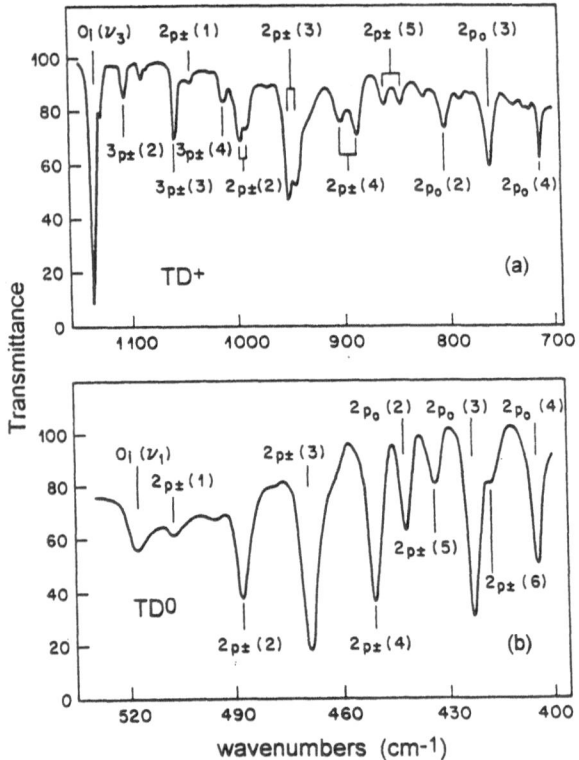

Figure 1. Spectra for several oxygen donor complexes. The spectral lines are labeled with their final state and donor complex number. (a) Spectra for the singly ionized charge state. (b) Spectra for the neutral charge state.

2.2. INFRARED ABSORPTION SPECTROSCOPY

In infrared absorption measurements, two groups of lines are observed for the oxygen donor [20-25] (Fig. 1). The temperature dependence and compensation behavior show that these groups of features can be assigned to the absorption lines that arise from the neutral (TD^0) and singly ionized (TD^+) charge states of a double donor [21]. The multiplicity of lines shows that the oxygen donor is not a unique defect but is a family of complexes with regularly spaced ground state energies. At least eleven distinct series of helium-like absorption features have been observed [3]. We follow the assignments and notation of Wagner and coworkers [3, 22] in this manuscript. Transitions in Fig. 1 are labeled by their final states. The number in parentheses denotes which of the family of complexes the transition arises from.

The different donor complexes form sequentially for continued annealing with ground state energies that get progressively shallower. The hydrogenic EMT prediction for the ground state energy of the neutral charge state is corrected by multiplying by the ratio of the atomic helium ionization energy to that of hydrogen [26]. For the singly ionized charge state, the hydrogenic EMT level energies are multiplied by a factor of 4 to account for the dipositive core. The ground state energies for the TD^0 complexes range from 69.3 to 49.9 meV. The ground state energies for TD^+ complexes range from 156.3 to 116.0 meV. Both ranges of binding energies are in reasonable agreement with the EMT predictions for a helium-like double donor, 56 and 126 meV, for the neutral and singly ionized charge states, respectively.

2.3. DEEP LEVEL TRANSIENT SPECTROSCOPY

In Czochralski-grown silicon that had been heat treated near 450°C for several hours, DLTS experiments [27] showed two new emission peaks, E(0.07) and E(0.15), whose emission energies are close to those expected for the oxygen donor. These features grow in together upon annealing at 450°C and similar peaks were not seen for Si with a low oxygen content, even when annealed for greater than 100 h at 450°C. The observation of a Poole-Frenkel effect confirmed that these emissions are due to donor levels. Thus, the E(0.07) and E(0.15) emission peaks were assigned to the (0/+) and (+/++) levels of the oxygen donor. The individual complexes in the family of oxygen donor complexes are not resolved by DLTS measurements which reflect their average properties.

2.4. ELECTRON PARAMAGNETIC RESONANCE

A number of new spectra were observed in EPR measurements made on oxygen-rich silicon that had been annealed near 450°C [28]. The dominant EPR spectra have g-tensor anisotropies that are consistent with C_{2v} symmetry. One of these spectra, NL8, was subsequently assigned to the paramagnetic charge state TD^+ of the oxygen donor from its behavior under stress that will be discussed in Sec. 5 below [12]. Ordinary EPR measurements do not resolve the individual members of the family of oxygen donor complexes. Another of the EPR spectra, NL10, has been the focus of recent attention [29]. NL10 has been assigned to an oxygen donor defect that has been partially passivated by hydrogen or complexed with an acceptor impurity to give a single donor. At present there are no stress data available for NL10, although such measurements might provide information about the NL10 defect's electronic structure and a means to correlate the results of different spectroscopic techniques.

Hyperfine interactions were not resolved by EPR measurements made on samples diffused with oxygen that was enriched with ^{17}O to 10% [28]. ENDOR measurements, discussed elsewhere in this volume, provide important structural information from hyperfine interactions with ^{17}O and ^{29}Si nuclei for the NL8 [30] and NL10 [31] spectra.

3. Infrared Spectroscopy with Uniaxial Stress

In partially compensated samples, the infrared absorption spectra of both charge states, TD^0 and TD^+, of the oxygen donor can be studied in the presence of applied uniaxial stress. From these results the symmetries of several of the different complexes in the family of oxygen donor complexes and the properties of the EMT states have been determined [5-7].

3.1. THE NEUTRAL CHARGE STATE, TD^0

Absorption spectra which were measured while uniaxial stress was applied are shown in Fig. 2 for several transitions of the neutral charge state of the family of oxygen donor complexes. The transition energies *are not split* for the [001] stress direction. For the [111] stress direction the absorption lines are split into doublets. For the [110] stress direction with a [001] viewing direction, the $1s \rightarrow 2p_\pm$ lines split into triplets while the $1s \rightarrow 2p_0$ lines are not split. These data are sufficient to determine the symmetry and electronic structure of TD^0.

3.1.1. *Effective mass theory description of the ground state*
The lack of splittings of the absorption lines for a [001] stress direction is in marked contrast to what would be expected for the $1s(A_1)$ ground state of a substitutional donor in silicon. For the magnitude of the stress used in Fig. 2, the

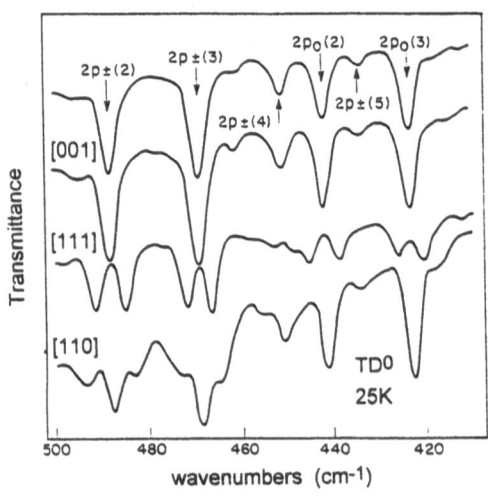

Figure 2. Infrared absorption spectra for TD^0 measured at T=25K. The upper spectrum was measured without applied stress. For the others, a stress of ~195 MPa was applied along the stress direction indicated. For the [110] stress direction the viewing direction was [001]. From ref. [7].

conduction band valleys (and hence the np excited states) can be estimated to be split by ~140 cm^{-1} from the shear deformation potential for Si [19]. Instead, much smaller splittings (~10 cm^{-1}) for the [111] and [110] stress directions are observed.

The lack of splittings for 1s→np absorption lines for the [001] stress direction can be explained within the framework of EMT and is similar to the behavior of transitions that originate from 1s(T$_2$) states for T$_d$ site symmetry [18]. In this case, *the ground state splits exactly like the excited states* and intervalley transitions are forbidden. Thus, the transition energies remain unchanged. Oxygen donor complexes have site symmetry lower than T$_d$ as will be discussed in the following sections. Equal ground and excited state splittings are preserved for lower symmetries than T$_d$ for a ground state constructed from wave functions associated with a pair of conduction band valleys that lie along the same direction in k-space.

3.1.2. *Thermal ionization of the ground state under stress*
Although the splitting of the effective-mass-like ground state cannot be detected as a splitting of the spectral features, the occupation statistics under stress can be used to probe the degeneracy of the ground state and the magnitude of the stress splitting [5-7]. In Fig. 3(a) is shown an energy level diagram with the allowed transitions for a 1s(T$_2$) initial state. For the polarization parallel to the stress, E//σ, the transitions 1s(-)→2p$_0$ and 1s(+)→2p$_\pm$ are allowed. Thus, measurements made with polarized light distinguish between transitions that originate from components of the ground state that shift differently under stress.

In Fig. 4 are shown a series of spectra taken with E//σ. The top spectrum was measured without applied stress and the spectrum labeled (a) was measured with a stress of σ = 196 MPa applied along the [001] direction at a temperature of T = 20K. Even though the stress splitting energy is 10 times the thermal energy kT, transitions originating from both 1s(-) and 1s(+) are observed. Further, the result in (a) is not obtained because relaxation from 1s(+) to 1s(-) is impeded by a kinetic barrier. In spectrum (b) of Fig. 4 the temperature is raised to 65K with σ = 196 MPa. Transitions that originate from the higher lying component, 1s(+), are reduced in intensity at elevated temperature. However, when the temperature is lowered to T = 20K with the stress maintained for spectrum (c), the population difference between 1s(+) and 1s(-) in (b) is not frozen in, but instead we regain spectrum (a). These results are not consistent with an electronically degenerate ground state [like a 1s(T$_2$) state] for which the 1s(+) level population would be reduced by a Boltzmann factor at elevated temperature. Thus, these results provide strong evidence that the oxygen donor has an *orbitally nondegenerate ground state* and that the applied stress shifts the energies of an *orientationally inequivalent set of defects*.

230

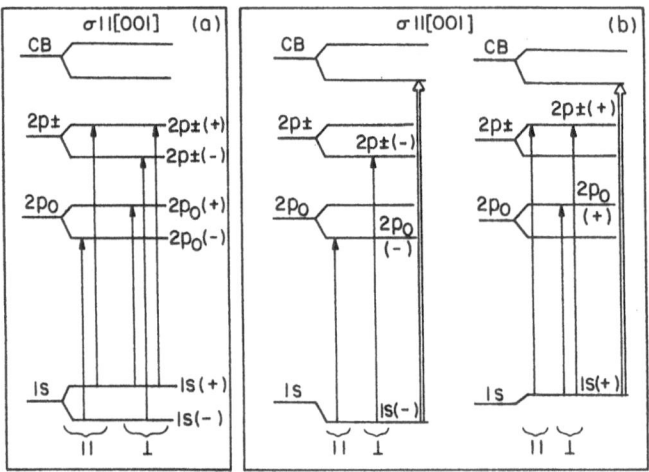

Figure 3. (a) Energy level diagram showing transitions originating from the 1s(T$_2$) level of a substitutional donor with stress applied along the [001] direction. (b) Energy level diagrams for an orientationally degenerate defect with a ground state constructed from the wave functions associated with a single pair of conduction band valleys. The levels and transitions for two different orientations of the defect's axes with respect to the direction of the applied [001] stress are shown. Single arrows show the allowed optical transitions. Double-lined arrows show thermal ionization transitions to the lowest conduction band valley. From ref. [7].

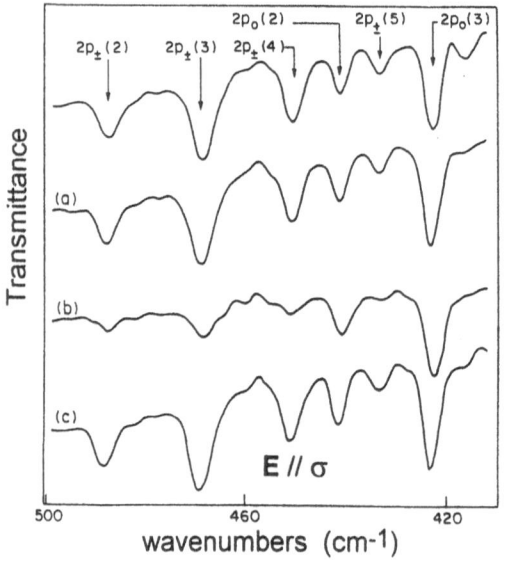

Figure 4. Infrared absorption spectra of TD^0. Upper: T=20K without stress. (a) T=20K, [001] stress of σ=196 MPa. (b) T raised to 65 K with the stress maintained. (c) Sample is cooled from 65K to 20K with the stress maintained. From ref. [7].

An energy level diagram is shown in Fig. 3(b) for a nondegenerate ground state that is constructed from a single pair of conduction band valleys. In this diagram an oxygen donor complex with a specific crystallographic orientation has a ground state made up from a particular pair of conduction band valleys; this valley pair is shifted up or down in energy depending upon the orientation of the center with respect to the applied stress.

The 1s(+) component of the ground state is preferentially depopulated at 65K in Fig. 4(b) because the 1s(+) state is shallower than the 1s(-) state with respect to the lowest conduction band valleys under stress. A rough estimate of the ground state energy shift for the orientationally degenerate defects can be made from this preferential depopulation of the 1s(+) state as its energy is raised by the stress to near the Fermi level. The shift of the 1s level, estimated from the stress-induced population changes, was found to be roughly consistent with the magnitude of the known shear deformation potential of Si [6, 7].

3.1.3. C_{2v} symmetry and the departure from effective mass theory

The splittings of the absorption lines shown in Fig. 2 for the [111] and [110] stress directions are due to a small departure from EMT that arises from the low symmetry of the oxygen donor's atomic structure. This orientational degeneracy gives rise to a small (in magnitude) splitting of the absorption lines under an applied stress that permits the differently oriented defects to be examined separately. The splitting patterns shown in Fig. 2 are consistent with C_{2v} symmetry [32]. The frequency shift, Δ, under uniaxial stress of an individual C_{2v} component is given by [32],

$$\Delta = A_2 \sigma_{xx} + A_2 \sigma_{yy} + A_1 \sigma_{zz} + 2A_3 \sigma_{xy}. \tag{2}$$

The σ_{ij} are the components of the stress tensor and the A_i's are components of the piezospectroscopic tensor. From our data we have determined that $A_1 = A_2 \approx 0$ and that $A_3(1) = 49$, $A_3(2) = 30$, $A_3(3) = 23$, and $A_3(4) = 18$ for oxygen donor complexes 1 through 4 in units of cm^{-1}/GPa. The splitting pattern is consistent with C_{2v} symmetry up to donor complex 4. Spectral congestion makes the stress splitting pattern difficult to determine for complexes beyond 4.

The departure from effective-mass-like behavior observed for the oxygen donor transition energies under stress should be due primarily to a departure of the ground state wave function from EMT because it is more compact than the excited states and will be affected more by the anisotropy of the defect's atomic structure. This supposition can be established better for TD^+ below.

3.1.4. *The C_2 axis conduction band valley pair*

Each of the differently oriented C_{2v} defects has its ground state constructed from a particular pair of conduction band valleys. Of the six possible 1s states, two of A_1 representation can be constructed from the pair of conduction band valleys along the k_z axis (where the z axis is the C_2 axis of the defect). The focus is on wave functions contructed from a pair of valleys because these have the appropriate stress response.

In Fig. 5, the EMT description and C_{2v} symmetry of the oxygen donor are brought together and related to the spectroscopic results [5-7]. In Fig. 5(a), a diagram of the set of six inequivalent C_{2v} defects is shown for a [110] stress and [001] optical viewing geometry. The four defect orientations labeled A are equivalent under stress and give rise to the main central component of the absorption line shown in Fig. 5(b). The defect orientations labeled B and C give rise to the side-band components of the absorption band. Figs. 5(c), (d) and (e)

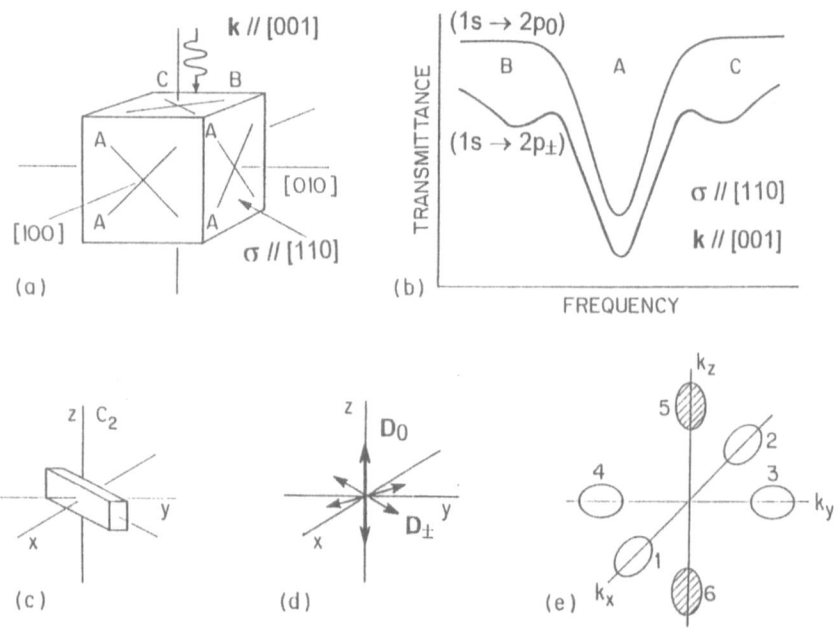

Figure 5. (a) The set of C_{2v} defects with stress applied along the [110] direction and with a [001] optical viewing direction. (b) Spectral lineshapes for 1s→2p0 and 1s→2p± transitions. The components of the bands are labeled with their corresponding defect orientations from (a). (c) A schematic C_{2v} defect with the orientation labeled C. (d) Transition moment directions for the 1s→2p0 and 1s→2p± transitions of defects C and B in (a). (e) The k_z axis pair of conduction band valleys for the C_{2v} defect in (c) is shown shaded. After ref. [7].

show a schematic of the defect labeled C; its transition moment directions, D_0 and $D_\pm$, for the $1s \to 2p_0$ and $1s \to 2p_\pm$ transitions, respectively; and the k_z axis valley pair from which the wave functions are constructed. For the $1s \to 2p_0$ transition, defects B and C have transition moments along [001] and therefore cannot absorb light that propagates along the [001] direction. The $1s \to 2p_\pm$ transitions for B and C have transition moments perpendicular to the [001] axis and therefore will give rise to side bands in the absorption spectrum. These expectations are confirmed in the spectrum shown in Fig. 2 for the [110] stress direction with a [001] viewing axis. The model is also consistent with spectra measured for a [110] stress direction and a [$1\bar{1}0$] viewing axis [6]. Thus, the polarization selection rules for [110] stress show that the *ground state is constructed from the 1s wave functions associated with the pair of conduction band valleys that lie along the k_z axis*, where z is the C_2 axis of the oxygen donor defect. There will be two 1s states that can be constructed from this pair of valleys which our results do not distinguish between.

3.2. THE SINGLY IONIZED CHARGE STATE, TD$^+$

Spectra are shown in Fig. 6 for the $1s \to 2p$ transitions of the singly ionized charge state of donor complexes 2 and 3 for several stress directions and $E//\sigma$ [6, 7]. For the [001] stress direction, conduction band-like splittings that are characteristic of substitutional donors are not observed. The small stress

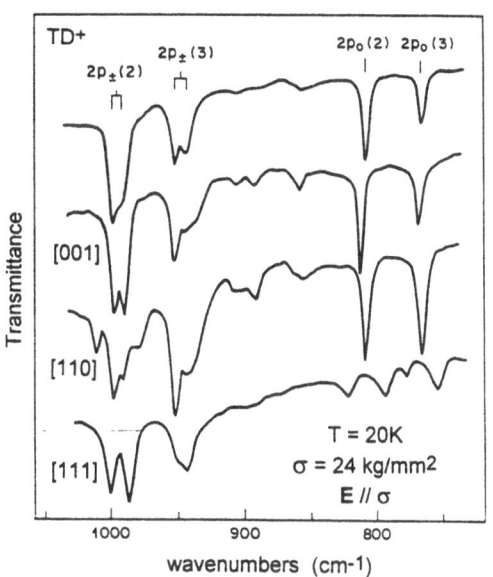

Figure 6. Infrared absorption spectra for TD$^+$ measured at T=20K and with light polarized with $E//\sigma$. The upper spectrum was measured without applied stress. For the others, a stress of ~235 MPa was applied along the stress direction indicated. For the [110] stress direction the viewing direction was [001]. From ref. [7].

splittings due to departures from EMT are more difficult to interpret than for TD^0 because of the splittings of the $1s \to np_\pm$ bands at zero stress which are of the same magnitude as the stress-induced splittings. For TD^+ it is helpful to measure spectra with polarized light, otherwise spectral congestion makes the splitting patterns difficult to systematize.

3.2.1. *Effective mass theory description of the ground state*
The data shown in Fig. 6 indicate that the TD^+ ground state is constructed from a single pair of conduction band valleys like for TD^0 because the spectral features are not split for the [001] stress direction [6, 7]. No splittings are observed because both the ground and excited states split exactly like the conduction band valleys. At 20 K, for stress splittings much greater than kT, transitions that originate from both $1s(-)$ and $1s(+)$ for $E//\sigma$ are observed, indicating that the TD^+ ground state is orbitally nondegenerate. These conclusions are supported by the preferential thermal ionization of the shallower $1s(+)$ state at elevated temperature and large stress values. Spectra measured with polarized light are shown in Fig. 7 for $T = 70$ K, and [001] stress with magnitude $\sigma = 284$ MPa. Only $1s(-) \to np_0$ transitions are observed strongly for $E//\sigma$ and only $1s(-) \to np_\pm$ transitions are observed for $E\perp\sigma$, in agreement with the polarization selection rules shown in Fig. 3(b).

3.2.2. *Deviations from effective mass theory*
The $1s \to np_\pm$ transitions for TD^+ are already split at zero stress. These splitting are of the $np_\pm$ excited states, rather than the ground state, because they are not equal for different principal quantum numbers, n, for the same donor complex.

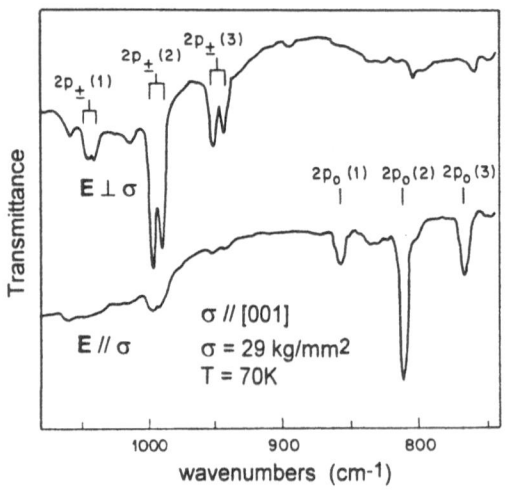

Figure 7. Infrared absorption spectra for TD^+ measured at T=70K and with a [001] stress of σ=284 MPa. From ref. [7].

(The $3p_\pm$ splittings are roughly a factor of 4 smaller than the $2p_\pm$ splittings for the same complex [22].) From the behavior of the [111] and [110] stress splittings, it has been determined that the symmetry of TD^+ is C_{2v} and that the $p_\pm$ splittings at zero stress also result from the C_{2v} symmetry of the oxygen donor. A brief discussion of the zero stress and stress-induced splittings follows. A more detailed discussion has been published elsewhere [6, 7].

The p-states observed for the oxygen donor that are consistent with its C_{2v} symmetry are constructed from the wave functions associated with the k_z-axis conduction-band-valley pair because the ground state wave functions are constructed from these valleys and intervalley optical transitions are forbidden. The p_0 state transforms like the A_1 representation of the C_{2v} point group. Linear combinations of the $p_\pm$ states that transform like the B_1 and B_2 representations give orthogonal p_1 and p_2 states. (These were called p_π and $p_{\pi'}$ in earlier publications [6, 7]. Here we follow the notation of ref. [14].) The p_1 and p_2 states are split by the C_{2v} anisotropy of the oxygen donor's extended "central cell", giving rise to the zero stress splittings that are observed. This conclusion was arrived at from studies of the stress-induced shifts and polarization selection rules for transitions to these states. Similar conclusions were arrived at independently from studies of the alignment of the oxygen donor by stress applied at the growth temperature [14]. (See Sec. 6 below.)

The following stress results are consistent with $1s \rightarrow np_1$ and $1s \rightarrow np_2$ transitions whose transition moments are directed along the principal axes of the C_{2v} defect. The stress-induced shifts of the $1s \rightarrow np$ transitions are the most straightforward to visualize for a [111] stress direction. Stress applied along the [111] direction divides the six orientationally degenerate C_{2v} defects into two inequivalent groups, each consisting of three defect orientations. Thus, the $1s \rightarrow np_0$ transitions are split into doublets by the [111] stress as can be seen in Fig. 6. For $E//\sigma$, only the $1s \rightarrow np_1$ transitions are allowed for one group of equivalent defects, and only the $1s \rightarrow np_2$ transitions are allowed for the other group. Thus the transitions to np_1 and np_2 final states (which are split already at zero stress) shift upon the application of stress but without the introduction of any new lines. The more complicated stress splitting patterns observed for other stress directions or polarizations are consistent with these results. The components of the piezospectroscopic tensor were found to be, $A_3(2) = 69 \text{ cm}^{-1}/\text{GPa}$ and $A_3(3) = 63 \text{ cm}^{-1}/\text{GPa}$ for donor complexes 2 and 3, respectively. For TD^+, A_1 and A_2 are smaller than A_3 but are nonzero and give rise to a slight asymmetry and center of mass shift of the lines.

The stress-induced splittings arise from a departure of the 1s ground state from EMT. This conclusion is supported by the nearly equal shifts under stress observed for transitions to different np excited states. If these were excited state

236

splittings, then the stress-induced splittings of transitions to different excited states of the same complex would not be equal. Stress splittings that arise primarily from a departure of the ground state from EMT are consistent with its being more localized that the excited states and thus more sensitive to the oxygen donor's anisotropic atomic structure, as was also noted for TD^0 above.

4. Deep Level Transient Spectroscopy with Uniaxial Stress

The DLTS feature, E(0.15), that has been assigned to TD^+ [27] is shown in Fig. 8 for uniaxial stresses applied along the [001], [110], and [111] directions. The DLTS feature is split into a doublet for [001] and [110] stress directions but not for [111] (refs. [8-11]). The ratios of peak heights of the shallower to deeper components are roughly 2:1 and 1:2 for [001] and [110] stress, respectively.

The DLTS results are consistent with an orientationally degenerate defect, for which each orientation has a ground state constructed from wave functions associated with a single pair of conduction band valleys [4, 7]. Unlike for optical transitions, the DLTS thermal ionization transitions are not subject to an intravalley selection rule. Thus the ground state splitting is observed directly as a splitting of the DLTS peak. The thermal ionization transitions (shown in Fig. 3b) are from a stress split ground state to the lowest conduction band valleys. Transitions originate from both components of the ground state, even for stress

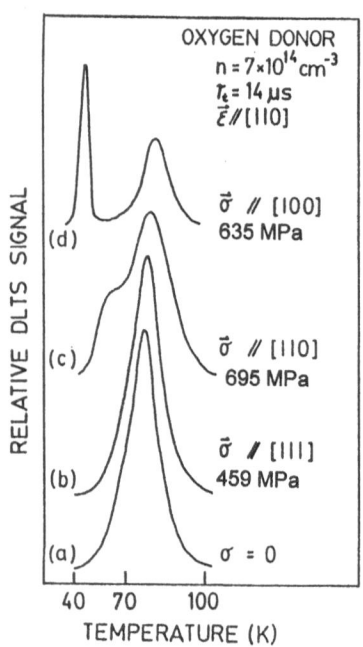

Figure 8. Effect of stress on the DLTS peak E(0.15) assigned to the (+/++) level of the oxygen donor. After ref. [9].

splittings that are many kT, which is consistent with an orientationally inequivalent set of defects.

The orientation dependence of the splittings and the ratios of the peak heights are also consistent with the conduction-band-like splittings that are expected for a ground state constructed from a single pair of valleys. If we ignore the possible stress dependence of the DLTS emission rate prefactor then the temperature difference for the components of the DLTS feature is proportional to the energy splitting. From the spectra shown in Fig. 8 one can estimate a shear deformation potential of 9 eV in agreement with the conduction band shear deformation potential (8.8 eV from optical spectra [19]).

5. Electron Paramagnetic Resonance with Uniaxial Stress

Lee *et al.* [12] have used the effect of uniaxial stress on the NL8 EPR spectrum to establish the assignment of NL8 to the oxygen donor and to probe the electronic structure of the oxygen donor's ground state. When the magnetic field **H** is parallel to the [110] direction, the EPR spectrum NL8 consists of 3 lines, each of which corresponds to defect orientations that remain equivalent in the presence of the field. The intensities are in the ratio 1:4:1 as can be seen in Fig. 9. The two weaker lines in Fig. 9 are due to the two orientations with their C_2 axes along the [001] direction. The strong line is due to the four orientations with their C_2 axes perpendicular to [001]. The advantage of the EPR measurements is that the orientational degeneracy and C_{2v} symmetry of the oxygen donor are apparent at the outset.

With the addition of uniaxial stress, the 1s ground-state energies of the orientationally inequivalent defects are split and the orientations whose binding energies are decreased by the applied stress can be preferentially depopulated [12]. When the level populations are modified by stress, there are consequent changes in the relative intensities of the components of the EPR spectrum because only the charge state TD^+ is paramagnetic and particular components will correspond to orientations that are preferentially in nonparamagnetic charge states. This apparent alignment is an electronic redistribution alignment and does not correspond to a reorientation of the oxygen donor defects. Thermal ionization cannot be used to depopulate the defects whose binding energies have been reduced (as it was in the infrared absorption experiments discussed above) because the sample becomes too conducting for EPR measurements. Instead, partial occupation of the oxygen donor defects and the preferential depopulation of orientations with reduced binding energy were achieved by compensation [12].

The stress-induced shifts of the ground state energy were studied for TD^+ and TD^0 [12]. If the compensation conditions are chosen so that the Fermi level is near the (+/++) level, orientations whose ground state energies are lowered by

238

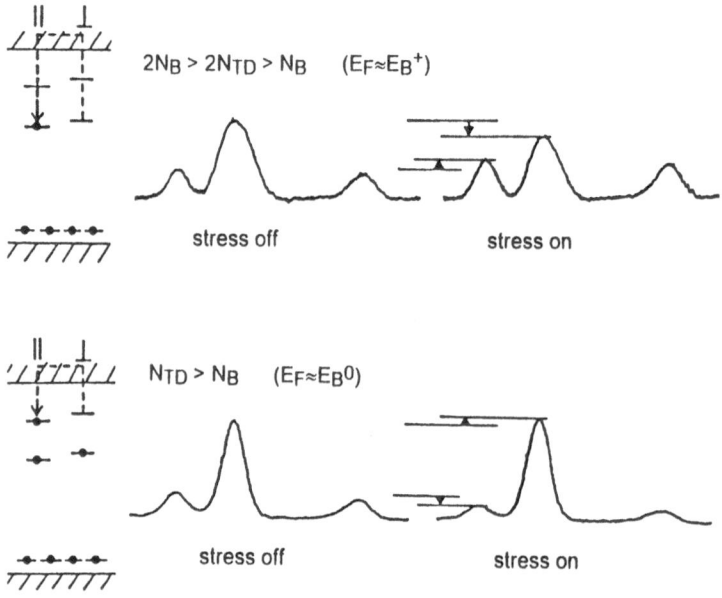

Figure 9. Electronic redistribution alignment for the NL8 EPR spectrum in partially compensated samples with H//[110] and σ//[001]. The energy level diagrams show the oxygen donor levels and the effect of stress on their occupation. After ref. [12].

stress will be preferentially in the paramagnetic charge state, TD^+. EPR spectra are shown in the upper part of Fig. 9 for σ//[001] with the corresponding level diagram. Those lines that correspond to orientations for which stress was applied along their C_2 axis (i.e., their z axis) were increased in intensity, showing that stress along the C_2 axis increases the binding energy of these defects. For stress applied perpendicular to the C_2 axis, the line intensity was reduced. From the orientation dependence and magnitude of the population changes, it was shown that the stress-induced energy shifts were consistent with the conduction band shear deformation potential. These results show that the ground state is constructed from the wave functions associated with the pair of conduction band valleys along the k_z axis.

For the EPR spectra shown in the lower part of Fig. 9, the compensation conditions were chosen so that the Fermi level was near the (0/+) level. For this case, the orientations whose ground state energies are lowered by stress will be preferentially in the *non*paramagnetic charge state, TD^0. In Fig. 9, the sense of the alignment is shown to be reversed from the case discussed in the above paragraph. This result confirms that TD^0 also has a ground state that is constructed from the wave functions associated with the pair of conduction band valleys along the k_z axis.

While most aspects of the EPR results under stress for NL8 are understood, an unexpected effect was also observed [12]. When the Fermi level was near the TD^+ level, the largest stress-induced population redistribution was observed for temperatures near 50K. As the temperature was lowered, the stress-induced redistribution unexpectedly disappeared, even when the samples were cooled under stress. This result remains a puzzle.

6. Localized Strain and Its Effect on the Electronic Structure

Coordinated infrared absorption and EPR experiments were performed by Wagner *et al.* [14, 15] on samples to which stress (600MPa) had been applied during the donor-formation heat treatments. In these experiments, the oxygen donors were found to be preferentially aligned along specific crystallographic directions. In this case, the alignment is a genuine atomic alignment; certain orientations of the C_{2v} defects have higher concentrations than others.

The results of EPR and infrared absorption experiments are shown in Fig. 10 for samples that had been stressed with $\sigma//[001]$ during a 460°C donor-formation anneal [14]. (Unlike most data shown in this paper, the spectra were recorded without stress applied during the measurements.) The changes in the relative intensities of the EPR lines for $H//[110]$ show that the [001] stress-alignment treatment causes the oxygen donor to be preferentially aligned with its C_2 axis perpendicular to the stress direction. The disappearance of the $1s{\rightarrow}2p_0$ transitions in the infrared absorption spectrum for $E//[001]$ shown in Fig. 10 also shows that the stress-alignment treatment has decreased the concentration of oxygen donor defects oriented with their C_2 axes along the [001] direction.

A study of the reorientation kinetics of the aligned oxygen donors leads to the conclusion that the oxygen donors reorient at high temperature and that the alignment that was produced during the donor-growth process represents the equilibrium distribution among the different orientations at the elevated temperature [15]. Qualitatively, the sense of the alignment of the oxygen donors shows that the local strain field that arises from the anisotropic core is compressive along the defect's C_2 axis [14]. This compressive strain causes the energy of an orientation with its C_2 axis along the external stress direction to be increased. And further, this localized compressive strain associated with the oxygen donor's atomic structure provides an natural explanation for the selection of the oxygen donor's ground state; it acts to lower the energy of the k_z-axis pair of conduction band valleys and their associated EMT-like states [14].

The EPR and infrared absorption experiments performed on samples containing stress-aligned oxygen donors have allowed the results from these spectroscopies to be correlated. For example, the relative intensities of transitions to the $np_\pm$ states that are split at zero stress for TD^+ were altered in

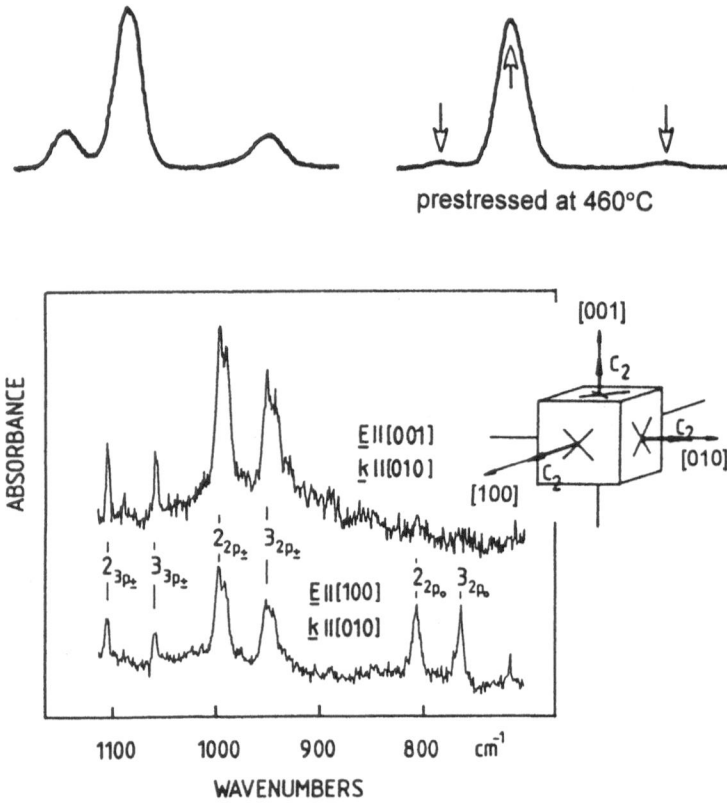

Figure 10. The NL8 EPR spectrum for a sample that had been stressed with σ//[001] during the 460°C donor-growth anneal, compared to a sample prepare without stress. Infrared spectra for a sample stressed during the donor-growth anneal along with a schematic showing the six C_{2v} defects and the crystal axes.

samples containing stress-aligned oxygen donors. It was concluded that the transitions are to p_1 and p_2 states, with representations B_1 and B_2, which are constructed from linear combinations of the $p_{\pm}$ states to be consistent with the C_{2v} symmetry of the oxygen donor. This conclusion is similar to that arrived at from stress studies of the infrared absorption bands (Sec 3.2.2.) but with the added bonus that the directions of the $1s \rightarrow np_1$ and $1s \rightarrow np_2$ transition moments could be correlated with the principal axes of the g-tensor [14].

7. Conclusion

Although the binding energies [3] for the oxygen donor complexes are close to those predicted by EMT, the low symmetry of the oxygen donor defect leads to electronic properties that reflect both the EMT character of the states and the oxygen donor's orientational degeneracy. The piezospectroscopic studies made of the infrared absorption [5-7], DLTS [8-11], and EPR [12] spectra are consistent with a ground state that is constructed from 1s wave functions associated with a single pair of conduction band valleys for charge states TD^0 and TD^+. The close similarity of the results confirm that the spectra measured by these different techniques can be assigned with confidence to the oxygen donor defect. Infrared absorption measurements show splittings for large stresses that indicate a small departure from EMT. The splitting pattern is characteristic of C_{2v} symmetry and is consistent with the C_{2v} g-tensor observed by EPR for the NL8 spectrum. Studies of the polarization properties of the infrared absorption features measured under stress and the stress-induced electronic redistribution among the components of the EPR line each show that the ground state is associated with the pair of conduction band valleys that lie along the k_z axis, where z is the C_2 axis of the defect.

Coordinated EPR and infrared absorption experiments, performed on samples that had been stressed during the growth to produce aligned oxygen donors, show that there is a local compressive strain field along the oxygen donor's C_2 axis [14]. An important consequence of this result is that this localized compressive strain acts to lower the energies of the EMT states associated with a particular pair of conduction band valleys, thereby relating the oxygen donor's electronic structure to its atomic structure.

Acknowledgements

I thank K.M. Lee and L.C. Kimerling for their contributions to our experiments on the oxygen donor's infrared absorption spectra. I also thank G.D. Watkins for several discussions and critical remarks that helped lead to the interpretation of our results. The preparation of this manuscript was supported by NSF Grant No. DMR-9415404.

References

1. Fuller, C.S., Ditzenberger, J.A., Hannay, N.B. and Buehler, E. (1954), *Phys. Rev.* **96,** 833.
2. Kaiser, W., Frisch, H.L. and Reiss, H. (1958), *Phys. Rev.* **112,** 1546.
3. Wagner, P. and Hage, J. (1989), *Appl. Phys. A* **49**, 123.
4. Michel, J. and Kimerling, L.C. (1994), in F. Shimura (ed.), *Oxygen in Silicon*, Academic Press, Boston, p. 251.
5. Stavola, M., Lee, K.M., Nabity, J.C., Freeland, P.E. and Kimerling, L.C. (1985), *Phys. Rev. Lett.* **54,** 2639.

242

6. Stavola, M. and Lee, K.M. (1986), in: J.C. Mikkelsen, Jr., S.J. Pearton, J.W. Corbett, and S.J. Pennycook (eds.), *Oxygen, Carbon, Hydrogen and Nitrogen in Crystalline Silicon,* Materials Research Soc., Pittsburgh, p. 95.
7. Stavola, M. (1987) *Physica* **146B**, 187.
8. Farmer, J.W., Meese, J.M., Henry, P.M. and Lamp, C.D. (1985), *J. Elect. Mat.* **14a**, p. 639.
9. Benton, J.L., Lee, K.M., Freeland, P.E. and Kimerling, L.C. (1985), *J. Elect. Mat.* **14a**, p. 647.
10. Henry, P.M., Farmer, J.W. and Meese (1984), *Appl. Phys. Lett.* **45**, 454.
11. Kimerling, L.C. (1986), in: J.C. Mikkelsen, Jr., S.J. Pearton, J.W. Corbett, and S.J. Pennycook (eds.), *Oxygen, Carbon, Hydrogen and Nitrogen in Crystalline Silicon,* Materials Research Soc., Pittsburgh, p. 83.
12. Lee, K.M., Watkins, G.D. and Trombetta, J. (1985), in N.M. Johnson, S.G. Bishop and G.D. Watkins (eds.), *Microscopic Identification of Electronic Defects in Semiconductors,* Materials Research Soc., Pittsburgh, p. 263.
13. Wagner, P. and Holm, C. (1985), *J. Elect. Mat.* **14a**, p. 677.
14. Wagner, P., Gottschalk, H., Trombetta, J. and Watkins, G.D. (1987), *J. Appl. Phys.* **61**, 346.
15. Watkins, G.D., in this volume; and Trombetta, J.M., Watkins, G.D., Hage, J. and Wagner, P., unpublished.
16. Kohn, W. and Luttinger, J.M. (1955), *Phys. Rev.* **98**, 915.
17. Kohn, W. (1957), in F. Seitz and D. Turnbull (eds.), *Solid State Physics,* Academic Press, New York, vol. 5, p. 257.
18. Ramdas, A.K. and Rodriguez, S. (1981), *Rep. Prog. Phys.* **44**, 1297.
19. Ramdas, A.K. and Rodriguez, S. (1992), in D.G. Seiler and C.L. Littler (eds.), *Spectroscopy of Semiconductors,* Academic Press, Boston,1992, p. 137.
20. Hrostowski, H.J. and Kaiser, R.H. (1958), *Phys. Rev. Lett.* **1**, 199.
21. Bean, A.R. and Newman, R.C. (1972), *J. Phys. Chem. Solids* **33**, 255.
22. Oeder, R. and Wagner, P. (1983), in S. Mahajan and J.W. Corbett (eds.), *Defects in Semiconductors II,* North-Holland, New York, p. 171.
23. Wruck, D. and Gaworzewski, P. (1979), *Phys. Stat. Sol. (a)* **56**, 557.
24. Pajot, B., Compain, H., Leroneille, J. and Clerjand, B. (1983), *Physica* **117B**, 110.
25. Suezawa, M. and Sumino, K. (1983), *Mater. Lett.* **2**, 85.
26. Ho, L.T. and Ramdas, A.K. (1972), *Phys. Rev. B* **5**, 462.
27. Kimerling, L.C. and Benton, J.L. (1981), Appl. Phys. Lett. **39**, 410.
28. Muller, S.N., Sprenger, M., Sieverts, E.G. and Ammerlaan, C.A.J. (1978), *Solid State Commun.* **25**, 987.
29. Martynov, Yu. V., Gregorkiewicz, T. and Ammerlaan, C.A.J. (1995), *Phys. Rev. Lett.* **74**, 2030.
30. Michael, J., Niklas, J.R. and Spaeth, J.-M. (1989), *Phys. Rev. B* **40**, 1732.
31. Gregorkiewicz, T., Bekman, H.H.P.Th. and Ammerlaan, C.A.J. (1998), *Phys. Rev. B* **38**, 3998; Bekman, H.H.P.Th., Gregorkiewicz, T. and Ammerlaan, C.A.J. (1989), *Phys. Rev. B* **39**, 1648.
32. Kaplyanskii, A.A. (1964), *Opt. Spectrosc. (USSR)* **16**, 329.

LOW-TEMPERATURE DIFFUSION AND AGGLOMERATION OF OXYGEN IN SILICON

U. GÖSELE, E. SCHROER, and P. WERNER
Max-Planck-Institute of Microstructure Physics
Weinberg 2, D-06120 Halle, Germany

T. Y. TAN
School of Engineering, Duke University
Durham, NC 27706, USA

In the last couple of years detailed information on the formation of thermal donors has become available which is so specific and unusual that at first sight understanding of the basic processes of thermal donor formation should have been a straightforward task. Contrary to expectations, this has not been the case. The present contribution tries to combine knowledge from thermal donor kinetics with features of observed enhanced long-range diffusion of oxygen to arrive at a set of likely conclusions and still open questions. Special emphasis will be placed on the question of possible processes rendering thermal donors electrically inactive. In a final part high temperature oxygen precipitation at grain boundaries will be dealt with which can lead to precipitates of a defined size in the range of 10-20 nm.

1. Introduction and Background

During Czochralski growth of single crystalline silicon, oxygen dissolved in the silicon melt from the quartz crucible is incorporated into the crystalline silicon in the form of electrically inactive oxygen interstitials in concentrations up to about 10^{18}cm^{-3}. At lower temperatures at which the dissolved oxygen is present in supersaturation it may precipitate out in various forms of silicon-oxide phases. Since in silicon based integrated circuits purposely introduced precipitates of oxygen are routinely used to getter away undesirable metal impurities from the active device region ("intrinsic gettering") the behavior of oxygen in silicon, including its diffusion has been investigated thoroughly over the last four decades. In condensed form the information on oxygen in silicon may be found in a recent book on "Oxygen in Silicon" [1] to which the reader is also referred for a wealth of valuable references. The diffusion of isolated interstitial oxygen O_i from about 350°C up to the melting point of silicon is well characterized by the diffusivity

$$D_i = 0.13 \exp(-2.53 \text{ eV}/kT) \text{ cm}^2/\text{s} \qquad (1)$$

R. Jones (ed.), Early Stages of Oxygen Precipitation in Silicon, 243–261.
© *1996 Kluwer Academic Publishers.*

based on many experiments and among those most notably the early low temperature measurements based on the stress-induced alignment of oxygen interstitials [2]. D_i based on expression (1) is shown in Fig. 1. The precipitation behavior of oxygen at temperatures above about 600°C is comparatively well understood and can accurately be described in terms of the diffusion of interstitial oxygen to the precipitates and the subsequent emission of one silicon self-interstitial for two oxygen interstitials incorporated (required by the volume increase associated with oxide formation) [1].

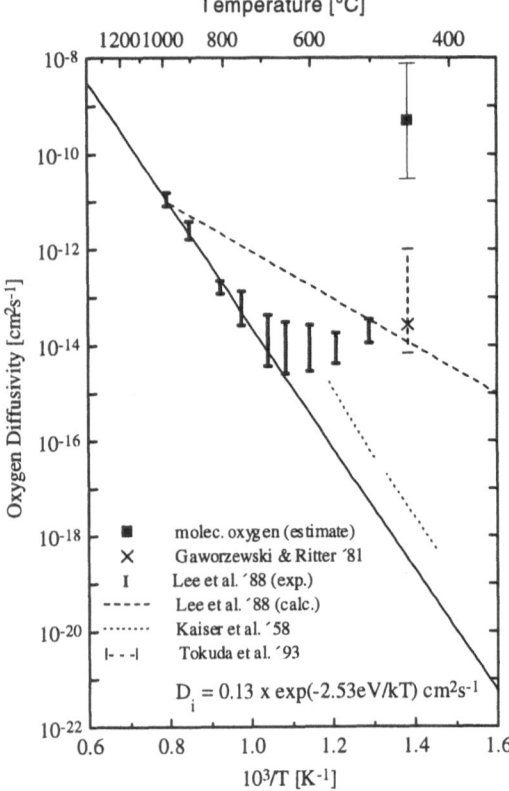

Fig. 1:
Accepted normal interstitial oxygen diffusion coefficient D_i in silicon based on eq. (1) as a function of inverse absolute temperature (e. g. [4]) together with estimates for D_i from thermal donor formation and decay [3] and observed enhanced out-diffusion data [28-30]. A calculated enhanced effective oxygen diffusion based on a dynamical equilibrium between molecular and interstitial oxygen [29,30] as well as an estimate of the diffusivity of the hypothetical oxygen molecule (dimer) [18,21] are also given.

At lower temperatures most prominently in the temperature range of about 400 - 500°C oxygen agglomerates are generated in the form of electrically active complexes which act as donors. They have been termed "thermal donors" since they are generated during a thermal treatment. The first extensive investigation of the formation and decay kinetics of thermal donors was presented by Kaiser, Frisch and Reiss in 1958 who also suggested the first detailed thermal donor model [3]. Even after almost forty years of research this papers clearly stands out and is probably matched in its thoroughness only by the recent paper by the group around Newman [4] which connected the thermal donor formation kinetics with the associated loss of interstitial oxygen over an extended range of temperatures.

In section 2 we will combine information from various earlier papers on thermal donors in silicon and germanium with later papers to point out the requirement any thermal donor model has to fulfill and indicate also the questions which still have to be answered. In section 3 we will discuss experimental indications of enhanced long-range diffusion at low temperatures. Finally, in section 4 we will turn to oxygen precipitation at high temperatures. We will deal with the surprising observation that oxygen precipitates at grain boundaries all appear to grow to an uniform defined size in the range of 10 - 20 nm.

2. Thermal Donor Formation and Deactivation

2.1. EXPERIMENTAL OBSERVATIONS AND SOME CONCLUSIONS

In the following we will combine experimental information from early papers on thermal donor formation and deactivation in silicon [3,5] and partly also germanium [6,7] with newer results obtained more recently [1,4,8-15]. We consider the following experimental observations and resulting conclusions as essential for any thermal donor model:

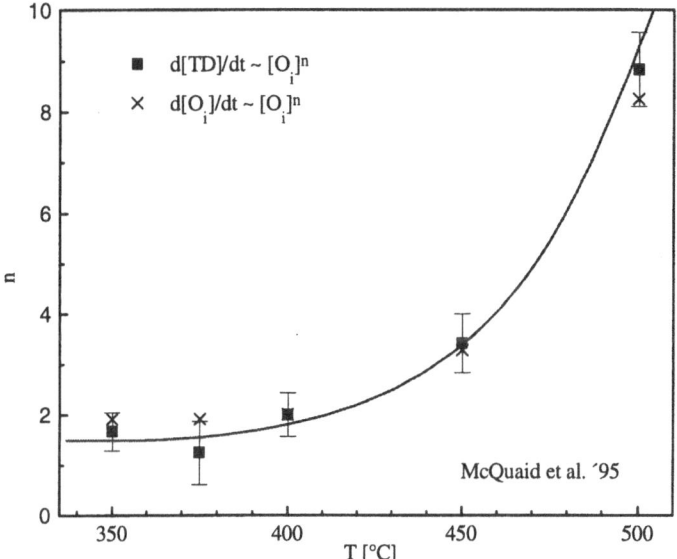

Fig. 2:   Reaction order n for thermal donor formation and the loss of oxygen interstitials as a function of temperature [4].

i) The initial thermal donor generation rate is strongly dependent on the interstitial oxygen concentration C_i via

$$(dC_{TD}/dt)_o = k_f (T) \, C_i^n \qquad (2)$$

where C_{TD} is the thermal donor concentration. The reaction order n at 450°C is around four. This has lead to the interpretation that four oxygen atoms are involved in the thermal donors as proposed by Kaiser et al. [3]. This reaction order of four has shaped the thinking about thermal donors over many decades. The observation of the

246

Newman group [4] that the reaction order actually changes from 2 at 350°C over 3.5 - 4 at 450°C to about 8 at 500°C (Fig. 2) is essential in terms of understanding thermal donor formation since it eliminates the fixation on the special role of the case of n = 4. A similar change-over from 2 at lower temperatures to 4 at higher temperatures was also found for the thermal donor formation kinetics in germanium [6,7]. The generation rate increases with increasing temperature with an approximate activation energy of about 1.8 eV [9,10]. At even higher temperatures the generation rate decreases with increasing temperature [15]

ii) The thermal donor concentration reaches a maximum concentration C_{TD}^{max} which is also associated with the oxygen concentration via

$$C_{TD}^{max} = k_{max} (T) C_i^m \tag{3}$$

where the power m is given by 3 at 450°C [3]. The maximum concentration decreases with increasing temperature, which is one of the reasons why at higher temperatures (say at 600°C) thermal donor concentrations are negligible. After reaching the maximum concentration the thermal donor concentration decreases again.

iii) Before reaching the maximum concentration the thermal donor generation rate may in good approximation be described by [5]

$$(d\ C_{TD}/\ dt) = k_d\ (T)\ (C_{TD}^{max} - C_{TD}) \tag{4}$$

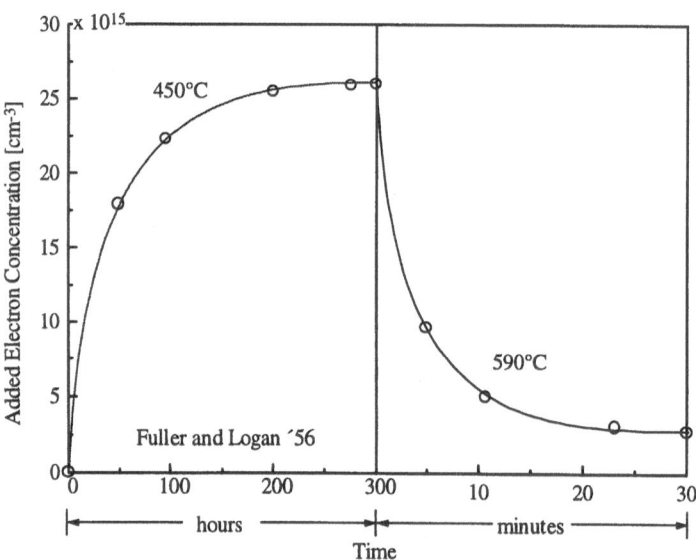

Fig. 3: Thermal donor concentration in terms of added electron concentration as a function of time first at 450°C (left) and then during a subsequent annealing at 590°C [5].

iv) When samples containing already thermal donors are annealed at a higher temperature than their previous generation temperature, the already introduced thermal donors decay with the same expression (4) containing now the lower maximum thermal donor concentration at the higher annealing temperature as schematically shown in Fig. 3 [3,5]. The coefficient k_d (T) describes a deactivation process of the thermal donor. Whether this deactivation process is a dissociation of the thermal donor back to dispersed oxygen interstitials (as it appears to be the case in germanium [6,7]) or a transformation into some other form of oxygen agglomerates, e.g. a normal precipitate is not known from the very beginning. The deactivation coefficient k_d(T) is thermally activated with an activation energy of about 2.8 eV (which is close to the activation energy of interstitial oxygen diffusion of 2.53 eV). From combination of equations (2), (3), and (4) at 450°C where it is known that n=4 and m=3 it follows that k_d is proportional to the interstitial oxygen concentration. Kaiser et al. [3] have attributed thermal donor deactivation to diffusion of oxygen interstitials to thermal donors due to their attachment. Within this interpretation the following expression for k_d should hold

$$k_d\ (T)\ =\ 4\ \pi\ r_{TD}\ D_i C_i \tag{5}$$

where r_{TD} is the radius of the thermal donor. The interstitial oxygen diffusivities calculated from the available data are 10-50 times higher than the known D_i from eq. (1). This discrepancy may possibly be explained in terms of some remaining hydrogen enhanced oxygen diffusivity as discussed by Newman and Jones [15]. Extending the results from 450°C to other temperatures based on (5) we suggest that the following approximate relation between the exponents n and m should hold

$$m = n - 1 \tag{6}$$

assuming that eq. (5) provides a valid description of the deactivation process.

v) The thermal donors consist of not just one configuration but of a number of configurations which develop successively [1,12]. By scaling the time and the concentration by temperature-dependent factors the same basic sequence may be obtained over a wide temperature range from 400 - 500°C [10]. During a subsequent treatment at higher temperatures than a previous thermal donor generation treatment all individual thermal donor species also appear to be deactivated exponentially [16].

vi) In the case of thermal donors in germanium it was shown very early that the loss of interstitial oxygen is closely related to the increase of the thermal donor concentration. At higher temperatures (>350°C) about four and at lower temperatures about two oxygen interstitials are lost for each thermal donor formed [6,7]. In addition, the reaction is reversible indicating a dynamical equilibrium between agglomeration and dissociation. In silicon the situation is different. In agreement with earlier experiments by Tan et al. [8] and Guse and Kleinhenz [13], the Newman group [4] showed that the loss of interstitial oxygen may approximately be described by the formation rate of oxygen dimers with the normal oxygen diffusivity, at least at temperatures below about 450°C. The key experimental observation, however, is that the reaction rate of the oxygen loss appears to be described by the same reaction order n as the thermal donor formation rate according to eq. (2)

$$- (d\ C_i/\ dt)/N_{TD} = (d\ C_{TD}/\ dt) = k_f\ (T)\ C_i^n \tag{7}$$

where the proportionality factor N_{TD} describes the number of oxygen atoms lost for each thermal donor generated (Fig. 2) [4]. This factor turns out to be around 8 - 12 independent of temperature in agreement with earlier measurements at 450°C [11]. Based on the presence of different forms of thermal donor species it is likely that N_{TD} describes the actual average number of oxygen atoms in thermal donor species and that at least initially practically all the lost oxygen interstitials will end up in thermal donors. This picture changes beyond the maximal donor concentration where the loss of oxygen interstitial proceeds without an increase in thermal donor concentration. This is different to the case of germanium where the loss of oxygen interstitials stops after a quasi-equilibrium concentration has been reached [6,7], which indicates a dynamical equilibrium between thermal donor formation by oxygen agglomeration and dissociation of thermal donors back to oxygen interstitials. In the case of oxygen it is more likely that the deactivation reaction limiting the thermal oxygen concentration is **not** the dissociation back to oxygen interstitials but rather the diffusion of interstitial oxygen to thermal donors (of any kind) and the triggering of a transformation to a nucleus of a normal oxygen precipitate according to

$$O_i + TD \rightarrow \text{oxygen precipitate} + I. \qquad (8a)$$

In (8a) the possible generation of one or more silicon self-interstitials (I) required for adapting the volume increase associated with the oxide formation is indicated. These self-interstitials will then agglomerate in terms of rod-like or similar point-defect clusters [17]. Alternatively, thermal donors could be deactivated by a thermally activated transformation to precipitates without the involvement of oxygen interstitials according to

$$TD \rightarrow \text{oxygen precipitate} + I. \qquad (8b)$$

Then instead of the proposed relation (6) $m = n$ should hold. In both cases the deactivated thermal donors do not emit oxygen interstitials back to the reservoir of oxygen interstitials in solution. The thermal donors are rather transformed into a precipitated form of oxygen in silicon. The same holds by annealing out of thermal donors at slightly higher temperatures. Already Fuller and Logan [5] pointed out that the thermal donor annealing at higher temperatures is not a reversible process since a subsequent thermal donor treatment at the lower temperature shows a much lower thermal donor formation rate indicating the permanent loss of oxygen interstitials. Nevertheless, a small percentage of the oxygen in the thermal donors deactivated during a higher temperature treatment goes back into solution [14].

vii) The observed loss of interstitial oxygen based on its normal diffusivity and the simultaneous generation of a whole sequence of oxygen containing thermal donors require the existence of a fast diffusing species which is most likely a complex of two oxygen atoms as first suggested in 1982 by two of the present authors [18]. Whether this oxygen dimer exists as an oxygen molecule hardly attached to the silicon lattice [18] or a complex of interacting oxygen interstitials [19] remains unclear.

ix) The presence of carbon does not increase the loss of oxygen interstitials but strongly increases the formation rate of thermal donors and at the same time leads to a decrease of the concentration of carbon on substitutional sites (1). All these observations may consistently be explained in terms of the trapping of fast diffusing oxygen dimers by substitutional carbon and pushing the carbon into interstitial sites [4].

x) Already Kaiser et al. [3] reported that thermal donor formation is enhanced in silicon crystals grown in a hydrogen atmosphere. In the meantime it has especially been shown by the Newman group [4] and others [1,20] that hydrogen enhances the diffusivity of interstitial oxygen. Consequently, the loss rate of oxygen interstitials and the thermal donor formation rate increases in the presence of hydrogen but hydrogen does not appear to influence the structure of thermal donors or the number of interstitial oxygen atoms lost for each thermal donor formed [4].

2.2. THERMAL DONOR MODEL: FORMATION AND DEACTIVATION

Kaiser et al. [3] suggested that thermal donors consist of four oxygen atoms and that these thermal donors could be deactivated by the attachment of an additional oxygen atom. Gösele and Tan [18] pointed out that the formation of a complex of four oxygen atoms requires a fast moving species and suggested oxygen molecules (O_2) which were assumed to be in a kind of steady state concentration due to a dynamical equilibrium between their formation and dissociation. Otherwise the suggested thermal donor model was similar insofar as the thermal donors consisted of four oxygen atoms (formed by two oxygen molecules) and that the deactivation occurred by the diffusion of oxygen interstitials to the thermal donors triggering a change into a normal precipitate according to reaction (8a). The required diffusivity D_2 of oxygen molecules was estimated [18] to be about nine orders of magnitude larger than D_i at 450°C (Fig. 1). The model in this form did not account for the sequentially occurring different thermal donors. In addition, the experimental observation that the loss rate of interstitial oxygen at 450°C could well be modeled by assuming that two oxygen atoms come together by the normal interstitial oxygen diffusivity and do not dissociate [8] was contradicting the assumption that oxygen molecules would dissociate easily. In a paper in 1989 Gösele et al. [21] explained the subsequent formation of a multiplicity of thermal donors by assuming that oxygen molecules could attach to existing thermal donor complexes to form the next species of thermal donors. In order to preserve the 4th order kinetics at 450°C it was assumed that the highly mobile oxygen molecules could become immobile by reacting with the silicon lattice under the emission of a silicon self-interstitial. These immobilized oxygen dimers were assumed to be part of a normal precipitation process and not to be involved in the thermal donor formation. Implicitly, it was assumed that thermal donor formation was a minor parallel reaction to an oxygen precipitation process independent of thermal donor formation. The results of the Newman group [4] comparing the loss rate of interstitial oxygen and the formation rate of thermal donors make it much more likely that basically the whole loss of oxygen interstitials is transformed into various species of thermal donors. Therefore, here we come back to the original thermal donor model suggested in 1982 and modified by the introduction of a multiplicity of donors by the attachment of additional oxygen molecules but without the assumption of an immobilization process of an oxygen molecule by the emission of a self-interstitial. The basic ingredients are as follows:

a) Oxygen interstitials come together to form fast diffusing oxygen dimers (in the form of molecules or some other form). The dissociation of dimers is temperature dependent and is so low at 400°C and lower temperatures that it can be neglected for thermal donor formation.

b) Fast diffusing oxygen dimers react with each other and subsequently with dimer agglomerates to form a sequence of O_4, O_6, O_8, O_{10}, O_{12}, O_{14}, and so on. These dimer aglomerates constitute the different species of thermal donors. From experimental results it can not be decided whether the thermal donor activity starts already at O_4, at O_4 or at even higher agglomerates. The result from thermal donors in germanium at low temperatures (300°C) that each thermal donor contains only two oxygen interstitials [6,7] point to the possibility that already O_2 might be a thermal donor species.

c) The observation that the formation of different thermal donor species is purely sequential indicates a chain-like linear arrangement of the dimers involved since otherwise different locations for attachment of additional oxygen dimers should be available and parallel sequences of thermal donors could develop.

d) The observation that the number of oxygen lost per thermal donor generated is approximately 10 independent of temperature and time indicates that the development of the various thermal donor species happens quickly. Medium size thermal donor clusters are relatively stable. Higher order thermal donors involving more than say 10 oxygen atoms become increasingly more likely either to be thermally unstable or not being able to capture additional oxygen molecules. The average number of oxygen per thermal donor is therefore limited to a certain number due to the inability of larger clusters to permanently incorporate further oxygen molecules.

e) The change of the reaction order from 2 to 8 with increasing temperature is associated with the increasing instability of smaller oxygen clusters, including that of oxygen dimers. Within this picture thermal donor species become energetically more favorable with increasing numbers of oxygen dimers involved up to a certain energy minimum for agglomerates with say 5-7 dimers (10-14 oxygen atoms). For higher numbers of dimers the agglomerates become increasingly energetically unfavorable.

f) The deactivation of the various thermal donor species is mainly due to reaction with diffusing oxygen interstitials which trigger a transformation to a normal precipitate possibly associated with the generation of a fast diffusing silicon self-interstitial. These self-interstitials may agglomerate to form rod-like defects. We consider it as less likely that the deactivation is due to either a spontaneous transformation (without the involvement of interstitial oxygen) to a precipitate or due to the dissociation of thermal donors, although a small contribution of dissociation when heating thermal donors to higher temperatures can not be excluded. Part of the reason why the oxygen diffusivity calculated from the deactivation process is 10-50 times higher than predicted by eq. (1) might be associated with the presence of some hydrogen enhancing the interstitial oxygen diffusion. Nevertheless, this discrepancy remains the weak point in the model presented

g) The deactivation mechanism described constitutes a way of generating nuclei of normal precipitates. This nucleation path exists besides the usual one which is not operating efficiently below about 500°C because of the low diffusivity of oxygen interstitials.

h) A determination of the proper reaction coefficients and the diffusivity D_2 of oxygen dimers and its dissociation constants and that of dimer agglomerates remains to be determined self-consistently. Attempts to model thermal donor formation with changing order of reaction from 2 to 8 with increasing temperature restricted the oxygen dimer diffusivity to values close to about 10^4 x D_i although the diffusivity suggested for molecular oxygen of 10^9 x D_i by two of the present authors [18] which corresponds to about 10^{-9} cm^2/s (shown in Fig. 1) is still orders of magnitude lower than the diffusivity of fast interstitial diffusers such as copper of about 10^{-5} cm^2/s. The trapping of fast diffusing oxygen dimers by oxygen interstitials should only occur in terms of a dynamical equilibrium since otherwise the efficient building up of higher thermal donor species would not be possible. There is also the possibility that oxygen dimers may be trapped by oxygen precipitates or point defect clusters.

In conclusion, we consider the thermal donor formation as the building up of a metastable oxygen dimer or molecular phase formed by the agglomeration of fast diffusing oxygen dimers in terms of a linear chain. This molecular phase can "leak" into the normal oxygen precipitation with a much higher rate than normal oxygen nucleation at this temperature would allow. Presently, there are no direct experimental indications for the presence of oxygen dimers with the exception of the most recent report of an infrared-active center introduced by hydrogenation of oxygen rich silicon at 275°C [22]. In the following section we will discuss some indications of enhanced long-range diffusion of oxygen at low temperatures.

3. Low-Temperature Long-Range Oxygen Diffusion

Although there are a number of reports on experimental indications of enhanced long-range oxygen diffusion in silicon no mechanism has been proposed which is consistent with all available data of enhanced long range oxygen diffusion and thermal donor formation. The present section will also rather point out some experimental results and open questions than offer a consistent solution.

Magee et al. [23] reported an increased concentration of oxygen in the abraded region of Cz silicon during annealing at 600°C and attributed this to enhanced long range diffusion of oxygen from the bulk to the abraded surface. It was later shown that the gettered oxygen in abraded and also arsenic implanted areas was due to diffusion of oxygen from the outside atmosphere [24,25]. The required diffusivity of the indiffusing oxygen was drastically enhanced compared to the diffusivity of interstitial oxygen. Since the silicon region in which this enhanced diffusion occurred was heavily damaged no further detailed interpretation of these results was suggested.

In 1981 Gaworzewski and Ritter [26] showed that the thermal donor formation rate decreased towards the surface. Interpreting this as due to oxygen out-diffusion they arrived at an effective oxygen diffusivity enhanced by more than four orders of magnitude compared to normal interstitial oxygen diffusivity. Gösele and Tan [18] interpreted this result in terms of fast diffusing oxygen molecules which were in dynamical equilibrium between formation and dissociation according to

$$O_i + O_i \underset{\leftarrow}{\overset{\rightarrow}{\rightleftharpoons}} O_2 \qquad (9)$$

which leads to an effective oxygen diffusivity given by

$$D_{eff} = D_i + 8\pi r_i D_i C_i \, D_2/(k_{diss}) \qquad (10)$$

where r_i is the reaction radius of oxygen interstitials and k_{diss} the ,dissociation coefficient of dimers. Eq. (10) predicts that the effective oxygen out-diffusion should increase with increasing oxygen concentration. Contrary to this expectation Hahn [27] found that the depth of the region of low donor generation rate decreases with increasing oxygen concentration. A similar result was obtained by Tokuda et al. [28] who also confirmed enhanced out-diffusion of oxygen at 450°C by secondary ion mass spectroscopy (SIMS). The effective out-diffusion coefficient decreases with increasing time (Fig. 4) and appears to decrease with increasing loss of interstitial oxygen by thermal donor formation and precipitation. All these observations indicate that the fast diffusing oxygen dimers may be trapped and/or decreased in concentration by the presence of thermal donors or precipitates.

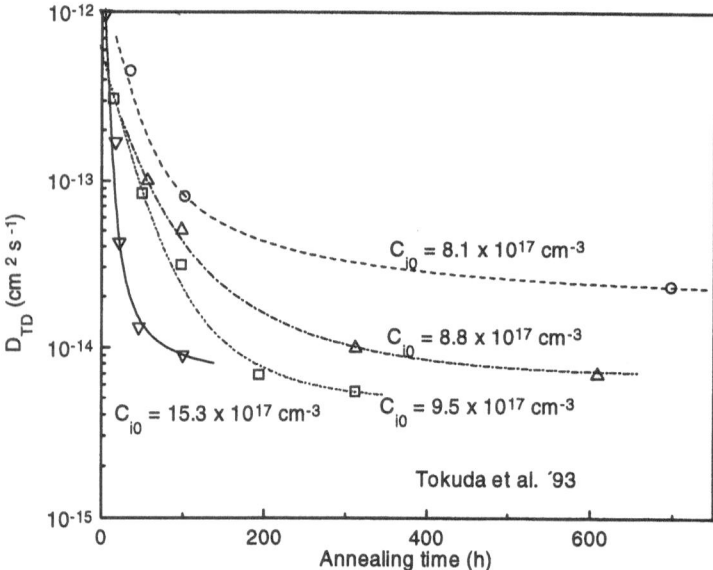

Fig. 4: Diffusivity D_{TD} as determined by the thermal donor concentration profile close to the surface for different interstitial oxygen concentrations in dependence of annealing time at 450°C [28].

Lee et al. [29,30] measured the out-diffusion of oxygen isotopes from silicon at temperatures from about 450°C to 1000°C. Below about 700°C they observed enhanced diffusion. One example of an out-diffusion profile is shown in Fig. 5. The obtained diffusivities are plotted in Fig. 1. The bars indicate the scatter in the data measured. The data do not show the expected concentration dependence derived from eq. (10) but rather an unsystematic variation from wafer to wafer. A local thermal donor generation rate which does not correlate with the local oxygen concentration but rather with the presence of defect agglomerates was observed by Rava et al. [31]. This behavior might be associated with interactions of oxygen dimers with grown-in defects. In analogy, enhanced oxygen out-diffusion might not only vary from wafer to wafer but also locally on a given wafer.

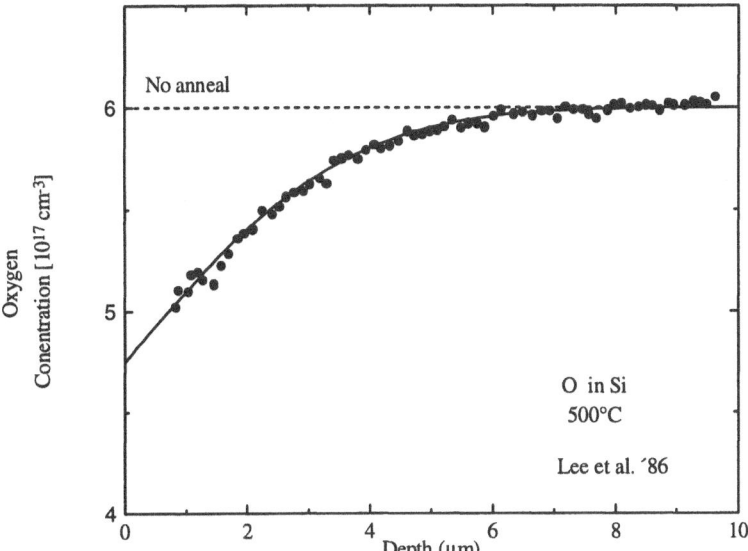

Fig. 5: ^{16}O concentration profile in silicon annealed at 500°C for 17 days as measured by SIMS. The solid curve is the best fit based on a constant effective diffusivity [29, 30].

Lee et al. [29,30] also observed many micrometer deep diffusion tails of implanted oxygen into silicon at annealing temperatures as low as 425°C as shown in Fig. 6. The exponentially decaying concentration tails were interpreted as fast diffusing oxygen molecules dissociating with a rather small dissociation coefficient. Of course, any radiation-induced and fast diffusing complex between oxygen and an intrinsic point defect decaying during the process of in-diffusion would lead to the same basic features. Within the interpretation of fast diffusing oxygen molecules the parameter $D_2/(k_{diss})$ may be determined from the exponentially decaying diffusion tails and used to calculate the effective diffusivity based on eq. (10). The resulting expression is [30]

$$D_{eff} = D_i + 4.3 \times 10^{-26} \exp(-0.88 \text{ eV}/kT) C_i \text{ cm}^4/\text{s} \qquad (11)$$

which is shown in Fig.1 for an interstitial oxygen concentration of 10^{18} cm^{-3}. Eq. (11) describes the out-diffusion results at around 500°C reasonably well but it clearly overestimates oxygen diffusion at higher temperatures considerably. Obviously, no clear-cut consistent description is available. It might be argued that at temperatures above 500°C already oxygen precipitation via thermal donor deactivation occurs. These precipitates might trap oxygen molecules and thus reduce the effective

254

diffusivity. Since eq. (11) would predict an increased effective oxygen diffusivity even at 800°C which is in clear contradiction with results from precipitation experiments [4] we may rather conclude that eq. (11) is not valid possibly because the in-diffusion results were interpreted incorrectly.

In conclusion, it is fair to state that there are a number of indications for enhanced low-temperature long-range diffusion of oxygen but that presently no self-consistent explanation in terms of fast diffusing oxygen molecules or any other fast diffusing species is available. We leave this subject by mentioning that in the case of nitrogen diffusion in silicon the enhanced long-range diffusion was observed and attributed to fast diffusing nitrogen molecules [32].

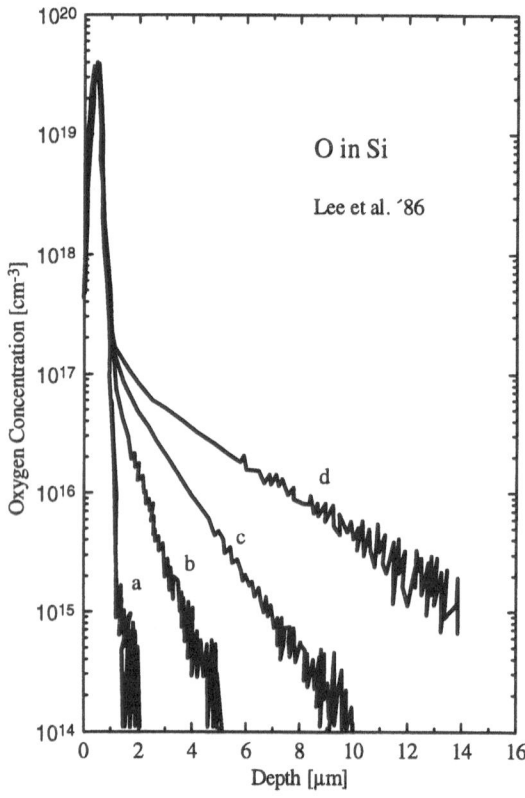

Fig. 6:
SIMS profiles of ^{18}O concentration in silicon. Curve a: as implanted; and annealed for 67 h at 525°C (curve b), at 480°C(curve c) or 425°C (curve d) [29].

4. Growth of Oxygen Precipitates to a Defined Size

So far we have discussed the behavior of oxygen in silicon at relatively low temperatures. This section deals with the final stage of oxygen precipitation at grain boundaries in crystalline silicon at high temperatures and long annealing times. In crystalline silicon oxygen precipitation is known to be governed by the diffusion of oxygen to the precipitates. Precipitates initially nucleated grow until the interstitial oxygen present in a supersaturated concentration is reduced nearly to the solubility value. The size distribution of the precipitates is spread out over a wide range. The final process of precipitation is governed by Ostwald ripening. Ostwald ripening describes the reduction of interface energy by increasing the volume of large precipitates and dissolution of smaller precipitates. For more information on oxygen precipitation in bulk silicon e. g. the recent review article by Borghesi et al. [33]. Finally the size distribution of the precipitates is spread out over a wide range. In the present paper we report on the special case of oxygen precipitation at silicon grain boundaries which results in precipitates within a narrow size distribution.

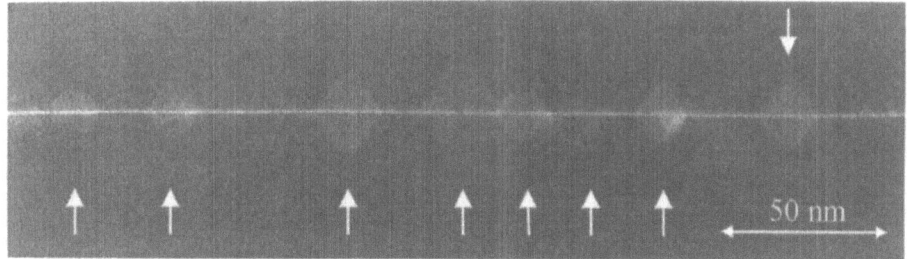

Fig. 7: TEM cross sectional micrograph of octahedron shaped SiO_2 precipitates at a silicon twist grain boundary. The white line from left to the right of the image is the twist boundary and the lighter areas indicated by the arrows are the SiO_2 precipitates. The mean width of the precipitates is $w_b = 15$ nm.

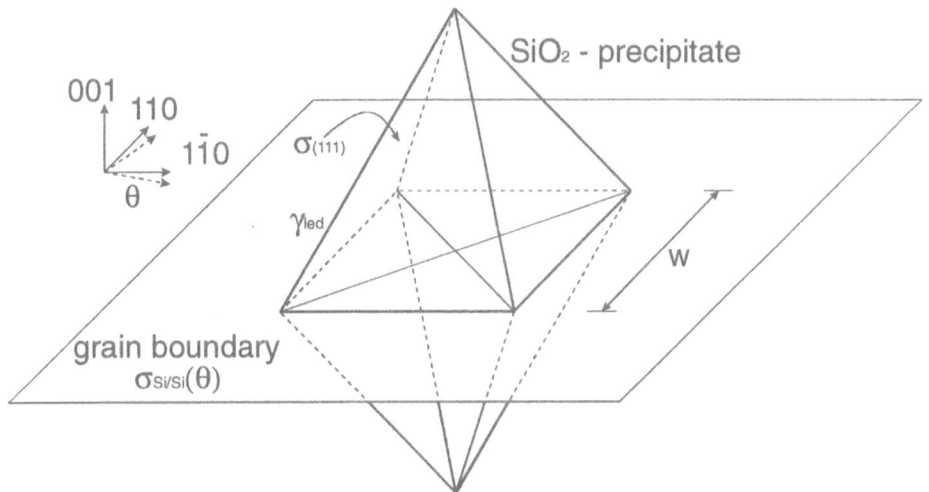

Fig. 8: Schematic picture of the octahedron shaped SiO_2 precipitates at the silicon grain boundary. The interface energy $\sigma_{(111)}$ and the ledge energy γ_{led} are depicted. The twist angle θ of the grain boundary is indicated by the dashed arrows of the crystallographic coordinate system.

We investigated the growth of SiO_2 precipitates at twist grain boundaries in otherwise single crystalline CZ-silicon. The twist grain boundaries were fabricated by hydrophobic wafer direct bonding of (001) oriented CZ-grown silicon wafers intentionally rotated by an angle of about 12° with respect to each other. Hydrophobic wafer direct bonding is known to produce a contact between the wafers without any interface layer which separates the two silicon crystals [34]. Subsequently, the bonded wafer pair containing the twist boundary was annealed at 1100°C for up to 5 days. Finally the SiO_2 precipitates generated during the high temperature annealing at the grain boundaries were investigated by conventional and high resolution transmission electron microscopy (HR)TEM. A cross sectional TEM micrograph of the SiO_2 precipitates after 5 days of annealing at 1100°C is shown in Fig. 7. The incident electron beam is parallel to the (011) orientation of one of the silicon wafers.

The interfaces between the SiO_2 precipitate and the silicon matrix are parallel to the {111} planes of the silicon. This is consistent with Jaccodine´s [35] argument that the specific interface energy σ of silicon is the lowest in (111) orientation because of the low bond density in this direction. The surface energy of the SiO_2 is not regarded in this argumentation because it is amorphous and thus has no preferred orientation. From these considerations the geometrical form of the SiO_2 precipitates can be determined to be of octahedron shape which is schematically shown in Fig. 8. In the TEM micrograph one can observe that the precipitate shape deviates slightly from the ideal octahedron shape. In most of the SiO_2 precipitates the top and the bottom of the octahedra is missing. The reason for this is the low surface to volume ratio at the peak ends of the octahedra. The deviation from ideal octahedron shape does not alter the principle of the following derivation. Therefore, it is neglected in the further discussion.

The statistical analysis of the precipitate size w_b distribution at the 12° twisted grain boundary is shown in Fig. 9. The mean size of the precipitates is $w_b = 15$ nm with a standard deviation of 3 nm. These values are obtained by fitting a Gaussian profile to the experimentally observed size distribution. However, there are a few larger SiO_2 precipitates at the grain boundary which were most likely formed by the coalescence of two or more octahedron shaped precipitates. We point out, that coalescence and Ostwald ripening are two principally different mechanisms. Coalescence needs a contact of the precipitates, whereas this is not necessary for ripening. Coalescence only occurs in the case of a high precipitate density. The SiO_2 precipitate density at the investigated twist boundary is fairly high as can be seen in Fig. 1 and therefore the coalescence of two precipitates is likely to occur.

As already mentioned the Ostwald ripening describing the minimization of total energy σ_{tot} by reducing the interface energy $\sigma_{int} = 2\sqrt{3}\sigma_{(111)}w^2$ does not account for a defined size of the SiO_2 precipitates even if the specific interface energy $\sigma_{Si/Si}$ at a grain boundary is considered. We could get either a varying precipitate size similar to the bulk case for low $\sigma_{Si/Si}$ or a continuous layer at the interface for sufficiently large $\sigma_{Si/Si}$. Therefore in order to account for the observed precipitate size we have to take into account an additional energy contribution in the minimization of the total energy of the system. We propose that this additional contribution is due to the energy of the ledges of the SiO_2 precipitate given by $\sigma_{led} = 12\gamma_{led}w$ where is the weighted averaged specific ledge energy γ_{led} per unit length and the factor 12 comes from the 12 ledges in the octahedron. Considering the ledge energy the total energy per unit area of grain boundary at the grain boundary is given by

$$\sigma_{tot} = n(2\sqrt{3}\sigma_{(111)}w^2 + 12\gamma_{led}w) + (1 - nw^2)\sigma_{Si/Si}, \tag{12}$$

where n is the density of the precipitates. A detailed analysis of the minimization procedure has been derived by Schroer et al. [36]. The result is a specific size w_b

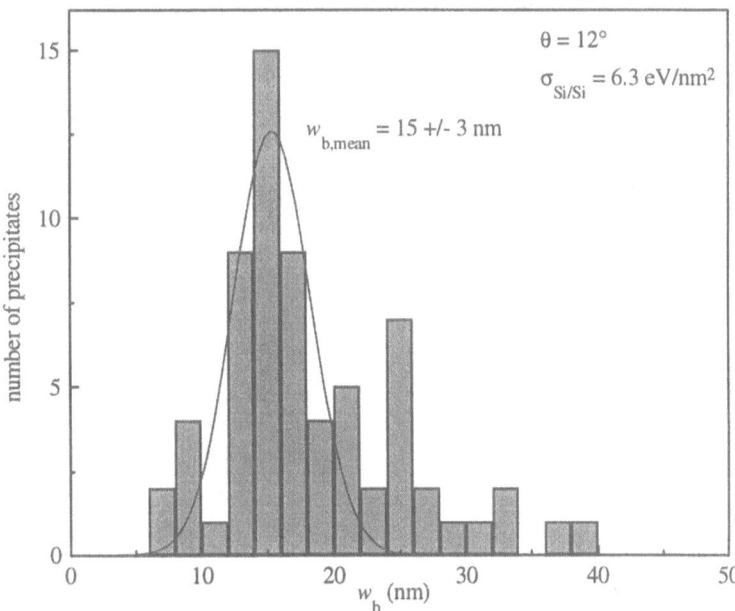

Fig. 9: Histogram of the distribution of the SiO_2 precipitate size w_b. The mean size of the SiO_2 precipitates is $w_b = 15$ nm with a standard deviation of 3 nm. The SiO_2 precipitates with the larger size are generated by coalescence of two or more octahedron shaped SiO_2 precipitates.

where the total energy attains a minimum. This size is given by

$$w_b = \frac{24\gamma_{led}}{\sigma_{Si/Si} - 2\sqrt{3}\sigma_{(111)}}. \tag{13}$$

The optimum size is independent of the amount of precipitated oxygen. It only depends on material parameters and energetic of the grain boundary determined by the twist angle θ in the case of the twist boundary used. An increased density n of the precipitates accounts for an increased overall volume of the precipitates. The volume $v = \sqrt{2}nw_b^3 / 3$ per unit area corresponds to the thickness of a continuous oxide layer with the same volume density as the layer of precipitates at the grain boundary.

For understanding the dependence of the total energy on the precipitate size in more detail we may rewrite eqn. (12) in a normalized form

$$\frac{\sigma_{tot} - \sigma_{Si/Si}}{12\gamma_{led}} = \frac{1}{w^2} - \frac{(\sigma_{Si/Si} - 2\sqrt{3}\sigma_{(111)})}{12\gamma_{led}}\frac{1}{w}. \tag{14}$$

This energy dependence is shown in Fig. 10 for various parameters $a = (\sigma_{Si/Si} - 2\sqrt{3}\sigma_{(111)})/\gamma_{led}$ as a function of w. From this drawing one can distinguish two ranges of a.

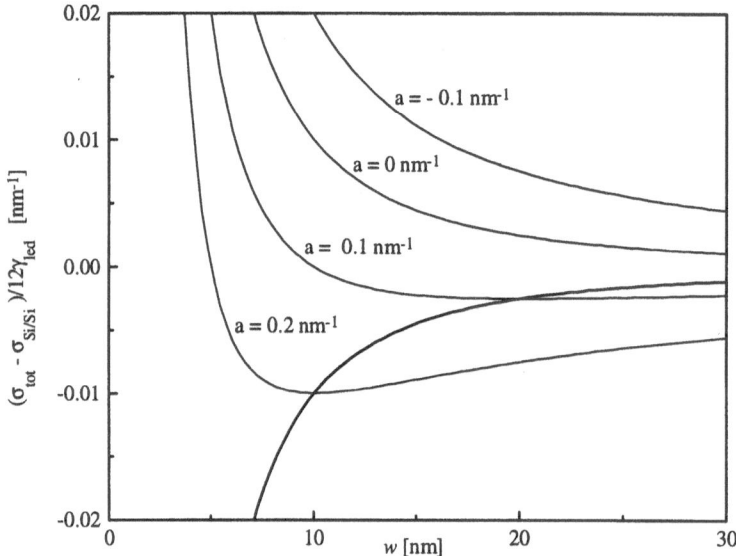

Fig. 10: Normalized total energy as a function of the precipitate size w for several values of the parameter a. In the case of parameters $a < 0$ the final stage of precipitation is governed by Ostwald ripening and for values of $a > 0$ the generation of precipitates of a defined size w_b is predicted. The bold curve connects the minima of the curves drawn.

For $-2\sqrt{3}\sigma_{(111)} \leq a \leq 0$ the value of w at the minimum of these energy curves corresponding to w_b for the respective value does not exist. In this range the energy minimization is achieved by increasing the size and reducing the density of the precipitates. Therefore, this is the range of normal Ostwald ripening. The lowest value of $a = -2\sqrt{3}\sigma_{(111)}$ is attained if the grain boundary energy is zero. This is the case for a silicon crystal without a grain boundary. The effective interface energy is reduced at the silicon grain boundary compared to the silicon matrix. Therefore, the critical size for nucleation of precipitates is smaller at the grain boundary. This leads to an enhanced nucleation rate due to the homogeneous nucleation at the grain boundary. Additionally, the density of nucleation sites for heterogeneous precipitation is most likely high at the grain boundary compared to the silicon matrix.

$a > 0$ corresponds to the case of energy minimization by the growth of SiO_2 precipitates to a defined size w_b given by eq. (13). In this case the effective surface energy of the precipitate is negative. Therefore, the precipitates will try to stay as small as possible, without a ledge energy this would lead to a thin continuous oxide layer at the grain boundary. Due to the presence of a ledge energy a defined maximum size of precipitates results. Further oxygen precipitation is not accomplished by further increase of the size of existing precipitates but rather by the nucleation of new precipitates. The increasing precipitate density accounts for an increasing volume of precipitated oxygen. Additionally, the nucleation rate is as well enhanced at the grain boundary.

Let us now apply these theoretical consideration to the presented experimental results. The grain boundary energy of a 12° twisted boundary is estimated to be $\sigma_{Si/Si} = 6.3 \text{eV} / \text{nm}^2$ and the maximal precipitate size measured is approximately $w_b = 15 \text{ nm}$. From this knowledge the largest value of the interface energy $\sigma_{(111)} < 1.8 \text{eV} / \text{nm}^2$ can be estimated from the fact that $a > 0$ holds in this case. This value is consistent with the interface energies determined from nucleation experiments which are reported to be in the range between $0.56 \text{ eV} / \text{nm}^2$ and $3.13 \text{ eV} / \text{nm}^2$.

Let us further consider the case of a continuous layer. Under this conditions $nw_b^2 = 1$ holds. From this relationship one can determine the required SiO_2 precipitate density at the grain boundary for the formation of a continuous layer to be $n = 4.3 \times 10^{11} \text{cm}^{-2}$. One can further calculate the effective thickness of this layer from $v = \sqrt{2} / (3a)$ to be $v = 3.6 \text{nm}$.

Finally we comment on the redistribution of an initially continuous SiO_2 layer by the generation of SiO_2 precipitates at a twist boundary. Ahn et al. [37] observed the redistribution of an initially continuous SiO_2 layer to a layer of precipitates under the condition that the twist angle between the wafers is less than $\theta = 3°$. For larger angles θ they did not observe any redistribution and the SiO_2 layer stays continuous and even increases in thickness when sufficient oxygen is transported to the layer. Ahn et al. [37] interpreted these results as a consequence of the fact that the interface energy $\sigma_{Si/Si}$ increases with increasing twist angle θ. For low twist angles the grain boundary energy is lower than the interface energy between the silicon and the SiO_2. In this case the total energy can be reduced by the redistribution of the SiO_2 from a continuous layer to precipitates. For larger twist angles θ the formation of the grain boundary requires more energy $\sigma_{Si/Si}$ per unit area than the gain of interface energy σ_{int} per unit area. Therefore, it is energetically more favorable for the SiO_2 layer to stay continuous rather than to redistribute into precipitates. From this point of view one expects the generation of a continuous interface layer at a grain boundary in the case of a rotation angle larger than $\theta = 3°$ during oxygen precipitation. However, as we have shown here this is not the case. This difference may be explained in terms of an energy barrier between a continuous SiO_2 layer and the precipitates at the grain boundary. Presently, it is not known whether the layer with precipitates thermodynamically is the metastable state and the continuous layer represents the lowest energy state or vice versa.

260

5. Conclusions

The behavior of oxygen in silicon at temperatures above about 700°C is well understood. Observations of unexpected features, such as that of a defined maximum size of oxygen precipitates at appropriate grain boundaries described in section 3, nowadays are rather rare. In contrast, the agglomeration behavior of oxygen at temperatures below about 500°C still remains a puzzle. It appears likely that due to the formation of fast diffusing oxygen dimers a metastable dimer phase develops consisting of a chain of agglomerated oxygen dimers. Dimer chains of different lengths constitute the different thermal donor species observed. A certain range of sizes of these dimer chains is energetically favored which leads to a typical average number of oxygen atoms involved in thermal donors. These chain-like agglomerates may be transformed to nuclei of normal oxygen precipitates via a thermally activated process possibly triggered by the diffusion of interstitial oxygen to the agglomerates. Thus thermal donor formation and subsequent transformation lead to a path to oxygen precipitation which for temperatures below about 500°C is much more efficient than normal oxygen precipitation.

6. References

1. Shimura, F. (ed.) (1994) *Oxygen in Silicon*, Academic Press, London.
2. Corbett, J. W., McDonald, R. S., Watkins, G. D. (1964) The configuration and diffusion of isolated oxygen in silicon and germanium, *J. Phys. Chem. Solids* **25**, 873-879.
3. Kaiser, K., Frisch, H. L., and Reiss, H. (1958) Mechanism of the formation of donor states in heat-treated silicon, *Physical Review* **112**, 1546-1546.
4. McQuaid, S. A., Binns, M. J., Londos, C. A., Tucker, J. H., Brown, A. R., and Newman, R.(1995) Oxygen loss during thermal donor formation in Czochralski silicon: new insights into oxygen diffusion mechanisms, *J. Appl. Phys.* **77**, 1427-1442.
5. Fuller, C. S. and Logan, R. A. (1957) Effect of heat treatment upon the electrical properties of silicon crystals, *J. Appl. Phys.* **28**, 1427-1436.
6. Fuller, C. S. (1961) Kinetics of donor formation in oxygen-doped germanium, *J. Phys. Chem. Solids* **19**, 18-28.
7. Kaiser, W. (1962) Electrical and optical investigations of donor formation in oxygen-doped germanium, *J. Phys. Chem. Solids* **23**, 255-260.
8. Tan, T. Y., Kleinhenz, R. L., and Schneider, C. P. (1986) On the kinetics of oxygen clustering and thermal donor formation in Czochralski silicon, in S. J. Pearton, J. W. Corbett, J. C. Mikkelsen, Jr., and S. J. Pennycook (eds.), *Oxygen, Carbon, Hydrogen and Nydrogen in Silicon*, Materials Research Society, Pittsburgh, PA, pp. 195-204.
9. Markevich, V. P., Makarenko, L. F., and Murin, L. I. (1986) Some new fearures of thermal donor formation in silicon at T < 800 K, *phys. stat. sol. a*, **97**, K173-K176.
10. Claybourn, M. and Newman, R. C. (1987) Activation energy for thermal donor formation in silicon, *Appl. Phys. Lett.* **51**, 2197-2199.
11. Schroder, D. K., Chen, C. S., Kang, J. S., and Song, X. D. (1987) Number of oxygen atoms in a thermal donor in silicon, *J. Appl. Phys.* **63**, 136-141.
12. Wagner, P. and Hage, J. (1989) Thermal double donors, *Appl. Phys. A* **49**, 123-138.
13. Guse, M. P. and Kleinhenz, R. (1992) Interstitial oxygen reduction in silicon at thermal donor temperatures, *J. Appl. Phys.* **72**, 4615-4618.
14. Götz, W., Pensl, G., and Zulehner, W. (1993) Observation of five additional thermal donors and kinetics of thermal donor formation and annihilation at temperatures above 500°C in Czochralski-grown Si, *Materials Science Forum* **117-118**, 213-218.

15. Newman, R. C. and Jones, R. (1994) Diffusion of oxygen in silicon, in ref. 1, pp. 290-352.

16. Michel, J. and Kimerling L. C. (1994) Electrical properties of oxygen in silicon, in ref. 1, pp. 251 -287.

17. Bergholz, W., Hutchison, J. L., and P. Pirouz (1985) Precipitation of oxygen at 485oC: Direct evidence for accelerated diffusion of oxygen in silicon?, *J. Appl. Phys.* **58**, 3419-3424.

18. Gösele, U. and Tan, T. Y. (1982) Oxygen diffusion and thermal donor formation, *Appl. Phys A* **28**, 79-92.

19. Snyder, L. C., Corbett,J. W., Deak, P., and Wu, R. (1988) On the diffusion of oxygen dimer in a silcon crystal , *Mater. Res. Soc. Symp. Proc.* **104**, 179 -184.

20. Stavola, M., Patel, J. R., Kimerling, L. C., and P. E. Freeland (1983) Diffusivity of oxygen in silicon at the donor formation temperature, *Appl. Phys. Lett.* **42**, 73-75.

21. Gösele, U., Ahn, K.-Y., Marioton, B. P. R., Tan, T. Y., and Lee, S.-T. (1989) Do oxygen molecules contribute to oxygen diffusion and thermal donor formation?, *Appl. Phys. A* **48**, 219-228.

22. Stein, H. J. and Medernach, J. W. (1996) Oxygen-related vibrational modes produced in Czochralski silicon by hydrogen plasma exposure, *J. Appl. Phys.* **79**, 2337-2342.

23. Magee, T. J., Leung, C., Kawayoshi, H., Furman, B. K., Evans, C. A. (1981) Gettering of mobile oxygen and defect stability within back-surface damage regions in Si, *Appl. Phys. Lett.* **38**, 891-893.

24. Schaake, H. F. (1982) Comment on "Gettering of mobile oxygen and defect stability within back-surface damage regions in Si", *J. Appl. Phys.* **53**, 1226.

25. Mikkelsen, Jr., J. C. (1983) Gettering of oxygen in Si damaged by ion implantation and mechanical abrasion, *Appl. Phys. Lett.* **42**, 695-697.

26. Gaworzewski, P. and Ritter, G. (1981) On the out-diffusion of oxygen from silicon, *phys. stat. sol. (a)* **67**, 511-516.

27. Hahn, S. (1986) Formation of a low thermal donor concentration layer in CZ Si wafer during 450°C/64h anneals, in S. J. Pearton, J. W. Corbett, J. C. Mikkelsen, Jr., and S. J. Pennycook (eds.), *Oxygen, Carbon, Hydrogen and Nydrogen in Silicon*, Materials Research Society, Pittsburgh, PA, pp. 181-186.

28 Tokuda, Y., Katayama, M., and Hattori, T. (1993) Depth profiles of thermal donors formed at 450°C in oxygen rich n-type silicon, *Semicond. Sci. Technol.* **8**, 163-166.

29. Lee, S.-Tong and Fellinger, P. (1986) Enhanced oxygen diffusion in silicon at thermal donor formation temperature, *Appl. Phys. Lett.* **49**, 1793 -1795.

30. Lee, S.-Tong, Fellinger, P., and Chen. S. (1988) Enhanced and wafer-dependent oxygen diffusion in CZ-Si at 500-700°C, *J. Appl. Phys.* **63**, 1924-1927.

31. Rava, P., Gatos, H. C., and Lagowski, J. (1982) Thermally activated oxygen donors in silicon, *J. Electrochem. Soc.* **129**, 2844-2849.

32. Itoh, T. and Abe, T. (1988) Diffusion coefficient of a pair of nitrogen atoms in float-zone silicon, *Appl. Phys. Lett.* **53**, 39-41.

33. Borghesi, A., Pivac, B., Sassella, A., and Stella, A., (1995) Oxygen precipitation in silicon, *J. Appl. Phys.* **77**, 4169-4244.

34. Tong Q.-Y., Schmid E., Reiche M., and Gösele U. (1994) Hydrophobic silicon wafer bonding, Appl. Phys. Lett. **64**, 625 - 627.

35. Jaccodine R. J. (1963) Surface energy of silicon and germanium, *J. Electrochem. Soc.* **110**, 524.

36. Schroer E., Hopfe S., Werner P., Gösele U., Duscher G., and Rühle M., (1996) Defined size of oxide precipitates at silicon grain boundaries, to be submitted to *Appl. Phys. Lett.*.

37. Ahn K.-Y., Stengl R., Tan T. Y., Gösele U., and Smith P. (1990) Growth, shrinkage, and stability of interfacial oxide layers between directly bonded silicon wafers *Appl. Phys. A* **50**, 85-94.

ROLES OF STRUCTURAL DEFECTS AND CONTAMINANTS IN OXYGEN PRECIPITATION IN SILICON

K. SUMINO
Nippon Steel Corporation
20-1 Shintomi, Futtsu City, Chiba Prefecture 293, Japan

Abstract

A review is given of recent works on the influence of structural defects and metallic impurities on the precipitation of oxygen in Czochralski-grown silicon crystals. Dislocations act as preferential nucleation centers of precipitates. Morphology of precipitates on dislocations depends on the temperature at which precipitation takes place. Precipitation of silicon oxide on dislocations immobilizes the dislocations and brings about high mechanical stability of Czochralski-grown silicon crystals. Fe impurities enhance the oxygen precipitation while Cu impurities not. A large affinity of Fe for oxygen seems to increase the density of active nuclei for the precipitation. Cu impurities develop colonies of Cu and do not affect oxygen precipitation. Grown-in defects in Czochralski-grown silicon seem to be divided into two kinds of defects. One causes the degradation of gate oxide integrity and the other acts as the nucleation centers for oxygen precipitation at relatively low temperatures. The generation processes of these kinds of defects are discussed.

1. Introduction

If integrated circuits could be produced on wafers of silicon crystals grown by the float-zone technique (abbreviated FZ-Si) with a high production yield, the precipitation of oxygen in silicon crystals would have called little attention of materials engineers and scientists in the field of microelectronics. Practically, integrated circuits can be produced with a high yield only on wafers of silicon crystals grown by the Czochralski technique (abbreviated CZ-Si). CZ-Si contains oxygen impurities at a concentration of about 20 ppm, which come from quartz crucibles for the melt at the time of crystal growth. Such oxygen impurities become supersaturated when wafers are brought to the temperature

263

R. Jones (ed.), Early Stages of Oxygen Precipitation in Silicon, 263–282.
© *1996 Kluwer Academic Publishers.*

264

range below about 1250 °C .

Oxygen impurities supersaturated in silicon give rise to two effects, which are opposite to each other (useful and harmful), on the performance of integrated circuits produced on CZ-Si wafers. The first merit of oxygen in CZ-Si is to increase the resistance to the occurrence of warping (plastic deformation) of wafers caused by thermal stress, when they are subjected to repeated thermal processing for device production. The strengthening of a silicon crystal due to the presence of supersaturated oxygen is caused by preferential precipitation of oxygen on dislocations [1,2]. Dislocations in the crystal are thus immobilized and are not activated by thermal stress. However, in an advance stage of precipitation, a rather high density of dislocations are generated around large-sized silicon oxide particles which precipitated in the matrix region. Silicon wafers are then softened, since such dislocations developed around the precipitates in a rather late stage of precipitation are free from pinning by oxygen and are easily activated by thermal stress [3]. They no longer have the merit of a high production yield for devices.

Since dislocations acts as preferential nucleation sites for any kind of supersaturated impurities, they can be used as effective sinks for contaminants which are harmful for device performance. Utilizing dislocations developed around silicon oxide precipitates in CZ-Si, harmful metallic impurities are effectively removed from the device-active region This technique for absorbing harmful impurities is known as *intrinsic gettering* [4].

Recently, a great deal of attention is being paid on the integrity of oxide films, as thin as of the order of 10 nm, formed on silicon wafer surfaces. It has been reported that the proof voltage of the film formed on a CZ-Si wafer is degraded at a few number of points on the wafer, where some kind of grown-in defects seem to be located [5]. There seems to be a good correlation between the distribution of such defective regions and that of silicon oxide precipitates [6].

It has long been known in the practical field that the precipitation of silicon oxides is remarkably enhanced by contamination of wafers with some kind of metallic impurities.

Control of precipitation of silicon oxide in CZ-Si can be achieved on the basis of control of structural defects and metallic impurities. The understanding of the role of structural defects or metallic impurities in the precipitation of silicon oxide in CZ-Si is not only important from the practical point of view but also interesting from the basic point of view.

This paper gives a review of the results of recent investigations on oxygen precipitation in CZ-Si as influenced by structural defects and metallic impurities. It contains three topics ; ① preferential precipitation on dislocations, ② effects of Fe and Cu impurities on the precipitation and ③ relation between grown-in defects and the precipitation.

2. Precipitation on Dislocations

2.1. STRUCTURE-SENSITIVENESS OF OXYGEN PRECIPITATION IN SILICON

When a CZ-Si crystal is subjected to plastic deformation at an elevated temperature, a considerable fraction of interstitially dissolved oxygen atoms are measured to be lost during deformation [7]. Practically, no appreciable loss of interstitial oxygen atoms is detected if a dislocation-free CZ-Si crystal is annealed under no load at the same temperature for the duration same as that spent for the deformation. This fact illustrates that the phenomenon of oxygen precipitation in silicon is very structure-sensitive and deformation-induced defects are very effective nucleation centers for oxygen precipitation.

The infrared absorption spectrum from a CZ-Si crystal deformed at $900°C$ shows a board peak in the wavenumber range $1000 - 1150$ cm^{-1}, as shown in Fig. 1(a). Transmission electron microscopy shows that many kinds of extended defects are present in the deformed crystal; namely, long dislocation lines, dislocation dipoles, small dislocation rings, faulted dipoles, and dark dots of which entities are not yet identified.

When a deformed CZ-Si crystal is annealed at temperature higher than 1300 °C , all the oxygen atoms lost during deformation are recovered and deformation-induced defects other than long dislocation lines are all eliminated. Survived dislocation lines are characterized by smoothness in the shape. If such a dissolution-treated crystal is annealed again at 900 °C , no characteristic absorption peaks are developed in the wavenumber range $900 - 1300$ cm^{-1}, as shown in Fig. 1(b), even though the concentration of interstitially dissolved oxygen is remarkably decreased at a high rate.

On the other hand, when a dislocation-free CZ-Si is annealed at $900°C$ for a long duration, a characteristic absorption peak centered around 1225 cm^{-1} is developed as seen in Fig. 1(c). This peak is believed to be related to silicon oxide precipitates.

The above observations suggest that various kinds of structural defects act as very effective nucleation centers for oxygen precipitation and also that the morphology and/or structure of precipitates seem to depend on the nature of defects on which they precipitate. Thus, details in precipitation of oxygen in silicon are greatly influenced by the kind of defects existing in the crystal.

2.2. PRECIPITATES ON DISLOCATIONS

As mentioned previously, preferential precipitation of oxygen on dislocations is the most important phenomenon underlying the fact that CZ-Si is almost exclusively used as the materials for microelectronics. The morphology of

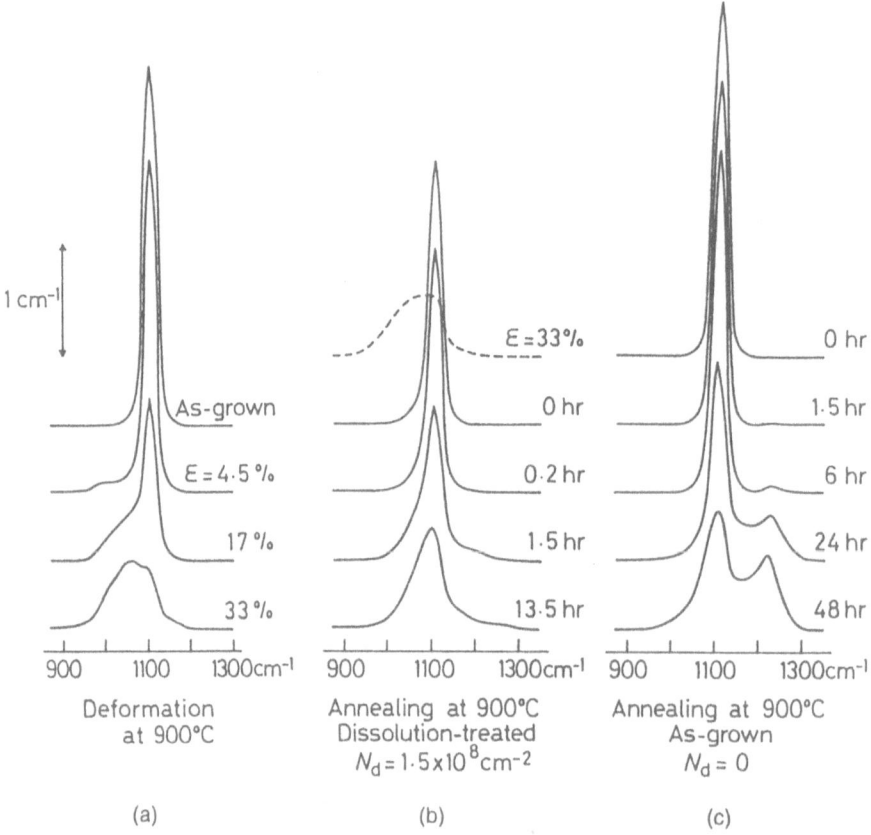

Figure 1. Infrared absorption spectra from CZ-Si crystals (a) after plastic deformation at 900 °C to strains shown in the figure, (b) containing only dislocations as defects, and (c) in as-grown state. Specimens in (b) and (c) are those after annealing at 900 °C for durations attached to the spectra [7].

silicon oxide precipitates depends on the temperature at which they precipitate even in the case that the precipitates are developed in the matrix region of CZ-Si.

Temperatures of precipitation for silicon oxide may be divided into three regions according to the morphology of precipitates formed in the matrix region of CZ-Si. The first is the temperature region lower than 650 °C (the low temperature range) where precipitates are of the shape of a small sphere with a diameter of order of 2 nm [8]. So-called rod-like defects and interstitial-type dislocation dipoles are seen together with such sphere-shaped precipitates. The second is the temperature range 850 − 1000 °C (the intermediate temperature range) where precipitates are of the shape of a square-shaped platelet with the

·habit planes {100} and the edges parallel to the <011> directions [9]. Such a platelet-like precipitate is accompanied by a large lattice strain in the matrix region around it and often emits punch-out dislocations. The third is the temperature range 1100 − 1200 °C (the high temperature range) where precipitates are in the shape of a octahedron with the habit planes {111} [10]. They are not accompanied by lattice strain around them but often by stacking faults of extrinsic type. The structure of precipitates developed in all the temperature range is amorphous according to transmission electron microscopy.

The dependence of the precipitate morphology on the temperature seen above has been well interpreted from an energy consideration. The formation of a silicon oxide particle causes a large volume expansion; the volume of a SiO_2 molecule is as large as 2.25 times that of a Si atom. This produces a very large strain in the matrix around a precipitate. In the low temperature range, the volume of the precipitate formed is so small because of a low diffusivity of oxygen that the interface energy of the precipitate dominates over the strain energy. As a consequence, the precipitate assumes a sphere shape. In the intermediate temperature range, the precipitate grows into a size that the strain around it is appreciable. However, the rate of self-diffusion is not high enough to release the strain around the precipitate by means of diffusion-related processes. Thus, the precipitate assumes the shape of a platelet which has the lowest strain energy among various shapes of precipitates of the same size. In the high temperature range, the volume of an individual precipitate is large and the self-diffusion of silicon takes place at high rates. The strain related to the volume expansion of the precipitates is effectively released by emission of interstitial silicon atoms from the precipitate or the absorption of vacancies from the matrix. The precipitate now assumes the shape of a sphere or a polyhedron with of habit planes of low energies.

The morphology of precipitates formed on dislocations is different from those developed in the matrix region. In a very early stage of precipitation, high resolution electron microscopic observations revealed that some kind of impurity clusters of a size of about 1 nm were formed on the dilatational side of 90 ° Shockley-type partial dislocations but not around 30 ° Shockley-type partial dislocations [11]. The size of the cluster was too small to determine its structure and composition. The author believes it to be a cluster of oxygen atoms before developing to silicon oxide.

Due to annealing in the low temperature range, a number of precipitates in an ellipsoidal shape, several tens nm in the length, are first closely aligned along dislocations, irrespective of the type of dislocations, as shown in Fig. 2 [11]. With an increase in the annealing duration, they grow along the dislocations and are connected to each other and, finally, all dislocation segments are continuously covered by precipitates of a cylindrical shape. The diameter of the precipitate cylinder is about 10 nm and the structure is

268

determined to be amorphous. Corresponding to such a change in the state of precipitates, the stress necessary to release the dislocations from the precipitates increases [12].

Precipitates developed on a dislocation in the intermediate temperature range are first distributed discretely with separations of about 100 nm. They are accompanied by strong lattice strain in the matrix and have the shape of a needle elongated along the direction of the dislocation as shown in Fig. 3 [11]. With an increase in the annealing duration, dislocations are displaced from the precipitates which were originally formed on themselves. 60 ° dislocations undergo parallel displacement, while screw dislocations become helical [13]. Such motions of dislocations are thought to be accomplished by absorption of interstitial silicon atoms which are emitted from the precipitates. This implies that precipitates now grow by themselves without the help of the dislocations.

In the high temperature range, precipitates are formed along dislocations but sparsely. The precipitate assume the shape of a polyhedron with the habit planes {111} and the edges parallel to the <110>, similar to those developed in the matrix region at the same temperature [11].

There are three reasons why a dislocation can act as a preferential nucleation site for precipitates : Namely, ① the release of the energy of the core region of dislocation due to the replacement by the precipitate, ② the interaction of the strain field of the dislocation with the precipitate, ③ enhancement of oxygen movement due to the pipe diffusion along the dislocation core. In the low temperature range, where the diffusion rate is low,

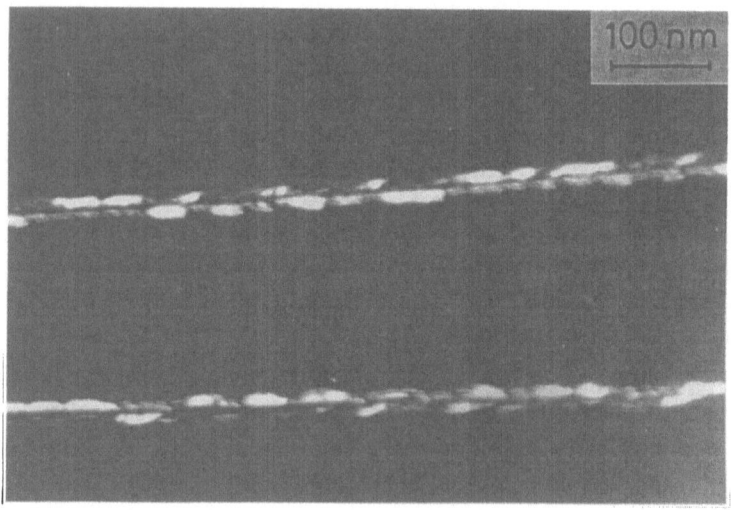

Figure 2. Transmission electron micrograph of silicon oxide precipitated on dislocations at 650 °C [11].

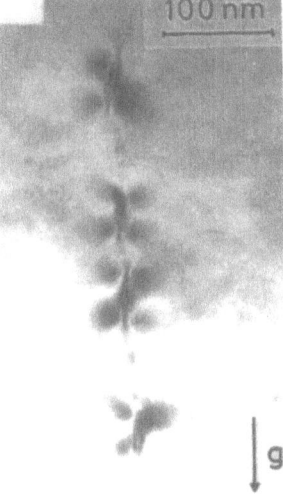

Figure 3. Transmission electron micrograph of needle-shaped silicon oxide particles precipitated on a dislocation at 900 °C [11].

the energy of the system may be the most efficiently released by replacing the core region of dislocation by the precipitates [14]. Thus, a cylinder of precipitate develops along the dislocation core. In the intermediate temperature range, the contribution of the energy of interface may play a role. The balance of the interfacial energy and the strain energy of the precipitate may lead to the development of needle-like precipitates aligned along the dislocation line. In the high temperature range the precipitate is free from the strain. Thus, the shape is determined in such a way that the interfaces takes the lowest energy.

3. Effects of Metallic Impurities on Precipitation

3.1. EFFECTS OF Cu AND Fe IMPURITIES ON OXYGEN PRECIPITATION

Cu and Fe are common metallic impurities in Si which are easily incorporated by contamination. They have the solubilities, diffusivities and precipitation characteristics very different from each other. They may give rise to different influences on oxygen precipitation in CZ-Si.

Cu was incorporated into CZ-Si specimens at temperatures between 700 and 1200 °C and Fe at temperatures between 900 and 1200 °C . Depending on the temperature of incorporation, the concentration of Cu ranged from 6.0×10^{16} to 1.0×10^{18} atoms/cm^3 while that of Fe from 6.0×10^{13} to 1.0×10^{16} atoms/cm^3.

It has been found that Cu impurities at any concentration in the above range give rise to no appreciable influence on the oxygen precipitation in precipitation treatment at any temperature. On the other hand, Fe impurities enhance oxygen precipitation at concentrations higher than 2×10^{14} atoms/cm^3, as shown in Fig. 4 [15]. The effect is observed in the temperature range higher than 700 °C and appears the most pronouncedly at the temperature

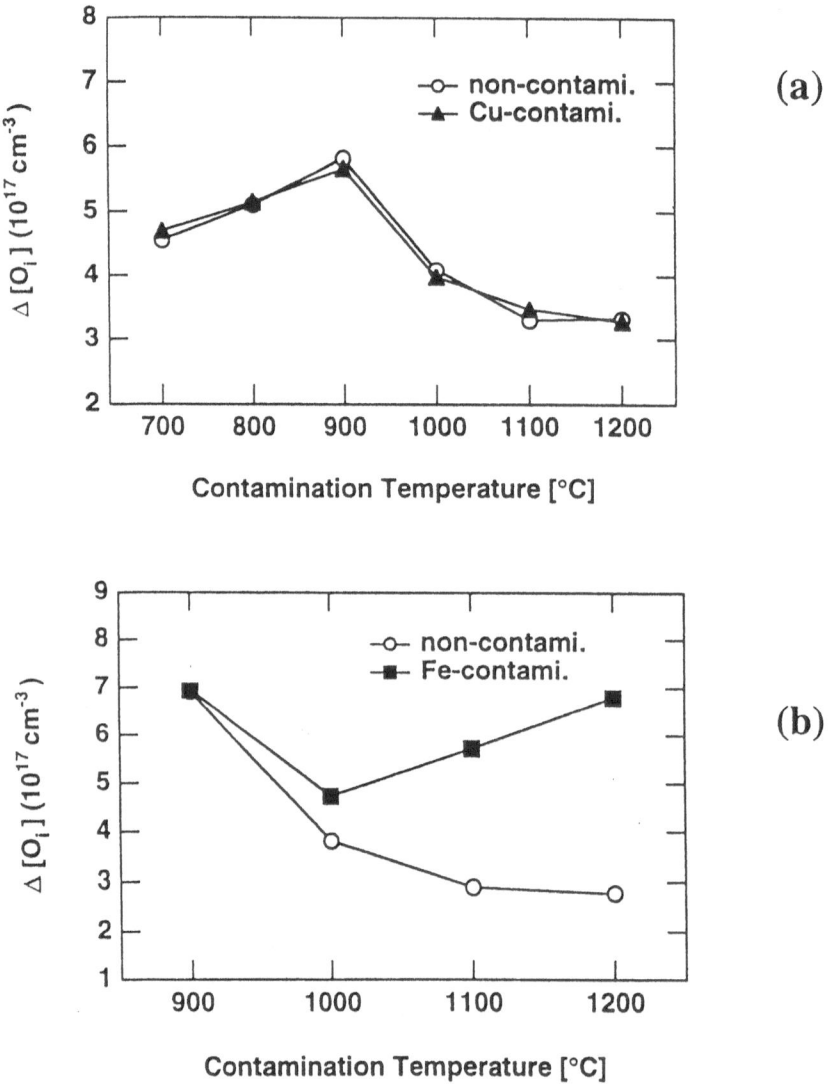

Figure 4. Amount of precipitated oxygen atoms $\triangle$ [O_i] due to annealing at 1000 °C for 20 h in specimens contaminated with (a) Cu and (b) Fe at temperatures shown in the abscissa. The data for non-contaminated specimens are also shown for the sake of comparison [15].

$900 - 1000\,^\circ\text{C}$. The enhancing effect becomes remarkable as the concentration of Fe increases.

3.2. CHARACTERISTICS IN PRECIPITATION ENHANCEMENT DUE TO Fe

The precipitation characteristics of oxygen in silicon can effectively be investigated by means of two step annealing. It consist of annealing at a rather low temperature $700 - 800\,^\circ\text{C}$ for the generation of precipitates which can grow at a high temperature and subsequent annealing at a rather high temperature of $1000 - 1100\,^\circ\text{C}$ for the growth of the precipitates. It is well established that no new precipitate particle appears during the second step annealing and only precipitates formed in the first step annealing grow. Using the two step annealing, we are able to find that whether Fe impurities play a role in the generation process or in the growth process of precipitates.

Figure 5 shows the concentrations of interstitially dissolved O atoms $[O_i]$ and the density of silicon oxide precipitates in specimens with and without doping of Fe, which are subjected to generation annealing (the first step annealing) at $750\,^\circ\text{C}$ for three different durations (4, 8, 12 h) and following growth annealing (the second step annealing) at $1050\,^\circ\text{C}$ for a fixed duration of 4 h [15]. The original concentration of interstitial O atoms in both kinds of specimens is 7.9×10^{17} atoms/cm^3. The density of precipitates in the Fe-doped specimen is high already after a rather short duration of generation annealing and increases slowly with increasing duration of generation annealing. Correspondingly, the amount of precipitated O atoms after growth annealing increases slowly with an increase in the duration of generation annealing. On the other hand, in non-doped specimens, the density of precipitates is first low and increases rapidly with the duration of generation annealing, reaching that of the Fe-doped specimen after annealing for 12 h. The amount of precipitated O atoms after growth annealing is rather small for a short duration of generation annealing and increases much more rapidly than in the Fe-doped specimen due to an increase in generation annealing.

When the duration of generation annealing is fixed at 12 h and that of growth annealing is varied, no essential differences are observed between the Fe-doped and non-doped specimens in the precipitation behaviour of O atoms as shown in Fig. 6 [15]. The densities of oxygen precipitates in both kinds of specimens are approximately equal to each other as have been seen in Fig. 5.

From above observations, we have the conclusion that Fe impurities enhance the generation rate of oxygen precipitates but do not affect their growth process, Namely, Fe impurities have no influence on the diffusion of O atoms in the temperature range around $1000\,^\circ\text{C}$.

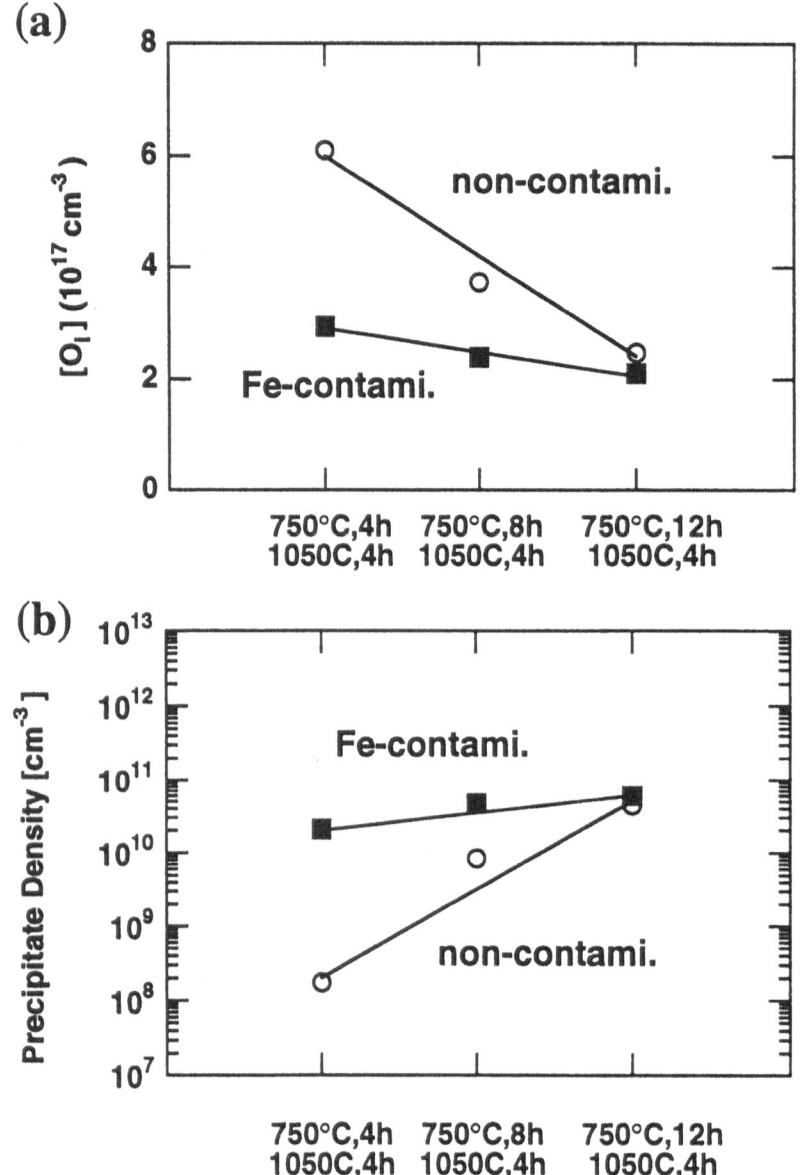

Figure 5. Variations in (a) the concentration of interstitial oxygen atoms [O_i] and the density of oxygen precipitates against the duration of generation annealing at 750 °C in specimens contaminated with Fe at 1200 °C and in non-contaminated specimens [15].

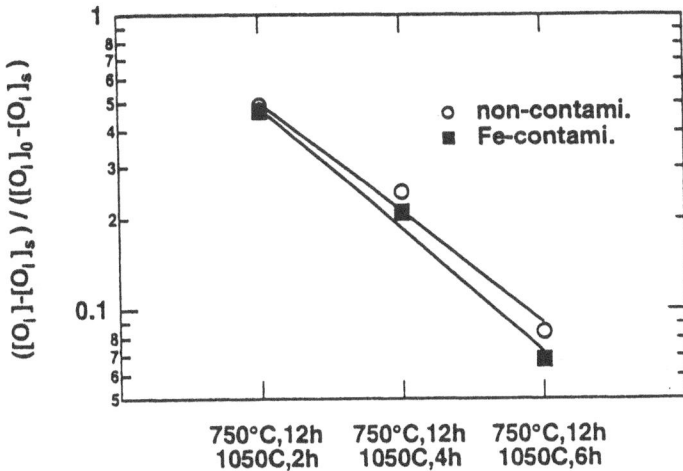

Figure 6. Variation in the normalized concentration of excess interstitial oxygen atoms against the duration of growth annealing at 1050 °C in specimens contaminated with Fe at 1200 °C and in non-contaminated specimens [15].

3.3. ORIGIN FOR DIFFERENT ROLES OF Fe AND Cu IMPURITIES

It is well accepted that an as-grown CZ-Si crystal contains a number of minute particles of silicon oxide [16]. The particles may have a size distribution which is determined by thermal history of the crystal. Any heat-treatment causes dissolution of some part of the particles and also growth of the other part, depending on the temperature. Since Fe oxide and Fe silicate are much more stable than Fe silicide [17,18], supersaturated Fe impurities may be incorporated into the minute silicon oxide particles rather than to precipitate in the Si matrix. The diffusivity of Fe is much higher than that of O in a Si crystal. Hence, silicon oxide particles incorporate Fe impurities much faster than to absorb O impurities to grow. We suppose that incorporation of Fe stabilizes minute silicon oxide particles in the as-grown Si crystal and brings about an increase in the density of the particles which can grow to precipitates at any temperature.

Transmission electron microscopy and electron beam induced current technique have revealed that Cu develops colonies of Cu_3Si during cooling from high temperatures at which doping was done [19]. The colonies are observed not to be affected by the precipitation of oxygen [15]. Owing to a very high diffusivity of Cu atoms in a silicon crystal, the colonies are developed very quickly by means of repeated precipitation of Cu_3Si particles on climbing edge dislocations [20]. Cu atoms incorporated in Cu_3Si are thought to be more stable than those incorporated in silicon oxide. Thus, oxygen precipitation proceeds independently of the presence of the colonies of Cu_3Si.

4. Correlation of Grown-in Defects with Oxygen Precipitation

4.1. HETEROGENEOUS PRECIPITATION OF OXYGEN IN AS-GROWN CZ-Si CRYSTALS

It is well known in the practical field that the amount of precipitated oxygen atoms is much larger at the top part than at the tail part in an as-grown CZ-Si crystal when they are subjected to the same two step annealing for oxygen precipitation. It is also reported that a large amount of oxygen precipitation is obtained in usual two step annealing of a CZ-Si crystal when the rate of cooling after solidification is slow in the temperature region lower than 650 °C [21]. On the one hand, the density of silicon oxide precipitates developed in a single step annealing in the temperature range higher than 800 °C does not depend on the duration of annealing [22]. These observations are reasonably interpreted in terms of the ideas that the oxygen precipitation in CZ-Si, at least, in the as-grown state takes place by means of heterogeneous nucleation and that the nuclei of silicon oxide precipitates are developed during the cooling after crystal growth. The classical nucleation theory of precipitation leads to a picture that large-sized nuclei with a low density is formed in a high temperature region, while small-sized ones with a high density in a low temperature region.

On the other hand, it has been found that the densities of precipitates in small specimens of CZ-Si which are quenched from about 1400 °C at a very high rate are almost the same after precipitation treatments in the temperature range 700~1000 °C , irrespective of the oxygen concentration [23]. This observation suggests that some kind of defects quenched-in from high temperature may coagulate rapidly into some clusters which act as the nuclei of silicon oxide precipitates. The defects may be intrinsic point defects such as vacancies and interstitials of Si atoms.

4.2. GROWN-IN DEFECTS IN CZ-Si

When the surface of a wafer of CZ-Si with a large diameter grown at a speed within a certain range is oxidized around 1100 °C , a well defined ring-shaped region with a high density of oxidation-induced stacking faults (OSF) is developed. Such a ring region is concentric with respect to the crystal axis as shown in Fig. 7 [24]. The density of OSF in the ring region is higher than those of the inner- and outer-side regions by orders of magnitude and the periphery of the ring is extremely sharp.

The ring region is not developed in a crystal grown at a high speed and starts to appear at the side surface of the crystal when the growth rate is reduced to a certain value. Upon further reducing the growth rate, the location of the ring region moves inward the crystal and finally disappears at the center of the crystal at a growth rate of 0.3 − 0.6 mm/min, depending on the structure of hot-zone of the CZ pulling furnace. The precipitation of oxygen due to a single step annealing at about 1100 °C is much enhanced in the ring region in comparison with the inner- and outer-side regions [24].

Habu *et al.* [25-28] have calculated the spatial distribution of frozen-in point defects in an as-grown CZ-Si crystal by solving phenomenological diffusion equations with some assumptions on the diffusivities and the thermal equilibrium concentrations of vacancies and interstitials in Si. An assumption has been also made that an entropy barrier exists for pair annihilation of a vacancy and an interstitial. The calculation has shown that a vacancy-rich region is developed in a cylindrical region centered at the crystal axis and an interstitial-rich region outside of it. It has been shown that, as the crystal growth rate is increased, the interstitial-rich region first appears at the center of the crystal at a certain low growth rate and expands toward the outer-surface of crystal. The OSF ring region may be regarded to be either the interstitial-rich region or the boundary region separating the vacancy-rich region and the interstitial-rich region. It has not yet been clarified how this picture is accommodated to the precipitation behaviour of oxygen observed in and outside the ring region. However, the calculation by Habu *et al.* demonstrates the possibility that the characteristic spatial pattern of defects observed in an as-grown CZ-Si crystal may be explained by taking into account of excess vacancies and interstitials frozen-in during crystal growth.

Until 1990, grown-in defects in Si crystals which are too small in the size to be observed by transmission electron microscopy (TEM) were detected by means of decoration by Cu or Li. Recently, much more convenient means for

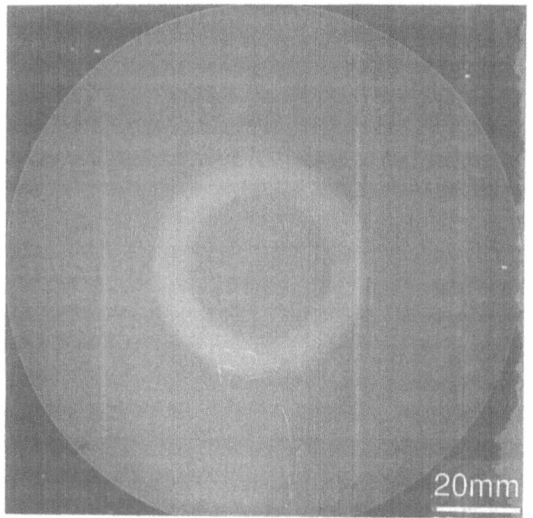

Figure 7. Ring-shaped region in a CZ-Si wafer observed with X-ray to-pography. White contrast shows that a high density of silicon oxide precipitates are there [24].

the detection have been developed. The defects are revealed as (1) etch pits accompanied by a wedge-shaped surface pattern (flow pattern) developed due to the Secco etching (FPD) [29], (2) small pits formed by SC-1 cleaning process (COP) [30], or (3) scattering centers for infrared lights (IRSC) [31]. There are also defects which are revealed as usual etch pits not accompanied by flow pattern by the Secco or Wright etching (EPD). It should be noted that the defects revealed as FPD, COP or EPD may not always be of the same atomic structure, since FPD, COP or EPD show only that etching or dissolution of Si surface is enhanced or irregular there. These defects in as-grown CZ-Si crystals recently call a great deal of attention in connection with the integrity of gate oxide films as thin as of the order of 10 nm formed on wafers.

FPD and COP are observed not only in an as-grown crystal of CZ-Si but also in that of FZ-Si [32,33] and are absent in epitaxially grown Si films. Their densities are reduced drastically if a CZ or a FZ-Si crystal is held at temperatures around 1200 °C for some duration during crystal growth [33,34]. These observations strongly suggest that FPD and COP are related to the same nature of defects which are characteristic of melt-grown Si crystals and that their essential constituents are some point defects which are frozen and agglomerate most efficiently at temperatures below 1200 °C during cooling of the crystal from melting point. On the other hand, IRSC are found only in CZ-Si and stable up to 1300 °C , while FPD and COP in a grown CZ-Si crystal up to 800 − 1000 °C [33,35]. IRSC have been identified to be tiny precipitates of silicon oxide 50 − 70 nm in the size which were developed during crystal growth [36]. TEM observation has also revealed that a high density of precipitates 10 nm in the size are involved in an as-grown CZ-Si [37].

The densities of FPD, COP and IRSC all increase with an increase in the growth rate of a CZ-Si crystal [32,35,38]. It is interesting that in a crystal with a OSF-ring the densities of the above kinds of defects in the outer-side region of the ring are much lower than those in the inner-side region, those in the outer-side region being lower than the detection limits [36]. Instead, large dislocation loops are observed in the outer-side region by TEM.

4.3. OXYGEN PRECIPITATION AS INFLUENCED BY GROWN-IN DEFECTS

In order to investigate the temperature range in which grown-in defects are generated, the so-called growth-halting experiments on CZ-Si crystals have been conducted by Iwasaki *et al.* [39,40,41]. In this type of experiments, a CZ-Si crystal is first grown at a growth rate of 1.0 mm/min until a suitable length of the crystal is obtained. Then, the growth rate is rapidly lowered to 0.2 mm/min. After keeping this growth rate for 100 min, the rate is raised rapidly again to 1.0 mm/min. Each portion of the crystal is held for a rather long time at some specific narrow temperature range which is determined by the position in the crystal. One can see how the holding at a specific temperature affects the generation of grown-in defects and precipitation behaviour of oxygen.

Figure 8(a) shows an X-ray topograph of a plate sliced parallel to the growth axis of a halted crystal after the two step precipitation annealing (800 °C for 4 h + 1000 °C for 16 h). Figure 8 (b) shows the concentrations of dissolved oxygen atoms at the portions along the central axis of the crystal before (open circles) and after (full circles) the precipitation annealing, plotted against the temperature at which the portion was held for 100 min [39]. It is seen that the oxygen precipitation is very scarce at the portion held at temperatures higher than 1350 °C . This portion is free from the OSF-ring and corresponds to the outer-side region of the ring. The amount of precipitated oxygen atoms is large at the portion held at 1350 − 1100 °C . This portion is termed A region. The precipitation of oxygen is the most pronounced at the portion held at 1100 −

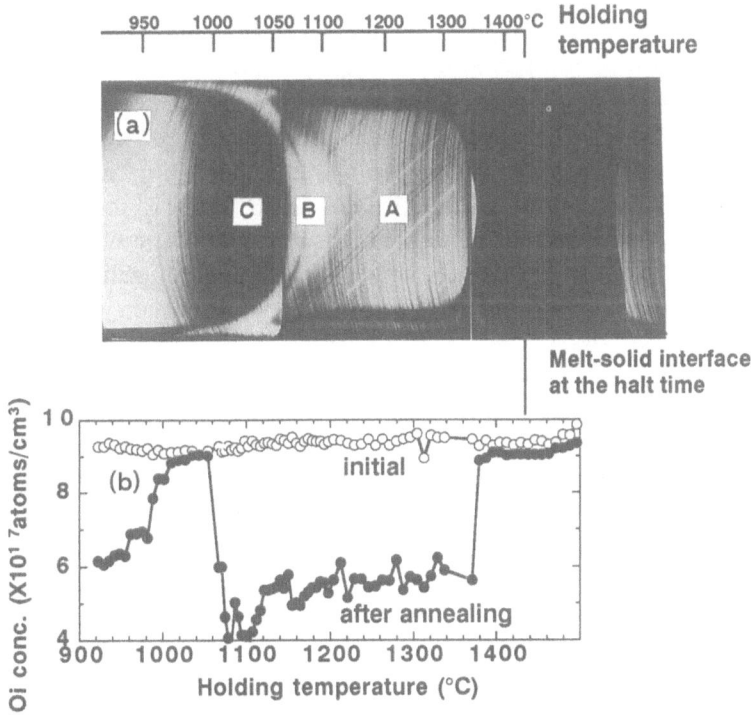

Figure 8. (a) X-ray topograph of a plate sliced parallel to the growth axis of a halted crystal after the two step annealing. (b) The concentration of interstitial oxygen atoms along the central axis before (open circles)and after (full circles) the two step annealing plotted against the temperature at which the crystal portion was held for 100 min [39].

1050 °C , which is termed B region. Oxygen precipitation is little at the portion held at 1050 − 980 °C . This portion is termed C region. A, B and C regions are all inner-side regions of the OSF ring. The integrity of gate oxide films has been measured to be high for the films formed on a wafer of the B region, while low for those on wafers of the A and C regions.

It is interesting to note that the precipitation in the C region is very little in comparison with the A and B regions. Figure 9 shows how the amount of precipitated oxygen atoms in the C region changes with the halting duration [39]. It is seen that the density of the nuclei of precipitates in the two step precipitation annealing is first high but decreases with the halting duration. This means that the nuclei of the precipitates are gradually eliminated during holding the crystal at 1050 − 980 °C .

Thermal stability of grown-in defects has been tested by means of annealing in the temperature range 1250 − 1350 °C . After annealing at these temperatures, the crystals have been subjected to the two step precipitation annealing. The defects become manifest to be revealed by the Wright etching.

Figures 10 (a), (b) and (c) show the densities of defects after annealing at 1250, 1300 and 1350 °C , respectively, plotted against the temperature at which the portion was held for 100 min [41]. It is to be noted that the B region is characterized by a high density of defects which survive at 1250 °C but disappear at 1300 °C . This kind of defects are probably related to the nuclei of precipitates developed at 800 °C in the two step annealing and are termed P defects. Another important characteristic of the B region is the low density of defects which survive at 1300 °C in comparison to the A and C regions. It is to be noted that the density of defects which survive after annealing at 1350 °C is very low in the B region and is the highest in the portion of the C region adjacent to the B region.

Since the B region shows a higher gate oxide integrity than the A and C regions, the defects which survive at 1300 °C seem to be the specific defects related to the degradation of the gate oxide integrity. This kind of defects are

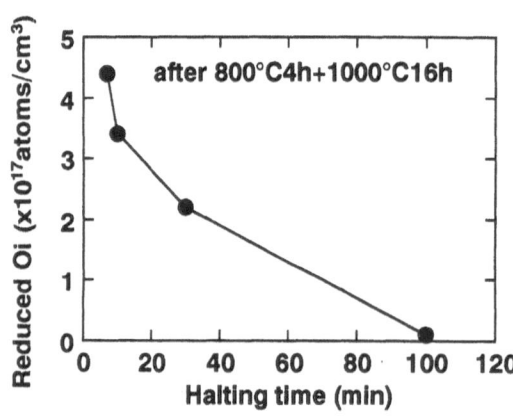

Figure 9. Reduction in the concentration O_i of interstitial oxygen atoms due to the two step annealing plotted against the duration of holding at a temperature of the C region [39].

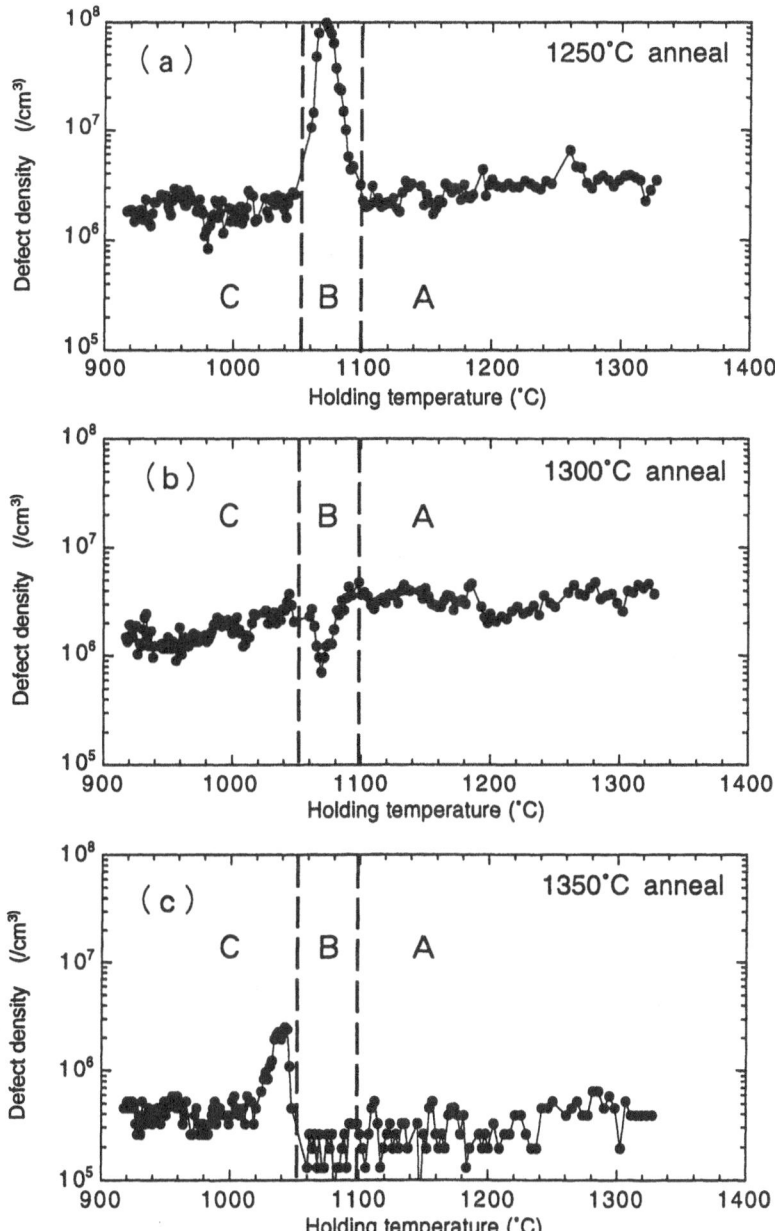

Figure 10. Densities of defects revealed by etching after annealing at (a) 1250 °C, (b) 1300 °C, and (c) 1350 °C plotted against the temperature at which the crystal portion was held for 100 min [41].

termed G defects. The density of G defects is the lowest in the B region among the A, B and C regions.

We have seen in the above that in the B region the formation of P defects dominates over the formation of G defects. On the other hand, the reverse holds in the C region. We may suppose that the formation of P defects and that of G defects are the processes competing each other, which consume the same kind of constituent defects. Iwasaki *et al.* have shown that once a crystal is held at a temperature in the B region (1100 - 1050 °C) for 100 min, the density of G defects no longer changes due to subsequent annealing in the temperature range of C region (1050 - 980 °C) for 100 min [40]. This fact implies that P defects formed in the B region are stable and are not converted to G defects by annealing at temperatures lower than 1050 °C .

We have the following picture : Formation processes of both G defects and P defects take place in a CZ-Si crystal at elevated temperatures. G defects and P defects both incorporate intrinsic point defects (probably, vacancies) at excess concentrations which are frozen-in during cooling of the crystal. On cooling a CZ-Si crystal from a high temperature, the formation of P defects is dominant in the temperature range of B region (1100 - 1050 °C). On the other hand, the reaction to form G defects becomes dominant in the temperature range of C region (1050 - 980 °C). The development of G defects consumes excess intrinsic point defects and suppresses the formation of P defects. A possible reaction to form a G defect may be the formation of some complex involving vacancies and oxygen atoms. The A region seems to contain another type of defects less stable than G defects, which diminish in the temperature range between 1300 °C and 1350 °C .

Iwasaki *et al.* have found that when a crystal is rapidly quenched from a temperature in the range 1380 - 1250 °C to the holding temperature, the boundary temperature between the B and C regions shifts to a low temperature as the quench temperature is lowered [40]. This observation supports the idea that one of the constituents of a G defect are excess intrinsic point defects.

The densities of FPD, EPD, COP and IRSC are all measured to be lower in the B region than in the A and C regions [42]. The densities of COP and large-sized IRSC show the maxima at the portion of the C region where the density of defects surviving at 1350 °C show the maximum. Thus, the defects giving rise to these pattern have some correlation with G defects.

5. Conclusion

Dislocations act as preferential nucleation centers for oxygen precipitation. It brings about high mechanical stability of CZ-Si compared with FZ-Si.

Fe impurities enhance the precipitation of oxygen in CZ-Si while Cu impurities not. This is attributed to the large affinity of Fe for oxygen. Contrarily,Cu impurities at an excess concentration develop Cu colonies and do not affect oxygen concentration.

Grown-in defects in CZ-Si are divided into two kinds of defects. One is the defects which cause the degradation of gate oxide integrity and the other the defects which act as nucleation centers for oxygen precipitation at relatively low temperatures. Generation processes of these defects have been discussed.

Acknowledgements

The author is deeply indebted to H. Harada, T. Iwasaki and H. Haga for their valuable discussion.

References

1. Sumino, K., Harada, H., and Yonenaga, I. (1980) Jpn. J. Appl. Phys. **19**, L49.
2. Sumino, K. (1986) in J.C. Mikkelsen Jr., S.J. Pearton, J.W. Corbett, and S.J. Pennycook (eds.), Mater. Res. Soc. Symp. Proc. Vol. **59**, *Oxygen, Carbon, Hydrogen and Nitrogen in Crystalline Silicon*, North-Holland, Amsterdam, p.369.
3. Yonenaga, I and Sumino, K. (1982) Jpn. J. Appl. Phys. **21**, 47.
4. Tan, T.Y. and Tice, W.K. (1976). Philos. Mag. **34**, 615.
5. Yamabe. K. and Taniguchi, K. (1985) IEEE Trans. Electron Devices, ED-**32**, 423.
6. Tachimori, K., Sakin, T., and Kaneko, T. (1990) Proc. 7th Symp. Crystal Technology, Jpn. Soc. Appl. Phys., 27.
7. Yonenaga, I. and Sumino, K. (1985) in H. Suzuki, T. Ninomiya, K. Sumino, and S. Takeuchi (eds.), *Proc. Yamada Conf. IX on Dislocations in Solids*, University of Tokyo Press, Tokyo, p.385.
8. Bourret, A., Thibault-Desseaux, J., and Seidman, D.N. (1984) J. Appl. Phys. **55**, 825.
9. Tempelhoff, K., Spiegelberg, F. Gleichmann, R., and Wruck, D. (1979) phys. stat. sol. (a) **56**, 213.
10. Bender, H. (1984) phys. stat. sol. (a) **86**, 245.
11. Minowa, K. and Sumino, K. (1993) unpublished work : Minowa K. PhD Thesis, Tohoku Univ. 1993.
12. Sumino, K. and Imai, M. (1983) Philos. Mag. A**47**, 753.
13. Minowa. K., Yonenaga, I. and Sumino, K. (1991) Materials Letters **11**, 164.
14. Cahn, J.W. (1957) Acta Metall. **5**, 169.
15. Shen, B., Jablonski, J., Sekiguchi, T., and Sumino, K. (1996) Jpn. J. Appl. Phys. in press.
16. Inoue, N., Osaka, J., and Wada, K. (1982) J. Electrochem. Soc. **129**, 2780.
17. Colas, G. and Weber, E.R. (1986) Appl. Phys. Lett. **48**, 1371.
18. Gallego, J.M. and Miranda, R. (1991) J. Appl. Phys. **69**, 1377.
19. Shen, B., Sekiguchi, T., Jablonski, J., and Sumino, K. (1994) J. Appl. Phys. **76**, 4540.

20. Solberg, J.K. and Nes, E. (1978) J. Mater. Sci. **13**, 2233.
21. Shinoyana, S., Hasebe, M., and Yamauchi, T. (1991) Oyo Butsuri **60**, 766.
22. Hasebe,M. (1992) PhD Thesis, State University of New York.
23. Ikari,A. (1995) PhD Thesis, Tsukuba University.
24. Hasebe, M., Takeoka, Y., Shinoyama, S., and Naito, S. (1990) in K. Sumino (ed.), *Defect Control in Semiconductors*, North-Holland, Amsterdam, p.157.
25. Habu, R., Yunoki, I. Saito, T., and Tomiura, A. (1993) Jpn. J. Appl. Phys. **32**, 1740.
26. Habu, R., Kojima, K., Harada, H., and Tomiura, A. (1993) Jpn. J. Appl. Phys. **32**, 1747.
27. Habu, R., Kojima, K., Harada, H., and Tomiura, A. (1993) Jpn. J. Appl. Phys. **32**, 1754.
28. Habu, R., Tomiura, A., and Harada, H. (1994) in H. R. Huff, W. Bergholz and K. Sumino (eds.), *Proc. 7th Internatl. Symp. Silicon Materials Science and Technology*, Electrochemical Society, Pennington, N.J., p.635.
29. Yamagishi, H., Fusegawa, I., Fujimaki, N., and Katayama, M. (1992) Semiconductor Science and Technology **7**, A 135.
30. Ryuta, J., Morita, E., Tanaka, T., and Shimanuki,Y. (1990) Jpn. J. Appl. Phys. **29**, L1947.
31. Gall, P., Fillard, J-P., Bonnafe, J. Rakotomavo, T., Rufer, H., and Schwenk, H. (1990) in K. Sumino (ed.), *Defect Control in Semiconductors*, North-Holland, Amsterdam, p.255.
32. Takeno, H., Ushio, S., and Takenaka, T. (1993) Abstracts of 17th Meeting, Jpn. Soc. Crystal Growth, p.15.
33. Shimanuki, Y. (1993) Abstracts of 17th Meeting, Jpn. Soc. Crystal Growth, p.1.
34. Shigematsu, T., Sano, M., and Sumita, S. (1992) Proc. 61st Meeting of 145 Committee, JSPS, p.13.
35. Kouno, M., Motoura, H., Uemura, K., and Nishimura, M. (1993) Abstracts of 17th Meeting, Jpn. Soc. Crystal Growth, p.5.
36. Sadamitsu, S., Umeno, S., Koike, Y., Hourai, M., Sumita, S., and Shigematsu, T. (1993) Jpn. J. Appl. Phys. **32**, 3675.
37. Nakai, K., Nakashizu, T. and Haga, H. (1993) in A.G. Cullis, A.E. Staton-Bevan, and J.L. Hutchison (eds.), *Microscopy of Semiconducting Materials 1993*, Institute of Physics Publishing, Bristol, p. 103.
38. Suga, H., Abe, H., Koya, H., Yoshimi, T., Suzuki, I., Yoshioka, H., Kagawa, N., and Akiyama, K. (1992) Proc. 61st Meeting of 145 Committee, JSPS, p.19.
39. Iwasaki, T., Tsumori, Y., Nakai, K., Haga, H., Kojima, K., and Nakashizu, T. (1994) in H.R. Huff, W. Bergholz, and K. Sumino (eds), *Proc. 7th Interntl. Symp. Silicon Materials Science and Technology*, Electrochemical Soc. Pennington, NJ., p.744.
40. Iwasaki,T., Harada,H. and Haga,H. (1996) in M. Suezawa and H. Katayama-Yoshida (eds.), *Proc. 18th Internatl. Conf. Defects in Semiconductors*, Trans Tech Publication, Switzerland, p.1731.
41. Iwasaki, T ., Nakai, K., Kojima, K., and Hasebe, M. (1996) Oyo Butsuri, in press.
42. Iwasaki, T., Nakai, K., Ohkubo, M., Harada, H., and Haga, H. (1995) Private Communication.

VARIOUS FORMS OF ISOLATED OXYGEN IN SEMICONDUCTORS

B. PAJOT
Groupe de Physique des Solides
Tour 23, Université Denis Diderot
2, place Jussieu
75251 Paris Cedex 05, FRANCE

1. Introduction

Oxygen has been investigated mostly in silicon because it is a residual pollutant with both detrimental and beneficial effects in this material. A wealth of information is therefore available on isolated oxygen in silicon, but it seems difficult to extrapolate to oxygen in other semiconductors because of differences in chemical reactivities and lattice relaxations. The aim of this review is to try to compare the situation in silicon with that in other semiconductors where isolated oxygen has been identified. I will present an overview of the evolution of our knowledge of the structure and dynamics of O_i in silicon, based on a comparison between the IR data and theoretical predictions and I will point out the similarities and differences with oxygen in germanium, gallium arsenide and gallium phosphide. I will make a parallel between the piezo-spectroscopic results on interstitial (O_i) and quasi-substitutional (off-centre or oxygen-vacancy) forms in GaAs and in Si. I will present a summary of the illumination effects observed for the oxygen-vacancy (OV_{As}) in semi-insulating (SI) GaAs in relation with changes in the charge states and metastability of this centre. I will discuss finally the properties of substitutional oxygen in GaP and other semiconductors.
I consider briefly the interaction of hydrogen with oxygen in GaAs, where at least another configuration of oxygen, bonded to hydrogen, is found, but where the microscopic structure is not yet resolved.

Abbreviations frequently used:

2DLFM:	2-dimensional low-frequency mode	IS:	Isotope shift
AM:	Average mass	LDOS:	Local density of states
BC:	Bond centre	LHeT:	Liquid helium temperature
FWHM:	Full width at half maximum	qmi:	quasi-monoisotopic

R. Jones (ed.), Early Stages of Oxygen Precipitation in Silicon, 283–302.

2. Interstitial Oxygen in Silicon

2.1 THE SITUATION

Oxygen is always present in silicon crystals, whatever the growth method used. The only difference is the actual content : from about 10^{15}/cm^3 in standard float-zoned (FZ) material to about 10^{18}/cm^3 in silicon pulled from a silica crucible by the Czochralski method (CZ silicon). This is a particularity of silicon, because the strength of the Si-O bond is larger than that of the Si-Si bond (~ 2.3 eV in silicon), despite the large relaxation energy required to accommodate an O atom in the silicon crystal [1].

Most of the oxygen present in CZ as-grown silicon is interstitial (O$_i$), with an O atom inserted between (and chemically bonded to) two Si atoms of the lattice, after breaking of the normal Si···Si bond. A direct and elegant proof of this geometry is given in [2]. A small concentration of oxygen (typically 0.2%) in as-grown material is also present in centres called thermal donors. These centres are electrically active and contain a small number of O atoms. They are usually suppressed from as-grown silicon by a short annealing near 600°C (this does not necessarily mean that the O atoms return in an interstitial location). There is no clear evidence of precipitated oxygen in as-grown silicon and in any case, it is assumed that the concentration of precipitated oxygen is small compared to that of O$_i$.

O$_i$ is electrically inactive and its concentration is too small to allow Raman scattering or extended X-rays absorption fine structure (EXAFS) measurements. Thus, infrared (IR) vibrational absorption has been used extensively for its measurement and study. Phonon resonance spectroscopy near 1 K has also been used to probe with a high sensitivity the low-frequency excitations of O$_i$.

2.2 MODELLING OF THE EXPERIMENTAL DATA

The presence of a broad IR absorption near 1106 cm^{-1} (the 9-micron band) in the room-temperature spectrum of CZ silicon was soon related to the presence of oxygen [3]. In the first IR studies at liquid helium temperature (LHeT), three vibrational lines at 518, 1136 and 1206 cm^{-1} were observed. They were attributed to the three modes of a restricted region of the silicon lattice, taken as a symmetric non-linear Si$_2$O quasi-molecule [4]. The strong line at 1136 cm^{-1} was ascribed to the asymmetric mode, traditionally labelled v_3 in molecular spectroscopy for a molecule with C_{2v} symmetry, while the lines at 518 and 1206 cm^{-1} were ascribed to v_2 and v_1, respectively. When temperature is raised above LHeT, new lines appear on the low-energy side of the 1136 cm^{-1} line, as shown in Figure 1 and the structure finally evolves into the 1106 cm^{-1} band at room temperature. These "hot" lines were attributed to transitions from thermalised levels close to the ground state, due to the splitting of the vibrational level, and the splitting explained by the orientational degeneracy of the O atom among equivalent sites in the crystal [5]. The v_3 mode is characterised at LHeT by isotopic satellites, due to the fact that natural Si contains ^{28}Si, ^{29}Si and ^{30}Si with respective abundances of 92.2, 4.7 and 3.1 %. The satellites due to ^{17}O and ^{18}O isotopes are more difficult to detect because their abundances with respect to ^{16}O are only 0.04 and 0.20 %, respectively [6]. When the full width at half maximum (FWHM) of the lines is

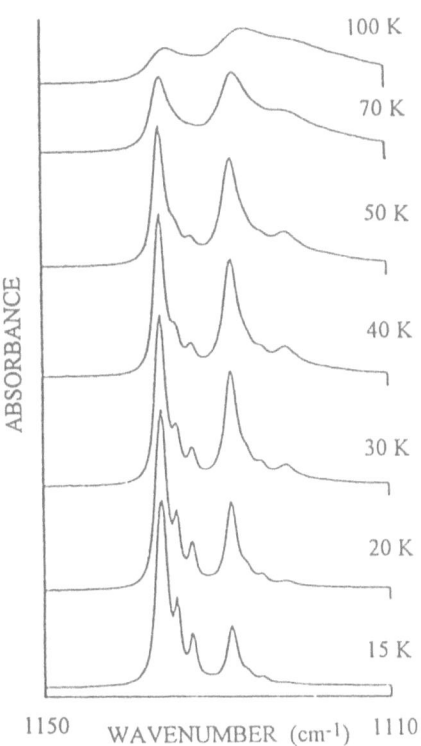

Figure 1. Absorption of $v_3(^{16}O)$ in silicon at between 15 and 100 K showing the progressive thermalisation of the low-frequency levels with temperature [7]

small, small isotope shifts (ISs) can be measured and they are a good test of the validity of the microscopic models.

The attribution of the 1206 cm^{-1} line to a symmetric mode was questioned [8] and it was suggested that it could be a combination of v_3 with the low frequency mode responsible for the thermalisation effects of v_3. A confirmation of this view came from a thorough investigation of the near and far IR absorption of O_i [9]. A far IR mode at 29 cm^{-1} was observed and ascribed to a vibration of the O atom in a plane perpendicular to the Si···Si broken bond. The motion of the O atom in this plane could be considered either as a perturbed 2-dimensional (2D) oscillator vibration or as the combination of a nearly free rotation about the Si···Si axis with a low-frequency vibration of the O atom in the (Si-O-Si) plane. At LHeT, only one line showed up, but at temperatures above LHeT, far IR lines corresponding to thermalised levels could also be observed (Figure 2). These transitions could be related with the ones near 1136 cm^{-1} and a unique energy level scheme derived. In this scheme, the low-frequency mode was taken as the v_2 symmetric mode of Si_2O and the line at 1206 cm^{-1} as a three-

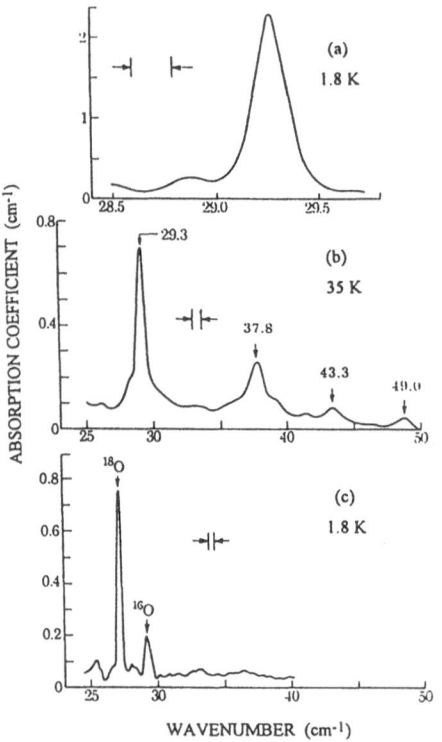

Figure 2. (a) Ground state low-frequency absorption of $^{16}O_i$ in silicon, (b) thermalised
transitions, (c) absorption of an ^{18}O-enriched sample [9]

quantum excitation involving a combination of v_3 with v_2 modes. The case of the 518 cm^{-1} line was not so clear and it was eventually considered as an oxygen-induced lattice absorption, involving a symmetric v_1-type motion. The models using the perturbed 2D oscillator or the combination of the low-frequency vibration with a nearly free rotator gave nearly identical values of the apex angle 2α of the Si-O-Si group (near 162°). A best fit analysis of the Si and O isotopic shift of the v_3 components had also been performed using a simple molecular model by adjusting the lattice interaction with the quasi-molecule through an adjustable mass M' added to that of the Si atoms and the Si-O-Si angle 2α. Values of 3 amu for M' and of 164° for 2α were obtained and the latter was comparable to the one derived from the far IR data [6]. Assuming an unaltered bond length of 1.6 Å for Si-O, the distance between the two bridged Si atoms deduced from an apex angle near 162° is about 3.2 Å, compared to a nearest neighbours distance of 2.35 Å in the silicon lattice. The local expansion due to oxygen can be partially compensated by a compression of the neighbouring Si-Si bonds, but an average increase of the lattice constant is none the less observed with oxygen concentration [10]. For $[O_i] = 10^{18}$ cm^{-3}, the relative volume change $\Delta V/V$ is + 1.3×10^{-5}. If the relative volume change is assumed to be $B^{-1}\sigma$, where B is the bulk modulus of silicon (~ 96 GPa), the above value of $\Delta V/V$ will produce a dilatation stress σ of - 1.3 MPa.

Later, another low-intensity line at 1749 cm^{-1} was found to be also related to O_i. It was ascribed to a combination of v_3 with a 2-phonon mode of the silicon lattice, but this attribution could not explain the large Si isotope effect observed (Figure 3) [6].

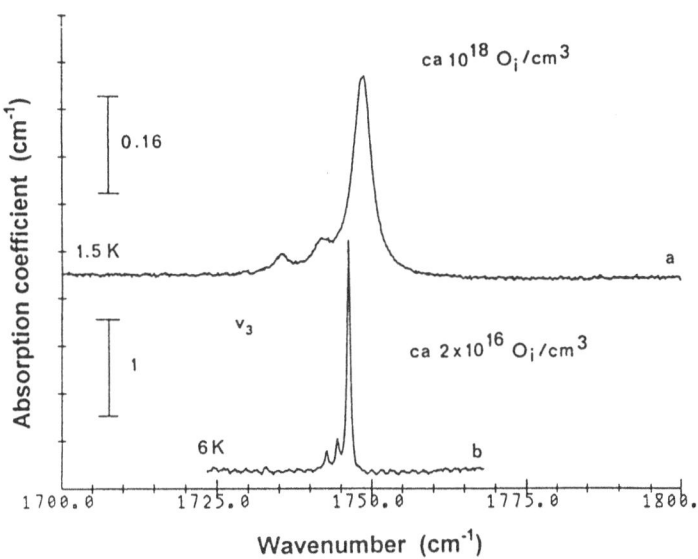

Figure 3. Comparison of the Si ISs of the 1749 cm^{-1} line (a) and v_3 mode (b) lines in silicon.
The abscissa scales are the same for (a) and (b), but the wavenumbers refer to (a) [6].

A quantum-mechanical calculation of the interaction between the asymmetric mode of O_i with the low frequency mode was performed in 1990 by Yamada-Kaneta *et al.* [11] using a $Si_3 \equiv Si\text{-}O\text{-}Si \equiv Si_3$ pseudo-molecule with point group symmetry D_{3d}. These authors started from an O atom bond-centred between two Si atoms. This yielded, as expected, two singly-degenerate modes A_{2u} and A_{1g} and a doubly-degenerate mode E_{2u}. They considered the Hamiltonians for the unperturbed motion of the O atom in the (111) plane and for the asymmetric mode A_{2u} along the <111> axis coupled by an interaction term. In the interaction scheme, a state is described by $|k, \ell, N >$ where k describes the radial dependence of the 2D anharmonic excitation, ℓ, its angular dependence and N, the high-frequency A_{2u} mode. Within the ground state manifold, the selection rules for electric dipole transitions are $\Delta k = 0, \pm 1, \Delta \ell = \pm 1$. For $\Delta N = 1$, they are the same for k with $\Delta \ell = 0$. In this description, the 29, 1136 and 1206 cm^{-1} lines correspond to transitions from the ground state $|0, 0, 0 >$ to the $|0, \pm 1, 0 >$,

$|0, 0, 1 >$ and $|1, 0, 1 >$ states, respectively. The calculation of the eigenvalues of the Hamiltonian as a function of the adjustable parameters was performed self-consistently using the experimental frequencies for ^{16}O and ^{18}O. The overall agreement with the experimental values of the frequencies of the transitions is good. For $N = 0$, the equilibrium positions of the O atom with respect to bond-centre (BC) location are 0.25 and 0.24 Å for ^{16}O and ^{18}O, respectively. For $N = 1$, these values increase to 0.34 and 0.32 Å. However, the potential barrier in the (111) plane between the equilibrium positions of the O atom were found to be small so that even at LHeT, the O atom could be thought of as <u>dynamically</u> located at a BC site between the two Si atoms, as shown in Figure 4.

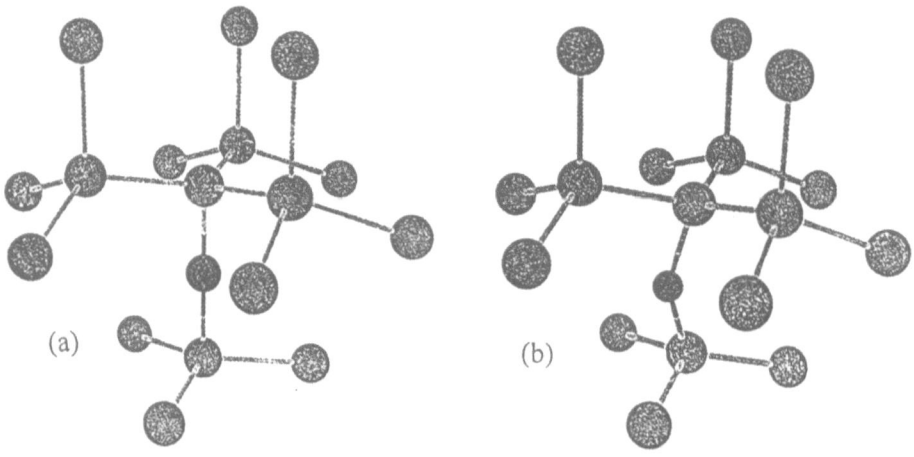

Figure 4. Representation of an interstitial O atom (dark grey) in the silicon (a) and germanium (b) lattices.

A result of these calculations is that the phenomenological coupling constant between the A_{2u} mode (labelled v_3 for convenience) and the 2D low-frequency motion (2DLFM), used to fit the data, is found to be negative. Now, in the interaction scheme, the radial dependence of the force constant of the v_3 mode is proportional to this coupling constant. It follows then that, contrary to an intuitive reasoning, when the distance of the O atom from the BC position increases, the frequency of v_3 should decrease. In this treatment, the 2DLFM is different from a symmetric modes of the pseudo-molecule and the 518 cm^{-1} line is ascribed to the E_{2u} mode. It was also suggested in this work that the missing A_{1g} mode could be one of the modes involved in the 1749 cm^{-1} line.

2.3 THE CONTRIBUTION OF THEORY

2.3.1. *Theoretical calculations*

First principles calculations have been performed on a Si$_3$≡Si-O-Si≡Si$_3$ cluster with a D_{3d} symmetry [12]. The calculation uses the Born approximation for the vibrational potential and the cluster-Bethe-lattice approximation for the infinite Si system. It predict that O_i is dynamically located at a BC site and that the potential experienced by the O atom in the direction perpendicular to the Si···Si axis is highly anharmonic and essentially flat near to the BC site. The relaxation of at least the first and second nearest neighbours of the O atom is shown to be essential, the final Si-O length being 1.56 Å. Based on these results, a phonon calculation is performed for O_i in an infinite Si system. The local densities of states (LDOS) of the O atom and of the two nearest neighbour Si atoms are shown in Figure 5. In the O and Si LDOSs, a peak is shown at 1150 cm^{-1}, associated with an asymmetric A_{2u} stretch mode of the unit.

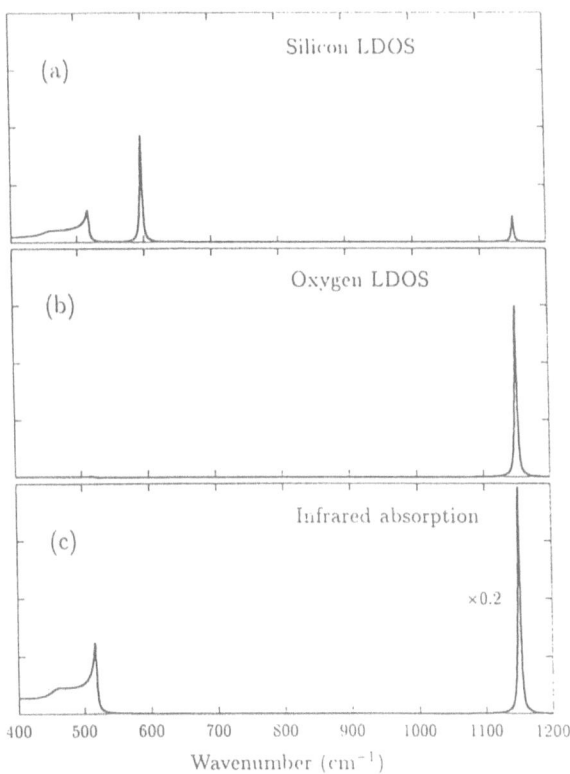

Figure 5. (a) Density of vibration modes projected on the Si atoms neighbours of the O atom. (b) Density of modes projected on the O atom. (c) Infrared absorption [12].

It is identified with the v_3 mode at 1136 cm^{-1}. The peak at 517 cm^{-1} in the Si LDOS corresponds to a doubly-degenerate transverse motion (E_u) of the Si atoms. The contribution at this frequency in the O LDOS is negligible. To remain consistent with the conventional spectroscopic notations, this mode must be labelled v_2. The experimental asymmetric line shape the 518 cm^{-1} mode is qualitatively reproduced in the calculated LDOS near 517 cm^{-1}. The dipole moment associated with this mode is perpendicular to the Si$\cdots$Si axis in agreement with the experimental results [13]. The strongest peak in the Si LDOS is at 594 cm^{-1} and it corresponds to a symmetric stretch mode (A_{1g}), or v_1 in the conventional spectroscopic notation. Such a mode is only Raman-active, but its combination with the A_{2u} or v_3 mode is IR active and should be observed near 1700 cm^{-1}. Figure 4 shows the dynamical location of interstitial oxygen in silicon. For a slightly non-linear Si-O-Si structure (or in the presence of anharmonicity), v_1 should become weakly IR-active. This mode, however, has not been detected yet, possibly due to the fact that the frequency expected makes this mode resonant with the two-phonon IR absorption of silicon, peaking at 613.6 cm^{-1}. By a first-principles method, it is not possible to compute accurately the potential perpendicular to the Si$\cdots$Si axis experienced by the O atom and this is also true from the low-frequency mode.

An *ab initio* calculation of the structure and dynamics of O_i in silicon has also been made using a crystalline molecular cluster $OSi_{44}H_{42}$ centred on an interstitial O atom [14]. After relaxation of the atoms of the $Si_3 \equiv Si$-O-$Si \equiv Si_3$ inner group, a minimum energy of the cluster was obtained for some values of the bond angles and interatomic distances. Frequencies at 1104 and 544 cm^{-1} are predicted for $^{28}Si_2{}^{16}O$. The calculated apex angle is slightly larger (172°) than the one derived from the IR data, but the Si-O and Si$\cdots$Si distances are very comparable. The agreement with experiment for the asymmetric mode is good and the 517 cm^{-1} line should correspond to the symmetric mode calculated to vibrate at 554 cm^{-1}, as the displacement of the O atom for this mode is found to be negligible. The frequency of the 2DLFM is very sensitive to the size of the cluster, and it is outside the range of applicability of the method used.

The main difference between the two theoretical analyses comes from the breaking of the D_{3d} symmetry in the latter. Thus, the symmetric mode at 554 cm^{-1} in [14] is an admixture of the A_{1g} and E_g modes of [12].

The stabilities of structures containing more than one O_i have been investigated by *ab initio* methods by Needels *et al.* [15]. They find that two or three O_i atoms in a staggered chain-like configuration could be stable. They suggest that v_3-like modes could be observed for the 2-O_i centre about 20 cm^{-1} above the mode for isolated O_i, but up to now such a mode has not been detected.

2.3.2. *Comparison with experiment*

O and Si ISs of the different O_i modes are a good test of theory. Under the assumption that the line at 1749 cm^{-1} line is a combination $A_{1g} + A_{2u}$, or ($v_1 + v_3$), of the ^{28}Si-^{16}O-^{28}Si bridge, it is possible to derive from the experimental frequencies of v_3 and of ($v_1 + v_3$) isotope-dependent frequencies of v_1. The O ISs have been measured in isotopically enriched samples [5,9,16]. The "measured" and calculated ISs of v_1 are listed in Table 1.

TABLE 1. Observed and calculated Si and O ISs (cm^{-1}) of O$_i$ modes, taken with respect to ^{28}Si$_2$^{16}O. The frequencies (cm^{-1}) in parentheses are calculated for ^{28}Si$_2$^{16}O. The ISs in brackets are determined indirectly [16].

Mode:	A_{1g} or ν_1 (596.3)		A_{2u} or ν_3 (1150.3)		$A_{1g} + A_{1u}$ (1746.6)	
Combination	Exp.	Calc.	Obs.	Calc.	Obs.	Calc.
^{28}Si^{16}O^{29}Si	[4.5]	5.0	1.9	2.3	6.4	[7.3]
^{28}Si^{16}O^{30}Si	[9.5]	9.7	3.7	4.5	13.2	[14.2]
^{28}Si$_2$^{17}O	[1.3]	0.0	26.9	26.0	28.2	[26.0]
^{28}Si$_2$^{18}O	[1.2]	0.0	51.4	49.7	52.6	[49.7]

For a linear Si-O-Si bridge, ν_1 is expected to be independent of the O isotope in the harmonic approximation. Experimentally, an IS of O is obtained for this mode, but smaller than that of Si by a factor of about two, indicating a small contribution of the O atom to ν_1, as a consequence of the anharmonicity of the potential. The large Si IS predicted by the calculations for ν_1 is also found experimentally and these two results confirm that the departure of Si-O-Si from a linear structure is indeed small. The fact that the O ISs experimentally derived for ν_1 are about the same is ascribed to a resonance effect of $\nu_3(^{17}$Si$_2$O) [16]. The combination $A_{1g} + E_u$ or $\nu_1 + \nu_2$ is IR-active, but, the expected intensity, FWHM and location (near 1129 cm^{-1}) of this combination makes it difficult to detect because it is too close to the strong absorption of the ν_3 mode. These comparisons seem to show that, even if not detected directly, the symmetric mode near 612 cm^{-1} predicted by the first-principles calculations do exist. A summary of the spectroscopic parameters at LHeT of the modes related to O$_i$ in silicon is given in Table 2.

TABLE 2. Experimental parameters of the O$_i$ modes at LHeT for ^{28}Si$_2$^{16}O. The relative intensity is given with respect to integrated absorption. In the combination mode, 2DLFM means only that the low frequency modes are involved.

Attribution	Position (cm^{-1})	Relative intensity	FWHM (cm^{-1})
2DLFM	29.3	0.05	0.1
E_u or ν_2	517.8	~0.25	5
A_{1g} or ν_1	(612.2)	?	~3
A_{2g} or ν_3	1136.36	1	0.6
$A_{2g} + 2$DLFM	1205.7	0.034	3.0
$A_{1g} + A_{2g}$	1748.6	0.015	3.7

The 2DLFM of O_i has been also detected by phonon resonance spectroscopy at 1 K [17]. Two absorptions are observed near 875 GHz, corresponding to the IR absorptions of ^{16}O and ^{18}O at 29.14 and 27.41 cm^{-1}. This kind of spectroscopy is very sensitive and it has been suggested that absorptions detected near 30 cm^{-1} are related to the shifts of the 2DLFM of O_i due to statistically distributed and interacting close O_i pairs.

3. Interstitial Oxygen in Germanium, GaAs and GaP

3.1 GERMANIUM

3.1.1 *The experimental background*
Oxygen is not normally present in CZ germanium, but in O-doped crystals, a line is seen at 862 cm^{-1} (LHeT) which shifts to 818 cm^{-1} with ^{18}O [18,19]. It is attributed to the ν_3 mode of O_i. Another broad feature (FWHM: 11 cm^{-1}) has been reported at 1269 cm^{-1} with an intensity proportional to that of ν_3 [20].

Natural Ge, with an average mass of 72.60 amu, contains isotopes ^{70}Ge, ^{72}Ge, ^{73}Ge, ^{74}Ge and ^{76}Ge with relative abundances of 20.50, 27.40, 7.80, 36.50 and 7.80 %, respectively. In silicon, at LHeT, only five Si_2O combinations are observed for ν_3 because for the same average mass (AM) of the Si atom, $^{29}Si_2O$ and $^{28}SiO^{30}Si$ lines nearly coincide and the former is 26 times more intense. Thus, by analogy, in germanium, 11 lines are expected for ν_3 in a spectral interval of 3 wavenumbers. High-resolution spectra of ν_3 at 6 K show 30 partially resolved components (Figure 6) with an average FWHM $\sim$ 0.04 cm^{-1} [21].

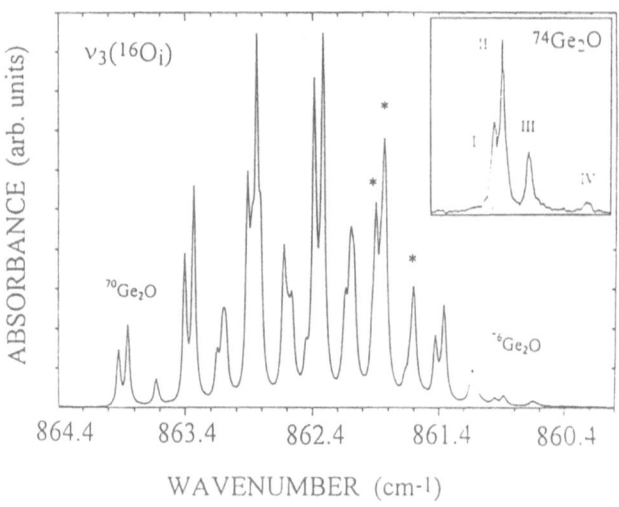

Figure 6. Isotope and thermal effects of $\nu_3(^{16}O)$ at 6 K in natural germanium [21]. In the inset, the absorption of qmi ^{74}Ge shows only thermal effects [22]. The asterisks denote the components of the inset.

The number of components indicates a very strong thermalisation of the levels at 6 K while a corresponding effect is found in silicon only near 100 K (Figure 1). In natural germanium, at 6 K, three components labelled I, II and III toward decreasing energies are more or less well seen for each AM. The I-II and II-III separations are ~ 0.07 and 0.23 cm⁻¹, respectively. In quasi-monoisotopic (qmi) ^{74}Ge, where extra components are only due to thermal effects, a fourth transition, IV, down-shifted from III by 0.48 cm⁻¹, is seen at 6 K [22] and shown in the inset of Figure 6.

When repeating the kind of fit of the ISs used for O_i in silicon [6], a value of the apex angle Ge-O-Ge of 140° and an interaction mass M' of 11.65 amu are obtained. Thus, assuming a Ge-O bond length 1.74 Å, a distance of 0.6 Å is found between O and the BC site (0.2 Å in silicon) and this results in a higher potential barrier to hindered rotation or tunnelling of the O atom. This explains qualitatively the small splitting of the vibrational levels in germanium. The ground state manifold has been explored by phonon spectroscopy [23]. Figure 7. shows the phonon spectrum of O_i in germanium at 1 K. At zero stress, above the 1.15 meV (9.7 cm⁻¹) onset of Al emitter-Sn detector combination, eight transitions are observed between ~ 10 and 22 cm⁻¹.

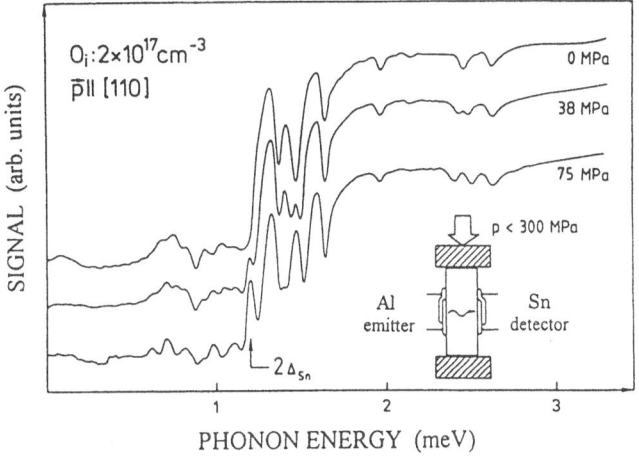

Figure 7. Phonon absorption spectrum of O_i in Ge at 1 K as a function of a [110] stress. The cut-off of the emitter/detector combination evaporated on the sample (bottom right) is at 1.15 meV [23].

An analysis of the results shows that even at 1 K, some of the transitions already arise from thermalised levels as the first two excited levels are 1.45 and 5.40 cm⁻¹ above the ground state. Similar conclusions have been reached from the temperature dependence of the IR absorption of qmi Ge samples [24]. The results can be interpreted in the hindered rotator framework already used with silicon [9,11]. Within the ground state manifold, all the transitions are allowed by phonon spectroscopy as long as the initial states are populated while by IR absorption, only $\Delta k = 0, \pm1$, $\Delta\ell = \pm1$ transitions are allowed. The first far IR lines should show up at 1.45, 3.95, 5 and near 7 cm⁻¹, but they have not yet been reported. The mid-IR lines labelled I, II, III and IV are attributed

294

respectively to transitions from the |0, 0>, |0, ±1>, |0, ±2> and |1, 0> initial $|k, \ell>$ states from $N = 0$ to the same final states for $N = 1$.

When temperature is lowered below 6 K, the IR spectrum simplifies because thermal population is reduced [24]. Then, the sharpness of the lines helps to resolve individual transitions for combinations with the same AM, but different isotopes, as shown in Figure 8 for a natural Ge sample at 1.6 K [25].

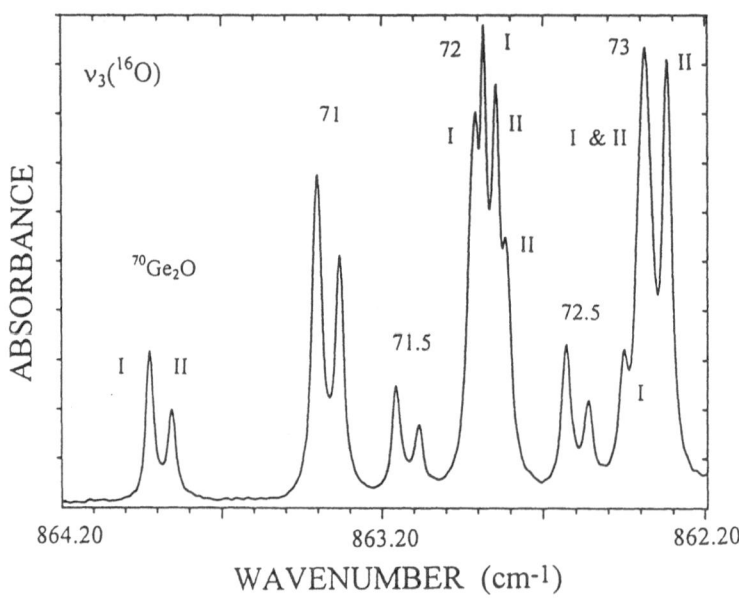

Figure 8. Part of the $v_3(^{16}O)$ structure in natural germanium at 1.6 K (resolution: 0.013 cm^{-1}) Note the absence of line III and the inversion of the intensities of lines I and II with respect to the 6 K spectrum [25].

For single combinations, a doublet due to lines I and II is seen, but for the AM 72, which is a superposition of $^{72}Ge_2O$ and $^{70}GeO^{74}Ge$, four partially resolved component are observed. This also occurs for AM 73 because the intensity of $^{73}Ge_2O$ is negligible compared to those of $^{70}GeO^{76}Ge$ and $^{72}GeO^{74}Ge$. In this latter case, two components nearly coincide and they cannot be resolved. From the relative intensities, it can be deduced that the most symmetric combination has the highest frequency. Preliminary calculations of the frequency shifts of component I of different combinations with the same AM agree well with experiment [26].

3.1.2 *Comparison between theory and experiment*
The results of first-principles calculations on O_i in germanium are [27]: a Ge-O-Ge angle of 140° and a Ge-O distance of 1.70 Å at the equilibrium, which localise the O atom at 0.58 Å from the BC location. The computed distance between the two Ge atoms is 3.20 Å, and it implies, as expected, a slightly smaller distortion due to O_i in germanium than in silicon (Figure 4). It is possible there to compute the splitting of the

ground state because the potential is larger and the agreement with experiment is very good. The calculated absorption spectrum of O_i predicts features at 0, 230, 416 and 877 cm^{-1}. The first one corresponds to the free rotation of oxygen and the other three ones to the v_2, v_1 and v_3 modes of Ge_2O. These latter modes are analogous to those listed in Table 2 for O_i in silicon. The rocking mode v_2 has never been reported but the asymmetric v_3 mode is well characterised experimentally (see above). While IR active because of the bent structure of Ge-O-Ge, the v_1 mode has not been observed so far, but its calculated combination with v_3 gives a feature at 1293 cm^{-1}, close to the band observed at 1270 cm^{-1}. With this attribution, the experimental position of v_1 should be 407 cm^{-1}. The Ge ISs of the $v_1 + v_3$ combination mode has been measured at LHeT with a resolution of 2 cm^{-1} using ^{16}O-doped qmi Ge samples. The results and their comparison with the calculations are summarised in Table 3 [28].

TABLE 3. Observed and calculated Ge ISs (cm^{-1}) of $^{16}O_i$ modes with respect to $^{70}Ge_2^{16}O$. ISs in brackets are determined indirectly. ISs in parentheses come from the high-resolution data of [24].

Mode:	v_1		v_3			$v_1 + v_3$	
Combination	Obs.	Calc.	Obs.		Calc.	Obs.	Calc.
$^{73}Ge_2O$	[4.5]	4.15	1.5	(1.54)	2.03	6.0	[6.18]
$^{74}Ge_2O$	[6.4]	5.47	2.0	(2.02)	2.68	8.4	[8.15]
$^{76}Ge_2O$	[9]	7.96	3.0	(2.94)	3.92	12	[11.88]
Natural Ge	[4.3]		1.5			6.0	

As in Table 2, a rather large IS due to the displacement of the two atoms of the crystal bonded to O is shown for v_1. The observation of the O ISs of the combination mode requires ^{17}O and ^{18}O-enriched samples which were not available.

3.2 GALLIUM ARSENIDE AND GALLIUM PHOSPHIDE

GaAs containing $^{16}O/^{18}O$ displays absorption features related to the presence of oxygen [29,30]. One features at 845 cm^{-1} have been attributed to the vibration of a Ga-^{16}O-As centre or O_i because the relative intensities of the two components of this doublet match the ^{69}Ga and ^{71}Ga relative abundances and ^{75}As is 100% abundant. [30]. The maximum peak absorption of this feature, the equivalent of v_3 in silicon and germanium, stays below 0.5 cm^{-1} so that no other feature related to O_i has been found up to now because it would escape detection. The FWHM of this mode depends on the homogeneity of the sample and 6 K values between 0.06 and 0.12 cm^{-1} have been measured, but no thermalised satellite bands are observed when raising the temperature [31]. This can be related to the fact that O is bonded to one As and one Ga atom, having different relaxation energies. At room temperature, the O_i mode is at 836 cm^{-1}

with a FWHM near 10 cm^{-1} and when ^{16}O is replaced by ^{18}O, the downward IS is 46 cm^{-1} [29]. A confirmation of the model comes from the stress-induced splitting of the mode, which shows that the Ga-O-As bridge is along a <111> axis [32].

Reorientation of O$_i$ leading to dichroism is observed when GaAs is cooled under a [110] stress from room temperature (Figure 9).

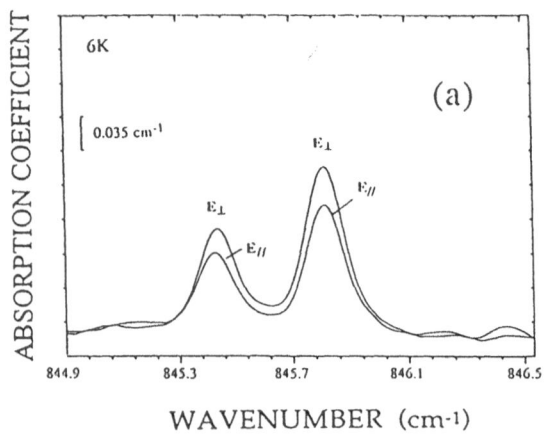

WAVENUMBER (cm^{-1})

Figure 9. Stress-induced dichroism of the v_3(^{16}O$_i$) doublet in GaAs at 8 K due to atomic reorientation. under a [110] stress of 120 MPa applied from 300 to 8 K. E$_{//}$ and E$_\perp$ are the polarisation direction with respect to stress direction. The spectra are taken at zero stress [31].

It shows that, like in silicon and germanium, O$_i$ aligns preferentially aside from the stress, but for silicon and germanium, starting temperatures above 300°C are required [2]. The reorientation energy in GaAs is between 0.8 and 1.0 eV, but it is not possible to know from the spectroscopic data whether the reorientation takes place preferentially about the Ga or As atoms. The dissociation energy is lower for diatomic GaO (285 kJ/mole) than for AsO (481 kJ/mole) and *ab initio* approximate Hartree-Fock calculations predict that the O atom should be slightly closer to As than to Ga [33]. On the other hand, cluster calculations indicate an inverse trend, with a reorientation energy of 0.5 eV about the Ga atom (breaking of an As-O bond) and of 1.84 eV about the As atom (breaking of a Ga-O bond) [34]. Under the simplifying assumption that the GaO and AsO binding energies in GaAs are the same, it can be guessed from the annealing temperature of the stress-induced dichroism that the diffusion coefficient of O$_i$ in GaAs is smaller than in germanium [2].

In O-doped GaP, lines at 1007 cm^{-1} and at slightly higher frequencies have been related to oxygen [35,36]. The one at 1007 cm^{-1} has been attributed, on the basis of a linear chain model, to the asymmetric mode of O$_i$ bonded to Ga and P atoms. However, no ^{18}O or Ga ISs have been reported for these modes, so that, if it seems very likely that the assignment is correct [37], direct proofs are still lacking. The linear chain model predicts also a symmetric mode of O$_i$ at 443 cm^{-1} which has not been observed up to now.

4. Quasi-substitutional oxygen in Gallium Arsenide

While O_i is electrically inactive, quasi-substitutional oxygen or OV (V stands for an atomic vacancy) is electrically active in silicon and GaAs. I restrict the discussion to GaAs because the situation in silicon is discussed in other parts of this book. In GaAs, the relevant vacancy is a missing As atom (V_{As}). Two IR features, A and B, observed at low temperature in as-grown high-resistivity GaAs grown from a quartz crucible, were ascribed to the asymmetric mode of two charge states of OV_{As} because each feature was a triplet involving two equivalent Ga atoms [38]. This model is confirmed by the observation of the $^{16}O/^{18}O$ IS of B [30]. In the accepted model, the O atom is bonded to two second-nearest neighbour Ga atoms as shown in Figure 10.

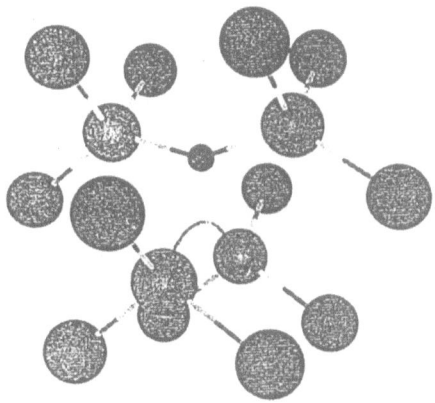

Figure 10. Model of the OV_{As} centre in GaAs. The four Ga atoms are shown in light grey and the O atom in dark grey

Modes A and B are observed under equilibrium, but in SI material, only A is observed as it corresponds to the most positive charge state, while in slightly n-type GaAs, A and B coexist [39,40]. In SI GaAs, illumination for a short time at LHeT with 1.25 eV photons ($\sim 1 \ \mu m$) produces a third mode, labelled C [40] or B' [41], which grows at the expense of A. The isotopic components of modes A, B and C are shown in Figure 10. Mode C is metastable and it correspond to a charge state intermediate between those of A and B. This implies that in GaAs, OV_{As} has a negative-U character and this has indeed been demonstrated [42,43]. In the SI GaAs samples where the deep centre EL2 and OV_{As} are both present, interconversion of A can occur under illumination at LHeT with 1.25 eV photons. FIG. shows the time dependence of the charge transfer from A to B, followed by the inverse. In both cases, C appears as in intermediate state. In the first part of the interconversion, the fast transfer from A to B is explained by the

298

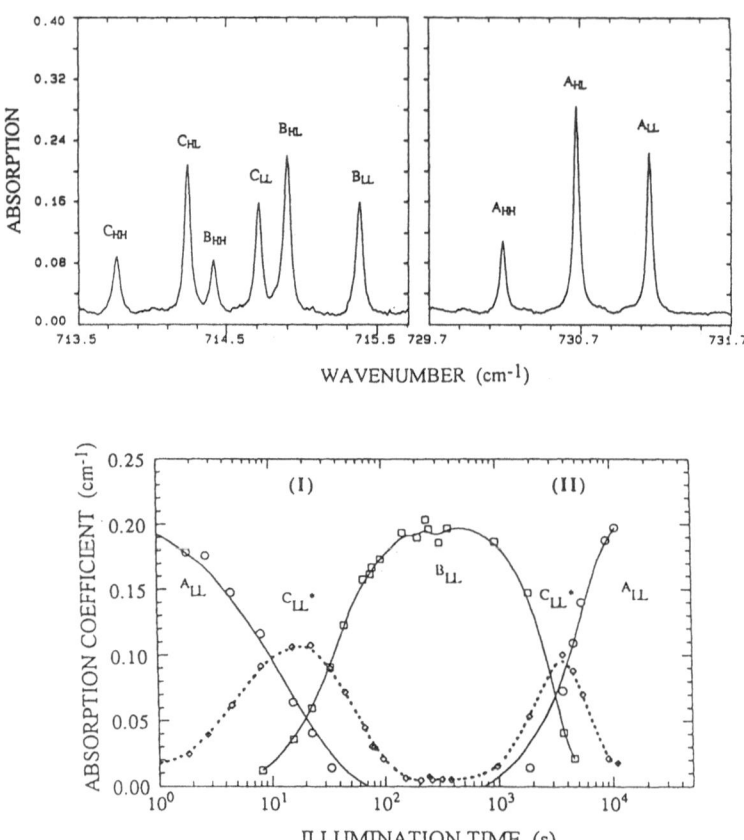

Figure 11. Above: Modes A, B and C of OV_{As} at 6 K in GaAs. HH, HL and LL are for $^{71}Ga_2O$, $^{71}GaO^{69}Ga$ and $^{69}Ga_2O$, respectively. Below: Change under illumination with 1.25 eV photons at 6 K of the concentration of the different charge states of OV_{As} in SI GaAs as a function of the illumination time (interconversion) [31,44].

photoionisation of EL2^0 in the conduction band and the preferential trapping of a first and a second photoelectron by OV_{As}. The second part is mediated by the photoionisation of EL2$^+$ in the valence band. The recombination of holes with EL2^0 competes with the phototransfer of EL2^0 into the metastable state at the same photon energy. The metastable state of EL2 is electrically inactive so that holes are available for recombination on the centres which have trapped electrons in the first step of the interconversion or on other hole traps [44]. At the end of this interconversion, GaAs is insensitive to further illumination. An attempt to determine the absolute values of the charge states of OV_{As} has been reported recently, based on optically detected electron paramagnetic resonance and electron nuclear double resonance measurements [45]. It seems to indicate that B' (or C) corresponds to a paramagnetic state with a single unpaired electron on the Ga-Ga reconstructed bond. This state is neutral with respect to the GaAs lattice, while A should be positive and B negative. The stress-induced splittings of modes A, B and C are consistent with a Ga-O-Ga dipole oriented along a <110> axis [31,32]. The symmetry of substitutional O in GaAs has been also investigated by *ab initio* calculations [46]. These calculations predict multiple charge states for this centre, with on-centre O^+ (isoelectronic to As) and O^0. In this scheme, mode A would correspond to O^-, B to O^{3-} and B' (or C) to O^{2-}. Like O, O^{2-} is paramagnetic, but the electronic density would be different.

The FWHM of A, B and C are near 0.05 cm^{-1} at LHeT and this contrasts with a value near 2 cm^{-1} for the OV modes in silicon. A main difference lies also in the fact that OV_{As} is a native defect in GaAs while OV must be produced by high-energy irradiation in silicon and germanium. In short term rapid thermal annealing, the A mode starts decreasing in intensity at ~ 600°C with a correlated increase of the O_i concentration, indicating that it is more stable than OV in silicon. Long-term annealing produces other O-related centres and finally the precipitation of oxygen [47].

5. Substitutional oxygen in GaP

Most of the information on substitutional O in GaP comes from luminescence and excitation spectra [37]. In GaP, because of a lattice constant smaller than that of GaAs, oxygen can replace a P atom. It acquires then a deep donor character with an ionisation energy near 0.9 eV, compared to ~ 0.1 eV for S, Se or Te. Both O^0 and O^+ display tetrahedral symmetry. Isoelectronic centres resulting from the pairing of Zn or Cd acceptor on a Ga site nearest neighbour from the O_P deep donor have also been identified. O^0 can bind an electron to give O^-, but the binding energy of this electron is much larger than the one extrapolated from the H^- model. Self-consistent semi-empirical calculations predict a small energy barrier between the off-centre location of the negatively charged O atom along a <100> direction and on-centre metastable O^-. The equilibrium configuration is obtained for a location 0.6 Å from the on-centre location [48]. O and Cd Isotope effects of the no-phonon lines of the excitons bonded to Cd,O isoelectronic pairs show a positive shift with increasing isotopic masses. This kind of IS is explained by a difference of mass-dependent zero-point energy between the initial and final state of the transition [49] and it has been observed for other no-phonon lines in other semiconductors. A local mode at 57.5 meV (464 cm^{-1}) has been ascribed to the symmetric vibration of O_P in GaP [50], to compare with 58.9 meV (475 cm^{-1}) for the linear chain model [35]. Unfortunately, no ^{18}O IS has been measured for this mode. In II-VI semiconductors, because of the large ionicity, O is generally assumed to locate on an anion site. In ZnTe, it behaves as a deep isoelectronic trap for electrons [51], but in other II-VI materials, its acceptor behaviour has been tentatively explained by an electron charge transfer from the lattice to substitutional O_{VI} [52].

6. OH-related centres in GaAs

In some O-containing CZ GaAs samples, fundamental vibrational modes are observed between 2900 and 3500 cm^{-1}. Such high frequencies imply the stretching of a NH or OH bond [53]. The strongest of these lines located at 3300 cm^{-1}, with a FWHM of 0.13 cm^{-1} has a weak satellite at 3298.54 cm^{-1} with a relative intensity of 0.03. The calculated frequency and relative intensity of an ^{18}OH oscillator compared to ^{16}OH at 3300 cm^{-1} are 3298.14 cm^{-1} and 0.037. Therefore, it is inferred that OH-related centres are present in these samples (CZ GaAs crystals are known to contain hydrogen as a residual impurity). The stress-induced splitting of this mode does not fit the usual pattern for <100>, <110> or <111>-oriented dipoles. The fact that more components are observed than expected does not mean here a dipole with a low symmetry because of the full polarisation of the components. This is interpreted as an OH radical weakly coupled to its environment, which can reorient under stress in directions different from the one at zero stress. This OH centre is electrically active and, under appropriate photo-excitation, it can trap a hole and give a new line at 3296 cm^{-1} (Figure 12).

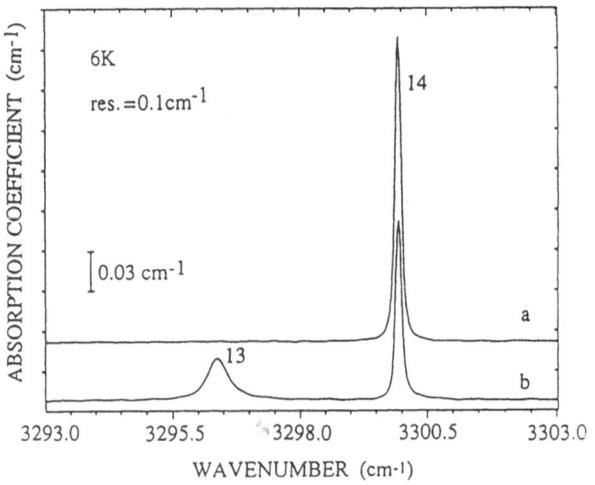

Figure 12. Absorption of an OH stretch mode in SI GaAs: (a) under equilibrium; (b) after 10 min. illumination with white light. Line 14 is due to the complex giving line 13 after the trapping of a hole.. The ^{18}O satellite is too weak to be seen at this scale [53].

The small change in frequency suggests that the hole is not trapped at the OH bond, but rather by its environment. Complexes involving O and H in germanium and in silicon are discussed in other parts of this book. The main differences with the one reported here is that they do not involve chemical bonding of O with a H atom.

Acknowledgements. I would like to thank E. Artacho, R. Jones and K. Laßmann for fruitful discussions and the communication of unpublished results. The loan of qmi Ge samples by E. E. Haller and by L. I. Khirunenko (within INTAS Project 93-320) is gratefully acknowledged.

301

7. References

1. For a review, see: Semiconductors and Semimetals, Vol. 42 (1994) F. Shimura (ed.) *Oxygen in Silicon*, Academic Press, San Diego.
2. Corbett, J. W., McDonald., R. S., and Watkins, G. D. (1964) The configuration and diffusion of isolated oxygen in silicon and germanium, *J. Phys. Chem. Solids* **25**, 873-879
3. Kaiser, W., Keck, P. H., and Lange, C. F. (1956) Infrared absorption and oxygen content in silicon and germanium, *Phys. Rev.* **101**, 1264-1268.
4. Hrostowski, H. J., and Kaiser, R. H. (1957) Infrared absorption of oxygen in silicon, *Phys. Rev.* **107**, 966-972.
5. Hrostowski, H. J., and Alder, B. J. (1960) Evidence for internal rotation in the fine structure of the infrared absorption of oxygen in silicon, *J. Chem. Phys.* **33**, 980-990.
6. B. Pajot and B. Cales (1986) Infrared spectroscopy of interstitial oxygen in silicon, *Mat. Res. Symp. Proc.* **59**, 39-44.
7. Krishnan, K., and Hill, S. L. (1981) Detailed Fourier transform (FTIR) study of the temperature dependence of the oxygen impurity in silicon, in H. Sakai (ed.), *Fourier Transform Infrared Spectroscopy (1981)* S.P.I.E. 289, S.P.I.E., Bellingham, Wash., pp. 27-31.
8. Chrenko, R. M., McDonald, R. S., and Pell, E. M. (1965) Vibrational spectra of lithium-oxygen and lithium-boron complexes in silicon, *Phys. Rev.* **138**, A1775-1784.
9. Bosomworth, D. R., Hayes, W., Spray, A. R. L., and Watkins, G. D. (1970) Absorption of oxygen in silicon in the near and far infrared, *Proc. Roy. Soc. Lond. A.* **317**, 133-152.
10. Windisch, D. and Becker, P. (1990) Silicon lattice parameters as an absolute scale of length for high precision measurements of fundamental constants, *phys. stat. sol (a)* **118**, 379-388 and references therein.
11. Yamada-Kaneta, H, Kaneta, C., and Ogawa, T. (1990) Theory of local-phonon-coupled low-energy anharmonic excitation of the interstitial oxygen in silicon, *Phys. Rev. B* **42**, 9650-9656.
12. Artacho, E., Lizón-Nordström, A., and Ynduráin, F. (1995) Geometry and quantum delocalization of interstitial oxygen in silicon, *Phys. Rev. B* **51**, 7862-7865.
13. Stavola, M. (1984) Infrared spectrum of interstitial oxygen in silicon, *Appl. Phys. Lett.* **44**, 514-516.
14. Jones, R., Umerski, and Öberg, S. (1992) *Ab initio* calculation of the local vibratory modes of interstitial oxygen in silicon, *Phys. Rev. B* **45**, 11 321-11 323.
15. Needels, M., Joannopoulos, J. D., Bar-Yam, Y., and Pantelides, S. T. (1991) Oxygen complexes in silicon, *Phys. Rev. B* **43**, 4208-4215.
16. Pajot, B., Artacho, E, Ammerlaan, C. A. J., and Spaeth, J-M. (1995) Interstitial O isotope effect in silicon, *J. Phys.: Condens. Matter* **7**, 7077-7085.
17. Dittrich, E, Scheitler, W., and Eisenmenger, W. (1987) Satellite phonon absorption lines above the 875 GHz resonance of interstitial oxygen in silicon, *Japan. J. Appl. Phys.* **39** (Suppl 26-3) 873-874.
18. Bloem, J, Haas, C., and Penning, P. (1959) Properties of oxygen in germanium, *J. Phys. Chem. Solids* **12**, 22-27.
19. Whan, R. E. (1965) Investigations of oxygen-defects in germanium between 25 and 700 K in irradiated germanium, *Phys. Rev.* **140**, A690-698.
20. Kaiser, W. (1962) Electrical and optical investigations of the donor formation in oxygen-doped germanium, *J. Phys. Chem. Solids* **23**, 255-260.
21. Pajot, B and Clauws, P. (1987) High-resolution local mode spectroscopy of oxygen in germanium in O. Engström (ed.), *Proc. of the 18th Internat. Conf. on the Physics of Semicond.*, World Scientific, Singapore, pp. 911-914.
22. Khirunenko, L. I., Shakovtsov, V. I., Shinkarenko, V. K., and Voroblako, F. M. (1990) Structure of infrared absorption of oxygen in germanium, *Sov. Phys. Semicond.* **24**, 663-665.
23. Gienger, M., Glaser, M., and Lassmann, K. (1993) Phonon spectroscopy of the low-energy vibrations of interstitial oxygen in germanium, *Solid State Commun.* **86**, 285-289.
24. Mayur, A. J.,Dean Sciacca, M., Udo, M. K., Ramdas, A. K., Itoh, K., Wolk, J, and Haller, E. E. (1994) Fine structure of the asymmetric stretching vibration of dispersed oxygen in monoisotopic germanium, *Phys. Rev. B* **49**, 16293-16299.

302

25. Pajot, B. (1995) Unpublished.
26. Artacho, E. (1995) Unpublished.
27. Artacho, E. and Ynduráin, F. (1996) Theory of interstitial oxygen in silicon and germanium, *Materials Science Forum* **196-201**, 103-108.
28. Artacho, E., Ynduráin, F., Pajot, B., Khirunenko, L. I., Ramirez, R., Herrero, C. P., Itoh, N, and Haller, E. E. (1996), to be published.
29. Akkerman, Z. L., Borisova, L. A., and Kravchenko, A. F. (1976) Infrared absorption spectra of oxygen-doped gallium arsenide, *Sov. Phys. Semicond* **10**, 590-591.
30. Schneider, J., Dischler, B., Seelewind, H., Mooney, P. M., Lagowski, J., Matsui, M., Beard, D. R., and Newman, R. C. (1989) Assessment of oxygen in gallium arsenide by infrared local vibrational mode spectroscopy, *Appl. Phys. Lett.* **54**, 1442-1444.
31. Song, C.Y. (1992) Etudes des centres présents dans GaAs:O semi-isolant par spectroscopie des modes localisés, *Doctoral Dissertation, Université Paris 7*.
32. Song, C., Pajot, B., and Porte, C. (1990) Piezospectroscopic study of interstitial oxygen in gallium arsenide, *Phys. Rev.* B **41**, 12 330-12 333.
33. Roberson, M. A., Estreicher, S. K., and Chu, C. H. (1993) Interstitial oxygen in elemental and compound semiconductors: fundamental properties and trends, *J. Phys. C.: Condens. Matter* **5** 8943-8954.
34. Jones, R. and Öberg, S. (1992) Unpublished.
35. Barker, A. S., Jr., Berman, R., and Verleur, H. W. (1973) Localized vibrational modes of interstitial oxygen and oxygen complexes in GaP, *J. Phys. Chem. Solids* **34**, 123-132.
36. Ulrici, B., Stedman, R., and Ulrici, W. (1987) Local vibrational mode absorption of hydrogen and oxygen centres in LEC-grown GaP and GaAs, *phys. stat. sol.* (b) **143**, K135-K139.
37. Dean, P. (1992) Oxygen in gallium phosphide, in S. T. Pantelides (ed.) *Deep Centers in Semiconductors: A State-of-the-Art Approach* (2nd edition), Gordon and Breach, Yverdon, pp. 215-377.
38. Zhong, X., Desheng, J., Ge, W., and Song, C. (1988) Model sudy of the local vibration center related to EL2 levels in GaAs, *Appl. Phys. Lett.* **52**, 628-630.
39. Alt, H. C. (1989) Photosensitivity of the 714 and 730 cm^{-1} absorption bands in semi-insulating GaAs: Evidence for a deep donor involving oxygen, *Appl. Phys. Lett.* **54**, 1445-1447.
40. Song, C., Pajot, B., and Gendron, F. (1990) Local mode spectroscopy and photo-induced effects of oxygen-related centers in semi-insulating gallium arsenide, *J. Appl. Phys.* **67**, 7307-7312.
41. Alt, H. C. (1989) Fine structure of the oxygen-related local mode at 714 cm^{-1} in GaAs, *Appl. Phys. Lett.* **55**, 2736-2738.
42. Alt, H. C. (1990) Experimental evidence for a negative-U center in gallium arsenide related to oxygen, *Phys. Rev. Lett.* **65**, 3421-3424.
43. Skowronski, M., Neild, S. T., and Kremer, R. E. (1990) Location of energy levels of oxygen-vacancy complexes in GaAs, *Appl. Phys. Lett.* **57**, 902-904.
44. Song, C. Y., Pajot, B., Ge, W. K., and Jiang, D. S. (1995) Relation between the metastability of EL2 and the photosensitivity of local vibrational modes in semi-insulating GaAs, *Phys. Rev.* B **52** 4864-4869.
45. Linde, M., Alt, H. Ch., and Spaeth, J.-M. (1996) Magneto-optical and ODENDOR investigations of the substitutional oxygen defect in gallium arsenide, *Materials Science Forum* **196-201**, 213-218.
46. Jones, R. and Öberg, S. (1992) Multiple charge states of substitutional oxygen in gallium arsenide, *Phys. Rev. Lett.* **69**, 136-139.
47. Skowronski, M. (1992) Complexes of oxygen and native defects in GaAs, *Phys. Rev.* B **46**, 9476-9481.
48. Khoo, G. S. and Ong, C. K. (1993) O⁻ in GaP: a negative-U centre, *J. Phys.: Condens. Matter* **5**, 3917-3924.
49. Heine, V. and Henry, C. H. (1975) Theory of the isotope shift of the zero-phonon optical transitions at traps in semiconductors, *Phys. Rev.* B **11**, 3795-3803.
50. Arai, T., Asanuma, N., Kudo, K, and Umemoto, S. (1972) The infrared absorption of GaP single crystals with silicon and oxygen impurities, *Japan. J. Appl. Phys.* **11**, 206-212.
51. Merz, J. L. (1968) Isoelectronic oxygen trap in ZnTe, *Phys. Rev.* **176**, 961-968.
52. Akimoto, K., Okuyama, H., Ikeda, M., and Mori, Y. (1992) Isoelectronic oxygen in II-VI semiconductors, *Appl. Phys. Lett.* **60**, 91-93.
53. Pajot, B and Song, C. (1992) OH bonds in gallium arsenide grown by the liquid-encapsulated Czochralski crystal-growth method, *Phys. Rev.* B **45**, 6484-6491.

OXYGEN-RELATED LUMINESCENCE CENTRES CREATED IN CZOCHRALSKI SILICON

E. C. LIGHTOWLERS AND GORDON DAVIES
Physics Department, King's College London
Strand, London WC2R 2LS, UK.

Abstract. A review is made of the major luminescence systems created in Czochralski (CZ) silicon by various thermal treatments, and by radiation damage with and without subsequent thermal treatments. Detailed investigations of these luminescence systems were carried out during the 1980s. More recent measurements have clarified some issues and shown that some other areas need to be revisited. Other recent measurements have shown that hydrogen has a dramatic effect on the luminescence centres created in both float-zone (FZ) and CZ Si and that several luminescence centres actually contain hydrogen. It has also been demonstrated that these and other centres can be made optically inactive if too much hydrogen is present.

1. Introduction

The purpose of this paper is to make a brief review of the major luminescence systems which are created in Czochralski (CZ) grown silicon by various thermal treatments, and radiation damage with and without subsequent thermal treatments. The creation of these luminescence systems certainly involves the high oxygen concentration present in CZ Si. However, although it is likely that most of the centres responsible for these luminescence systems actually contain oxygen, in many cases no direct evidence has been reported. Many of these systems were first discovered between the late 1960s and the early 1980s [1] and detailed investigations were carried out in various laboratories throughout the 1980s; a review is included in reference 3. We have carried out some further investigations during the 1990s [3, 4]. Our recent discoveries that hydrogen has a dramatic effect on the luminescence centres created in both float-zone (FZ) and CZ Si and that several centres actually contain hydrogen [5-7], not originally envisaged, have recently increased our activity in this area. In order to produce some original figures for this paper, some new spectra were very recently recorded using samples prepared up to 15 years ago as well as new material currently being produced in connection with our hydrogen studies. It is clear from examining these spectra, and other recent measurements, that some areas investigated during the 1980s need revisiting using better technology, better samples and recently developed ideas concerning data interpretation.

2. Experimental

The spectra shown in this paper were all recorded using Fourier transform photoluminescence (PL) spectroscopy, first developed in the mid 1980s [8]. The instruments employed were either a Nicolet 60SX or Bomem DA8 FTIR spectrometer, modified for PL and fitted with a North Coast cooled Ge diode detector. Since some important spectral features have energies below the spectral range of a Ge detector, ie < 0.7 eV, a modified Cincinnati Electronics cooled InSb diode detector, fitted with a 2.8

303

R. Jones (ed.), Early Stages of Oxygen Precipitation in Silicon, 303–318.
© 1996 *Kluwer Academic Publishers*.

μm short wavelength pass cold filter, was also employed. This is about 50 times less sensitive than a Ge detector because of room temperature black body radiation, which introduces a factor of about 2500 in data acquisition time for the same signal/noise ratio and resolution. Most of the photoluminescence measurements were carried out with the samples immersed in liquid helium at 4.2 K, and the luminescence was excited by a 514 nm Ar+ laser.

3. Radiation damage

Figure 1 shows a spectrum obtained from a CZ Si sample with $[^{12}C] = 4.5 \times 10^{17} \, \mathrm{cm}^{-3}$ and $[^{14}C] = 2 \times 10^{17} \, \mathrm{cm}^{-3}$ irradiated with a dose of $4 \times 10^{17} \, \mathrm{cm}^{-2}$ 2 MeV electrons. Only the G-line (0.9695 eV) and C-line (0.7896 eV) luminescence systems can be observed, which is generally the case for either FZ or CZ Si after irradiation at, or close to, room temperature with no subsequent thermal treatment; their absolute and relative intensities are dependent on the carbon and oxygen concentrations and the radiation dose [9, 10]. The zero-phonon G-line shows no carbon isotope shift or splitting, but the doublet structure of the S local mode (figure 1b) would suggest that the centre contains one carbon atom. However, optically detected magnetic resonance (ODMR) measurements [11] have demonstrated that the centre responsible for the G-line luminescence system contains two carbon atoms, and this is confirmed by a closer examination of the S local mode [12]. The isotope splitting of the zero-phonon C-line implies that the centre responsible for this luminescence system contains at least one carbon atom. Radiation damage and other processes, create vacancies and self-interstitials, and the latter interact with substitutional carbon to produce interstitial carbon which is mobile at room temperature. This results in the creation of the G-line and C-line and a range of other luminescence centres, particularly after various thermal treatments [2, 10].

The local mode structure of the C-line luminescence system is shown in figure 2. This has been investigated several times during the 1980s [9, 13, 14, 15] and the labelling employed is that used in the last report [15]. The centre responsible for this

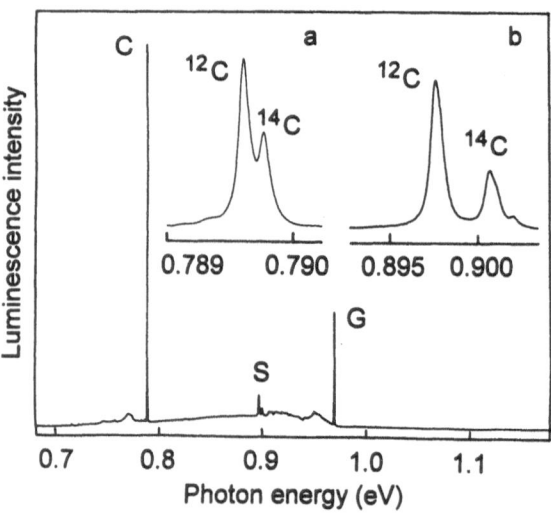

Figure 1. PL spectrum from a CZ Si sample with $[^{12}C] = 4.5 \times 10^{17} \, \mathrm{cm}^{-3}$ and $[^{14}C] = 2 \times 10^{17} \, \mathrm{cm}^{-3}$, irradiated with $4 \times 10^{17} \, \mathrm{cm}^{-2}$ 2 MeV electrons. Inset a: isotope structure of the C-line. Inset b: isotope structure of the S local mode of the G-line. Resolution: 0.065 meV.

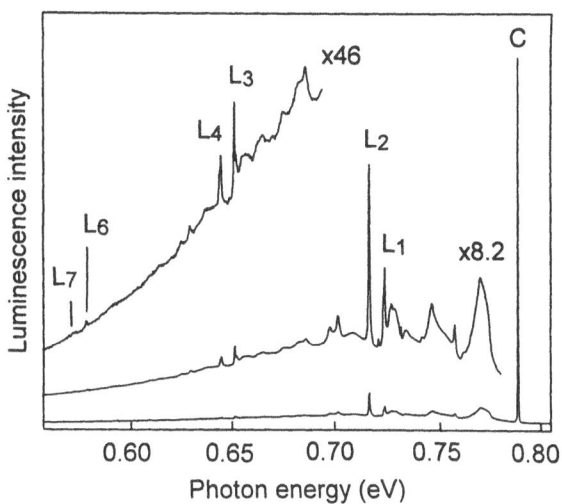

Figure 2. PL spectrum of the C-line luminescence system obtained from a FZ Si sample with [C] = 1.7 x 10^{17} cm^{-3} indiffused with ^{18}O for 48 hours at 1300°C, and then irradiated with 4 x 10^{17} cm^{-2} 2 MeV electrons. ^{18}O:^{16}O~ 2:1. Resolution: 0.25 meV.

luminescence system is considered to be an interstitial carbon-oxygen pair and two different structures have been proposed: one based on electron paramagnetic resonance (EPR) measurements [16], and one based on theoretical modelling of the local vibrational modes derived from both PL and mid-infrared absorption spectroscopy [17]. A quick analysis of the spectra recorded for this paper has yielded information which is not in complete agreement with that derived from previous PL measurements.

Figure 3 shows an expansion of the region containing the L_1 and L_2 local mode features obtained from the same material as figure 2, with an ^{18}O:^{16}O isotope ratio of ~2:1, at a resolution of 0.065 meV. The isotope structures of L_1 and L_2 are both consistent with the presence of one oxygen atom, and the silicon isotope structure can be seen for L_2 as reported previously [13]. Figure 4a shows a lower resolution spectrum

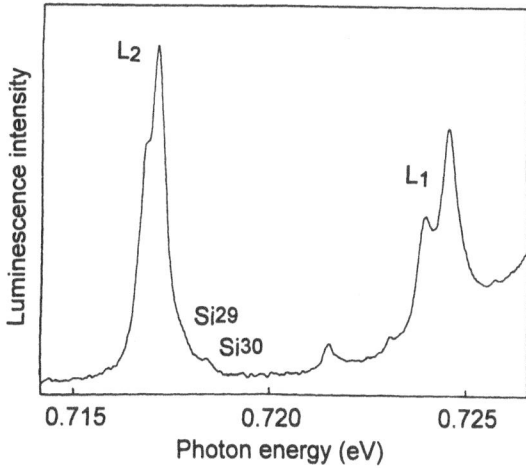

Figure 3. PL spectrum of the L_1 and L_2 local modes obtained from the same material as figure 2, at a resolution of 0.065 meV.

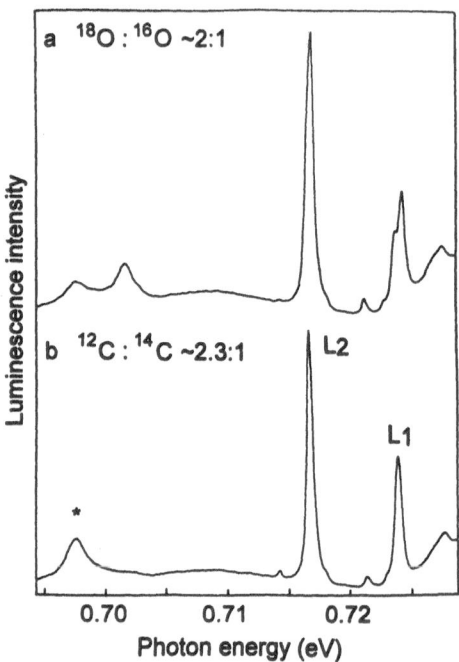

Figure 4. PL spectra of the L_1, L_2 and (*) local modes obtained from (a) the same material as figure 2 and (b) the same material as figure 1. Resolution: 0.25 meV.

from the same material extended to a lower energy using an InSb detector. Figure 4b shows a spectrum obtained from the same material as figure 1. There is no isotope structure associated with carbon or any measurable shift in L_1, L_2 or the feature labelled with an asterisk [15]. However, the latter shows a large oxygen isotope splitting compared with L_1 and L_2. The local mode frequencies are 741.4 cm^{-1} (91.92 meV) for ^{16}O and 709.1 cm^{-1} (87.92 meV) for ^{18}O, and the ratio of 1.046 is close to but somewhat lower than $\sqrt{18/16} = 1.061$ expected for a pure oxygen vibrational mode.

Figure 5 shows an expansion of the region containing the features labelled L_3, L_4 and L_5 obtained from (a) the same material as figure 2, (b) the same material as figure 1, and (c) electron irradiated CZ Si with $^{13}C:^{12}C \sim 20:1$. L_4 coincides almost exactly with $2L_2$ and has the same oxygen isotope structure as L_2. L_5 is very close to the sum of L_2 and the feature labelled with an asterisk, and shows the same oxygen isotope structure as the latter. L_3 coincides almost exactly with $L_1 + L_2$, but the shape and size in figures 5a and 5b would appear to be inconsistent with this interpretation. However, in the ^{13}C doped material (figure 5c) a new feature appears at higher energy, labelled C^{13}, and an additional feature also appears in the material containing ^{12}C and ^{14}C, labelled C^{14} in figure 5b. It is clear from these data that there is a carbon vibrational mode which for ^{12}C coincides with L_3, made up also of $L_1 + L_2$, but shifts to higher energies for ^{13}C and ^{14}C. The local mode frequencies are 1116 cm^{-1} (138.4 meV) for ^{12}C, 1079 cm^{-1} (133.8 meV) for ^{13}C and 1047 cm^{-1} (129.8 meV) for ^{14}C, and the ratios are 1.034 and 1.066 compared with $\sqrt{13/12} = 1.041$ and $\sqrt{14/12} = 1.080$ expected for pure carbon vibrational modes.

Careful examination of the C-line spectra has shown that there are other weak features which have not been identified in this or previous work. A more detailed

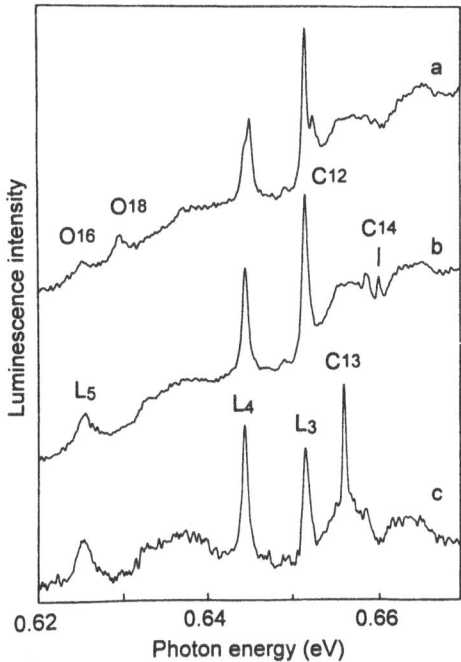

Figure 5. PL spectra of the L_3, L_4 and L_5 local modes obtained from (a) the same material as figure 2, (b) the same material as figure 1, and (c) electron irradiated CZ Si with $^{13}C:^{12}C \sim 20:1$. Resolution: 0.25 meV.

investigation using higher resolution and improved signal/noise ratio might lead to a more definitive analysis of the structure of the centre. New mid-infrared absorption measurements using better samples, in particular ^{18}O doped material, and modern equipment might also be helpful.

Another unexpected phenomenon was observed during recording spectra for this paper. Some ^{18}O doped material which had been prepared for the work reported in reference 13 was reinvestigated. This was FZ Si with [C] $\sim 2 \times 10^{17}$ cm^{-3}, ^{18}O diffused yielding $^{16}O:^{18}O \sim 2:1$, which had been irradiated with 1×10^{18} cm^{-2} 2 MeV electrons. This material originally had a strong G-line and strong C-line, similar to the spectrum shown in figure 1. However, after being at room temperature for about 12 years after irradiation, the G-line had completely disappeared and replaced by a line at 0.9499 eV, labelled F in reference 2. This is now the dominant luminescence system as shown in figure 6. This luminescence system had previously been observed in neutron irradiated CZ Si annealed at temperatures above 250°C and was destroyed at temperatures above 450°C [18]. In reference 18 it is called the G-line and it was considered to be associated with a vacancy-oxygen complex. However, the zero-phonon line has been shown to have carbon isotope structure [13]. A doublet feature in the local mode structure observed in this material has a 2:1 ratio which suggests that the centre does contain oxygen. It would appear to be another carbon-oxygen complex created as the centre responsible for the G-line is destroyed.

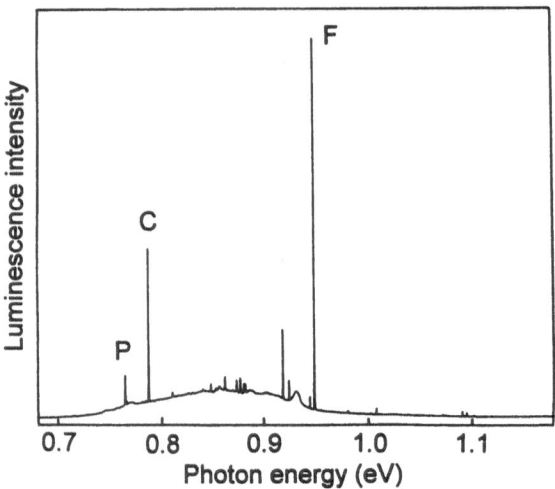

Figure 6. PL spectrum from a FZ Si sample with [C] ~ 2 x 10^{17} cm^{-3}, indiffused with ^{18}O yielding ^{16}O:^{18}O ~ 2:1, irradiated with 1 x 10^{18} cm^{-2} 2 MeV electrons and left at room temperature for ~ 12 years.

4. Thermal treatment

We will now consider the effects of thermal treatment at 470°C on 50 Ω cm p-type carbon-lean CZ Si, with [O] ~ 8 x 10^{17} cm^{-3} and [C] < 10^{16} cm^{-3}, which creates and eventually destroys thermal donors. Figure 7 shows a spectrum obtained before any thermal treatment. This is dominated by boron bound-exciton luminescence, though small traces of phosphorus and arsenic can also be detected in the no-phonon region [8], and there is a very weak P-line (0.7671 eV). Figure 8 shows a spectrum from the same material after heating at 470°C for 48 hours in flowing Ar gas. There is now a strong P-

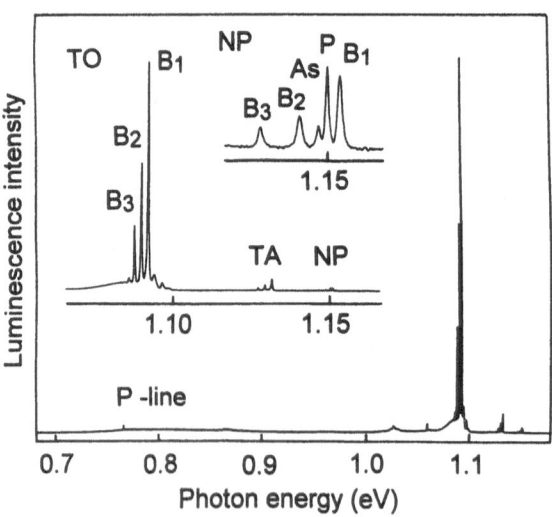

Figure 7. PL spectrum from 50 Ω cm p-type CZ Si with [O]~8 x 10^{17} cm^{-3} and [C] <10^{16} cm^{-3}, as grown with no thermal treatment.

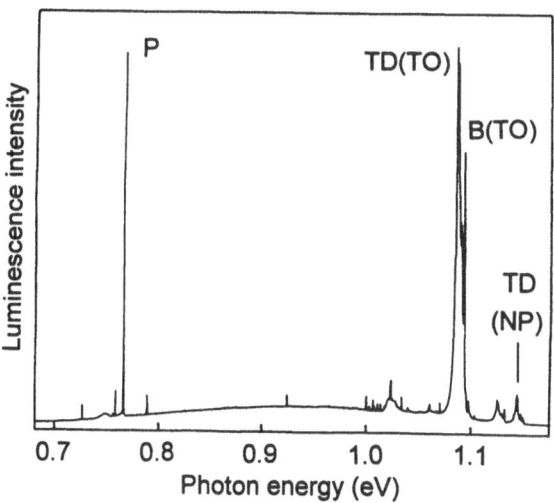

Figure 8. PL spectrum from the same material as figure 7 after heating for 48 hours at 470°C in flowing Ar gas.

line and features associated with the decay of excitons bound to thermal donors, together with a range of other weaker features. The overall luminescence efficiency has fallen by an order of magnitude because of the strong non-radiative Auger effect in excitons captured by deep donors.

The shallow bound-exciton region in figure 8 is expanded in figure 9. The decay of excitons bound to shallow donors and acceptors can occur with the emission of momentum-conserving transverse acoustic (TA) and transverse optic (TO) phonons with the same wavevector as the conduction band minimum or with no phonon (NP) emission [19]. The no-phonon (NP) features are weak, but they are not broadened by phonon-assisted decay. However, the width of the phosphorus NP line in figure 7 is 0.19 meV, although in high purity FZ Si the natural line width is 0.005 meV [20] and the spectral resolution employed here is 0.065 meV. Therefore, the spectral resolution of the thermal

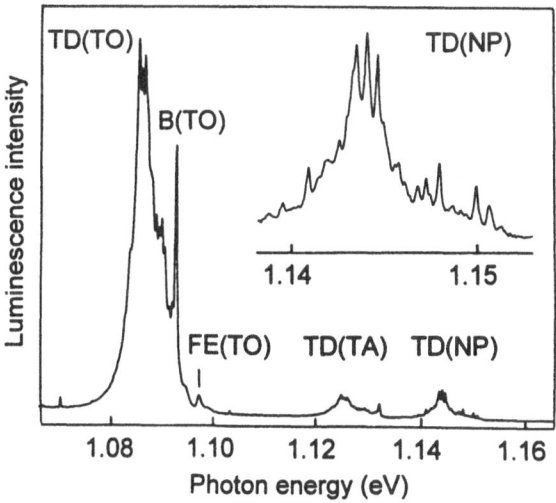

Figure 9. Expansion of the shallow bound exciton region of figure 8.

donor no-phonon TD(NP) features in figure 9 is limited by the broadening due to the high oxygen concentration in CZ Si. The two highest energy features in the TD(NP) region are due to boron and phosphorus, and the large number of overlapping features at lower energy are due to the range of thermal donor species present; the spectrum changes continuously with annealing time as the different thermal donor species are produced. There has been no attempt to analyse this structure because of the broadening caused by the high oxygen concentration and the low luminescence efficiency due to the competing non-radiative Auger effect.

The P-line is always created in the early stages of thermal donor formation, and subsequently destroyed as the process continues. It has been investigated in some detail as described in the paper by Kürner et al [15] and references therein. The centre responsible for the P-line has been shown to contain carbon from the isotope shift of the zero-phonon line [13, 15], and oxygen from the isotope shift of one of the local modes [15]. It is considered to contain two oxygen atoms and one carbon atom, and to be created by the interaction between an oxygen dimer and an interstitial carbon atom released by a silicon self-interstitial created by oxygen dimer formation [15]. Because of the competition from the non-radiative Auger effect in material containing thermal donors and the fact that all the local mode structure is outside the range of the Ge detector, no detailed examination has been made of the local mode structure. The P-line can be created by radiation damage and subsequent thermal treatment without the simultaneous creation of thermal donors but, as is the case with thermally treated carbon-rich CZ Si, a range of other deep level centres is created simultaneously, and often a broad underlying continuum, with the P-line luminescence relatively weak. Figure 10 shows a low resolution (0.5 meV) spectrum recorded for this paper, using the same material as that for figure 6, heated for 30 minutes at 450°C. The major local mode features previously identified [15] are shown here with the same labelling. A detailed high resolution investigation of the isotope structure of the local modes identified, and possibly weaker features suspected to be present, will require first of all a detailed investigation of the optimum oxygen and carbon concentrations, radiation dose and thermal treatment to maximise the intensity of the P-line and minimise the intensities of other systems. When this has been achieved, suitable isotope-enriched material can be produced for PL and possibly also mid-infrared absorption measurements.

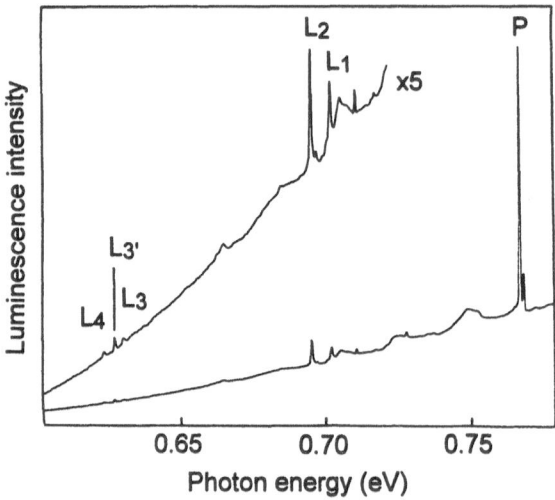

Figure 10. PL spectrum showing the P-line luminescence system from the same material as figure 6, heated at 450°C for 30 minutes. Resolution: 0.5 meV.

When the thermal treatment at 470°C is carried on for more than a few 10s of hours, the P-line disappears and the thermal donors eventually vanish. As the latter occurs, the O^j and later the $O^{j'}$ luminescence systems appear, then a feature at 0.885 eV and eventually the "903" or 9-line luminescence system. The spectrum shown in figure 11 was obtained after heating at 470°C for 480 hours. The O^j and $O^{j'}$ systems have been investigated at several institutions since the mid 1980s, giving rise to very different interpretations. Weber and Queisser [21] considered that the O^j system was associated with the recombination of free holes with electrons bound to thermal donors, and that $O^{j'}$ was a phonon replica. Later they suggested [22] that the strongest feature might be associated with an oxygen precipitate rather than a thermal donor. Dörnen and Hangleiter [23] considered that the O^j features were associated with donor-acceptor pair recombination between electrons on thermal donors and holes on boron acceptors. These interpretations are inconsistent with the fact that the O^j and $O^{j'}$ luminescence systems reach a maximum intensity when thermal donors have been completely destroyed. Thewalt et. al. [24, 25] reported a very detailed investigation leading to the conclusion that two isoelectronic centres had been created as the thermal donors disappeared, and that these centres could bind up to four excitons at high excitation densities. More recently, Leisert et al [26] suggested that the O^j and $O^{j'}$ lines are caused by the decay of an exciton bound to a thermal donor, with the internal excitation of an electron at the donor, ie a two-electron transition.

In order to clarify this issue some magnetic field and uniaxial stress perturbation studies have been carried out [4]. These have confirmed that the defects responsible for the O^j and $O^{j'}$ luminescence systems are isoelectronic centres created as the thermal donors are destroyed. These centres can bind one or two excitons with split singlet-triplet states, rather than four excitons as originally suggested [24, 25]. It is not clear why thermal donors should eventually become neutral isoelectronic centres. Also, the broad asymmetric line shapes of the O^j and $O^{j'}$ features require some further thoughts.

After a very long treatment at 470°C, or by raising the temperature to accelerate the process, the major luminescence system observed is the so-called "903" system or 9-line [3, 27-29], shown in figure 12. The major feature is at 902.2 meV and there is additional structure at lower energies which varies with different materials and different thermal treatments. It was suggested by Tajima et al [29] that this luminescence system

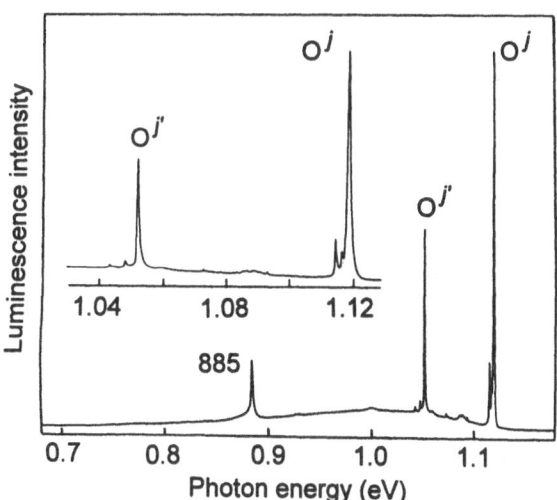

Figure 11. PL spectrum from the same material as figure 7, after heating for 480 hours at 470°C in flowing Ar gas.

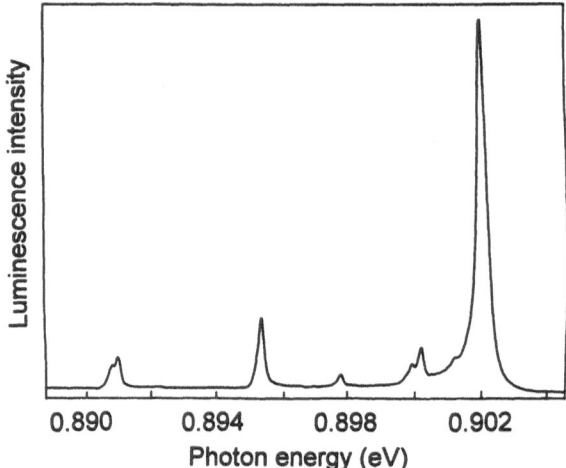

Figure 12. PL spectrum from the same material as figure 7, after heating for 188 hours at 470°C and 10 hours at 650°C.

might be associated with rod-like defects, which are created by various thermal treatments of CZ Si and by heavy radiation damage in both FZ and CZ Si. A detailed investigation [3] has shown a strong correlation between the formation of this luminescence system and the creation of etch pits thought to be associated with rod-like defects. This investigation has led to the proposal that the optical transitions observed are associated with the decay of excitons bound to specific structures within rod-like defects.

It has long been established that heating CZ Si with a high carbon concentration at ~450°C does not immediately give rise to the production of thermal donors; they are eventually produced but at much lower concentrations than in carbon-lean CZ Si. In the carbon-rich material, as well as the P-line, the H-line (0.9256 eV), T-line (0.9351 eV) and I-line (0.9650 eV) are also produced together with several other weaker luminescence features. This is illustrated in figure 13 which shows luminescence spectra obtained from

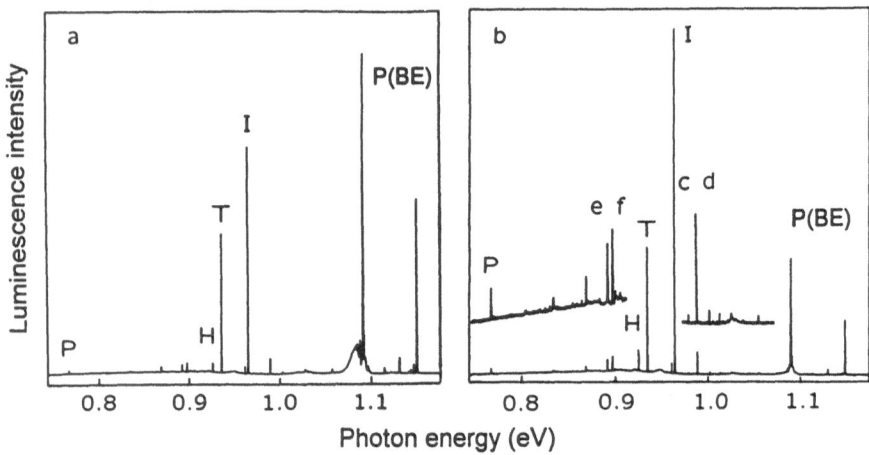

Figure 13. PL spectra from carbon-rich CZ Si with [O] ~ 1.8 x 10^{18} cm^{-3} and [C] ~3.4 x 10^{17} cm^{-3}, (a) heated for 15h at 450°C in flowing Ar gas, (b) heated for 15h at 450°C in air.

high carbon CZ Si, with [O] ~1.8 x 10^{18} cm^{-3} and [C] ~ 3.4 x 10^{17} cm^{-3}, heated for 15 hours at 450°C (a) in flowing Ar gas and (b) in air. It is clear that heating in air produces stronger deep-level luminescence features with respect to the phosphorus bound-exciton (P(BE)) luminescence than heating in flowing Ar gas. Because of this observation [6] and other recent discoveries [5], we considered that this must be due to hydrogen indiffusion, the source being water vapour in air.

To prove this point, a sample of the same material was heated for 15 hours at 450°C in static Ar gas containing a partial pressure of D_2O equivalent to 100% saturation at ~22°C. From the upper spectrum (a) in figure 14 it can be seen that the T-line, I-line and other smaller features labelled a, b, c and d all have deuterium-hydrogen isotope structure, but not the H-line or the P-line, not shown here. The lower spectrum (b) was recorded after etching off 26 μm, giving some indication of how far the deuterium had indiffused at 450°C for 15 hours. It is clear that hydrogen is a common contaminant in silicon and can be readily introduced simply by heating in air and several other procedures [6]. Recent work has shown that the centre responsible for the T-line contains two carbon atoms and one hydrogen atom [30] and that the centre responsible for the I-line contains carbon, oxygen and hydrogen [31]. The H-line is known to contain carbon [13] and since it is only produced in CZ Si, it almost certainly also contains oxygen, though this has not been investigated.

5. Radiation damage and thermal treatment

Many luminescence centres are created in silicon by radiation damage with subsequent thermal treatment at various temperatures. Recent measurements have shown that hydrogen can have a dramatic effect on the centres produced and that many actually

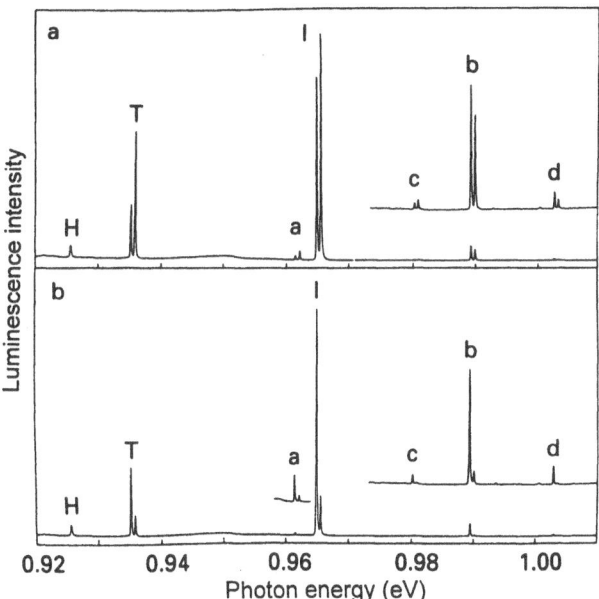

Figure 14. PL spectra, showing the hydrogen-deuterium isotope structure of the zero-phonon lines, obtained from the same material as figure 13 heated for 15h at 450°C in static Ar gas containing a partial pressure of D_2O equivalent to 100% saturation at ~22°C: (a) as treated, (b) after the removal of 26 μm.

314

contain hydrogen [5, 7]. In CZ Si with a moderate or high carbon concentration, the dominant species produced by radiation damage and subsequent annealing at ~450°C are the same as those produced by thermal treatment alone at ~450°C in CZ Si with a high carbon concentration, namely, the P-, H-, I- and T-lines. An additional feature produced by radiation damage and subsequent thermal treatment, but not by thermal treatment alone, is the M-line. We will concentrate here on material deliberately doped with hydrogen to illustrate two very important points, namely, that some centres contain hydrogen but that the presence of too much hydrogen eliminates all the deep level luminescence centres.

Figure 15 shows PL spectra obtained, under identical conditions, from the same CZ Si with [O] = 1.2 x 10^{18} cm^{-3} and [C] = 2 x 10^{17} cm^{-3}, with hydrogen deliberately incorporated [32] at concentrations of (a) 6.7 x 10^{14} cm^{-3}, (b) 2.2 x 10^{15} cm^{-3}, (c) 6.1 x 10^{15} cm^{-3} and (d) 1.5 x 10^{16} cm^{-3}, irradiated at a moderate dose of 5 x 10^{16} cm^{-2} 2 MeV electrons and then heated at 450°C for 30 minutes. At this radiation dose, the number of self-interstitials and vacancies released will be ~ 5 x 10^{15} cm^{-3} [2] and the number of luminescence centres eventually created after subsequent thermal treatment most likely <10^{15} cm^{-3}. In figure 15a, the M-, P-, H-, I- and T-lines, together with two other strong lines at 0.7817 eV and 0.7983 eV, have intensities similar to the boron bound-exciton (B(BE)) features. In figure 15b, with ~3.3 x higher hydrogen concentration, the low energy features are all still present but about a factor of 2 weaker relative to the boron bound exciton features. With much higher hydrogen concentrations of 6.1 x 10^{15} cm^{-3} and 1.5 x 10^{16} cm^{-3} in figures 15b and c, all the low energy features have disappeared except for a marginal T-line. It is clear that the presence of too much hydrogen eliminates the optical activity of all these centres. This has been shown to be the case for a range of other materials investigated, and by complementary experiments in which the radiation dose is increased at a constant level of hydrogen incorporation [7].

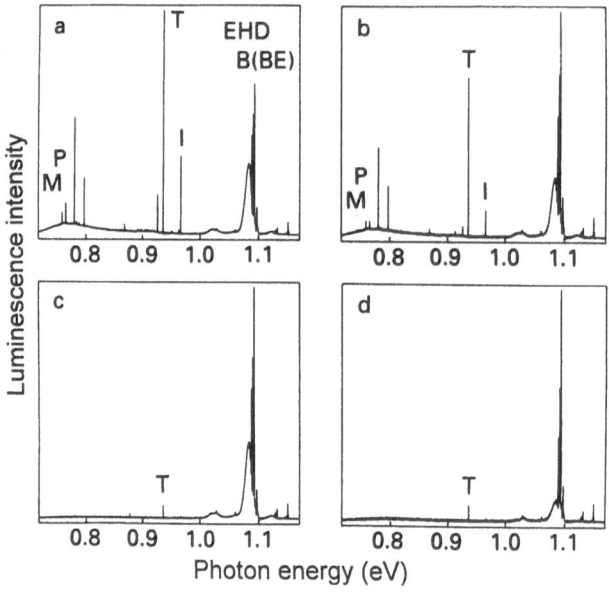

Figure 15. PL spectra obtained from CZ Si with [O] = 1.2 x 10^{18} cm^{-3} and [C] = 2 x 10^{17} cm^{-3}, with hydrogen deliberately incorporated at concentrations of (a) 6.7 x 10^{14} cm^{-3}, (b) 2.2 x 10^{15} cm^{-3}, (c) 6.1 x 10^{15} cm^{-3} and (d) 1.5 x 10^{16} cm^{-3}, all irradiated with 5 x 10^{16} cm^{-2} 2 MeV electrons and subsequently annealed at 450°C for 30 min.

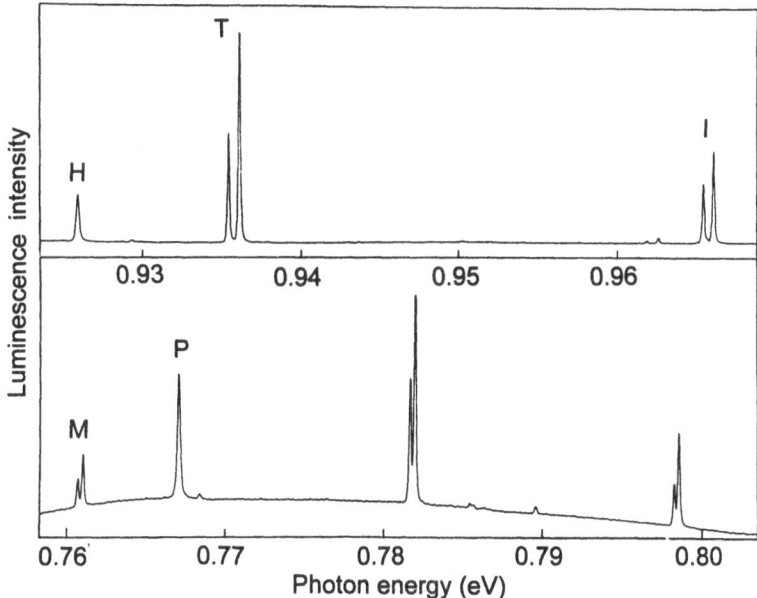

Figure 16. PL spectrum showing the hydrogen-deuterium isotope structure of the zero-phonon lines, obtained from CZ Si with [O] ~ 8 x 10^{17} cm^{-3}, [C] ~ 2 x 10^{17} cm^{-3}, deliberately doped with a nominally 1:1 H_2:D_2 mixture at a concentration of ~1.5 x 10^{16} cm^{-3}, irradiated with 4 x 10^{17} cm^{-2} 2 MeV electrons then heated for 30 min. at 450°C.

Figure 16 shows a PL spectrum obtained from a CZ Si sample with [O] ~ 8 x 10^{17} cm^{-3} and [C] ~ 2 x 10^{17} cm^{-3} deliberately doped with a nominal H_2:D_2 = 1:1 mixture at a concentration of 1.5 x 10^{16} cm^{-3}, irradiated with 4 x 10^{17} cm^{-2} 2 MeV electrons and then heated at 450°C for 30 minutes. It is clear that the centres responsible for the M-, T- and I-lines and the features at 0.7817 eV and 0.7983 eV all contain, almost certainly, one hydrogen atom, whereas those responsible for the P- and H-lines are not hydrogen related. It would appear, therefore, that some centres are made optically active by the presence of one hydrogen atom but become optically inactive if they capture a second hydrogen atom. Other centres become optically inactive by capturing one hydrogen atom.

6.   Conclusions

It is clear from the work reported here that, in terms of the luminescence centres produced, the behaviour of oxygen in silicon subjected to thermal treatment, or radiation damage and subsequent thermal treatment, is strongly dependent on the other major contaminants, carbon and hydrogen. An extension of the photoluminescence studies and other investigations carried out mainly in the 1980s could provide a more detailed understanding of the complicated behaviour of oxygen, carbon and hydrogen in silicon.

7.   Acknowledgements

This work was supported by the Engineering and Physical Sciences Research Council (UK).

316

8. References

1. See, for example: Jones, C.E., Johnson, E.S., Dale Compton, W., Noonan, J.R. and Streetman, B.G. (1973) Temperature, stress, and annealing effects on the luminescence from electron-irradiated silicon, *J. Appl. Phys.* **44**, 5402-5410; Minaev, N.S. and Mudryi, A.V. (1981) Thermally-induced defects in silicon containing oxygen and carbon, *Phys. Stat. Sol. (a)* **68**, 561-565, and references therein.

2. Davies. G. (1989) The optical properties of luminescence centres in silicon, *Physics Reports* **176**, 83-188.

3. Jeyanathan, L., Lightowlers, E.C., Higgs, V. and Davies, G. (1994) Luminescence associated with rod-like defects in Czochralski silicon, *Mat. Sci. Forum* **143-147**, 1499-1504.

4. Jeyanathan, L., Davies, G. and Lightowlers, E.C. (1995) Characterisation of the 1117 meV and 1052 meV optical transitions in heat-treated Si, *Phys. Rev. B* **52**, 10923-10931.

5. Safonov, A.N. and Lightowlers, E.C. (1994) Hydrogen related optical centres in radiation damaged silicon, *Mat. Sci. Forum* **143-147**, 903-908.

6. Lightowlers, E.C., Newman, R.C. and Tucker, J.H. (1994) Hydrogen-related luminescence centres in thermally treated Czochralski silicon, *Semicond. Sci. Technol.* **9**, 1370-1374.

7. Lightowlers, E.C. (1995) Hydrogen incorporation and interaction with impurities and defects in silicon investigated by photoluminescence spectroscopy, *Mat. Sci. Forum* **196-201**, 817-824.

8. Colley, P. McL. and Lightowlers, E.C. (1987) Calibration of the photoluminescence technique for measuring B, P and Al concentrations in Si in the range 10^{12} to 10^{15} cm^{-3} using Fourier transform spectroscopy, *Semicond. Sci. Technol.* **2** 157-166.

9. Davies, G., Oates, A.S., Newman, R.C., Woolley, R., Lightowlers, E.C., Binns, M.T. and Wilkes, J.G. (1986) Carbon-related radiation damage centres in Czochralski silicon, *J. Phys. C: Solid State Phys.* **19**, 841-855.

10. Davies, G. and Newman, R.C. (1994) Carbon in Monocrystalline Silicon, in *Handbook on Semiconductors, Vol. 3*, ed. Mahahjan, S., North Holland, Amsterdam.

11. O'Donnell, K.P., Lee, K.M. and Watkins, G.D. (1983) Origin of the 0.97 eV luminescence in irradiated silicon, *Physica B* **116**, 258-263.

12. Davies, G., Lightowlers, E.C. and do Carmo, M.C. (1983) Carbon-related vibronic bands in electron-irradiated silicon, *J. Phys. C: Solid State Phys:* **16**, 5503-5515.

13. Davies, G., Lightowlers, E.C., Woolley, R., Newman, R.C. and Oates, A.S. (1984) Carbon in radiation damage centres in Czochralski silicon, *J. Phys. C: Solid State Phys.* **17**, L499-L503.

14. Thonke, K., Watkins, G.D. and Sauer, R. (1984) Carbon and oxygen isotope effects in the 0.79 eV defect photoluminescence spectrum in irradiated silicon, *Solid State Commun.* **51**, 127-130.

15. Kürner, W., Sauer, R., Dörnen, A. and Thonke, K. (1989) Structure of the 0.767 eV oxygen-carbon luminescence defect in 450°C thermally annealed Czochralski-grown silicon, *Phys. Rev. B* **39**, 13327-13337.

16. Trombetta, J.M. and Watkins, G.D. (1987) Identification of an interstitial carbon-interstitial oxygen complex in silicon, *Appl. Phys. Lett.* **51**, 1102-1105.

17. Jones, R. and Öberg, S. (1992) Oxygen frustration and the interstitial carbon-oxygen complex in Si, *Phys. Rev. Letters* **68**, 86-89.

18. Tkachev, V.D. and Mudryi, A.V. (1977) Radiative recombination centres in silicon, irradiated by fast neutrons and ions, *Inst. Phys. Conf. Ser. No. 31*, 231-243.

19. Thewalt, M.L.W. (1982) *Exitons,* ed. Rashba, E.I. and Sturge, M.D. North Holland, Amsterdam, p393.

20. Thewalt, M.L.W. and Brake, D.M. (1990) Ultra-high resolution photoluminescence studies of bound excitons and multi-bound exciton complexes in silicon, *Mat. Sci. Forum* **65-66**, 187-198.

21. Weber, J. and Queisser, H.J. (1986) New optical transitions at thermal donors in silicon, *Mat. Res. Soc. Proc.* **59**, 147-152.

22. Weber, J., Köhler, K., Stützler, F.J. and Queisser, H.J. (1986) Evidence for an inhomogeneous distribution of thermal donors in silicon from electrical and optical measurements, *Mat. Sci. Forum* **10-12**, 979-984.

23. Dörnen, A. and Hangleiter, A. (1986) Time resolved study of thermal donor related luminescence lines in silicon, *Mat. Sci. Forum* **10-12**, 967-972.

24. Thewalt, M.L.W., Steele, A.G., Watkins, S.P. and Lightowlers, E.C. (1986) Thermal-donor-related isoelectronic center in silicon which can bind up to four excitons, *Phys. Rev. Lett.* **57**, 1939-1942.

25. Steele, A.G., Thewalt, M.L.W. and Watkins, S.P. (1987) A second isoelectronic multiexciton center in annealed Czochralski silicon, *Solid State Commmun.* **63**, 81-84.

26. Liesert, B.J.Heijmink, Gregorkiewicz, T. and Ammerlaan, A.J. (1993) Photoluminescence of silicon thermal donors, *Phys. Rev. B.* **47**, 7005-7012.

27. Weber, J and Sauer, R (1983) Photoluminescence study of thermally treated silicon crystals, *Mat. Res. Soc. Proc.* **14**, 165-169.

28. Weber, J. and Watkins, G.D. (1985) Photoluminescence and optically detected magnetic resonance (ODMR) from thermally treated silicon samples, *Proc. 13th Int. Conf. on Defects in Semiconductors,* The Metallurgical Society of AIME, p661-667.

29. Tajima, M., Gösele, U., Weber, J. and Sauer, R. (1983) Photoluminescence associated with thermally induced microdefects in Czochralski-grown silicon crystals *Appl. Phys. Lett.* **43**, 270-274.

30. Safonov, A.N., Lightowlers, E.C., Davies, G., Leary, P., Jones, R. and Öberg, S. (1996) Interstitial carbon-hydrogen interaction in silicon. To be submitted for publication.

31. Gower, J.E., Safonov, A.N., Lightowlers, E.C. and Davies, G. (1996). Carbon-hydrogen-oxygen related centre responsible for the photoluminescence I-line (0.965 eV). Published in this book.

32. Binns, M.J., McQuaid, S.A., Newman, R.C. and Lightowlers, E.C. (1993). Hydrogen solubility in silicon and hydrogen defects present after quenching *Semicond. Sci. Technol.* **8**, 1908-1911.

THE NITROGEN-PAIR OXYGEN DEFECT IN SILICON

F. BERG RASMUSSEN

Van der Waals-Zeeman Institute, University of Amsterdam,
Valckenierstraat 65-67, NL-1018 XE Amsterdam,
The Netherlands

S. ÖBERG

Department of Mathematics, University of Luleå,
Luleå, S-95187, Sweden

R. JONES, C. EWELS AND J. GOSS

Department of Physics, University of Exeter,
Exeter, EX4 4QL, United Kingdom

AND

J. MIRO AND P. DEÁK

Department of Atomic Physics,
Technical University of Budapest,
Budafoki út 8, H-1111 Budapest, Hungary

Abstract.

The nitrogen-pair oxygen defect in silicon has been studied by infrared absorption spectroscopy on samples implanted with various combinations of ^{14}N, ^{15}N, ^{16}O and ^{17}O. The measurements give direct evidence for the involvement of nitrogen and oxygen in the defect and show that the impurity atoms comprising the defect are only weakly coupled. *Ab initio* cluster calculation on several models of the nitrogen-pair oxygen defect have been performed and are compared with experiment. Based on these investigations a model consisting of a bridging oxygen atom adjacent to the nitrogen pair is suggested.

R. Jones (ed.), Early Stages of Oxygen Precipitation in Silicon, 319–327.

1. Introduction

The properties of N in Si has gained considerable interest in recent years for several reasons. N is often used in various processing steps of Si technology but has important effects on the material. For example, N impurities can pin dislocations [1, 2] and form electrically active defects [3]. Furthermore, N readily interacts with other light impurities as C and O [3].

Although the interaction between N and O in Si-hosts is of great interest, very little information on the structure of N-O defects is available. It has been suggested that the so called *shallow thermal donors* [4] are due to N-O complexes [5, 6] but even the incorporation of N in these defects remains controversial [7, 8, 9]. More recent work [10] suggests that N inhibits the formation of thermal donors [11] but leads to an increase in the concentration of shallow donors.

The dominant N defect in O-free Si is a N-pair, consisting of two neighbouring $\langle 100 \rangle$ oriented Si-N split interstitials bonded together[3, 13]. The structure of the N-pair is shown in Fig. 2 if the O-atom is disregarded. In float-zone Si implanted with ^{14}N and ^{16}O, Stein [12] observed local vibrational modes (LVMs) at 805, 1000 and 1030 cm^{-1} in addition to modes at 770 and 967 cm^{-1} due to the N-pair [3, 13]. Hereafter these three modes are referred to as NNO-modes. The NNO-modes have also been observed in CZ-material grown in an N-atmosphere [14] where their intensities have been shown to be correlated with each other [15]. The NNO-modes were ascribed to a complex consisting of a N-pair perturbed by a nearby O-atom, firstly because the modes are only observed in samples containing both N and O, and secondly because the observed LVM frequencies at 805 and 1000 cm^{-1} are close to those of the N-pair.

The NNO-defect is the dominant N-O defect observed by LVM-spectroscopy in Si and hence is probably the simplest N-O complex. Hence, studies of the NNO-defect form the starting point for understanding larger N-O complexes and their possible relation to shallow donor formation in Si. In spite of this the microscopic structure of the NNO-defect has remained unknown. We have therefore initiated a combined infrared absorption and theoretical study of NNO-defects in Si. Previously [16], we have discussed the effect of mixed N isotopes on the LVMs of the defect but here we investigate the effect of mixed O isotopes.

2. Experimental details and results

Float-zone, phosphorus doped, n-type Si-samples cut from a wafer (residual impurity concentration $< 10^{16}$ cm^{-3}) were implanted at 300 K with various combinations of ^{14}N, ^{15}N, ^{16}O and ^{17}O. The samples had a typical size of $10 \times 10 \times 0.3$ mm^3 and were polished and implanted on both sides. The

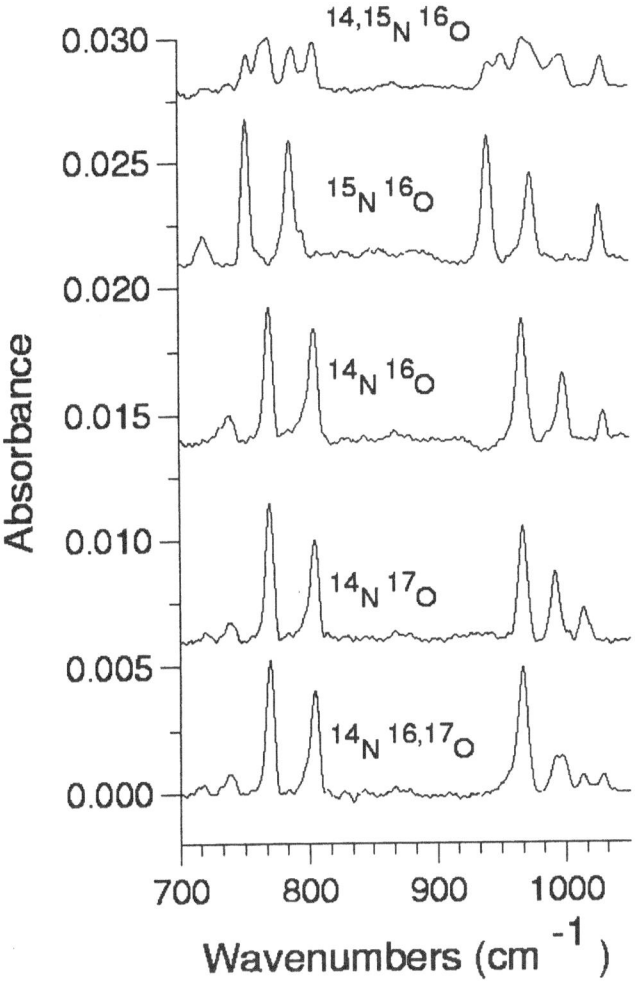

Figure 1. Absorbance spectrum observed after annealing at 650°C for one h of samples implanted with various combinations of ^{14}N, ^{15}N, ^{16}O and ^{17}O. In all cases the total N and O doses are equal

energies and doses [16] were chosen to obtain a nearly uniform O and N profile from 0.21 to 0.82 μm below the surface with a concentration of 6.5×10^{19} cm^{-3} of both O and N.

All samples were annealed under flowing N$_2$ ambient for one hour in the temperature range from 500 to 700°C. Infrared absorption measurements were carried out at 77 K using a Nicolet System 800 Fourier Transform Infrared spectrometer equipped with a HgCdTe detector and a Ge-KBr beamsplitter. Typically 1000 scans were taken with a resolution of 6 cm^{-1}.

Figure 1 shows infrared absorbance spectra after annealing at 650°C. In the case of ^{14}N and ^{16}O five strong modes are observed at 770, 805, 967, 999 and 1030 cm^{-1} as well as a weak feature at 739 cm^{-1}. The two modes at 770 and 967 cm^{-1} belong to the N-pair whereas the 805, 999 and 1030 cm^{-1} modes coincide with those reported previously for the NNO-defect [12, 14, 15]. The observed LVM frequencies for all measured isotope combinations are given in Table 1. All the NNO-modes shift with ^{15}N establishing that N is involved in the defect. Similarly, the observed shift with ^{17}O for the two top modes at 999 and 1030 cm^{-1} confirms the involvement of O. The modes at 805 and 999 cm^{-1} shift downwards by 19 and 25 cm^{-1} respectively when ^{14}N is substituted by ^{15}N. These shifts are close to the shifts of 18 and 26 cm^{-1} observed with the N-pair [13]. This gives further indication that the NNO defect is a N-pair with a nearby O-atom compressing the bonds of the pair and hence increasing the frequencies of the N-related modes.

The NNO-mode at 1030 cm^{-1} only shows a shift of 2 cm^{-1} with ^{15}N but a substantial shift with ^{17}O (17 cm^{-1}) indicating that this mode is predominantly O-related. Isolated interstitial O (O_i) is the dominant O-defect observed by LVM-spectroscopy in as-grown CZ-material where it gives rise to absorption at 1136 cm^{-1} [17], about 100 cm^{-1} above the 1030 cm^{-1} NNO-mode. This mode shifts by 29 cm^{-1} when ^{16}O is replaced with ^{17}O, much more than the 17 cm^{-1} shift of the NNO mode, suggesting that the NNO mode is strongly correlated with other neighbouring modes. This suggests that the NNO-defect basically consists of an O_i entering a dilated Si-Si bond close to a N-pair. A recent calculation on the interaction between O_i and dislocations in Si found a strong preference for O to bridge dilated Si-Si bonds [18].

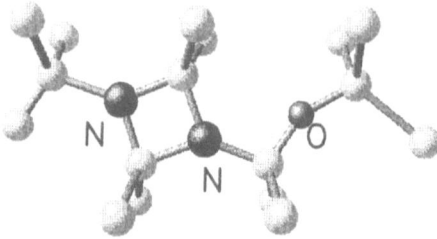

Figure 2. Structure of the nitrogen-pair oxygen defect shown in the (110) mirror plane of the defect. The N pair has a similar structure but does not contain the bridging O atom.

In samples implanted with ^{14}N, ^{15}N and ^{16}O, extra N-pair modes are observed at 762 and 951 cm^{-1} due to the coupling between the two N-atoms in the pair [13]. In direct contrast, no new modes related to the NNO-defects are observed in samples containing both N-isotopes (Fig. 1, Table 2) indicating that the N atoms of the NNO-defect are only weakly coupled dynamically. A weak coupling between the two N-atoms is expected if the O-atom lies much closer to one N-atom than the other, making the two N-atoms inequivalent.

Finally, the small mode at 739 cm^{-1} shifts downwards to 721 with ^{15}N whereas no shift with the O-isotope is observed showing that the mode is related to N. The 739 cm^{-1} mode is only observed in samples implanted with both N and O and it disappears together with the NNO-modes if the samples are annealed above $\sim$700°C. This indicates that the 739 cm^{-1} mode also belongs to the NNO-defect.

3. Cluster calculations and discussion

Due to the weak coupling between the impurity atoms comprising the NNO-defect the number of N and O atoms cannot be determined from the number of observed modes (see e.g. [13]). However, based on the discussion above a defect involving one N-pair and one O_i is considered by far the most likely, as also suggested by previous authors [12, 14, 15].

There are a number of ways an O_i can interact with a N-pair. One possibility is shown in Fig. 2 where O_i bridges the nearest Si(1)-Si(5) bond to the N-pair defect lying in the (110)-mirror plane of the pair. This is a very likely model as previous cluster calculations on the N-pair defects [13] found the Si-atoms numbered 1 and 5 to be displaced by 4 % of their bond length in defect-free Si, i.e. making the Si(1)-Si(5) bond dilated and hence a favourable site for O_i. However, other possibilities exist. For example, the O_i could enter the second nearest Si(6)-Si(5) bond, the direct Si(1)-N(1) bond or a bond between e.g. Si(3) and one of its nearest neighbour Si-atoms sticking out of the (110)-mirror plane of the N-pair. This last model however is rather unlikely as the Si-Si bonds bordering the N-pair along [100] were found to be compressed and not dilated [13].

Ab initio local density functional calculations [20] have been carried out using norm-conserving pseudopotentials and large H-terminated clusters in order to investigate these models [16]. The method gives Si-Si bond lengths to within 1.5 % and accounts successfully for the structures and vibrational modes of both isolated O_i and the N-pair in Si [19, 13]. A 134 atom cluster, $N_2OSi_{71}H_{60}$, was used for these calculations in which the O-atom occupied one of the sites given above and the N-atoms were placed as found for the N-pair [13]. The self-consistent energy and forces on all the atoms were found, and the whole cluster relaxed by a conjugate gradient algorithm. Second derivatives of the energy were evaluated for the impurities and their Si neighbours. These were then used, together with a Musgrave-Pople potential [20], to construct the dynamical matrix of the cluster from which the vibrational frequencies were found.

TABLE 1. Calculated and observed local vibrational modes (cm^{-1}) due to the nitrogen-pair oxygen defect in silicon, see the text. For modes where basically only one impurity atom is involved in the motion; this atom is given in the first column of the table. The numbering refers to Fig. 2.

	$^{14}N\ ^{14}N\ ^{16}O$	$^{15}N\ ^{15}N\ ^{16}O$	$^{14}N\ ^{15}N\ ^{16}O$	$^{15}N\ ^{14}N\ ^{16}O$	$^{14}N\ ^{14}N\ ^{17}O$
	Calculated				
N(1)	1070.4	1039.6	1069.8	1040.3	1070.3
O	861.1	860.1	860.7	860.4	840.5
N(2)	808.2	788.1	788.2	808.1	806.9
	723.9	705.5	708.3	722.2	723.5
	671.5	655.8	669.5	656.7	669.3
	Displaced Si(1)				
O	962.9	962.7	962.8	962.8	939.5
N(1)	948.9	920.1	948.0	921.3	949.4
N(2)	813.3	792.6	792.7	813.0	813.3
	727.0	708.2	714.0	723.9	727.0
	688.1	672.3	683.5	674.2	687.7
	Observed				
O	1030	1028	1028		1013
N(1)	999	974	999, 974		991
N(2)	805	786	805, 786		804
	739	721			739

None of the models considered here, except the one shown in Fig. 2, were consistent with the positions and isotopic shifts of the LVMs observed (Fig. 1). Thus, focus will be given to the model of Fig. 2. This structure contains only saturated bonds and is electrically inert. The calculated LVMs are given in Table 2. The two modes at 1070 and 808 cm^{-1} are clearly N-related, as they barely shift with the O-isotope, whereas the 861 cm^{-1} mode is O-related. As evidenced by the mixed isotopic calculations (Table 2) there is very little coupling between the two N-related modes in agreement with the fact that no new NNO-modes are observed in samples containing both N-isotopes (Fig. 1). Hence, although the calculated modes are in the correct region and account for the observed weak coupling between the N-related modes, the O-mode is found between the two N-modes contrary to observation. However, the ordering and position of the modes are very sensitive to the position of the Si(1)-atom being bonded to both the O-atom and the N-pair (Fig. 2). If the Si(1)-atom is displaced by 0.07 Å along ⟨110⟩, increasing the N(1)-Si(1) bond length by 0.07 Å and decreasing the Si(1)-O bond by 0.05 Å, the displaced Si modes of Table 2 are obtained. In this case the calculated and observed frequencies agree within 70 cm^{-1} and the N atoms stay decoupled. The changes in bond lengths and differences between observed and calculated modes are within the accuracy of the calculation.

The highest mode (963 cm^{-1}) is now simply an antisymmetric stretch in the Si-O bonds. It shifts downwards by 23 cm^{-1} when ^{16}O is substituted by ^{17}O. This shift is somewhat larger than the observed shift of 17 cm^{-1}. The calculation found no O-shift of the 949 cm^{-1} mode whereas a downwards shift of 8 cm^{-1} is observed (Table 2). Note however, that with the displaced Si(1)-atom the calculated O-mode and highest N-mode cross when ^{16}O is substituted by ^{17}O. Such a crossing is not observed experimentally (Fig. 1). It is considered that the extreme sensitivity of the position of the common Si-atom (Si(1)) and the possibility of crossing of the two highest modes makes it rather difficult to calculate the isotopic shifts for these modes. In the calculations almost no interaction between the two highest modes were found. However, just a small interaction would automatically cause a smaller O-shift of the 963 cm^{-1} mode and a (larger) O-shift of the 949 cm^{-1} mode. Contrary to this, the mode at 813 cm^{-1} shows no shift with ^{17}O in close agreement with the observed shift of 1 cm^{-1}.

Consider now the shifts with N isotopes. Substituting ^{14}N by ^{15}N, the calculated shifts of the 963, 949 and 813 cm^{-1} modes are 0, 29 and 21 respectively in good agreement with the observed shifts of 2, 25 and 19 cm^{-1}. Finally, a mode involving motion of both N-atoms were found at 727 cm^{-1} in the calculation. The 727 cm^{-1} shifts 19 cm^{-1} with ^{15}N but does not show any shift with O. The position and shift of the 727 cm^{-1} calculated mode is astonishingly close to the observed mode at 739 cm^{-1} that displays a 18 cm^{-1} shift with ^{15}N and no resolvable shift with O. Furthermore, the 727 cm^{-1} mode is expected to split into four components when both N-isotopes are present (Table 2). Considering the small intensity of the observed mode at 739 cm^{-1}, this explains why this mode is not observed in samples implanted with both ^{14}N and ^{15}N.

Based upon the above discussion we suggest the structure in Fig. 2 as a model of the NNO-defect in Si. This model accounts for the dynamical properties of the defect with fair agreement between calculated and observed positions and isotopic shifts of the local vibrational modes.

4. Conclusions

Infrared absorption experiments on the NNO-defect in Si have been carried out. The involvement of both N and O in the defect is shown directly by observing N and O-related isotopic shifts. The measurements were compared with *ab initio* calculations on various models of the defect. The model shown in Fig. 2 was found to account for the dynamical properties of the NNO-defect when allowances for the extreme sensitivity of the position of the common Si-atom (Si(1) in Fig. 2) were taking into account. No other model investigated was consistent with observation.

The NNO-defect has no deep or shallow donor levels in the configuration shown in Fig. 2. This indicates that if N-related shallow donors exists [5, 6] they must involve larger numbers of N or O-atoms, or possibly Si-interstitials — as recently proposed [21] — or H.

Acknowledgements

This project has been supported by The Danish National Research Foundation through the Aarhus Center for Advanced Physics (ACAP). RJ thanks the British Council, PD and JM thank the Hungarian National Scientific Research Fund (OTKA 2744) and SÖ thanks the Swedish National Research Council, for financial support. Finally, FBR gratefully acknowledges a grant from The Danish Natural Science Research Council (jr. no.11-1226-1)

References

1. Sumino K., Yonenaga I., Imai M. and Abe T. (1983) Effects of nitrogen on dislocation behavior and mechanical strength in silicon crystals, *J. Appl. Phys.* **54**, 5016-5020.
2. Heggie M. I. H., Jones R. and Umerski A. (1993) Ab Initio Total Energy Calculations of Impurity Pinning in Silicon, *Phys. Status Solidi a* **138**, 383-387.
3. Stein H. J. (1985) Nitrogen in Crystalline Silicon, in J. C. Mikkelsen, Jr., S. J. Pearton, J. W. Corbett and S. J. Pennycook (eds), *Oxygen, Carbon, Hydrogen and Nitrogen in Crystalline Silicon*, MRS, Pittsburgh, PA, Vol. 59, pp. 523-535.
4. Navarro H., Griffin J., Weber J. and Genzel L. (1986) New Oxygen Related Shallow Thermal Donor Centres in Czochralski-grown Silicon, *Solid Stat. Commun.* **58**, 151-155.
5. Suezawa M., Sumino K., Harada H and Abe T. (1988) The Nature of Nitrogen-Oxygen Complexes in Silicon, *Jpn. J. Appl. Phys.* **27**, 62-67.
6. Hara A., Aoki M., Koizuka M. and Fukuda T (1994) Model for NL10 thermal donors formed in annealed oxygen-rich silicon crystals, *J. Appl. Phys.* **75**, 2929-2935.
7. Griffin J., Hartung J., Weber J., Navarro H. and Genzel L (1989) Photothermal Ionisation Spectroscopy of Oxygen-Related Shallow Defects in Crystalline Silicon, *Appl. Phys. A* **48**, 41-47.
8. Heijmink Liesert B. J., Gregorkiewicz T. and Ammerlaan C. A. J. (1993) Photoluminescence of silicon thermal donors, *Phys. Rev. B* **47**, 7005-7012.
9. Martinov Yu. V., Gregorkiewicz T. and Ammerlaan C. A. J. (1995) Role of hydrogen in formation and structure of the Si-NL10 thermal donor, *Phys. Rev. Lett.* **74**, 2030-2033.
10. Yang D., Que D. L. and Sumino K. (1995) Nitrogen effects on thermal donor and shallow thermal donor in silicon,*J. Appl. Phys.* **77**, 943-944 (1995).
11. Wagner P. and Hage J., (1989) Thermal Double Donors in Silicon ,*Appl. Phys. A* **49**, 123-138.

12. Stein H. J. (1985) Oxygen-Nitrogen Interactions in Ion-Implanted Silicon, in John-son N. M., Bishop S. G. and Watkins G. D. (eds), *Microscopic Identification of Electronic Defects in Semiconductors*, MRS, Pittsburgh, PA, pp. 287-291.

13. Jones R., Öberg S, Berg Rasmussen F. and Bech Nielsen B. (1994) Identification of the Dominant Nitrogen Defects in Silicon, *Phys. Rev. Lett.* **72**, 1882-1885.

14. Wagner P., Oeder R. and Zulehner W. (1988) Nitrogen-Oxygen Complexes in Czochralski-Silicon,*Appl. Phys. A* **46**, 73-76.

15. Qi M. W., Tan S. S., Zhu B., Cai P. X., Gu W. F. , Xu X. M., Shi T. S. , Que D. L. and Li L. B. (1991) The evidence for interaction of the N-N pair with oxygen in Czochralski silicon,*J. Appl. Phys.* **69**, 3775-3777.

16. Jones R., Ewels C., Goss J., Miro J., Deák P., Öberg S. and Berg Rasmussen F. (1994) Theoretical and isotopic infrared absorption investigations of nitrogen-oxygen defects in silicon, *Semicond. Sci. Technol.* **9**, 2145-2148.

17. Newman R. C. (1973) *Infrared Studies of Crystal Defects*, Taylor and Francis, London.

18. Umerski A. and Jones R. (1993) The interaction of oxygen with dislocation cores in silicon, *Phil. Mag. A* **67**, 905-915.

19. Jones R., Umerski A. and Öberg S. (1992) *Ab initio* calculation of the local vibratory modes of interstitial oxygen in silicon , *Phys. Rev. B* **45**, 11321-11323.

20. Jones R. (1992) *Ab initio* cluster calculations of defects in solids,*Phil. Trans. R. Soc. Lond. A* **341**, 351-360.

21. Deák P., Snyder L. C. and Corbett J. W. (1992) Theoretical studies on the core structure of the 450°C oxygen thermal donor in silicon, *Phys. Rev. B* **45**, 11612-11626.

THERMAL DOUBLE DONORS IN SILICON:
A NEW INSIGHT INTO THE PROBLEM

L.I. MURIN and V.P. MARKEVICH
Institute of Solid State and Semiconductor Physics,
Academy of Sciences of Belarus
P. Brovki str. 17, Minsk 220072, Belarus

Abstract

The initial stages of the thermal double donor (TDD) formation in Cz-Si crystals are considered. Some basic parameters (binding energy, rate of dissociation, steady-state concentration, diffusivity) of the most abundant oxygen cluster - oxygen dimer are estimated. A good correlation between the behaviours of oxygen dimers and trimers, and the IR absorption bands at 1012 cm^{-1} and 1006 cm^{-1} is found. It is shown that some features of the enhanced oxygen diffusion and TDD generation are not consistent with a conception of the very fast diffusing oxygen dimers. Along with dimers the mobility of trimers and the first TDD species is suggested.

1. Introduction

For a long time it has been recognized that the kinetics of thermal double donor (TDD) formation at 300-500 °C in Czochralski-grown Si crystals can not be described within the framework of oxygen clustering limited by the normal diffusion rate of interstitial oxygen atoms. In fact all the modern TDD kinetic models assume the existence of a fast diffusing oxygen-related species (FDS). However, the origin of such a species continues to be one of the main puzzles in the problem of TDD formation or, in general, in the problem of oxygen diffusion and precipitation at T $\leq$ 500 °C. Various kinds of FDSs have been suggested including complexes of oxygen with an intrinsic defect or impurity, self-interstitials created by O_2 formation, quasi-free or unbonded oxygen atoms, oxygen molecules or dimers. At present the latter suggestion seems to be the most attractive because it allows to explain consistently a number of facts concerning enhanced in- and out-diffusion of oxygen at T $\leq$ 600 °C as well as kinetics of oxygen loss and TDD formation at T = 300-500 °C [1-3].

On the other hand it is well known that at T $\geq$ 700 °C all the processes related to oxygen diffusion (in- and out- mass-transport, precipitation etc.) imply the normal oxygen diffusivity. Therefore the existence of any fast moving oxygen-containing

329

R. Jones (ed.), Early Stages of Oxygen Precipitation in Silicon, 329–336.
© 1996 *Kluwer Academic Publishers.*

species should not effect noticeably the oxygen diffusion at high temperatures. In the present work we try to examine how the conception of the fast diffusing dimers successively forming a TDD series is consistent with the above-stated requirement as well as with some peculiarities of the initial TDD generation kinetics.

2. Some characteristic features of dimers and trimers

2.1. ON THE STEADY-STATE CONCENTRATION OF DIMERS AND UPPER LIMIT FOR THEIR DIFFUSIVITY

In general, the kinetics of dimer (O_2) and trimer (O_3) formation in bulk crystals assuming O_2 mobility can be described by the following set of differential equations

$$\frac{dN_2}{dt} = k_1 N_1^2 - k_{-2} N_2 - N_2 \sum_{n \geq 1} k_{dn} N_n - k_{dx} N_2 N_x + \sum_{n \geq 1} k_{-(n+2)} N_{(n+2)} + k_{-dx} N_{dx},$$

$$\frac{dN_3}{dt} = k_{d1} N_2 N_1 - k_{d3} N_2 N_3 - k_{-3} N_3 + k_{-5} N_5 , \qquad (1)$$

where N_n are the concentrations of centres containing n oxygen atoms , N_x is the concentration of some additional traps for dimers (e. g., carbon atoms), k_1, k_{dn}, k_{dx} and k_{-2}, k_{-n} ($n \geq 3$), k_{-dx} are the appropriate formation and dissociation rate constants.

It is reasonable to assume that at temperatures $\geq$550 °C the dimer dissociation is a dominant process as compared to trapping. Otherwise the rapid loss of interstitial oxygen should occur which has never been observed experimentally in perfect and carbon-lean crystals. For the same reason we neglect the emission of self-interstitials by the dimerization process and do not include an appropriate term in Eqs. (1). Thus at $T \geq 550$ °C the steady-state concentration of dimers can be simply found as

$$N_2 = k_1 N_1^2 / k_{-2} , \qquad (2)$$

with $k_1 = 4\pi r_1 D_1$ and $k_{-2} = v_2 \exp[-(E_{m1} + E_{b2})/kT]$, where r_1 is the capture radius for O_2 formation, D_1 is the normal O_i diffusivity, E_{m1} is the O_i migration energy, E_{b2} is the dimer binding energy, and v_2 is the frequency factor. Since the O_2 dissociation should involve the O_i reorientation (usual diffusion jump) we have suggested that $k_{-2} = 2(\tau_{Oi})^{-1} \exp(-E_{b2}/kT)$, where $(\tau_{Oi})^{-1}$ is the rate of oxygen reorientation. Assuming $r_1 = 5 \times 10^{-8}$ cm, $D_1 = 0.17 \exp(-2.54 \text{ eV}/kT)$, $\tau_{Oi} = 2.12 \times 10^{-15} \exp(2.54 \text{ eV}/kT)$ [4] and $N_2 \ll N_1$ we obtained $N_2 = 1.1 \times 10^{-22} N_1^2 \exp(E_{b2}/kT)$. Fig. 1a illustrates the temperature dependence of N_2 for some values of E_{b2}. It should be pointed out that practically the same expression for dimer equilibrium concentration can be deduced from the statistical consideration of oxygen atoms distribution.

In order to estimate a range of real values of E_{b2} the following information seems to be useful. In accordance with the data of [2] at $T = 450$ °C the rates of dimers dissociation and their transformation into stable TDD-related clusters are comparable. On the other hand the times usually required for reaching the steady-state TDD generation rate at this temperature are about several hours [5]. So some limits for the

E_{b2} values can be estimated if we assume that 1 h $< (k_{-2})^{-1} <$ 4 h at T = 450 °C. It results in 0.1 eV $\leq E_{b2} \leq$ 0.2 eV. Of course this estimation is rough but it seems to be realistic. E. g., if we take E_{b2} = 0.5 eV then an incubation stage for the TDD generation or oxygen loss should be ~ 500 hours at 450 °C which is not consistent with the experimental observations.

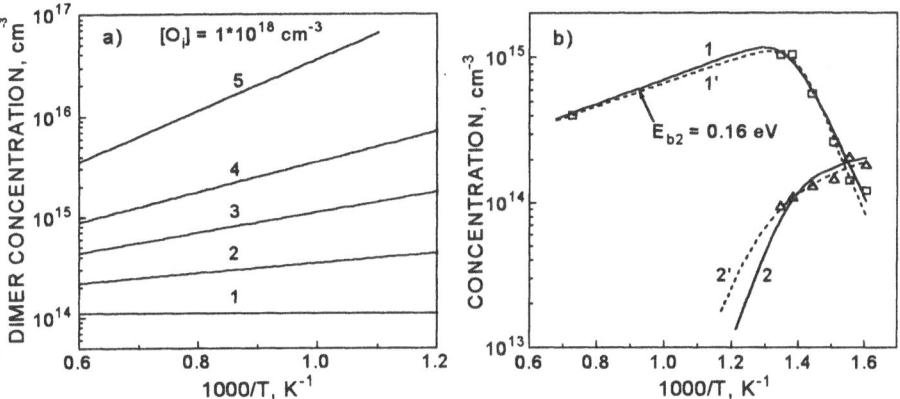

Figure 1. Temperature dependence of quasi-equilibrium concentrations of a) dimers, calculated for T >550 °C at given different values of binding energy E_{b2}, (eV): 1 - 0, 2 - 0.1, 3 - 0.2, 4 - 0.3, 5 - 0.5 and b) dimers (1, 1') and trimers (2, 2') calculated by fitting to the experimental data of [7] ($\square$ - 1012 cm^{-1}-cluster, Δ - 1006 cm^{-1}-cluster), 1', 2' -assuming the mobility of trimers.

Thus one can expect that the dimer density at high temperatures lies between curves 2 and 3 in Fig.1a. Then it is possible to estimate the upper limit for dimer diffusivity. In the presence of mobile dimers the effective oxygen diffusivity can be expressed as

$$D_O^{eff} = D_1(1 + 2N_2 D_2 / N_1 D_1). \tag{3}$$

To be consistent with the absence of enhanced oxygen diffusion at T $\geq$ 700 °C the second term in Eq.(3) should not exceed 1, i.e., $D_2^{max} \leq N_1 D_1 / 2N_2$. For $N_1 = 10^{18}$ cm^{-3}, E_{b2} = 0.1 eV, and, accordingly, $N_2 = 4 \times 10^{14}$ cm^{-3} at T =700 °C the upper limit for the dimer diffusivity is 1.5×10^{-11} cm^{-2}s^{-1}. It stands to reason that the prefactor of the diffusion coefficient of dimers can not be much lower (not more than by 100 times) than that in D_1. So the temperature dependence of the maximum allowable diffusivity of dimers can be deduced as $D_2 = 1.7 \times 10^{-3} \exp(-1.56$ eV/kT). At T = 400 °C it results in $D_2^{max} = 3.5 \times 10^{-15}$ cm^{-2}s^{-1} or $\approx 2 \times 10^5 D_1$.

It is more difficult to estimate the concentration of O_2 complexes at the TDD formation temperatures because a number of different processes can occur simultaneously there (see Eqs. (1)). But this problem can be solved taking into account the recent results of Lindstrom and Hallberg [6,7].

2.2. LINKAGE OF THE IR ABSORPTION BANDS AT 1012 cm^{-1} and 1006 cm^{-1} WITH DIMERS AND TRIMERS

Recently the infrared vibrational bands related to TDDs in silicon were discovered [6,7]. Important new information about the behaviours of the earlier known oxygen-related bands at 1012 cm^{-1} and 1006 cm^{-1} was obtained as well. It was shown that there are two kinds of oxygen complexes giving rise to these bands: a) the most abundant oxygen complexes which usually exist in as-grown crystals and can be rapidly transformed into the first types of TDDs during the initial stages of heat-treatment (HT) at T = 350-450 °C; b) oxygen complexes related to the higher order TDDs formed after more prolonged anneals. Further we shall consider only the first kind of complexes.

An unique feature of the band at 1012 cm^{-1} is a weak dependence of its intensity on the HT temperature. This band was observed after heat-treatments at 650 °C [6,8], 1100 °C [7] and even after a 1350 °C dispersion treatment [8]. On the other hand rapid variations in its intensity occurred at HT temperatures 350-450 °C [7] and relatively short times were needed to attain the quasi-equilibrium level. Besides this band was not found in carbon-rich Si:O crystals [6].

In our opinion all these features of the 1012 cm^{-1} band behaviour can correlate with the properties of only one kind of oxygen cluster, namely, oxygen dimer. Only this complex is in a position to exist in noticeable concentration at high temperatures up to the Si melting point (see Fig. 1a) and to undergo rapid variations in the density at elevated temperatures due to its high formation and/or trapping rates.

It is most likely that the main traps for mobile dimers in carbon-lean Si crystals are the interstitial oxygen atoms because their concentration is much higher than that of any other oxygen-related complex. The oxygen trimers formed in such a way can be considered as the second kind of abundant oxygen clusters. Earlier [5] the oxygen trimers were assigned by us to the TDD precursors or nuclei. The latter centres were found to exist in as-grown crystals, annihilate at T $\geq$ 550 °C and be generated again at the TDD formation temperatures. The correspondence of these features to those of the electrically inactive centre giving rise to the band at 1006 cm^{-1} makes it possible to identify this centre with the oxygen trimer.

In [7] the temperature dependence of the quasi-equilibrium levels (integrated absorption coefficients) of the 1012 and 1006 cm^{-1} bands was obtained. Using the calibration coefficient $k_m = 6 \times 10^{15}$ cm^{-1} [7] we plotted in Fig. 1b the data of [7] as the temperature dependence of the concentrations of the 1012- and 1006-clusters. It is evident that a comparison of this dependence with the expected one for dimers and trimers may be an important checking for the above-stated assignments. To do this we have to evaluate the concentrations of O_2 and O_3 complexes at the TDD formation temperatures. To simplify the task we have considered only the initial stages of the TDD generation when it is possible to neglect the processes of TDD dissociation [5] and capture of dimers into TDDs. Under these assumptions the analytical expressions for N_2^{eq} and N_3^{eq} can be easily deduced by solving in quasi-equilibrium approximation the set of Eqs. (1) and the simulation of the temperature dependence of N_2^{eq} and N_3^{eq} can be performed.

The results of an appropriate fitting to the experimental data are shown in Fig. 1b, curves 1 and 2. The details of this procedure will be published elsewhere. Herein we only mention that a satisfactory agreement between calculated and experimental data could be achieved only with the following relationship for the rate constants: $k_{-3} \gg k_{d3}N_3$ and $k_{d3} \gg k_{d1}$. The fitting parameter k_{d3} was found to be $5 \times 10^{-8} \exp(-1.6 \text{ eV}/kT)$. For $r_3 = 5 \times 10^{-8}$ cm and at T = 400 °C it results in $D_2 = 8.5 \times 10^{-14} \text{cm}^{-2}\text{s}^{-1}$. It is evident that this value is significantly higher than the estimated value of D_2^{max}.

For the suggested scheme of oxygen clustering and TDD formation the value of k_{d3} can be also estimated from the experimentally obtained values of the stationary rate of TDD formation $dN_{TDD}^{st}/dt = k_{d3}N_2N_3$. At T = 400 °C and $N_1 = 10^{18}$ cm^{-3} $dN_{TDD}^{st}/dt \approx 3 \times 10^9$ cm^{-3}s^{-1} [5]. Taking $N_2 = 4 \times 10^{14}$ cm^{-3} and $N_3 = 1.5 \times 10^{14}$ cm^{-3} (see Fig. 1b) the value of k_{d3} was found to be 4.5×10^{-20} cm^{3}s^{-1} and $D_2 = 7 \times 10^{-14}$ cm^{2}s^{-1} or $20D_2^{max}$. Thus too high dimer diffusivity is required again.

To escape this contradiction we have to imply the mobility not only of dimers but of trimers too. In such a case the kinetics of trimer formation can be expressed as

$$\frac{dN_3}{dt} = k_{d1}N_2N_1 - (k_{d3}N_2 + k_3N_1)N_3 - k_{-3}N_3, \qquad (4)$$

where $k_3 = 4\pi r_3 D_3$, and D_3 is the diffusivity of trimers.

So far as $N_1 \gg N_2$ then even at $k_3 < k_{d3}$, i. e. $D_3 < D_2$, the relation $k_3N_1 \gg k_{d3}N_2$ can be realized. The fitting curves 1' and 2' in Fig. 1b were obtained for the latter case. The values of fitting parameters k_{d1} and k_3 were found to be comparable and the deduced diffusivities D_2 and D_3 were lower than D_2^{max}. It should be emphasized that there are no limitations on the magnitude of trimer diffusivity since their concentration drops rapidly at T > 550 °C.

In the next section we shall show that some features of the formation kinetics of the first types of TDDs indicate on their mobility as well.

3. A new mechanism of the consecutive formation of TDDs

There are a great number of experimental facts and observations which demonstrate that the formation of the whole family of TDDs occurs successively by the addition of some structural unit. An opinion has been widely spread that this unit is a fast diffusing species (FDS). It was shown earlier [5] that important information on the behaviour of FDS as well as on some reaction rate constants can be obtained from the dependence of the total generation rate of TDDs of a higher type on the concentration of TDDs of the preceding type:

$$\sum_{m>n} \frac{dN_{TDD-m}}{dt} = k_n N_{FDS} N_{TDD-n} - k_{-(n+1)} N_{TDD-(n+1)} - \sum_{m>n} k_{pm} N_{TDD-m}, \quad (4)$$

where k_{pm} are the rate constants of TDD-m transformation into electrically inactive centres.

It was demonstrated [5] that during the initial stages of TDD formation the probabilities of thermal donor dissociation or their transformation into electrically inactive complexes were negligible in the case of Si:O crystals with $N_1 \geq 8 \times 10^{17}$ cm^{-3} at HT temperatures T $\leq$ 430 °C. It allowed the direct determination of the values of $k_n N_{FDS}$ for the first types of TDDs (Fig. 2a.) from the kinetics of their formation. It was found that the value of $k_1 N_{FDS}$ is directly proportional to N_1 in a wide range of temperatures and oxygen concentrations. Besides it was established that this product is independent of the HT time and the total TDD concentration in a rather wide range of variation of these parameters. Moreover these values were found to be the same in carbon-lean and carbon-rich Si crystals [9] in spite of a big difference in the TDD generation rate observed. In the present paper we have examined how these findings can be consistent with the possible assignment of FDS to dimers.

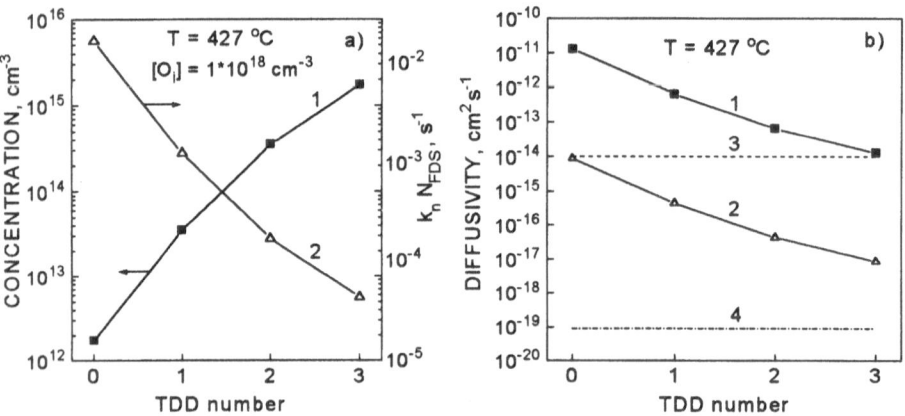

Figure 2. TDD number dependence of a) TDD-n (n = 0,1,2,3) steady-state concentration (curve 1) and $k_n N_{FDS}$ values (curve 2), and b) diffusivities of dimers (curve 1) and TDD-n species (curve 2), required for the consecutive TDD formation. Curves 3 and 4 correspond to D_2^{max} and normal oxygen diffusivity at T = 427 °C, respectively.

i) Assuming $N_{FDS} = N_2 = 7 \times 10^{14}$ cm^{-3} and $r_n = 5 \times 10^{-8}$ cm the diffusivity of dimers can be easily evaluated from the determined values of $k_n N_{FDS}$. It follows from the data plotted in Fig. 2b that extremely high values of dimer diffusivity $D_2 >> D_2^{max}$ are required to account for the successive transformation of the first types of TDDs.

ii) Simulation has shown that the quasi-equilibrium density of dimers can not be directly proportional to the concentration of interstitial oxygen even at the HT temperatures at which the dimer dissociation is negligible (the similar conclusion was made earlier in [3]). Besides an increase in the total TDD concentration is found to result in a reduction in dimer density due to the increasing rate of their capture.

iii) A remarkable reduction in the TDD generation rate for carbon-rich Si:O crystals can be explained simply by the capture of dimers by the substitutional carbon atoms. An occurrence of such a process is also indicated by a number of other facts [1,2,6]. So in the Si:O,C crystals an essential decrease in N_2^{eq} should occur and, if dimers are FDSs, the corresponding changes should be observed for the $k_n N_{FDS}$ values but it is not the case.

Thus it is evident that the assignment of FDS to dimer is not consistent with the above-stated features of the TDD formation kinetics. Our attempts to explain these features by invoking the FDS of other nature (self-interstitial, quasi-free oxygen etc.) have failed too. So we would have to come to the conclusion that not only the dimers and trimers but the first types of TDDs also are mobile and the rapid transformation of TDD-n into TDD-(n+1) occurs through their capture by interstitial oxygen. This allows all the features of the TDD formation initial kinetics to be explained naturally, i. e., if $k_n N_{FDS} = 4\pi r_n D_{TDD-n} N_1$ then the rates of TDD transformation from one species to another should be directly proportional to oxygen concentration, be HT time- and carbon concentration-independent. Besides, in this case the anomalously high diffusivities for oxygen complexes are not required (see Fig. 2b.).

We believe that the driving force for the migration of the first TDD species may be associated with their bistability and the existence of deep levels in the forbidden gap, i. e., the recombination-enhanced diffusion of these centres can occur at the TDD formation temperatures. Perhaps such an enhancement can account for the rapid reorientation of the first TDD species observed earlier in the stress-induced alignment studies [10].

4. Concluding remarks

The consideration of the early stages of the TDD formation performed in the present paper has led us to the following scheme of the process.

The initial aggregation of O_2 complexes is controlled by the normal oxygen diffusion rate. These centres have relatively low binding energy of about 0.2 eV but nevertheless they are the most abundant oxygen clusters even at high temperatures (>1000 °C). They give rise to the IR absorption band at 1012 cm^{-1}. At the TDD formation temperatures the diffusivity of dimers is higher than that of interstitial oxygen and the effective generation of trimers can occur as a result of capture of dimers by O_i. The oxygen trimers can be considered as the second type of the more abundant oxygen clusters. They give rise to the IR absorption band at 1006 cm^{-1}.

The formation of the subsequent oxygen clusters corresponding to the TDD species can go in two parallel ways: a) through the capture of dimers into O_n ($n \geq 2$) complexes, i. e. $O_2 + O_2 \rightarrow$ TDD-0 + $O_2 \rightarrow$ TDD-2 etc., $O_3 + O_2 \rightarrow$ TDD-1 + $O_2 \rightarrow$ TDD-3 etc. and b) through the capture of mobile O_n ($n \geq 3$) complexes by interstitial oxygen atoms, i. e. $O_1 + O_3 \rightarrow$ TDD-0, $O_1 +$ TDD-0 $\rightarrow$ TDD-1, $O_1 +$ TDD-1 $\rightarrow$ TDD-2, etc. For the first TDD species the second way of generation is a dominant one. Since the mobility of O_n clusters is reduced significantly with an increase in their size it results in the growth of their concentration and beginning from a certain TDD-number the contribution of dimers in successive transformation of thermal donors becomes

336

essential. The high migration ability of the first TDD species (bistable TDDs) may be associated with some kind of the electronically enhanced defect reactions [11].

According to this scheme all the small oxygen clusters being mobile should contribute to the effective oxygen diffusivity and their existence at the TDD formation temperatures should result in enhanced oxygen loss [2], in- and out-diffusion [1,12], formation of the TDD depleted layers [12] etc. On the other hand at T > 600 °C the quasi-equilibrium concentrations of these clusters (excepting dimers) are negligible and the normal oxygen diffusion should occur.

We believe that the suggested mechanism of the TDD generation can be tested by a careful study of the peculiarities of the TDD depleted layer formation in Si crystals with different oxygen and carbon content and at different HT temperatures. Such experiments and modelling are now in progress and some preliminary results confirm our suggestion.

5. References

1. Gosele, U., Ahn, K.-Y., Marioton, B.P.R., Tan, T.Y., and Lee, S.-T. (1989) Do oxygen molecules contribute to oxygen diffusion and thermal donor formation in silicon?, *Appl. Phys.* **A48**, 219-228.
2. McQuaid, S.A., Binns, M.J., Londos, C.A., Tucker, J.H., Brown, A.R., and Newman R.C. (1995) Oxygen loss during thermal donor formation in Czochralski silicon: New insights into oxygen diffusion mechanisms, *J. Appl. Phys.* 77, 1427-1442.
3. McQuaid, S.A., Newman, R.C., and Munoz, E. (1995) Models of oxygen loss and thermal donor formation in silicon by clustering of rapidly diffusing oxygen dimers, *Mater. Sci. Forum* **196-201**, 1309-1314.
4. Stavola, M., Patel, J.R., Kimerling, L.C., and Freeland, P.E. (1983) Diffusivity of oxygen in silicon at the donor formation temperature, *Appl. Phys. Lett.* **42**, 73-75.
5. Murin, L.I. and Markevich, V.P. (1990) Effects of various pre-treatments and impurity content on thermal donor formation in silicon, in K. Sumino (ed.), *Defect Control in Semiconductors*, Elsevier Science Publishers B.V., North-Holland, pp. 199-210.
6. Lindstrom, J.L. and Hallberg, T. (1994) Clustering of oxygen atoms in silicon: a new approach to thermal donor formation, *Phys. Rev. Lett.* **72**, 2729-2732.
7. Hallberg, T. (1995) Optical Properties of Oxygen-related Thermal Donors and Precipitates in Silicon, Dissertation, Linkoping.
8. Pajot, B., Stein, H.J., Cales, B., and Naud, D. (1985) Quantitative spectroscopy of interstitial oxygen in silicon, *J. Electrochem. Soc.* **132**, 3034-3037.
9. Murin, L.I. and Markevich, V.P. (1995) Effect of carbon on thermal double donor formation in silicon, *Mater. Sci. Forum* **196-201**, 1315-1320.
10. Wagner, P., Hage, J., Trombetta, J.M., and Watkins, G.D. (1992) Reorientation of stress induced alignment of thermal donors in silicon, *Mater. Sci. Forum* **83-87**, 401-406.
11. Sheinkman, M.K. and Kimerling, L.C. (1990) The mechanisms of electronically enhanced defect reactions in semiconductors, in K. Sumino (ed.), *Defect Control in Semiconductors*, Elsevier Science Publishers B.V., North-Holland, pp. 97-105.
12. Tokuda, Y., Katayama, M., and Hattori, T. (1993) Depth profiles of thermal donors formed at 450 °C in oxygen-rich n-type silicon, *Semicond. Sci. Technol.* **8**, 163-166.

INTERACTION OF POSITRONS WITH VACANCY-OXYGEN COMPLEXES AND OXYGEN CLUSTERS IN SILICON

M. FUJINAMI
Advanced Technology Research Laboratories
Nippon Steel Corporation
1618 Ida, Nakahara-ku, Kawasaki 211, Japan

The isochronal annealing behavior in O-ion-implanted Si has been investigated using variable-energy positron annihilation spectroscopy. For the $2x10^{15}$ O^+/cm^2 implanted Si, multivacancy-oxygen complexes induced by room temperature implantation are transformed into multivacancy-multioxygen ones around 600°C. Annealing at 800°C results in the formation of oxygen microclusters, whose concentration increases sublinearly with the dose over the range from $2x10^{14}$ $/cm^2$ to $2x10^{15}$ $/cm^2$. It is shown that the positron technique is very useful for the study of the oxygen-defect complexes in Si, which are not observable by transmission electron microscope, as well as the open-volume type defects.

1. Introduction

The initial stage in formation for oxygen precipitates in Si is under continuous study to control the crystal defects in fabrication of Czochralski (CZ)-grown Si. A basic understanding of behavior of the oxygen-related defects is fundamentally important and extended studies have been done. In this paper, variable-energy positron annihilation spectroscopy is applied to the study of defects, formed during O^+-ion implantation of Si and annealing, and the applicability of the positron probe is investigated through the interaction between excess oxygen atoms and vacancies induced. Positron annihilation spectroscopy is well-known as the valuable tool for the open-volume type defects such as monovacancies, divacancies and their agglomerates[1]. In this decade oxygen related defects in Si have been studied, and

337

R. Jones (ed.), Early Stages of Oxygen Precipitation in Silicon, 337–344.
© *1996 Kluwer Academic Publishers.*

some groups reported that the positrons could be trapped by even oxygen microclusters in a crystal Si[2-4].

In the conventional positron annihilation spectroscopy the radioisotopes such as ^{22}Na are used as a positron source. Positrons with energies up to several hundred keV are projected into μm range in the sample and, as a result, defect information in the bulk can be obtained. Variable-energy positron annihilation spectroscopy has been developed recently, and utilized for the determination of the defect depth profile near the surface region[5]. The mean penetration depth of the positron can be controlled by tuning the positron energy at the desired value and many applications to the study of defects in ion-implanted Si wafers have been performed[6-9].

2. Experimental

The substrates used in this study were p-type 5-10 Ωcm CZ-Si(100) wafers. The samples were implanted with 180 keV O^+ ions to doses from 2×10^{14} /cm^2 to 2×10^{15} /cm^2 at room temperature. The implantation angle was 7° off-normal to the (100) face for suppression of channeling effects. According to the results of TRIM calculation, the mean projected range and its standard deviation are 370 nm and 83 nm, respectively. Annealing was carried out at 300 to 800°C for 30 min in a standard furnace in flowing nitrogen.

Positron annihilation measurements were done in an ultrahigh vacuum chamber by the monoenergetic positron beam installed in Nippon Steel Corporation. The samples were subjected to an aqueous 4 vol% HF etch before measurements in order to remove the surface oxide. The Doppler broadening of the 511 keV annihilation γ-rays, characterized in terms of the S parameter, was measured by a Ge detector with respect to incident positron energy in the range of 0.1 to 29 keV. The mean penetration depth, z (nm), of the positron in Si target follows a power-law dependence on incident energy, E (keV), $z=17.4E^{1.6}$. The S parameter is defined as the ratio of the counts in a central region of the annihilation photopeak to those in the whole of the photopeak. Throughout this work, the S parameter was normalized to the value, S_b, for bulk silicon. The Doppler broadening of the annihilation γ-rays reflects the momentum distribution of electrons which positrons annihilate. Generally speaking, as positrons trapped in open-volume type defects annihilate valence electrons with a high probability, the degree of the Doppler broadening becomes small. This results in an increase in the value of S because the momentum of valence electrons is lower than that of core electrons. However, this scenario is inapplicable in the case that positrons are trapped at the vacancies coupled with impurities such as oxygen. Here the the electrons in the impurity atoms contribute to the annihilation process and the value of S is strongly affected by their momentum distribution rather than that of the electrons in the matrix atoms[10, 11]. The program POSTRAP4[12] was employed in order to solve the diffusion equation for positrons and to determine the defect depth profile from the S parameter versus the positron incident energy (S-E curve).

3. Results and Discussion

Figure 1 shows the S-E curves for unirradiated CZ-Si(100) and the Si samples implanted with 2×10^{14} O^+ /cm^2 and after subsequent annealing at different temperatures. The S-E curve for the unirradiated Si becomes a criterion, in which the defect density is less than the sensitivity limit (5×10^{15} /cm^3) of the positrons. The measured S parameter decreases slowly with the incident positron energy until the value of S becomes S_b, equal to 1. This is due to the fact that the H-terminated Si surface gives a higher characteristic S value and that the positrons thermalized after implantation easily diffuse towards the surface in the perfect crystal Si.

In the as-implanted sample, the measured value of S is significantly higher than S_b and evidently indicative of the presence of the open-volume type defects. The room temperature ion implantation of Si is known to yield the formation of divacancies, V_2, as the predominant defects[13]. Assuming that only V_2 were induced, the fits of a

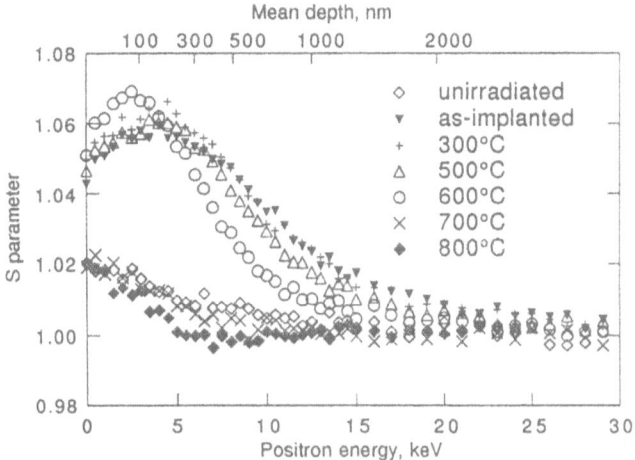

Figure 1.   The S parameter vs. incident energy (E) curves for the samples implanted with 2×10^{14} O^+ /cm^2 and after subsequent annealing.

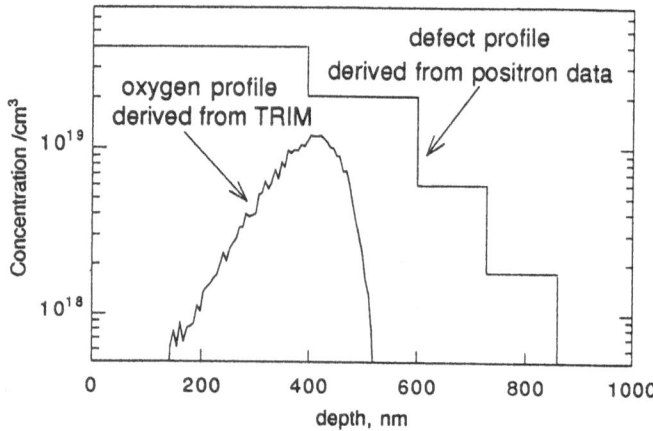

Figure 2. The oxygen profile and the histogram vacancy profile for the sample implanted with 2×10^{14} O^+ /cm².

diffusion and annihilation model to the S-E curve were done. As shown in Fig.2, the result indicates that the defects, whose distribution extends to depth of 900 nm from the surface, are induced not only in the implanted region but also in a deeper region.

Annealing at 300°C leads to a slight increase in the value of S around 5 keV(depth 230 nm). As an increase in defects content is difficult, this is a consequence of the fact that some agglomerates of vacancies are formed. Divacancies become mobile and anneal around 200°C, so that this consideration is consistent to the above assumption that the V_2 are dominant vacancy type defects in the as-implanted sample. A lowering in S at higher incident energy (> 5 keV) is initiated in the sample annealed at 500°C. By annealing at 600°C, S values at low incident energy (< 4 keV) increase, whereas they decrease at higher energy (> 4 keV). This shape of the S-E curve indicates that the diffusion of defects towards the surface and their disappearance in the deeper region take place in this temperature range. The S-E curve of the sample annealed at 700°C is almost identical to that of the unirradiated sample, suggesting that the open-volume type defects are completely removed. This annealing behavior is very similar to that of common dopant (B, P) ion implantation to Si, and can be explained by the introduction of only vacancy-based defects[7]. However, after annealing at 800°C, the S values measured from 3 keV to 10 keV are slightly below those of the unirradiated material. It seems that some defects related to oxygen are formed.

Figure 3 shows the S-E curves for the Si samples implanted with 2×10^{15} O^+ /cm² and after subsequent annealing. Even in the high dose sample, the measured values of

S are almost unchanged, compared with the 2×10^{14} O^+ /cm^2 implanted sample. But the annealing behavior is completely different between them. Up to 400°C, no variation in the S-E curve is observed. If the dominant defects induced by ion implantation are simple vacancy-type defects, the value of S should be changed by annealing below 400°C. On the other hand, some multivacancy-oxygen complexes such as V_3O are stable up to relatively high temperature[14]. It is reported that multivacancy-multioxygen complexes (V_xO_y) gives the higher S parameter than S_b if the value of x is larger than that of y[15]. It is, hence, considered that the open-volume type defects coupled with oxygen atoms are formed in 2×10^{15} O^+ /cm^2 implantations.

The first annealing stage can be seen at 500°C and a lowering in S is observed at more than incident energy 3 keV (> 100 nm depths). The shape of the S-E curve for the sample annealed at 600°C is characteristic. The S value has a maximum at the incident energy 2 keV (50 nm depth), showing that the vacancy-type defects still remain in the vicinity of the surface. It decreases rapidly with the incident energy until a minimum value below the S_b is reached at 7 keV (390 nm depth), and after that, it returns to the S_b. The meaning of the positron traps having the S value below S_b in the O ion-implanted region is that they are related to oxygen defects. It is, hence, proposed that the multivacancy-oxygen complexes and the interstitial oxygen atoms migrate in the implanted region and are stabilized to form more complicated V_xO_y, in which the value of x is less than that of y.

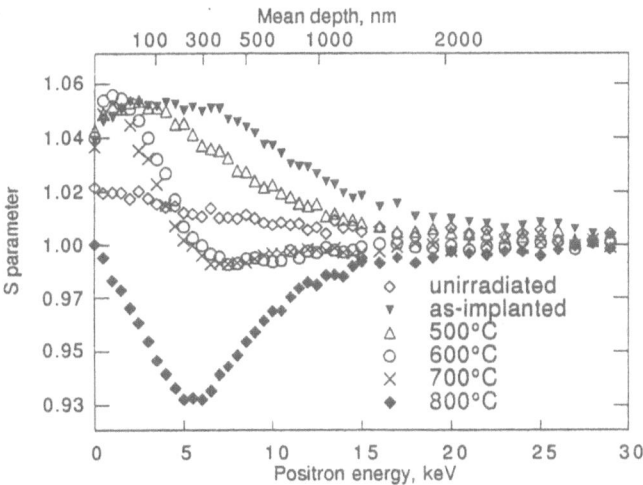

Figure 3. The S-E curves for the samples implanted with 2×10^{15} O^+ /cm^2 and after subsequent annealing.

Annealing at 800°C gives rise to a large lowering in S, indicating that the open-volume type defects disappear, but new positron traps appear. The minimum value of S is estimated to be 0.93 at the incident energy 5.5 keV (260 nm depth) and it is impossible to give it by annihilating positrons with electrons involved in Si atoms. The electrons of oxygen atoms have high momentum, and the positrons which annihilate them result in a low S value. Hence, the positron traps formed at 800°C are considered to be the defects including some oxygen atoms associated with open spaces, oxygen microclusters. The S-E curves for the samples annealed at 800°C after O implantations with doses of $2x10^{14}$, $5x10^{14}$, and $1x10^{15}$, $2x10^{15}$ /cm^2 are displayed in Fig. 4. The minimum S value in each S-E curve is strongly dependent on the oxygen implantation dose in this range. Assuming that the positron traps are distributed uniformly in the 180-500 nm depth range, or the implanted-oxygen profile, and that the specific trapping rate is equal to that of divacancies, $3x10^{14}$ /s, fits to the S-E curves have been done to derive their concentration. The results are summarized in Table 1, indicating that the mean ratio of the oxygen content to the positron traps is several tens. It is, therefore, speculated that several tens of oxygen atoms are involved in the oxygen microclusters formed at 800°C.

The pictures of cross-sectional transmission electron microscope (TEM) for the Si samples implanted with $2x10^{15}$ O$^+$/cm^2 and after subsequent annealing at 600°C and 800°C is exhibited in Fig. 5. No damage can be observed in the as-implanted sample and the sample annealed at 600°C. In the sample annealed at 800°C, the extended defects such as dislocations and stacking faults are recognized in the 350-600 nm depth range, whereas TEM observation shows no damage in the region to 350 nm to depth from the surface. The location of these extended defects does

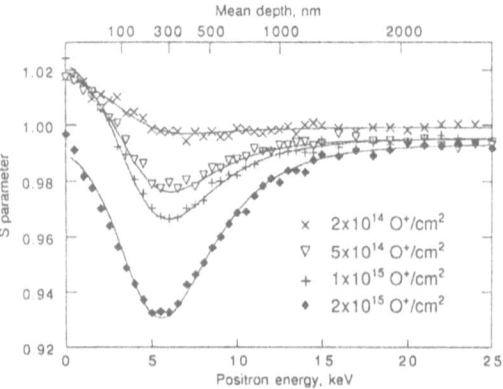

Figure 4.   The evolution of the S-E curves for the samples annealed at 800°C after O implantations with $2x10^{14}$, $5x10^{14}$, $1x10^{15}$, and $2x10^{15}$ O$^+$/cm^2. The solid lines are the fitted curves.

343

TABLE 1. The average atomic oxygen concentration calculated from TRIM and the density of the oxygen clusters derived from the positron S-E curve at the 180 to 500 nm depth for the samples implanted with different doses.

Samples (O$^+$/cm^2)	O conc. (atoms/cm^3)	Positron traps (/cm^3)
2x10^{14}	5.9x10^{18}	1.5x10^{17}
5x10^{14}	1.4x10^{18}	7.0x10^{17}
1x10^{15}	3.0x10^{19}	1.3x10^{18}
2x10^{15}	5.9x10^{19}	3.0x10^{18}

not always correspond to that of the positron traps, suggesting that the oxygen microclusters detected by a positron probe are indistinguishable by TEM.

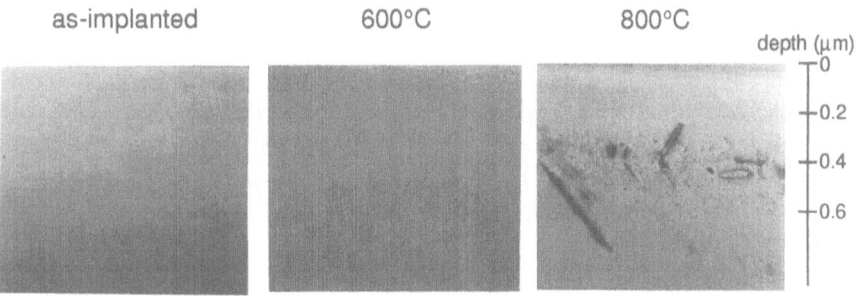

Figure 5.   Cross-sectional TEM photograph of the samples implanted with 2x10^{15} O$^+$ /cm^2 and after subsequent annealing at 600°C and 800°C.

4. Conclusions

The behavior of the defects induced in the O ion implanted Si has been studied by the variable-energy positron annihilation spectroscopy. There is a clear difference in the isochronal annealing behavior of the 2x10^{14} /cm^2 and the 2x10^{15} /cm^2 ion implanted Si. In the case of the 2x10^{14} /cm^2 sample, the annealing behavior is almost similar manner to that of common dopant, such as B and P, ion implantation, and can be explained by the behavior of the vacancy-based defects. In the case of the 2x10^{15} /cm^2 sample, the vacancies coupled with oxygen atoms are formed and it is impossible to remove the induced defects even by annealing up to 800°C. Further, annealing at 800°C leads to the formation of the positron traps, or oxygen microclusters, having very low characteristic S value. It has been proved that the positron is valuable probe to the study of oxygen-related defects in Si.

5. References

1. Asoka-Kumar, P., Lynn, K.G., and Welch, D.O. (1994) Characterization of defects in Si and SiO_2-Si using positrons, *J. Appl. Phys.* **76**, 4935-4982.
2. Dannefaer, S. and Kerr, D. (1986) Oxygen in silicon: A positron annihilation investigation, *J. Appl. Phys.* **60**, 1313-1321.
3. Uedono, A., Kawano, T., Wei, L., Tanigawa, S., Ikari, A., Kawakami, K., and Itoh, H. (1994) Oxygen microclusters in Czochralski-grown Si probed by positron annihilation, *Jpn. J. Appl. Phys.* **33**, L1131-L1134.
4. Coleman, P.G., Chilton, N.B., and Baker, J.A. (1990) Positron implantation studies of oxygen in P^+-silicon epilayers, *J. Phys.: Condens. Matt.* **2**, 9355-9361.
5. Schultz, P. J. and Lynn, K. G. (1988) Interaction of positron beams with surfaces, thin films, and interfaces, *Rev. Mod. Phy.* **60**, 701-779.
6. Fujinami, M. and Chilton, N.B. (1993) A slow positron beam study of vacancy formation in fluorine-implanted silicon, *J. Appl. Phys.* **73**, 3242-3245.
7. Fujinami, M. and Hayashi, S. (1995) Study of defect behavior in ion-implanted Si wafers by slow positron annihilation spectroscopy, *Mater. Sci. Forum* **196-201**, 1165-1170.
8. Nielsen, B., Holand, O.W., Leung, T.C., and Lynn, K.G. (1993) Defects in MeV Si-implanted Si probed with positrons, *J. Appl. Phys.* **74**, 1636-1639.
9. Uedono, A., Wei, L., Dosho, C., Kondo, H., Tanigawa, S., Sugiura, J., and Ogasawara, M. (1991) Defect production in phosphorus ion-implanted SiO_2(43nm)/Si studied by a variable-energy positron beam, *Jpn. J. Appl. Phys.* **30**, 201-206.
10. Ikari, A., Haga, H., Uedono, A., Ujihira, Y., and Yoda, O. (1994) Oxygen clusters in quenched Czochralski-Si studied by infrared spectroscopy and positron annihilation, *Jpn. J. Appl. Phys.* **33**, 1723-1727.
11. Uedono, A., Cho, Y.K., Tanigawa, S., and Ikari, A. (1994) Positron annihilation in proton irradiated Czochralski-grown Si, *Jpn. J. Appl. Phys.* **33**, 1-5.
12. Aers, G.C. (1990) An algorithm for defect profiling using variable energy positrons, in by P.J. Schultz, G.R. Massoumi, and P.J. Simpson (eds.), *Positron Beams for Solids and Surfaces, Proceedings of the 4th International Workshop on Slow-Positron Beam Techniques for Solids and Surfaces*, American Institute of Physics, New York, pp.162-170.
13. Stein, H.J., Vook, F.L., and Borders, J.A. (1969) Direct evidence of divacancy formation in silicon by ion implantation, *Appl. Phys. Lett.* **14**, 328-329.
14. Kawasuso, A., Hasegawa, M., Suezawa, M., Yamaguchi, S., and Sumino, K. (1995) An annealing study of defects induced by electron irradiation of Czochralski-grown Si using a positron *Appl. Surf. Sci.* **85**, 280-286.
15. Uedono, A., Ujihira, Y., Ikari,, A., Haga, H., and Yoda, O. (1993) Positron annihilation in electron irradiated Cz-Si, *Hyperfine Interactions* **79**, 615-619.

FORMATION OF THERMAL DONORS IN CZOCHRALSKI GROWN SILICON UNDER HYDROSTATIC PRESSURE UP TO 1 GPa

V.V. EMTSEV
Ioffe Physicotechnical Institute, RAS
194021 St.Petersburg, Russia
B.A. ANDREEV
Institute of Chemistry of High-Purity Substances, RAS
49 Tropinin St., GSP-445, 603600 Nizhnyi Novgorod,
Russia
A. MISIUK
Institute of Electron Technology, S8
Al.Lotnikόw 32/46, 02-668 Warsaw, Poland
K. SCHMALZ
Institute of Semiconductor Physics
Walter-Korsing-Str. 2, D-15230 Frankfurt(Oder),
Germany

Abstract. Effects of compressive stress on the formation of thermal donors in Cz-Si have been investigated at T=450°C and 600°C. Whereas the formation of Thermal Double Donors under stress is suppressed at T=450°C, other thermal donors are formed with increasing rates. At T=600°C under stress all families of thermal donors are produced with higher rates than that in heat-treated Cz-Si without stress. Some possible causes of the difference in the behavior of thermal donors under stress are briefly discussed.

1. Introduction

Formation processes of thermal donors in Czochralski grown silicon (Cz-Si) have been studied in considerable detail. There are several families of thermal donors formed at T=450°C. Among them, a family of Thermal Double Donors (TDD) consisting of nearly twenty spieces [1-4] has been investigated most intensively. Along with this family of thermal donors, two other kinds of donor states are observed in Cz-Si after prolonged heat treatment at T=450°C (t>50 hours): Shallow Donors with the ionisation energy less than 40 meV (SD) and Donors at $\approx E_C - 0.09$ eV ($D_{0.09}$) [5,6]. All these thermal donors can be produced at higher temperatures T≤600°C as well [7,8], though their formation kinetics are strongly changed compared to those at

345

R. Jones (ed.), Early Stages of Oxygen Precipitation in Silicon, 345–353.
© *1996 Kluwer Academic Publishers.*

T=450°C [9]. Further increasing in the temperature of heat treatment up to T=700°C results in a substantial decrease in the contributions of SD, TDD, and $D_{0.09}$ against the background of rapidly increasing concentrations of deep donors over the energy interval of $\approx E_C - 0.1$ eV to $\approx E_C - 0.25$ eV (DD) [9] which are attributed to rodlike defects [10]. Amongst the various factors that have a strong effect on the formation of thermal donors (heat treatment regimes, atmospheres, impurities etc), the impact of mechanical stress is not well understood. Bending Cz-Si samples during heat treatment at T=450°C Wang and Corbett [11] observed a remarkable effect of mechanical stress (up to 2 MPa) on the formation of thermal donors whose total concentration was twice as large as in the reference samples without stress. At the other extreme, some dramatic effects associated with the behaviour of oxygen in Cz-Si subjected to plastic deformation due to heat treatment under hydrostatic pressure up to 8 GPa have been reported in [12]. In the present paper we discuss the influence of moderate mechanical stress on the formation of thermal donors in Cz-Si at T=450° and 600°C.

2. Experimental

Wafers of high-resistivity Cz-Si of p-type were used for heat treatment. The hole concentration before heat treatment was (2 to 5) x 10^{13} cm^{-3} and the compensation ratio of shallow acceptors (boron) was less than 0.5. The oxygen and carbon contents were estimated on the basis of the well known IR absorption bands at 1108 cm^{-1} and 604.9 cm^{-1}, respectively. The oxygen content in the initial material was 8 x 10^{17} cm^{-3} (DIN 50435/1) and the carbon content was less than 5 x 10^{16} cm^{-3}. Samples were subjected to a hydrostatic pressure of 3 MPa or 1.07 GPa in an atmosphere of pure argon during heat treatment at T=450 and 600°C. The duration of heat treatment was 10 hours for all samples studied in order to draw a comparison between our data and the data obtained in [11]. After heat treatment, a layer of nearly 20 μm was removed from the sample surface by polishing and etching. Infrared photoconductivity spectra were recorded at low temperatures (T= 4 - 20 K) by a contactless method over the range 200 - 800 cm^{-1} using a Bruker IFS-113 V Fourier transform spectrometer. Measurements of the electron concentration vs temperature for all heat-treated samples were taken over the temperature span of 20 to 300 K. A computer analysis of the n (T) curves was carried out with the help of the relevant electrical neutrality equation; see for instance [10].

3. Results and Discussion

3.1. HEAT TREATMENT AT T=450°C

In Figure 1 we plot the temperature dependence of the electron concentration n(T) for a sample heat-treated at 450°C under the low stress. The electron concentration at room temperature usually evaluated from conductivity data is widely used as a measure of the

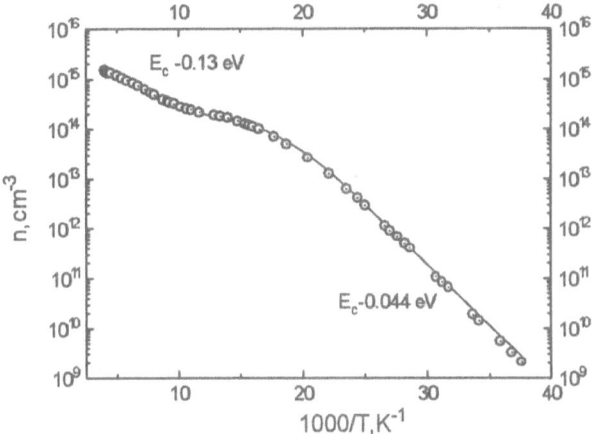

Figure 1. Temperature dependence of the electron concentration in Cz-Si heat-treated at 450°C for 10 h under a hydrostatic pressure of 3 MPa. Points, experimental; curve, calculated. The positions of shallow and deep donors are indicated.

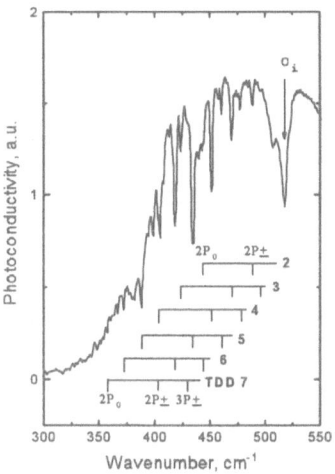

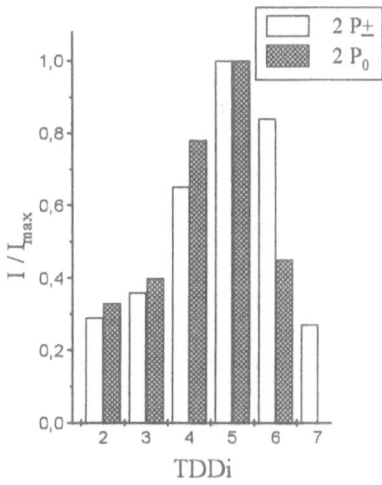

Figure 2. Photoconductivity spectrum of Cz-Si heat-treated at 450°C for 10 h under a hydrostatic pressure of 3 MPa. The spectrum was recorded in darkness at 11 K with a resolution of 1 cm^{-1}.

Figure 3. Relative distribution of the Double Thermal Donors for the photoconductivity spectrum shown in Figure 2. The intensities of the absorption lines due to the $1s \rightarrow 2p_0$ and $1s \rightarrow 2p\pm$ transitions were used for qualitative estimates of the TDD distribution.

TDD concentration; see for instance [11]. In our case $n(T=300$ K$)$ is equal to 1.6×10^{15} cm^{-3} and this value is quite close to what is observed for Cz-Si heat-treated at the same conditions without the stress. At first sight, there is no remarkable effect of the stress applied. However, analysis of the electrical and optical data indicates conclusively that the formation processes of thermal donors are strongly changed under compressive stress. The photoconductivity spectrum for the same sample shown in Figure 2 revealed the presence of several species of the TDD after heat treatment. Based on the intensities of the relevant spectral lines, one can estimate qualitatively the relative distribution of TDD (Figure 3). As is seen from this figure, the TDD distribution peaks at TDD5, whereas without stress a similar distribution can be achieved only for prolonged heat treatment ($t>50$ h); see also [2]. Moreover, the absolute concentrations of TDD were found to differ greatly in both cases, with and without stress. Without stress, Thermal Double Donors are the only family of thermal donors produced in Cz-Si during heat treatment at $T=450\degree$C for $t<100$ h and their total concentration N_{TDD} is of about 8×10^{14} cm^{-3} for $t=10$ h. In a sharp contrast, the formation of TDD is remarkably suppressed ($N_{TDD} \leq 3 \times 10^{14}$ cm^{-3}) under the stress applied. Along with the TDD, there are two other donor families at $\approx E_C -0.09$ eV ($N_{0.09}=2.5 \times 10^{14}$ cm^{-3}) and $\approx E_C -0.13$ eV ($N_{0.13}=1 \times 10^{15}$ cm^{-3}). From this observation, it is evident that the formation processes of thermal donors at 450°C are rather sensitive to compressive stress. Therfore, one cannot rely on measurements of the electron concentration at room temperature for a true evaluation of the TDD contribution. Nevertheless, this concentration can be used for general characterization of thermal donors. After heat treatment of Cz-Si at 450°C for 10 h under the low stress ($P=2.4$ MPa), Wang and Corbett [11] observed an increase in the electron concentration at room temperature by a factor of 1.6 as compared to the reference sample heat-treated without the stress. In our case the electron concentrations at room temperature turned out to be practically the same in both samples. The difference between two observations may be due to different oxygen contents in the samples studied (1.4×10^{18} cm^{-3} and 8×10^{17} cm^{-3}, respectively). Changes in the formation of thermal donors are much more pronounced with increasing stress ($P=1.07$ GPa). First of all, the electron concentration at room temperature in Cz-Si heat-treated at $450\degree$C for 10 h under the high stress is increased by an order-of-magnitude as compared to the reference sample without the stress (Figure 4). Still, optical measurements revealed the presence of TDD (Figure 5) and their relative distribution is similar to that observed for lower stress; cf Figures 3 and 6. Judging from our analysis of $n(T)$, one can make only a rough estimate of the total concentration of TDD (a few 10^{14} cm^{-3}), since their contribution is partially overlapped with the contribution of shallow thermal donors SD with much higher concentrations ($N_{SD} \approx 2 \times 10^{15}$ cm^{-3}). In any case, it is clear that the formation of TDD under the high stress is also suppressed. It is interesting to note that without stress shallow thermal donors distributed over the energy range of $\approx E_C -0.03$ eV to $\approx E_C -0.04$ eV make their appearance only for very long heat treatment of Cz-Si ($t \geq 100$ h) [6]. What is more, from Figure 4 it is evident that this high hydrostatic pressure leads to the enhanced formation of deep thermal donors at

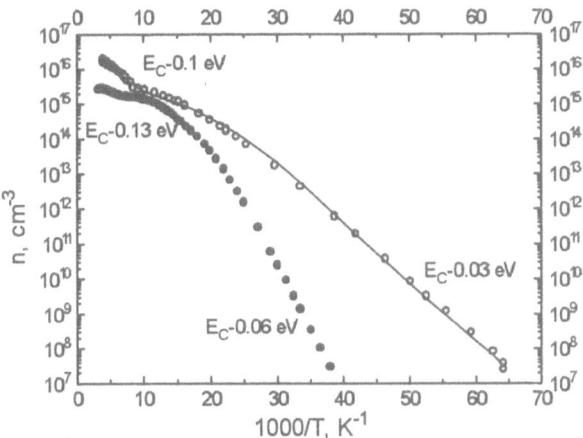

Figure 4. Temperature dependence of the electron concentration in Cz-Si heat-treated at 450°C for 10 h under a hydrostatic pressure of 1.07 GPa (open circles). Cf n(T) in the same material heat-treated at 450°C for 16 h under atmospheric pressure (black circles). Points, experimental; curve, calculated. The positions of shallow and deep donors are indicated.

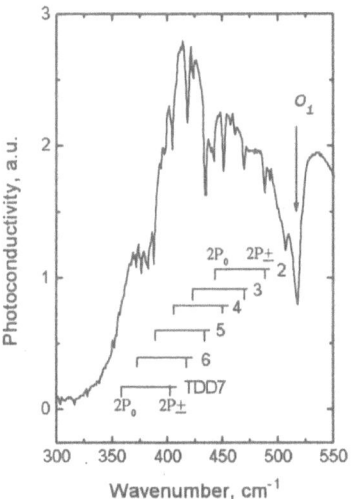

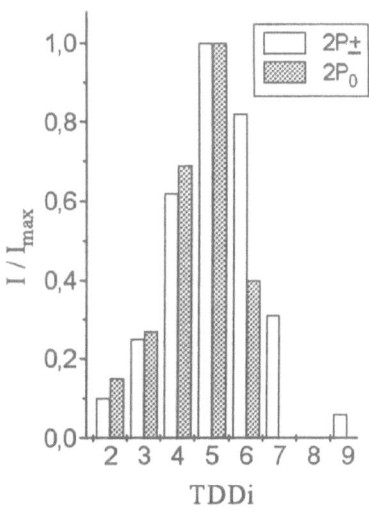

Figure 5. Photoconductivity spectrum of Cz-Si heat-treated at 450°C for 10 h under a hydrostatic pressure of 1.07 GPa. The spectrum was recorded in darkness at 6.2 K with a resolution of 1 cm^{-1}.

Figure 6. Relative distribution of the Double Thermal Donors for the photoconductivity spectrum shown in Figure 5. The intensities of the absorption lines due to the $1s \rightarrow 2p_0$ and $1s \rightarrow 2p_\pm$ transitions were used for qualitative estimates of the TDD distribution.

$\approx E_C - 0.1$ eV, so they turned out to be dominant ($N_{0.1} \geq 1.3 \times 10^{16}$ cm^{-3}) among other thermal donors.

3.2. HEAT TREATMENT AT T=600°C

From our earlier studies of Cz-Si samples cut from the same ingot [6,7,9], we have learnt that heat treatment at this temperature features a latent period of about 20 h in the formation of thermal donors, so one can start electrical measurements only after annealing for 24 h. As will be apparent from Figures 7 and 8, under the stress thermal donors are formed in sizable concentrations even for t=10 h. Optical measurements showed unambiguously that the lower-numbered species of TDD (TDD1-TDD3) are available in Cz-Si after heat treatment under the stress. In a sharp contrast to heat treatment at T=450°C, the formation of TDD at T=600°C was found not to be impeded at all. Their total concentrations are equal to $N_{TDD} \approx 3 \times 10^{13}$ cm^{-3} and $\approx 8 \times 10^{13}$ cm^{-3} at 3 MPa and 1.07 GPa, respectively. This should be compared with $N_{TDD} \leq 2 \times 10^{13}$ cm^{-3} in the reference sample heat-treated at T=600°C for a longer period (t=24 h) without the stress. Besides, it has been established that the total concentration of shallow thermal donors formed under the high stress is increased by an order-of-magnitude as against the reference sample, being $N_{SD} \approx 3 \times 10^{14}$ cm^{-3} and $\approx 3 \times 10^{13}$, respectively. In other words, the known components of "New Donors" including $D_{0.09}$ appear to be sensitive to compressive stress and no additional component with donor states at $E \leq E_C - 0.2$ eV is introduced as a result of heat treatment under stress. The formation rates and concentrations of all families of thermal donors increase by several times with increasing stress. It is important that in all cases the total concentration of compensating acceptors did not vary significantly after heat treatment of Cz-Si at both temperatures under stress. It means that even the highest stress applied does not lead to the appearance of new deep acceptor centers in considerable concentrations. This observation together with the conclusion that the formation processes of thermal donors at T=600°C remain qualitatively unchanged permits us to exclude remarkable effects due to plastic deformation; for effect of plastic deformation of heat-treated Cz-Si see [12]. One can also neglect a difference in the initial state of oxide precipitates in as-grown Cz-Si, since all samples studied were cut from the central part of the same wafers. A substantial decrease in the TDD formation at T=450°C under compressive stress means that it is inviting to conclude that the basic model of the electrically active core of TDD involving the self-interstitial as a constituent [13] is true; in the case of a vacancy-type core a reverse effect could be expected. At the same time, under stress the growing oxide precipitates can be less effective in emitting excess native interstitials and in this way the TDD production can be slown down as well. On the other hand, this compressive stress may be responsible for the shift of the TDD distribution to the higher-numbered species. To our mind, the enhanced formation of other shallow and

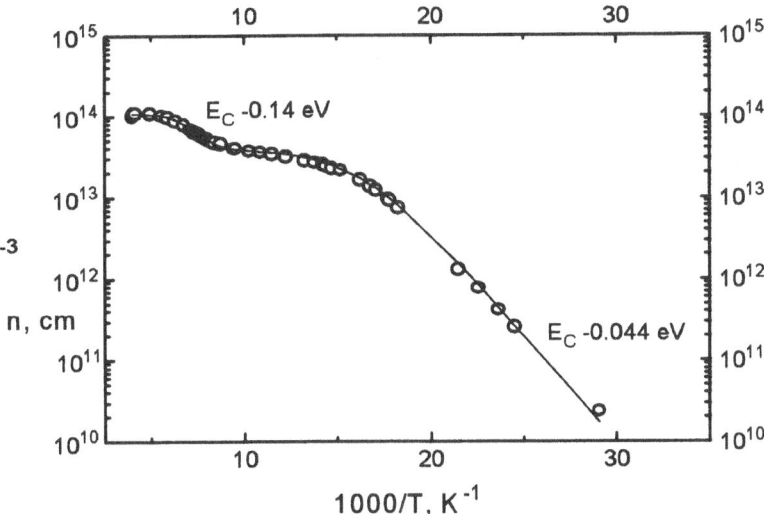

Figure 7. Temperature dependence of the electron concentration in Cz-Si heat-treated at 600°C for 10 h under a hydrostatic pressure of 3 MPa. Points, experimental; curve, calculated. The positions of shallow and deep donors are indicated.

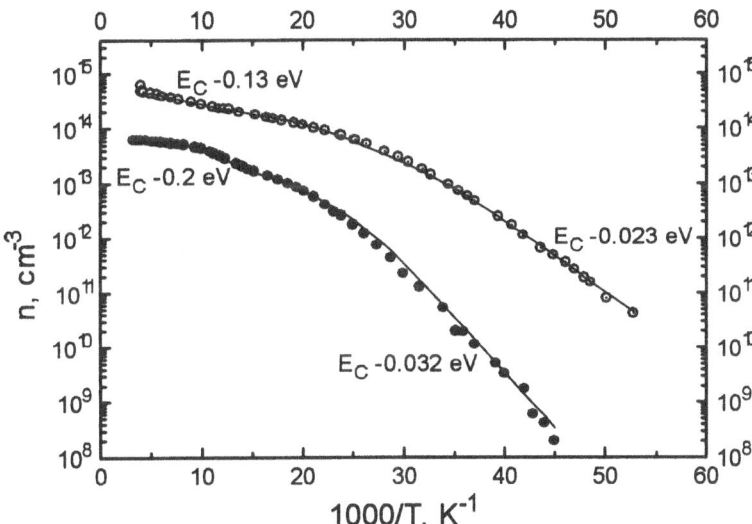

Figure 8. Temperature dependence of the electron concentration in Cz-Si heat-treated at 600°C for 10 h under a hydrostatic pressure of 1.07 GPa (open circles). Cf n(T) in the same material heat-treated at 600°C for 24 h under atmospheric pressure (black circles). Points, experimental; curves, calculated. The positions of shallow and deep donors are indicated.

deep thermal donors, different from the TDD, can be attributed to a rise in the oxygen diffusivity at T=450°C under the high stress. With increasing temperature of heat treatment some of the above factors can fall off, thus leading to a substantial increase in concentrations of all components of "New Donors" as a result of increasing diffusivity of oxygen at T=600°C due to combined effects of the temperature and stress.

4. Conclusions

The impact of compressive stress on the formation processes of thermal donors in Cz-Si at T=450°C and 600°C has been studied. We observed that the formation of Thermal Double Donors is remarkably suppressed at T=450°C under stress, being accompanied by shifting their distribution to the higher-numbered species. At the same time the formation of shallow and deep thermal donors are strongly enhanced. With increasing temperature of heat treatment all families of thermal donors were found to be formed with increasing production rates and the latent period in their formation becomes shorter. Some possible causes for the changes in the formation processes of thermal donors under stress are briefly discussed.

References

1. Makarenko, L.F. and Murin, L.I. (1986) On the nature of α-traps in heat-treated Si<O> *Fiz. Tekh. Poluprovodn.* **20**, 1530–1533.
2. Wagner, P. and Hage, J. (1989) Thermal Double Donors in Silicon, *Appl. Phys. A* **49**, 123–138.
3. Götz, W., Pensl, G., and Zulehner, W. (1992) Observation of five additional thermal donors species TD12 to TD16 of regrowth of thermal donors at initial stages of the oxygen donor formation in Czochralski-grown silicon, *Phys. Rev. B.* **46**, 4312–4315.
4. Liesert, B.J.H., Gregorkiewicz, T., and Ammerlaan, C.A.J. (1992) Silicon thermal donors: photoluminescence and magnetic resonance study of boron- and aluminium-doped silicon, in G. Davies, G.G. DeLeo, and M. Stavola (eds.), *Materials Science Forum*, Trans Tech Publications, Switzerland, **83–87**, part I, pp.407–412.
5. Kamiura, Y., Hashimoto, F., and Yoneta M. (1989) Novel thermal donors generated in Cz silicon by prolonged annealing at 470°C, in G. Ferenczi (ed.), *Materials Science Forum*, Trans Tech Publications, Switzerland, pp. 673–678.
6. Emtsev, V.V., Oganesyan, G.A., and Schmalz, K., (1993) Oxygen-related thermal donors in heat-treated Cz-Si, in N.T. Bagrayev (ed.), *Defect and Diffusion Forum*, Scitec Publication Ltd, Switzerland, **103–105**, pp. 505–516.
7. Emtsev, V.V., Daluda, Yu.N., and Schmalz, K. (1991) "New Donors" in heat-treated Cz-Si, in M. Kittler and H. Richter (eds.), *Solid State Phenomena*, Trans Tech Publications, Switzerland, **19&20**, pp. 229–234.
8. Andreev, B.A., Golubev, V.G., Emtsev, V.V., Kropotov, G.I., Oganesyan, G.A., and Schmalz, K., (1993) Formation of "new" donors as a result of heat treatment of silicon with different oxyge concentrations, *Fiz. I Tekh. Poluprovodn.*, **27**, 567–582 (*Semiconductors*, **27**, pp. 315–323).
9. Emtsev, V.V., Oganesyan, G.A., and Schmalz, K. (1996) New Donors in heat-treated Cz-Si their components and formation kinetics, in H. Richter, M. Kittler, and C. Clayes (eds.), *Solid State Phenomena*, Sci-Tech Publications Ltd, Switzerland, **47–48**, pp. 259–266.

10. Emtsev, V.V. and Oganesyan, G.A. (1996) Formation of deep thermal donors in heat-treated Czochralski silicon, *Appl.Phys. Letts* (in print).

11 Wang, P.W. and Corbett, J.W. (1986) Studies of oxygen thermal donor formation under stress, in J.C. Mikkelsen, Jr., S.J. Pearlton, J.W. Corbett, and S.J. Pennycock (eds.), *MRS Symposia Proc.*, MRS, Pittsburgh, Pennsylvania, **59**, pp. 167–172.

12 Vitman, R.F. (1986) Changes in the phase states of oxygen in silicon subjected to heat treatment and high pressure, *PhD Thesis*, Ioffe Physicotechnical Institute, Academy of Sciences of the USSR, Leningrad, pp. 140–175.

13. Deák, P., Snyder, L.C., and Corbett, J.W. (1992) Theoretical studies on the core structure of the 450°C oxygen thermal donors in silicon, *Phys. Rev. B*, **45**, 11612–11626.

COMPLEXES OF OXYGEN AND GROUP II IMPURITIES IN SILICON

E. McGLYNN*, M.O. HENRY*, and S.E. DALY*
*School of Physical Sciences
Dublin City University, Dublin 9, Ireland
K.G. McGUIGAN$
$Department of Physics,
Royal College of Surgeons in Ireland, Dublin 2, Ireland

Abstract

A detailed photoluminescence (PL) study of oxygen-rich silicon doped with the group II impurities Be, Zn and Cd is reported. Uniaxial stress and Zeeman measurements show that the defects responsible for the PL are of C_{2v} or C_{1h} symmetry and pseudo-donor in nature. Isotope substitution experiments indicate the involvement of one metal atom in each case. Strong circumstantial evidence for direct oxygen involvement in the defects is also presented.

1. Introduction

Oxygen reacts with group II elements to form stable compounds with strong ionic bonding between the group II atoms and the neighbouring oxygen atoms. Similarly bonded impurity pairs incorporating oxygen and a group II atom have been recognised for decades as excellent examples of molecular type defects in semiconductors, e.g. Zn-O and Cd-O in GaP [1]. These well-established results for GaP motivated us to examine whether corresponding impurity pairs might form in silicon. More particularly, the prospect of obtaining new information on oxygen-related donors via such pairing encouraged this study. The results we describe below confirm that the presence of the group II impurities Be, Zn and Cd in oxygen-rich silicon leads to the formation of a variety of defects which are not produced if oxygen concentrations are low. The defects are shown to possess rather similar properties, in general. Some important variations do occur, however, and considerable additional study is required in order to provide a full analysis.

2. Results

Comprehensive PL data were recorded for oxygen-rich silicon samples containing Be, Zn and Cd. The experimental details are described by Henry et al [2]. The spectra of the as-received implanted samples included the common damage-related defects. These

355

R. Jones (ed.), Early Stages of Oxygen Precipitation in Silicon, 355–362.
© 1996 Kluwer Academic Publishers.

356

vanished during annealing in the range from room temperature to 500°C. The defects characteristic of the group II impurities appear with optimum signal levels for annealing in the range 550 to 750°C. For the case of Be, the cooling rate is an important factor - rapid cooling favours the production of Be pair defects [3]. In all cases, annealing above 800°C results in the loss of the characteristic luminescence. Subsequent anneals at lower temperatures generally results in the reappearance of the spectra, albeit with reduced intensity.

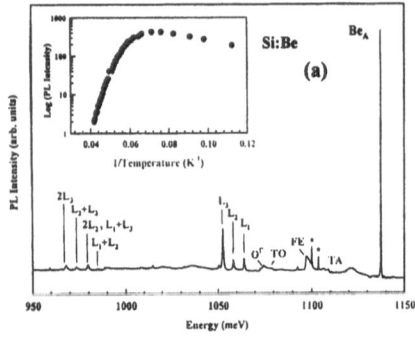 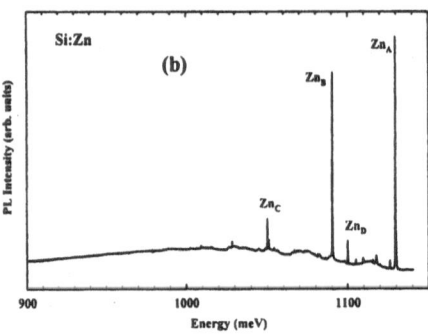

Figure 1(a).   A PL spectrum recorded at 12K of Si: Be. The temperature dependence of the intensity of the Be_A line is shown in the inset.

(b).   A PL spectrum recorded at 6K of Si:Zn.

2.1 BERYLLIUM

Figure 1(a) contains a PL spectrum, obtained at 12K, of a Si:Be sample annealed for 30 minutes at 600°C. An intense zero-phonon line occurs at 1137.98(5) meV which has been labelled Be_A. An extensive sideband occurs in the spectrum with acoustic, optic and local mode features clearly evident. The local modes, labelled L_1, L_2 and L_3, are especially prominent; they are also clearly seen in the two-phonon region and are labelled accordingly in the figure. Note that the intensity of the two-phonon line labelled $2L_3$ is somewhat weaker than would be expected. We have not studied this anomaly in any detail. The full sideband, with the exception of line $2L_3$, can be reconstructed successfully assuming linear coupling to a spectrum of phonon modes where the weighted Huang-Rhys parameter value is 3.7±0.4. This implies a large relaxation for the defect in the optical transition. The sharp lines at 1100 and 1104 meV are found to occur with different intensities relative to Be_A in different samples and they are not considered in this paper. It should be noted that the Be_A line was reported previously by Gerasimenko et al [4]. These authors also identified the line L_3 in their spectra and they associated it with Be_A. However, their spectra were considerably more complex than those we have recorded presumably due to contamination in their samples.

The Be$_A$ spectrum remains essentially unchanged over the temperature range 5-20K for which it is observed. No evidence of excited states has been found in the spectra. We show the temperature dependence of the total Be$_A$ PL intensity in the inset of Figure 1(a). The behaviour is very characteristic of recombination at defects in silicon - an initial increase in intensity as the temperature is raised from liquid helium temperatures followed by a rapid loss of the intensity as the excited state of the defect becomes thermally unstable. This effect is well-understood [5] and the fit to the data in Figure 1 is obtained for a thermal binding energy of ~27 meV. The spectroscopic binding energy relative to the silicon energy gap is approximately 31 meV.

2.2 ZINC

A PL spectrum of (CZ)Si:Zn is shown in Figure 1(b). The spectrum is dominated by two lines at 1129.57 (5) meV and 1090.47 meV, labelled Zn$_A$ and Zn$_B$ respectively, each of which is accompanied by a broad sideband. Several weaker lines are also evident in addition to Zn$_A$ and Zn$_B$, most notably those at ~1100 meV and 1050 meV labelled Zn$_D$ and Zn$_C$, respectively. The broad sidebands indicate coupling to a large range of lattice phonons. Sharp intensity changes at ~1065 meV and ~1025 meV correspond to the O$^\Gamma$ cut-off for Zn$_A$ and Zn$_B$ respectively. Detailed measurements reveal that the A and B defects possess characteristic in-band phonon modes of energy ~12 meV and 8 meV, respectively [2]. The value of the Huang-Rhys parameter estimated from the fraction of the total intensity in the zero-phonon lines is ~3 for both Zn$_A$ and Zn$_B$. The weakness of Zn$_C$ and Zn$_D$ preclude making a similar calculation for all four cases. For Zn$_A$ and Zn$_B$, the temperatures dependence of the PL spectrum also shows that both the ground and excited states are split, with separations in the range 1 to 4 meV observed between the levels [2].

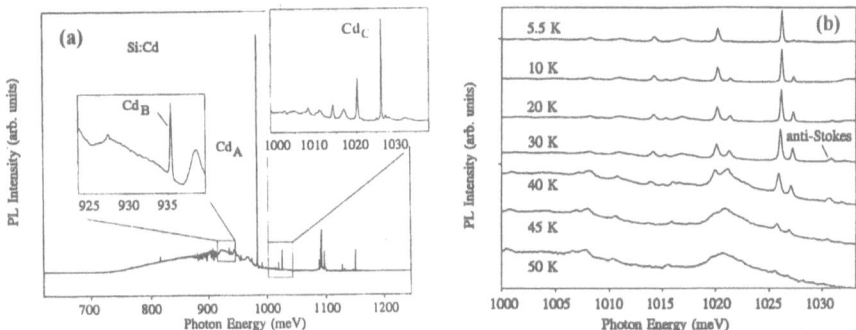

Figure 2(a). A PL spectrum recorded at 5K of Si:Cd; (b) Details of the Cd$_C$ spectrum.

2.3 CADMIUM

The case of Cd differs in several respects from those of Be and Zn. Whereas only one defect is observed for Be and four very similar defects for Zn, we find that for Cd more complex effects occur. A PL spectrum of cadmium-implanted CZ silicon annealed at 550°C for 30 minutes is shown in Figure 2(a). The spectrum is dominated by a zero phonon line (ZPL) at 983.21(5) meV (referred to as Cd_A), and smaller ZPLs can be seen at 1026.10(5) meV - Cd_C - and 935.35(5) meV - Cd_B. The relative intensities of the three ZPLs vary from sample to sample and therefore arise from recombination at different centres. Our results indicate that the intensity of Cd_B relative to Cd_A decreases with anneal temperature between 500-700°C, while the intensity Cd_C relative to Cd_A increases with anneal temperature in the same range.

The Cd_A sideband is quite broad, indicating coupling to a range of lattice modes, in addition to the weak in-band local mode at ~973 meV. The Cd_B sideband overlaps the much stronger Cd_A sideband and consequently it is very difficult to distinguish between the two bands; however, a weak in-band local mode of the Cd_B line can be seen at ~928 meV. In addition, both the Cd_A and Cd_B vibronic bands lie on the high energy side of a broad band which survives to higher temperatures (>50K), the presence of which makes accurate determination of the correct baseline very difficult. Allowing for this background we estimate the value for the Huang-Rhys factor to be ~3 and the relaxation energy to be ~50 meV for Cd_A.

The Cd_C ZPL and sideband are shown in Figure 2(b). It is immediately apparent that this defect is quite different to both Cd_A and Cd_B, with in-band local modes of ~6 meV(L_1) and ~9 meV(L_2) being dominant. The Huang-Rhys parameters estimated for these phonon modes are 0.65 and 0.4, respectively.

The distinctive nature of Cd_C is reinforced when the temperature dependence of the spectrum is studied. An additional line due to transitions from an excited state is observed together with L_1 and L_2 sidebands and with Huang-Rhys parameter values which are essentially the same as those for the main Cd_C line. The thermal activation energy equals the spectroscopic separation indicating a common ground state. The anti-Stokes L_1 phonon is considerably lower in energy than the Stokes - 4.7 meV compared to 6 meV.

2.4 UNIAXIAL STRESS

Uniaxial stress measurements were made in all three cases for samples oriented along <001>, <111> and <110> crystal directions. We begin with the case of Si:Zn since this provided the clearest set of data and best polarisation information. The data are summarised in the fan diagrams of Figure 3. Note that for Zn_A and Zn_C, data pertaining to nearby lines are shown by open circles. The solid lines show theoretical fits to the data. For Zn_A, transitions between non-degenerate states at rhombic I (C_{2v}) symmetry defects provide the best fit to the data. For all other cases the symmetry must be monoclinic I (C_{1h}) in order to account for the observed splitting and shift rates.

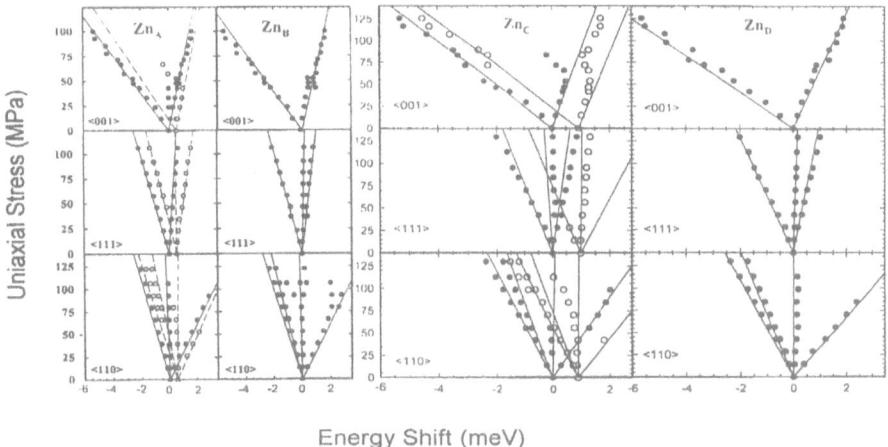

Figure 3.   Fan diagrams of the uniaxial stress data for Si:Zn.

The data obtained for Be_A are shown in Figure 4.   Two components are observed for <001> and <111> stresses and three for <110> stresses, as is the case for Zn_A.   A good fit to the data can be obtained in this case also (at low stresses) if the Be_A ZPL is assigned to transitions at rhombic I (C_{2v}) symmetry defects. At stresses greater than 40 MPa, substantial curvature is observed for <001> stress, and to a lesser extent for <110> stresses, whereas the shift rates are linear for all <111> stresses. This has been analysed in detail by Daly et al. [3] and the fits shown in Figure 4 are based on including interactions with an excited state of the Be_A defect which cannot be observed in the PL measurements.

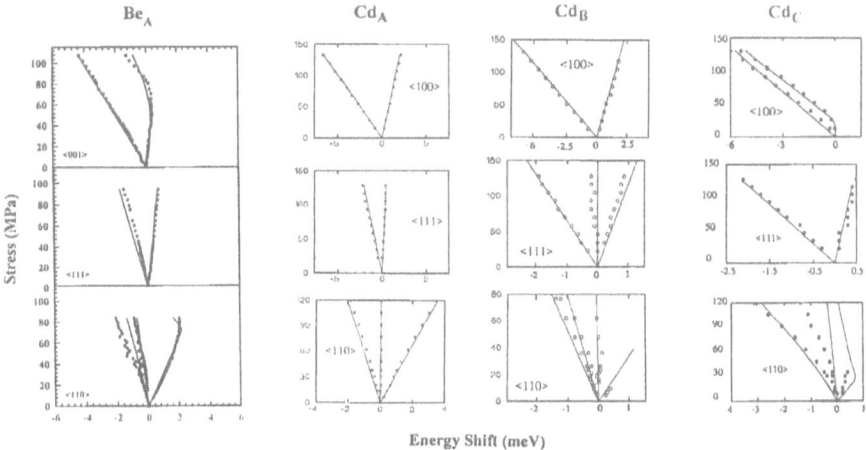

Figure 4.   Fan diagrams of the uniaxial stress data for Si:Be and Si:Cd.

The uniaxial stress data for the three Cd-related defects are also shown in Figure 4. Here, too, the Cd_A data are seen to be consistent with rhombic I (C_{2v}) symmetry, whereas monoclinic I (C_{1h}) symmetry is required to account for the Cd_B data. The case of Cd_C is much more complex, with a rapid non-linear shift to lower energy occurring under small <001> stresses. The solid lines show fits obtained assuming rhombic I (C_{2v}) symmetry including interactions with the adjacent excited state observed in the PL measurements. Detailed measurements of the effects of stress on the excited state are required in order to confirm and refine the fit.

A full summary of the theoretical fits for all defects is given in Table 1. In all cases, the initial and final states are orbitally non-degenerate A or B states. The uniaxial stress data are not sufficient to discriminate between A_1 or A_2, B_1 or B_2 states.

Line	Energy (meV)	Stress Parameters (meV/GPa)				Defect symmetry
		A_1	A_2	A_3	A_4	
Be_A	1138.00	-35.3	16.3	-15.9	-	Rhombic I
Zn_A	1129.57	-52.0	15.0	-17.2	-	Rhombic I
Zn_B	1090.47	-51.2	15.4	-16.8	±2.2	Monoclinic I
Zn_C	1050.31	-51.0	13.5	-15.8	±2.5	Monoclinic I
Zn_D	1100.42	-55.0	17.6	-16.5	±2.5	Monoclinic I
Cd_A	983.15	-49.0	15.0	-14.6	-	Rhombic I
Cd_B	935.30	-51.0	16.0	-16.6	±3.5	Monoclinic I
Cd_C	1026.10	-44.2	12.0	-15.2	-	Rhombic I

Table 1: The best fit parameter values obtained by applying the theory of Kaplyanski [7] to the uniaxial stress data.

2.5 ZEEMAN MEASUREMENTS

Magnetic fields in the range 0-5T are found to have no effect in all cases studied to-date, with neither the position nor linewidth of any of the lines being altered in the presence of the field.

2.6 ISOTOPE SUBSTITUTION MEASUREMENTS

The availability of a selection of naturally occurring isotopes in the case of Zn and Cd enabled the effects of isotope substitution to be studied. For Be, there exists only one stable isotope. Clear and unambiguous confirmation of the involvement of Zn in the corresponding A and B defects, and of Cd in the corresponding A, B and C defects is obtained. The data are shown in Figure 5. In all cases, the evidence indicates that there is only one group II atom involved. The data for the low energy in-band local modes are particularly clear, showing that the energies are inversely proportional to the square root of the isotope mass, i.e. indicating that the vibration is predominately that of the group II atom.

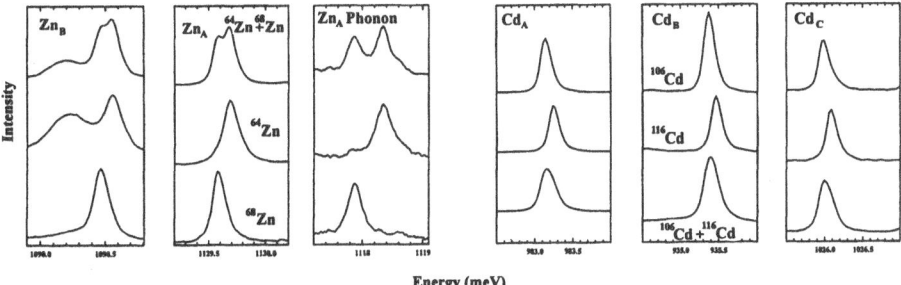

Figure 5.   Effects of isotope substitution on the main lines in Si:Zn. and Si:Cd.

A doublet structure is not observed in the case of samples implanted with two Cd isotopes. This is due partly to the small shift involved and also to the incorporation of fewer atoms of the heavier isotope (the second to be implanted) in the defects [6].

3. Discussion

The uniaxial stress data show that there are only small variations in the best fit parameter values, except for Be_A and Cd_C for which interaction effects are important. Defects are found to be either rhombic I (C_{2V}) or monoclinic I (C_{1h}) and in the case of the latter, the departure from rhombic I (given by the value of the A_4 parameter [7]) is small. In all cases, magnetic fields have no effect on the positions or linewidths of the zero-phonon lines (at the resolutions employed of ~ 0.1 meV). The absence of any magnetic field effects and the trends in the uniaxial stress data enable two important factors to be established. Firstly, electron-hole overlap, which would be manifested by the occurrence of magnetic field induced splittings of exchange-coupled electron-hole pair states [8], is negligible. Consequently, the observed radiation may be attributed to the recombination of electron - hole pairs with one of the carriers highly localised at the defect and the other in a delocalised shallow level. We may also conclude, because of the absence of magnetic field effects, that the delocalised carrier cannot be a hole, since such a carrier should show the characteristic splitting of a j=3/2 hole [8]. The uniaxial stress data also support the view that the delocalised carrier is the electron. The observed shift rates (for the majority of the defects) under <001> stress are very similar to those of the conduction band valleys [9]. The highly localised hole behaves as a S = 1/2 particle. The recombination of a S = 1/2 electron with this hole can account for the absence of magnetic field effects if we assume equal g - values for the two carriers. The term pseudo-donor is now generally used to describe such defects. This description enables the origin of the A and B states indicated by the fits to the stress data to be readily explained. These states emerge from the parent E and T of a tetrahedral donor impurity, with the electronic degeneracy being lifted by the C_{2v} or C_{1h} symmetry of the defects.

The question of direct oxygen involvement in the defects is still open despite the strong circumstantial evidence. Although co-implantation of oxygen with Be and Zn into oxygen-lean material results in the appearance of the luminescence, proof of oxygen involvement via shifts in the PL line energies for different oxygen isotopes has not been obtained. Despite this, it is highly likely that oxygen atoms are directly involved. The occurrence of a variety of defects, in the case of Zn in particular, with relative intensities which depend on the annealing treatment has not yet been studied in detail. This case seems to be the most promising for exploring possible connections with oxygen-related donor formation, since Be and Cd present less varied characteristics.

4. Conclusions

The group II impurities Be, Zn and Cd are found to produce distinctive luminescent defects in oxygen-rich silicon. In the case of four defects for Si:Zn and two for Si:Cd, the PL can be attributed to recombination at pseudo-donor centres. For Be and for one defect in the case of Cd, more complex defect characteristics are observed although these, too, show pseudo-donor nature. In all cases, the presence of oxygen at minimum concentrations of approximately 10^{17} cm^{-3} is required in order to produce the luminescence centres. For Zn and Cd, isotope substitution experiments indicate that only one group II impurity is involved in the defects. Local-mode vibrations are found to involve the motion of the metal atom with respect to an essentially static lattice. The nature and degree of oxygen involvement in the defects remain to be established.

5. References

1. Henry, C.H., Dean, P.J and Cuthbert, J.D. (1968) New Red pair luminescence from GaP, *Phys.Rev.* **166**, 754-6.
2. Henry, M.O., Campion, J.D., McGuigan, K.G., Lightowlers, E.C., doCarmo, M.C. and Nazaré, M.H. (1994) A photoluminescence study of Zn-O complexes in silicon, *Semicond. Sci. Technol.* **9**, 1375.
3. Daly, S.E., Henry, M.O., McGuigan, K.G. and doCarmo, M.C. (1996) A complex luminescent centre in Beryllium-doped oxygen-rich silicon, *Semicond. Sci. Technol.*, **11**, 6903-8.
4. Gerasimenko, N.N., Zaltser, B. A., Safronov, L.N. and Smirnov, L.S. (1985) Radiative recombination centres in silicon irradiated with beryllium ions, *Sov. Phys. Semicond.* **19**, 762-5.
5. Davies, G. (1989) The optical properties of luminescence centres in silicon, *Physics Reports* **3** and **4**, 83-188.
6. McGlynn, E. (1996) A photoluminescence study of cadmium related defects in silicon, Ph.D. Thesis, Dublin City University
7. Kaplyanski, A.A. (1969) Non-cubic centres in cubic crystals and their piezospectroscopic investigation, *Opt. Spectrosc.* **16**, 329-37.
8. Dean, P.J. and Herbert, D.C. (1979) Bound Excitons in Semiconductors, in K.Cho(ed) *Excitons* Springer-Verlag, 55-182.
9. Laude, L.D., Pollak, F.H. and Cardona, M. (1971) Effects of uniaxial stress on the indirect exciton spectrum of silicon, *Phys. Rev. B* **3**, 2623.

COPPER AND OXYGEN PRECIPITATION DURING THERMAL OXIDATION OF SILICON: A TEM AND EBIC STUDY

A. CORREIA and D. BALLUTAUD,
Laboratoire de Physique des Solides, 1 Place Aristide Briand 92195 Meudon Cedex (France)

A. BOUTRY-FORVEILLE
Laboratoire d'Electrochimie Interfaciale, 1 Place Aristide Briand 92195 Meudon Cedex (France)

1. Introduction

Oxygen in silicon has been shown, in low temperature experiments, to act as a trap for interstitial copper , and in high temperature precipitation to promote copper intrinsic gettering and colony generation [1][2]. The precipitation of oxygen alone has been invoked to explain lifetime degradations observed after annealing at T>700°C [2]. Moreover, the presence of oxygen has also been shown to be at the origin of the electrical activation of dislocations in a number of examples [3]. In these cases, it is interesting to know if oxygen was solely responsible for such degradations or if some metallic fast diffuser also played a role by decorating the oxide-based defects [4]. For instance, an investigation of annealing-induced oxide-based point-like defects has been carried out by Ho and Wald [5] by comparing the minority carrier diffusion lengths in similarly treated float-zone (FZ) and Czochralski (Cz) silicon samples. The diffusion length was seen to decrease with annealing in both materials, but more strongly in the Cz silicon. These authors concluded that both degradations was related to a metal contamination, and that the bigger effect in the Cz silicon was due to additional centers corresponding to SiO_x clusters having gettered metallic impurities [5].

In the present study, we correlate electrical measurements and microstructure characterization in silicon, thermally treated in conditions where oxygen, or oxygen and copper are introduced into the samples. The electrical activity produced by oxygen clusters, copper precipitates, and copper-decorated oxygen clusters is then quantitatively evaluated. As the oxygen concentration is playing a role in both the precipitation of oxygen and the gettering of metallic impurities, we compare Cz and FZ silicon.

The microstructure of the defects is checked at a microscopic scale by transmission electron microscopy (TEM) combined with energy dispersive x-ray spectroscopy (EDS), and at a macroscopic scale by secondary ion mass spectrometry (SIMS) profiling of oxygen and copper. Local value of the minority carrier diffusion

363

R. Jones (ed.), Early Stages of Oxygen Precipitation in Silicon, 363–369.
© *1996 Kluwer Academic Publishers.*

length is determined using an electron-beam-induced current (EBIC) in the scanning electron microscope.

2. Experimental

Starting material was boron-doped ([B] = 3×10^{16} cm^{-3}) p-type silicon with a (001) surface orientation and 500 μm thick. The oxygen content was respectively about 8×10^{17} cm^{-3} in Czochralski (Cz) silicon and less than 10^{16} cm^{-3} in Float-zone (FZ) silicon (measured by infra-red absorption with a calibration coefficient of 3×10^{16} cm^{-2} at room temperature). The samples were submitted to mechanical polishing, to chemical etching, then to thermal oxidation in dry oxygen in sealed silica ampoules at one atmosphere and 1000°C for one hour in a clean furnace. ^{18}O was used as a tracer for SIMS analysis. The samples were cooled to room temperature at a rate of about 3K/s. Some samples were contaminated with copper prior to the thermal oxidation by mechano-chemical etching of the sample backside for 3 hours using a [Cu(NO$_3$)$_2$, NH$_4$F] aqueous solution [6].

The copper concentration in the starting silicon bulk, before the thermal annealing, was checked to be lower than 10^{-13} cm^{-3} by neutron activation analysis [7]. In the samples contaminated first with copper, the dose appeared to be about 5×10^{14} cm^2 after thermal oxidation, while the copper dose remained lower than 10^{-13} cm^{-2} in the other samples.

EBIC was performed before and after hydrogenation using a JEOL JSM 840 scanning electron microscope. The collecting junctions were of the metal-insulator-semiconductor type, prepared by removing the oxide layer by HF, reoxiding the surface at 400°C in one atmosphere of oxygen during 20 minutes, and depositing under ultra-high-vacuum a 100 nm thick aluminum layer. The minority carrier diffusion length L_n was deduced from the variation of the collection efficiency as a function of beam voltage. For L_n determination, fitting curves were calculated using a model based on that of Wu and Wittry [8].

Plan-view and cross-sectional specimens for TEM high resolution observations were prepared using respectively chemical etching of the sample backside and ion milling. The microscope used (JEOL 2000FX) was equipped with an energy dispersive x-ray spectrometer.

SIMS (Cameca IMS-4F instrument) profiling with calibration (concentration and abrasion depth) was used to analyze the profiles of ^{18}O and ^{63}Cu in some of the samples.

3. Results

The EBIC collection efficiency as a function of beam voltage has been measured in regions far from extended defects. The lifetime of minority carriers are lowered for all the samples after thermal oxidation. The diffusion length values for the different samples are displayed in Table I.

TABLE 1. Diffusion lengths (Ln) in the different samples after oxidation.

Sample	Ln (μm)		
	Before treatment	After oxidation	After copper contamination and oxidation
Cz	100	32	16
FZ	200	85	45

The degradation of the diffusion lengths is more important for the Cz silicon (high initial oxygen concentration), and also more important in the case of copper contamination.

In the case of the Cz samples, SIMS profiling performed in regions free of extended defects (checked by ionic SIMS imaging) indicates that ^{18}O (coming from the oxidation) is present in the bulk up to 10^{17} cm^{-3} (Fig. 1, curve 1), i.e. above the solubility at 1000°C (7.5×10^{16} cm^{-3}). This suggest the presence of oxygen traps in the Cz material, which can be considered as small SiO_x clusters.

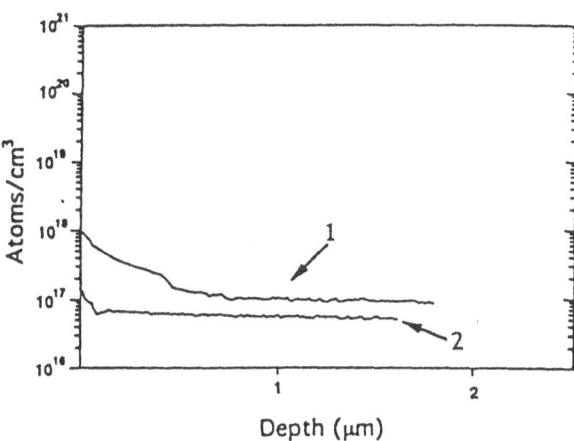

Figure 1. SIMS profiles of ^{18}O concentration after thermal oxidation: (1) Cz sample; (2) FZ sample.

TEM cross-section views of the oxidized Cz samples exhibit point-like defects in the bulk to a density of approximately 10^{15} cm^{-3}. One of these defects could be observed in the high resolution mode (figure 2): it appears to be plate-like shaped and lying in a {111} plane. A similar shape has been reported by Ponce and Hahn [9] for oxide precipitates. But these were extended by a stacking fault, which is not seen here. On the contrary, a noticeable strainfield around the particle (the dark field in figure 2) due to the precipitate volume expansion is observed showing an unrelaxed stress.

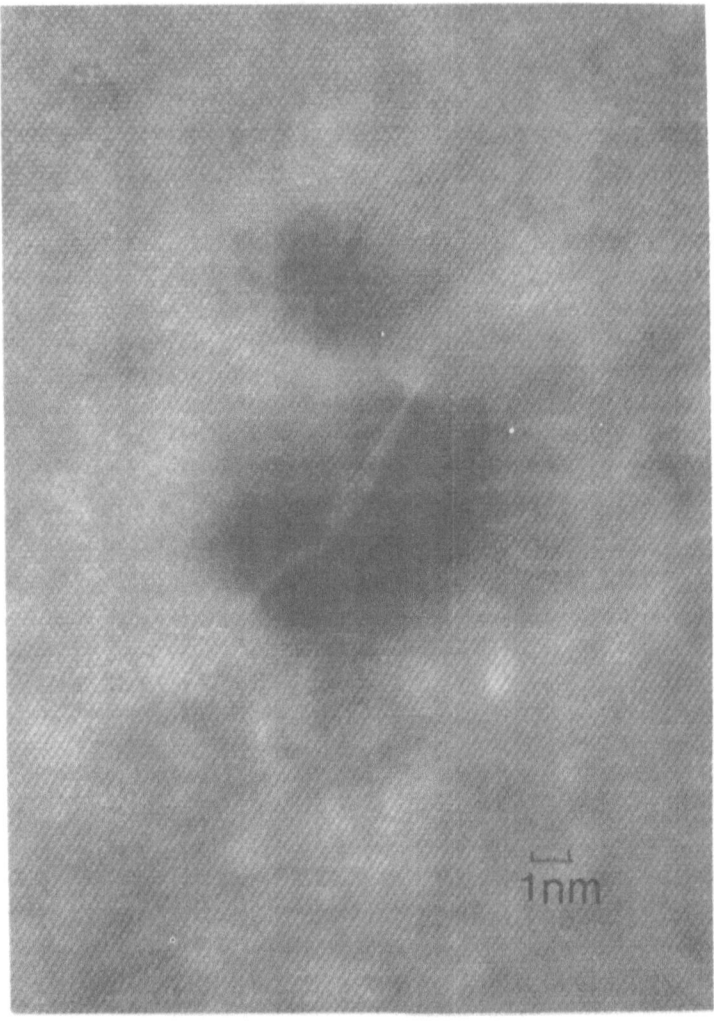

Figure 2. High resolution TEM cross-sectional view, in [110] zone axis, of a precipitate in the oxidized Cz sample, about 2 μm below the oxide-silicon interface

After oxidation, the concentration of oxygen atoms above the solubility is essentially made of the ^{16}O atoms already present before treatments (8×10^{17} cm^{-3}), as the ^{18}O concentration (10^{17} cm^{-3}) can be neglected compared to the ^{16}O one. If we divide this quantity by the density of precipitates mentioned above, we obtain a mean number of 1000 oxygen atoms per precipitate. Supposing that these particles possess the SiO_2 composition and density, this gives them a volume of about 25 nm^3. Reporting this volume for the platelet seen on figure 3, gives this precipitate a dimension of 5nm perpendicular to the plane of the micrograph, i.e. similar to its largest dimension in this plane. In the absence of more precise microanalytical data, this is a good indication that these small defects are essentially due to oxide precipitation.

The ^{18}O SIMS profile recorded in a FZ sample (figure 1 curve 2) shows that in contrast to the Cz case, the oxygen concentration remains lower than the solubility at 1000°C. This confirms the absence of oxygen precipitation in this case, as it has been checked also by TEM observations.

Copper SIMS profiles where performed in regions free of extended defects (Fig. 3). Neglecting the far end of the profiles, where copper concentration is below the SIMS detection limit (8×10^{-15} cm^{-3}), copper doses appear to be, respectively, 5×10^{14} and 6×10^{13} cm^{-2} in the Cz and FZ cases. To average these doses over the sample thickness (500µm) leads to respective concentrations of approximately 10^{15} and 10^{16} cm^{-3} in the two cases analyzed. These values are more than one order of magnitude below the solid solubility of copper at 1000°C (3×10^{17} cm^{-3})[10]. Considering the copper solubility in silicon, 10^{15} cm^{-3} at 540°C and 10^{16} cm^{-3} at 700°C, the precipitation process during cooling began at a mean starting temperature of 630°C. At this temperature, the copper diffusion coefficient ($\sim10^{-5}$ cm^{2}s^{-1}) still allows a very rapid migration of the soluble copper. A calculation of the diffusion length, $L_{cu}=(D\tau)^{1/2}$, of copper atoms during cooling, may be carried out using either a mean coefficient, or the concept of equivalent time of diffusion at the start temperature T_0 [11]. Taking T_0=630°C and a cooling rate of 3 K s^{-1}, the calculation gives L_{Cu} = 350µm. In a classical error-function out diffusion

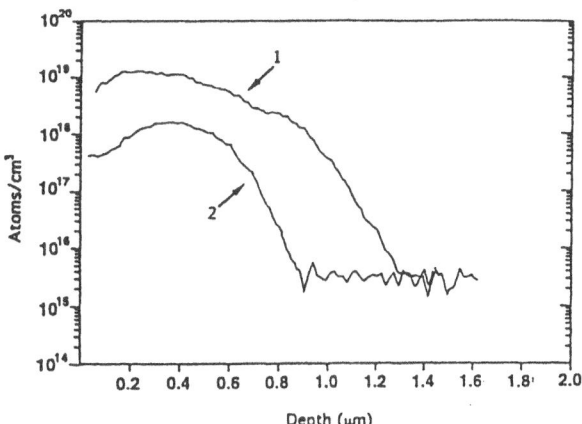

Figure 3. SIMS profile of ^{63}copper concentration: (1) Cz sample; (2) FZ sample.

curve, L_{Cu} corresponds to the depth where one-half of the starting concentration remains. As the samples are 500 µm thick, this qualitatively indicates that a non-negligible part of the copper stays in the bulk of the samples after cooling. If it finds trapping sites, this copper will be at the origin of small clusters, more likely appearing as point-like recombination centers in EBIC than actual precipitates of a new phase, due to the difficulty of relaxing the silicon matrix at low temperature.

The oxide-silicon interface and oxide precipitates are known to getter copper at high temperature[12]. We therefore make the hypothesis that the oxide-based precipitates that we observe in the Cz sample are actually surrounded by copper, situated at the particle-silicon interface and/or in the highly strained region in the nearby matrix. If we suppose that only one type of these centers is responsible for the recombinations, the minority carrier diffusion length $L_n=(D\tau)^{1/2}$ can be expressed in terms of density N_t and minority carrier cross section σ_n of this deep trap:

$$L_n = (D/N_t \ \sigma_n \ v_n)^{1/2},$$

where v_n is the electron thermal velocity, and one has employed, as low-injection formalism implies in the particular experimental conditions used, the simplified relation $\tau=(N_t \ \sigma_n \ v_n)^{-1}$. In this equation, the only unknowns are N_t and σ_n. In the case of the Cz samples, if one supposes that the precipitates counted in TEM (i.e. $N_t=10^{15}$ cm^{-3}), are at the origin of the decreased value of L_n, the only unknown left is the capture cross section σ_n. Taking $D=30$ cm^2 s^{-1} and $v_n=10^7$ cm s^{-1}, σ_n comes out to be $\sim 10^{-15}$ cm^2. This value must be compared to the largest of those found by Brotherton *et al.* by DLTS [13] (i.e. 3×10^{-15} cm^2).

In the FZ sample, copper probably aggregates without involving oxygen, i.e. differently from the precipitate shown on figure 2. Figure 4 shows the small copper precipitates which are detected by TEM in FZ silicon and analyzed as Cu_3Si particles (density $\sim 10^{14}$ cm^{-3}). The density of copper precipitates is one order of magnitude higher in the Cz than in the FZ silicon. In the frame work of oxygen based nuclei in the Cz silicon, this effect can be understood in terms of more efficient nucleation on these than on point-like defects existing in the FZ. However, it is not sure that these precipitates are similar to those already well known in FZ material through their DLTS signature [13].

Acknowledgments

This work is part of a research project on silicon for high efficiency solar cells, funded by the EEC JOULE programme

4. References

1. Fiermans, L. and Vennick, J. (1965) *Phys. Status Solidi* **12**, 277.
2. Shimura, F. (1991) *Solid State Phenomena* **19-20**, 1.
3. Pichaud, B, Minari F. and Martinuzzi, S., (1991) *J. Phys. (France) IV*, **1-C6**, 187.
4. Maurice, J.-L. (1993) *Phil. Mag. A* **68**, 951.
5. Ho, C. T. and Wald, F.V. (1981) *Phys. Status Solidi A* **67**, 103.
6. De Mierry, P., Ballutaud, D., Aucouturier, M. and Etcheberry, A. (1990) *J. Electrochem. Soc.* **137**, 2966.
7. Revel, G., Deschamps,N., Dardenne, C. and Pastol, J.-L. (1984) *Nucl.Chem. Lett.* **85**, 137.
8. Wu, C. J. and Wittry, D. B. (1978) *J. Appl.Phys.* **49** 2827.
9. Ponce, F. A. and Hahn, S. (1989) *Mater. Sci. Eng. B* **4**, 11.
10. Weber, E. R. (1983) *Appl. Phys. A* **30**, 1.
11. Philibert, J. (1985) *Diffusion et transport de matière dans les solides*, Les Editions de Physique, Paris, p. 337.
12. Falster, R.J., Fisher, G.R. and Ferrero, G. (1991) *Appl. Phys. Lett.* **59**, 809.
13. Brotherton, D., Ayres, J. R. and Gill, A. (1987) *J. Appl. Phys.* **62**, 1826.

Figure 4. High resolution TEM plan-view, in [100] zone axis, of a precipitate in the oxidized FZ sample.

COMPUTER SIMULATED DISTRIBUTION OF DEFECTS FORMED DURING CZ-SI CRYSTAL GROWTH

K. KAWAKAMI, H. HARADA AND T. IWASAKI
Advanced Technology Research Labs., Nippon Steel Co., 3434 Shimata, Hikari 743, JAPAN

AND

R. HABU
SKY Aluminium Co.,Ltd, 1351 Uenodai, Fukaya 366, JAPAN

1. Introduction

Oxygen precipitation in Cz-silicon single crystals depends not only on the initial oxygen concentration and annealing conditions but also on the crystal growth conditions. The R-OSF (Ring-likely distributed Oxidation induced Stacking Fault) reported by M. Hasebe[1] is observed after oxidation treatment only in the crystals grown at limited pull rates. Fig. 1 is an X-ray topograph showing the radius increase of the R-OSF with pull speed. In the region where the pull rate is lower than a critical rate (v_{crit}), the R-OSF shrinks out and cannot be seen (E-zone). v_{crit} differs with different thermal growing conditions, but W. von Ammon showed recently that v_{crit} is proportional to the axial temperature gradient (G) in crystals at the melt/solid interface[2]. The oxygen precipitation in the R-OSF region after two-step annealing (800°C, 4 hours and 1000°C, 16 hours) is much lower than that in its neighbourhood. The E-zone also shows low oxygen precipitation in the two-step annealing. Conversely, it is known that precipitation in the R-OSF region after a single anneal at 1000°C is higher than in other regions[1].

AOP (Anomalous Oxygen Precipitation) is also reported in the literature[3] where it is found in crystals detached abruptly during the growth.

T. Iwasaki[4] reported oxygen precipitation behaviour in crystals grown with a pull rate change (initially 1.0 mm/min, then abruptly changed to 0.2 mm/min, and finally returned to 1.0 mm/min). Fig. 2 shows the oxy-

R. Jones (ed.), Early Stages of Oxygen Precipitation in Silicon, 371–379.
© *1996 Kluwer Academic Publishers.*

gen precipitation distribution in one of these crystals. The B-zone, which exhibits the highest oxygen precipitation, is almost the same shape and at the same location as the top of the AOP region seen in the previously described detached-crystals. The C-zone contains much less oxygen precipitation compared to the neighbouring zones. The E-zone is another low oxygen precipitation region and is bounded by upper and lower R-OSF regions. The positions of the boundaries (R-OSF regions) are symmetrically located with respect to the slower pull-rate region.

When the pull rate is varied and the radius of the R-OSF is changed, small regions with enhanced oxygen precipitation are seen near the boundaries of the E-zone. We call these enhanced precipitation regions "Dimples" as indicated in Figs. 1 and 2.

Many nonuniform oxygen precipitation phenomena are observed in Cz-silicon crystals, as described above. Although they have been studied by many people, their formation mechanisms are still not well understood. A-defects and D-defects were observed in Fz-silicon and were referred to as interstitial and vacancy-type defects respectively[5]. The relationship between the defect distribution and the growth rate for the Cz-silicon case is quite similar to that for Fz-silicon. In other words, in Fz-crystals, A and D defects are formed when the growth rate is low and high respectively. Therefore we considered that the origin of the E-zone formation in Cz-crystals is same as that of the A-defect region in Fz-crystals. Then we assumed that the E-zone is self-interstitial (I) rich and the other zones are vacancy (V) rich, and numerically simulated the concentration changes of V and I during crystal growth to investigate the formation mechanisms of these distributed defects.

2. Calculation method

The equation for the time-dependent change in I-concentration (N_I) in a growing crystal is given by Habu[6] as the following:

$$\frac{dN_I}{dt} = div(D_I \, grad(N_I)) \quad - \quad div\left[D_I(e_{fI} - Q_I)\frac{grad(T)}{kT^2}N_I\right]$$

$$- \quad V_g\frac{\partial N_I}{\partial z} - F_{pair} - F_{reac,I}, \tag{1}$$

where D_I is the diffusion coefficient, e_{fI} denotes the formation enthalpy, Q_I represents the effect of heat flow upon the diffusion flow, which is negligibly small compared to e_{fI}, v_g is the pull rate, F_{pair} represents the pair annihilation rate and $F_{reac,I}$ represents a total effect of other reactions. The equation for V can be obtained by replacing I with V in Eq. 1. We ignored $F_{reac,I}$ and $F_{reac,V}$ in this simulation, because these terms are not important at a high temperature; the macro scale distribution of secondary defects would be determined at a higher temperature than the temperature at which these reactions become effective. We simulated distributions of N_I and N_V in a cylindrical crystal 450 mm in height with a 130mm diameter. The boundary conditions, the treatment of temperature fields and F_{pair} are the same as described by Habu[6], but physical constants for point defects are different. In this calculation the following are used: the thermal equilibrium concentration of V: $N_{Veq} = V_0 \, exp(-4.3eV/kT)$, the thermal equilibrium concentration of I: $N_{Ieq} = 0.0057 \, V_0 \, exp(-3.6eV/kT)$, and $D_V = D_I = 0.55 \, exp(-1.0eV/kT)$ cm^2/sec, where V_0 is an arbitrary constant.

3. Simulation results

We investigated the influence of growth conditions on the distributions of N_I and N_V.

3.1. INFLUENCE OF PULL RATE (V_G)

Both where v_g is constant and where it changes, the difference between N_V and N_I ($DVI = N_V - N_I$) is smallest and first becomes negative in edge regions where v_g is being reduced, as shown in Figs. 3 and 4 (a). This behaviour is in agreement with the experimental results based on the assumption that the I-rich region is the E-zone. The range of v_g where both I-rich and V-rich regions coexist in the same height is between 0.73 mm/min and 0.53 mm/min in the case where v_g is gradually changed. The range is wider than that in the case where v_g is constant. Chemical potentials (μ_I, μ_V) of V and I in the crystal whose pull rate is gradually reduced are shown in Fig. 4 (b). The contour of $\mu_V = 12000K$ in the I-rich region near the boundary resembles the shape of a "Dimple", and the contour of $\mu_I = 8000K$ in the V-rich region is located at almost the same place as the R-OSF region.

3.2. INFLUENCE OF AXIAL TEMPERATURE GRADIENT (G)

We simulated N_V and N_I in crystals grown in three different temperature patterns. The temperature gradient (G) above 1320K was varied (low=4.24K/mm, medium=6.96K/mm, high=9.67K/mm) while kept constant at lower temperatures. Fig. 5 shows the relationship between DVI and v_g/G at the centre of the top position. The DVI's are almost the same in the high and medium G cases as those in the experiment, but DVI in the low G case is lower because the out diffusion length is too long. The x-axis intercept, where DVI=0, corresponds to v_{crit}/G, and the mean value of v_g/G in high and medium G cases is about two thirds of that obtained from the experiment [3].

3.3. INFLUENCE OF ABRUPT CHANGE OF PULL RATE

We simulated point-defect distributions in crystals whose pull rates are changed abruptly to from 1.0 mm/min to 0.1 mm/min, and returned to 1.0 mm/min after 150 min. Fig. 6 (a) shows the contours of DVI after the pull rates are returned to 1.0 mm/min. The I-rich region is formed symmetrically to and wider than the region where the pull rate is lowered, which is in good agreement with the experimental results, although the shapes are different. It can bee seen in Fig. 6 (b) that the contour of μ_I = 12000K resembles a "Dimple", and that the contour of μ_V = 8000K resembles the R-OSF region. To investigate the effects of the low pull rate, we calculated the defect distributions during growth stopping. Contours of μ_V and μ_I before stopping and after stopping for 60 min are shown in Fig. 6. All contours of μ_I become lower after stopping, but the contours of μ_V higher than 14000K are not changed. This is due to the concentration of the minority defect, I, being too small to annihilate with the high concentration of the majority defect, V, in the region where μ_V is over 14000K. Therefore μ_V is kept high in this region. The shape of the μ_V = 14000K contour resembles that of the B-zone when compared with those of lower μ_V. The relative position of the region where μ_I > 8000K to the position where μ_V = 14000K, which is assumed to be the boundary of the B-zone, is in accordance with the R-OSF region. The difference of the shapes of the contours is caused by the amount of I-V pair annihilation.

4. Discussion

Here we set the following three assumptions. Firstly, no reactions among V, I, and oxygen occurs near the melting point. Secondly, the radius of crystal is larger than the diffusion length of the point-defects at high temperature. Thirdly, the pull rate is constant and $dT/dz(=G)$ is also constant near the melt/solid interface. Then Eq. 1 is reduced to the following one-dimensional equation applicable to both V and I;

$$0 = \frac{d}{dT}\left[D\left(\frac{dN}{dT} - \frac{D(e_f - Q)}{kT^2}N\right)\right] - \left\{\frac{v_g}{(dT/dz)}\right\}\frac{dN}{dT} \qquad (2)$$

N equals the thermal equilibrium concentration at the melt/solid interface and $dN/dT = 0$ at the point where the diffusion coefficient is sufficiently low. Then Eq. 2 depends only upon one parameter, v_g/G. This indicates that the quantity of point-defects introduced from melt is determined only by v_g/G and that the critical pull rate (v_{crit}) is proportional to the temperature gradient (G) at the interface. This explanation is unchanged in more general cases where the boundary conditions at the melt/solid interface are functions of v_g/G. The disappearance of the R-OSF at the edge of the crystal ingots will be also determined by the parameter v_g/G. Usually, commercial silicon crystals are pulled at high speeds so as not to have the R-OSF. In particular, the top position of the crystals has to be pulled at higher speeds than other positions are pulled since this region does not have the thermal resistance of the upper crystal part, so that it possesses a larger temperature gradient and cools faster than other positions.

The dependence on pull rates of the boundaries between V-rich and I-rich regions is explained as follows. N_V is higher than N_I at the melt/solid interface. This causes the crystals pulled at high speeds to have only V-rich regions. The larger the formation enthalpy or the diffusion coefficient is, the more uphill diffusion is effective. The enthalpy of V is larger than that of I and the diffusion coefficients are the same in these simulations. Therefore, N_V is more effectively reduced at lower pull rates than N_I. Out diffusion also reduces these concentrations especially near edges. The difference between N_{Ieq} and N_{Veq} implies different boundary conditions for out diffusion. Therefore, the I-rich region appears at higher pull rates near the edge than near centre. The other combination of these physical constants can also explain these phenomena.

The simulations show that the pull rate change influences the far position from the melt/solid interface as shown in Fig. 6 (a). Also, they show that the pull rate range where the I-rich and the V-rich regions coexist in the same height is different between in a constant pull rate condition and in a changing pull rate condition. These results are also explained by the effect of macro scale diffusions at high temperature.

After sufficient pair annihilation occurs, the minor point-defects have higher concentrations at boundaries between V-rich and I-rich regions or near crystal edges than in other positions. This is because pair annihilation reduces the product $N_I N_V$ to $N_{Ieq} N_{Veq}$, keeping DVI constant. This statement can be also made for the other reaction where both V and I are concerned and almost the same numbers of V and I are annihilated.

Considering the correspondence between simulation and experimental results with regard to various oxygen precipitation phenomena, we propose models of defect distribution and the formation mechanisms in the crystals grown in two conditions of pull rates ($v_g \gg v_{crit}$ and $v_g \approx v_{crit}$), as schematically shown in Fig. 8. The defect distribution and its formation mechanism are explained as follows.

Firstly, we will discuss each defect zone in the crystal shown in Fig. 2. It should be noted that the crystal is kept in the pulling furnace for a longer time than usual because of the very slow growth rate and that the defect distribution may be intensified. Here we consider the nuclei for oxygen precipitates will be voids formed by V agglomeration. In the E-zone, I and V are reduced mainly by uphill diffusion and I becomes dominant due to the larger enthalpy of V. The lower concentration of V reduces the number of nucleation centres. In the A-zone, the amount of pair-annihilation is a little larger. In the B-zone, many medium sized voids are nucleated and cause oxygen precipitation enhancement. In the C-zone, voids and oxygen precipitates grow larger, absorbing more V and oxygen than usual. Therefore N_V is reduced and the precipitation at lower temperatures is suppressed.

Secondly, we consider the defect distribution in V dominated crystal regions grown at the pull rate greater than v_{crit}. Near the crystal edge or the boundary between V- and I-rich regions, the number density of voids is large and the size is small. The voids absorb V, I, and oxygen, and form oxygen precipitates. The remaining concentration after pair annihilation or absorption to the voids is larger than that far from the edge or the boundary. The remaining I form stacking faults around the preformed oxygen precipitates. These are the R-OSF seen in the V-rich region near the edge or the boundary. The remaining V are effectively absorbed into the stacking faults, and N_V is reduced further. In the case where the pull rate is much larger than v_{crit}, DVI becomes larger, and hence N_I is too small to form stacking faults.

Thirdly, the defect formation in I-rich crystal regions grown at a pull rate a little larger than and nearly equal to v_{crit} is examined. Interstitials form a greater number of small dislocations at lower temperatures near the crystal edge, or at the boundary of I- and V-rich regions, rather than in other I-rich regions, and N_I will be reduced nearly to zero. The concentration of V becomes higher there than the other areas, and V form voids or oxygen precipitation nuclei at low temperatures. These form the Dimple. Far from the edge or the boundary, since the concentration of V in these regions is very low, the Dimple is not formed.

Fourthly, just on the boundary between I-rich and V-rich regions, N_I and N_V are comparable and small. The temperature at which voids are formed becomes lower, and these voids will vanish by faster absorption of I than that of oxygen because of the difference in diffusibility. Therefore no nuclei are formed.

Finally, we consider AOP formation. The AOP arises in the region under the B-zone, where I and V not yet agglomerated. During rapid cooling caused by the abrupt detaching, I and V are annihilated and the remaining V forms voids in the same way as in the case of usual growth. The difference between the detached crystal and one cooled in the normal way is the degree of supersaturation of V. Rapid cooling produces high supersaturation and a

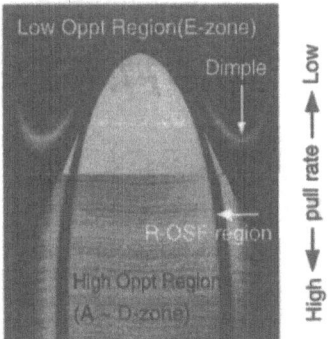

Fig. 1. X-ray topograph of a wafer vertically sliced from a crystal whose pull speed is increased gradually.

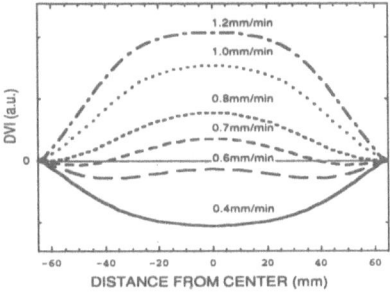

Fig. 3. Radius distribution of DVI at 400mm in height at the crystals grown at different pull speeds

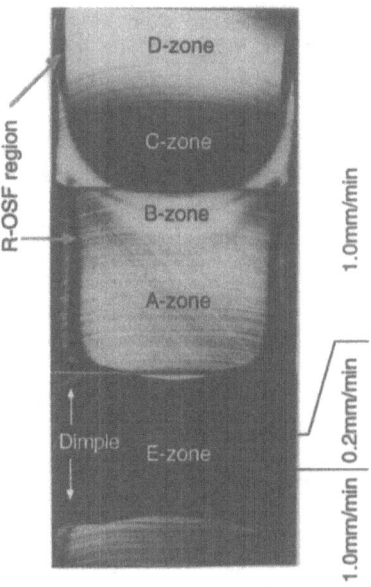

Fig. 2. X-ray topograph of a wafer vertically sliced from a crystal whose pull speed is first 1.0mm/min ,changed to 0.2mm/min for 100min and returned to 1.0mm/min.

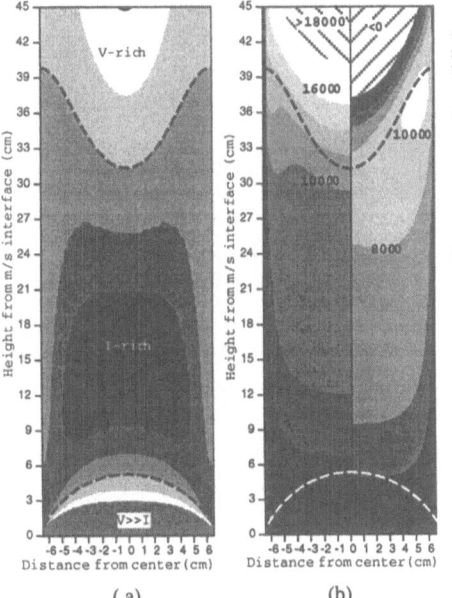

Fig. 4. Contours of DVI (a), μ_v (left half of (b)) and μ_i (right half of (b)) in a crystal whose pull rate is reduced gradually to 0.1mm/min from 1.0mm/min for 600min.

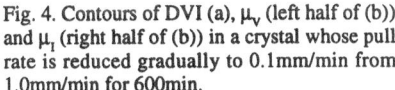

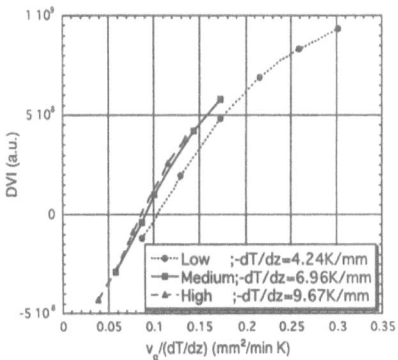

Fig. 5. Relation between DVI at the center of the topmost position and Vg/G

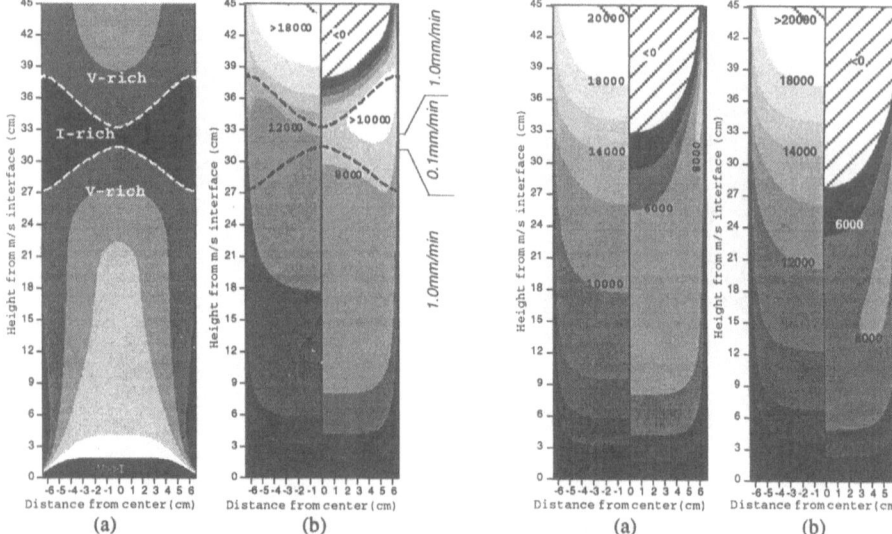

Fig. 6. Contours of DVI (a),μ_v(left half of (b)) ,and μ_I (right half of (b)) in the crystal whose pull rate is changed abruptly as shown on right side of the figure.

Fig. 7. Contours of μ_v(left half) and μ_I (right half) in each crystal; (a) the pull rate is 1.0mm/min, (b) the pull rate is zero for 60min after 1.0mm/min.

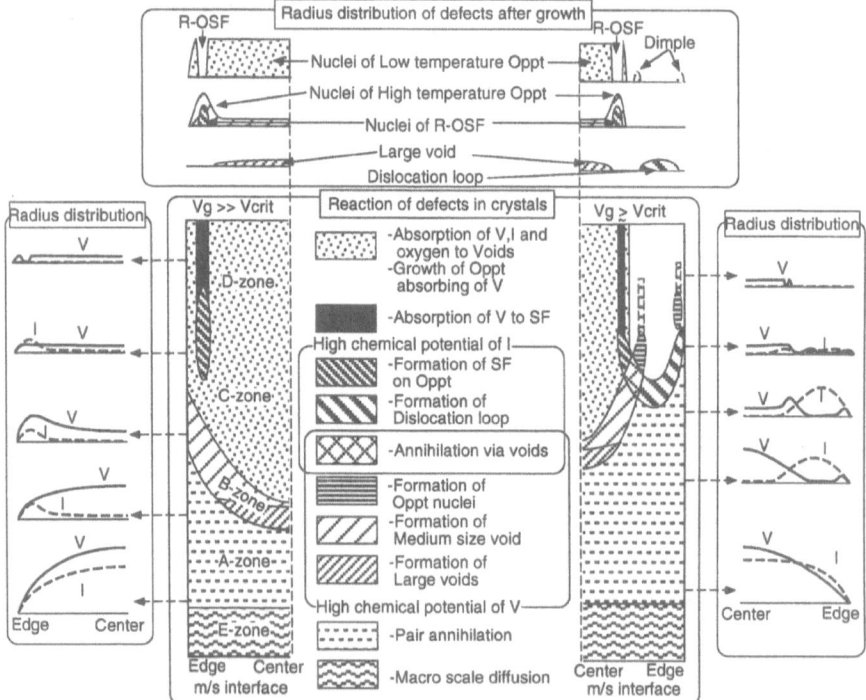

Fig. 8. Schematic diagrams of defect formation mechnism in Cz-Si.
Left: Defect distribution in V dominant crystals grown at the pull rate $v_g \gg v_{crit}$.
Right: Defect distribution in V and I coexisted crystals grown at pull rate $v_g > v_{crit}$.

high number density of voids whose size is small. In particular, the number density is very high near the boundaries of low-temperature void formation. These are nuclei for the anomalous oxygen precipitation. The V-rich region near the melt/solid interface in Fig. 4 (a) corresponds to the AOP region in the case that the pull rate before detaching is sufficiently low.

The combination used in these calculations provides not quantitative, but qualitative agreement of simulation with experiment. There is some disagreement in that the I-rich regions appear near the edge at high temperatures as shown in Fig. 6, and with the changing v_{crit}/G as shown in Fig. 5. These differences with experimental results could be removed by using more appropriate combination of parameters and further consideration for the figure of the melt/solid interface and the radius distribution of temperature.

5. Conclusions

We simulated the distribution of I and V formed during Cz-silicon crystal growth based on equations considering diffusion and pair annihilation. The results of these calculations agree qualitatively with the experimental results very well. We proposed the formation mechanisms of the R-OSF, the Dimple, the AOP and the B-zone. The R-OSF and Dimples are formed near the boundary of I- and V-rich regions by agglomeration of highly supersaturated minor point defects to the nuclei formed from majorities and oxygen atoms. The B-zone is formed in V-rich region and caused by high supersaturation of V remaining after pair annihilation. The AOP is formed under the B-zone and is caused by V before reacting in the B-zone.

References

1. Hasebe M.,Takeoka Y., Shinoyama S. and Naito S. (1990) Ring-likely distributed stacking faults in CZ-Si Wafers, Defect Control in Semiconductors, Elsevier, North-Holland, 157-161.
2. Abe T., Takeno H. (1992) Dynamic behavior of intrinsic point defects in FZ and CZ silicon srystals, Mat. Res. Soc. Sym. Proc. vol 262, Mat. Res. Soc., Pittsburgh, 3-14 ·
3. von Ammon W., Dornbrtger E., Oelkrug H. and Weidner H. (1995) The dependence of bulk defects on the axial temperature gradient of silicon crystals during Czochralski growth, J. Crystal Growth 151, 273-277
4. Iwasaki T.,Tsumori Y.,Nakai K.,Haga H.,Kojima K. and Nakashizu T.(1994) Influence of cooling condition during crystal growth of CZ-Si on oxide breakdown property, Semiconductor Silicon 1994, Electrochem. Soc., Pennington, 744-755
5. Abe T., Kimura M. (1990) Behavior of point defects in Fz silicon crystals, Semiconductor Silicon 1990, Electrochem. Soc., Pennington, 105-116
6. Habu R., Tomiura A., Harada H.,(1994) Distribution of grown-in crystal defects formed by point defect diffusion, Semiconductor Silicon 1994, Electrochem. Soc., Pennington, 635-646

A SMALL ANGLE NEUTRON SCATTERING STUDY OF OXYGEN PRECIPITATION IN SILICON

R.J.STEWART, S.MESSOLORAS AND S.RYCROFT
J.J.Thomson Physical Laboratory, University of Reading,
Whiteknights, Reading RG6 6AF, England

1. Introduction

The importance of small angle neutron scattering (SANS) in the study of inhomogeneities in materials [1,2], such as SiO_2 precipitates in silicon, is that the technique provides structural information on a scale ranging from a few Ångstroms up to inhomogeneities with dimensions of the order of 5000 Å and in addition such studies can be made on bulk crystals. As a result of the scattering of neutrons by the inhomogeneities in a crystal a diffraction pattern of the average inhomogeneity structure is determined. This diffraction pattern cannot be unequivocally translated into real space, but this is balanced by the ability to investigate the inhomogeneity structure throughout a large volume of a crystal (up to more than 10 cm^3 for silicon) in a relatively short time (from a few minutes to about 1h on a high flux neutron source for silicon studies, the time depending on inhomogeneity size and number density).

Neutrons are uncharged and suffer little absorption in silicon and thus can easily penetrate deep into a crystal. The principal interaction with atoms is via the short range, strong nuclear interactions which are independent of the atomic number of the nucleus and vary in an apparently random way from nucleus to nucleus. The short range nature of the interaction (about 10^{-15} m) compared to the wavelength of the neutrons used in this study (about 10^{-9} m) means that the scattering from a single nucleus is spherically symmetric, which simplifies the analysis of the diffraction data. Agglomerations of impurities (eg SiO_2 precipitates) can thus be considered as particles made up of regularly spaced isotropic scattering centres for neutrons. The amplitude of the wave scattered by a single nucleus (the scattering centre) is denoted by b and is called the scattering length. Consider one inhomogeneity of this type bathed in a neutron beam of wavelength λ, all the scattering centres are then sources of scattered waves. When the scattering direction is the same as that of the incident beam, these scattered waves are all in phase (ie independent of the mutual disposition of the scattering centres). However, as the scattering angle increases, the difference in phase between the various scattered waves also increases. The amplitude of the resultant wave

R. Jones (ed.), Early Stages of Oxygen Precipitation in Silicon, 381–388.
© *1996 Kluwer Academic Publishers.*

thus decreases with increasing angle because of the increasing destructive interference, it becomes zero when there are as many waves with phases between 0 and π as there are between π and 2π. This will occur when the scattering angle 2θ is of the order of λ/L, L being the average dimension of the particle. For n such particles, neglecting interparticle interference effects, the scattering would be n times that for a single particle. Thus the scattering from agglomerates of impurity atoms in an otherwise perfect lattice varies with the 2θ (or the magnitude of the scattering vector $|Q| = 4\pi Sin\theta/\lambda$). From the experimentally determined Q dependence of the scattering it is possible to determine the size and shape of the inhomogeneity present. For scattering events the probability of a neutron being scattered into unit solid angle is referred to as the differential scattering cross-section ($d\sigma/d\Omega$), which has units of barns/steradian. From the magnitude of this cross-section the number of inhomogeneities and their composition can be determined [1].

2. Experimental

The D11 SANS instrument [3] at the Institut Laue-Langevin in Grenoble, France was used for the studies in this paper. The long wavelength neutrons from the cold source, are monochromated by a helical slot velocity selector ($\delta\lambda/\lambda = 8\%$). These are then passed down a guide tube, with movable guides allowing collimation lengths of 2, 5, 10, 20 and 40 m to be selected. The beam then strikes the sample and is then detected with a position sensitive BF_3 detector, with a 64×64 grid of 10 mm^2 elements. The sample to detector distance can be varied continuously from 1.1 to 35 metres giving access to a Q range from $5x10^{-3}$ to 4 nm^{-1}. The entire instrument including the sample area was under vacuum to avoid scattering from window materials and the air, which would otherwise add to the background. In order to avoid Bragg scattering from the silicon matrix, which might mask the SANS from the inhomogeneities, a neutron wavelength longer than twice the maximum interplanar spacing in silicon ($2d_{max} = 6.26$Å) was employed, typically a wavelength of 9Å. The Czochralski-grown single crystals used have a [100] growth direction and the neutron beam was incident along this direction for most of the SANS measurements. All the measured SANS data were corrected for the background, sample holder scattering and transmission. The corrected counts were then converted into absolute scattering cross sections by using the incoherent scattering from a Vanadium single crystal which is used as a standard [1].

3. Theoretical Background of Small Angle Neutron Scattering

The observed SANS from a precipitate is the Fourier transform of the geometrical shape of the precipitate and it is given by [1]

$$\frac{1}{NV}\left[\frac{d\sigma}{d\Omega}\right] = \frac{n_p}{N}\left[(<c_p> - <c_m>)\int_{V_p}e^{iQ\cdot r}d^3r\right]^2 \qquad (1)$$

where n_p is the number of precipitates per unit volume, N is the number of atoms per unit volume, V is the volume of crystal bathed by the neutron beam, V_p is the volume of one precipitate, $<c_p>$ and $<c_m>$ are the mean scattering length densities of the precipitate and matrix respectively and r is a radial vector. Under the Guinier Approximation the small angle scattering cross section can be approximated by the expression [1]

$$\frac{1}{NV}\left[\frac{d\sigma}{d\Omega}\right] = \left[\frac{n_p V_p^2}{N}(<c_p> - <c_m>)^2\right] \exp-(Q^2 R_g^2/3) \qquad (2)$$

the term in square brackets is called the *Guinier constant* (C_g) and R_g is the *Guinier Radius* or radius of Gyration of the inhomogeneity. For spherical objects R_g is related to the sphere radius R_s by the expression $R_s = \sqrt{(5/3)}\, R_g$. Using the above equation, the linear region of a plot of $\log(d\sigma/d\Omega)$ versus Q^2 will yield R_g from the slope and C_g from the intercept. If we know $\langle c_p \rangle - \langle c_m \rangle$ then from the experimentally determined value of C_g we can determine the product $V_p^2.n_p$. Now for a precipitate containing oxygen in a silicon matrix we have [1]

$$<c_p> - <c_m> = (1/v_p)(b_o m_p - b_{Si}(1 - m_p)) - (1/v_m)(b_o m_m - b_{Si}(1 - m_m)) \qquad (3)$$

Now b_O and b_{Si} the scattering lengths of oxygen and silicon are known as is v_m the volume occupied by one atom in a silicon matrix. The precipitates are also known to be a form of SiO_2, thus the concentration of oxygen in a precipitate m_p is 2/3. In addition the total oxygen concentration is only 10^{-5} and thus the concentration of oxygen in the matrix m_m can be taken to be zero. Thus we know all the terms on the right hand side of (3) except v_p, the volume per atom in the precipitate. This will depend on the form of SiO_2, combined IR, SANS and HRTEM measurements have established a value of $14.4 Å^3$ for v_p [9]. Thus assuming this value for v_p a value for $\langle c_p \rangle - \langle c_m \rangle$ can be found and hence from C_g a value for $V_p^2.n_p$. Now if the shape and size of the precipitate can be established from the data V_p will be known and thus n_p can be determined.

4. Data analysis

Two basic types of scattering have been observed from SiO_2 precipitates in silicon, either isotropic scattering (no azimuthal dependence around the incident beam direction on the detector plane) or anisotropic scattering, some typical examples of anisotropic scattering are shown in figure 1. The bright cross in figure 1a represents the scattered intensity observed on the area detector from a heat treated silicon single crystal. In the case of isotropic scattering, the data were radially averaged and a Guinier analysis was used to estimate the radius of gyration and number density of the SiO_2 precipitates. The anisotropic data was analyzed by simulating the scattering patterns for model precipitates

384

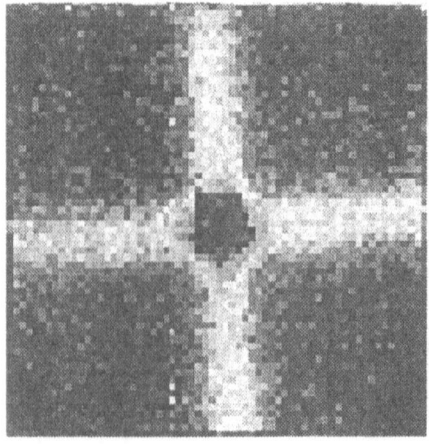

Figure 1a. 216 hrs at 750^0 C.
<100> orientation.

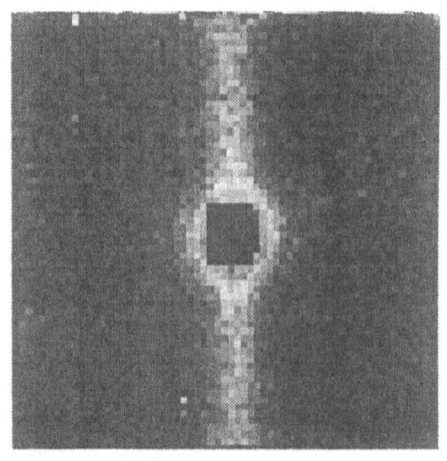

Figure 1b. 216 hrs at 750^0 C.
<110> orientation + 45^0 rotation.

Figure 1c. Simulated scattering from cuboids.
L=475Å width factor w = 0.08.

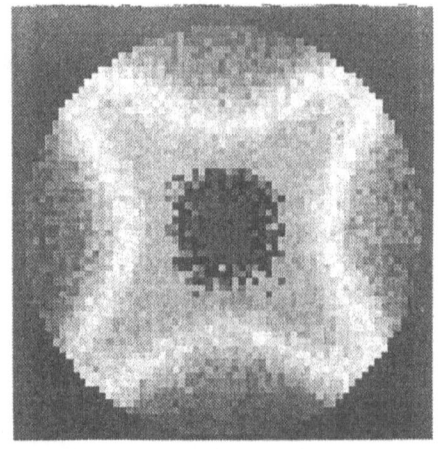

Figure 1d. 12 hrs at 950^0 C.
<100> orientation + 45^0 rotation.

using equation (1) and comparing these with the measured scattering. Model calculations using *cuboids* to simulate the *cushion shaped* precipitates have been found to work well [8,9,10,11]. Sections of the detector patterns along the symmetry axes, and perpendicular to them at an offset of 20 cm from the beam centre were compared to refine the simulations. For a cuboid of dimension (L × L × wL), it can be shown that $L = R_g \sqrt{\{12/(2+w^2)\}}$, where $0 < w < 1$ ($w=1$ would be a cube). Thus the R_g values calculated from the radial averages can be used to further refine the simulations.

5. Results

As grown Czochralski single crystals show almost no SANS except for a temperature dependent contribution of as yet unknown origin which disappears on lowering the measurement temperature to 100°K [4]. Oxygen which typically occurs at a level of 10^{18} atoms/cm^3 exists predominantly in bonded interstitial sites. During subsequent heat treatments the oxygen can diffuse and agglomerate. Structural studies using SANS and electron microscopy have established that oxygen segregates in silicon as SiO_2 [5,6,7,8] in well defined cushion shaped precipitates [9,10]. The orientation of individual precipitates observed by electron microcopy was such that the cushions lay in {100} plane with their edges along the <110> directions [10]. Our SANS studies with the neutron beam incident along different crystallographic directions and using crystal volumes in excess of 1cm^3 confirmed that most if not all such SiO_2 precipitates have this orientation [11], the anisotropic scattering observed for two orientations is shown in figures 1a and 1b. The effect of carbon (12), boron (13) and antimony (14) impurities on the kinetics and form of oxygen precipitation in silicon heated at 750°C has also been studied by SANS. In general boron and carbon when present in high concentrations (greater than 5×10^{17} atoms per cm^3) result in a larger number of smaller SiO_2 precipitates than observed in undoped silicon. In contrast in heavily antimony doped silicon larger and less numerous SiO_2 precipitates are found in comparison with undoped silicon. The SiO_2 precipitate sizes, number densities and oxygen content, resulting from heat treatment at 750°C are listed in tables 1 and 2. The marked differences observed in the shape of the anisotropic scattering for precipitates with different w values (cuboid dimensions LxLxwL) is shown in figures 1a and 1d. The simulated scattering (fig 1c) which was found to be the best fit for the data from a sample heated for 72h at 750°C predicts w=0.08 whereas that for a sample heated for 12h at 950°C (fig 1d) predicts w=0.25. In large diameter silicon crystals the oxygen concentration can often vary across the crystal diameter and since the form of the SiO_2 precipitates is dependent on the oxygen concentration [8] different forms are found across the diameter of some crystals [15].

SANS measurements on SiO_2 precipitates in silicon become progressively more difficult as the size of the precipitates get smaller since the magnitude of the scattering cross-section depends on $V.n_p.(N_p)^2$, where N_p is the average number of atoms in a precipitate. Now $V.n_p$ is the number of precipitates bathed by the incident neutron

386

Table 1. Precipitate Parameters after 750°C treatment

Si Crystal Spec & Oxy Conc cm^{-3}	750°C 24h	750°C 72h	750°C 216h	Parameters of precipitates
9.6e17[O] Undoped	Weak Isot 17 0.7 7.2e13 1.5e17	370×60 140 111 3.4e11 1.3e17	760×76 315 2910 3.0e11 6.2e17	LxwL(Å) R_g(Å) C_g(mb sr^{-1} atom^{-1}) n_p(cm^{-3}) n_pN_p[O] (cm^{-3})
1.2e18[O] Undoped	110x72 44.2 91 2.4e13 9.8e17	130x65 55.2 127 2.1e13 1.1e18	250×75 103.5 419 3.9e12 8.4e17	LxwL(Å) R_g(Å) C_g(mb sr^{-1} atom^{-1}) n_p(cm^{-3}) n_pN_p[O] (cm^{-3})
2.3e18[O] Undoped	Weak Isot 14.2 3.3 9.9e14 1.2e18	Weak Isot 14.6 4.1 1.0e15 1.4e18	Weak Isot 18.93 7.36 4.0e14 1.1e18	LxwL(Å) R_g(Å) C_g(mb sr^{-1} atom^{-1}) n_p(cm^{-3}) n_pN_p[O] (cm^{-3})
1.3e18[Sb] 7.7e17[O]	No SANS observed	No SANS observed	No SANS observed	LxwL(Å) R_g(Å) C_g(mb sr^{-1} atom^{-1}) n_p(cm^{-3}) n_pN_p[O] (cm^{-3})
1.3e18[Sb] 1.1e18[O]	No SANS observed	Isotropic 25.9 0.39 3.2e12 2.3e16	650x60 810 2.7e11 3.1e17	LxwL(Å) R_g(Å) C_g(mb sr^{-1} atom^{-1}) n_p(cm^{-3}) n_pN_p[O] (cm^{-3})
2.6e19[B] 9.1e17[O]	Weak Isot 17.7 25 2.0e15 4.7e18*	Weak Isot 16.9 23 2.5e15 4.9e18*	Weak Isot 17.1 24 2.4e15 5.0e18*	LxwL(Å) R_g(Å) C_g(mb sr^{-1} atom^{-1}) n_p(cm^{-3}) n_pN_p[O] (cm^{-3})

* This value is higher than the oxygen concentration initially present, however if the precipitates involve B_2O_3 and SiO_2 as postulated in [13] then the amount of oxygen in precipitates is reduced to an acceptable value. n_pN_p[O] is the concentration of oxygen atoms in SiO_2 precipitates. The oxygen concentrations in the as grown crystals were measured using IR absorption measurements using the calibration from [16].

beam and since the solubility of oxygen in silicon is only about 50 ppm at the melting point $V.n_p.N_p$ is very small. If large samples are used V can be increased but at most by a factor of 10, however in the limit of all the available oxygen being involved in SiO_2 precipitates of a given size the cross-section will fall proportional to N_p as the precipitate size reduces (i.e. as R_s^3 for spherical precipitates), thus if the precipitate average size reduces by a factor of 10 the cross-section would reduce in magnitude by a factor of 1000.   For a given oxygen concentration our results show that as the temperature is reduced a larger number of smaller SiO_2 precipitates are formed. It thus becomes progressively more difficult to resolve the SANS as the temperature of treatment is reduced. We have observed well resolved SANS in samples heated for 300h at 600°C [17]. However a heat treatment of 300h at 550°C shows only a very tiny change in the SANS compared to an as grown crystal (Table 3). This scattering is at

Table 2. Precipitate parameters for carbon doped silicon

Si Crystal Spec & Oxy Conc cm^{-3}	750°C 0.5,1&2h	750°C 4h	750°C 8h	Parameters of precipitates
9.5e17[O] 5.1e17[C]	No SANS observed	Isotropic 17.7 0.6 4.9e13 1.1e17	Isotropic 21.6 4.7 1.1e14 4.8e17	LxwL(Å) R_g(Å) C_g(mb sr^{-1} atom^{-1}) n_p(cm^{-3}) n_pN_p[O] (cm^{-3})
	750°C 12h	750°C 24h	750°C 72h	
9.5e17[O] 5.1e17[C]	Isotropic 43.1 9.1e13 1.3e18 1.2e18	Isotropic 34.6 70.6 1.0e14 1.8e18	Isotropic 34.4 73.7 1.1e14 1.9e18	LxwL(Å) R_g(Å) C_g(mb sr^{-1} atom^{-1}) n_p(cm^{-3}) n_pN_p[O] (cm^{-3})

The oxygen concentrations in the precipitates are too high for SiO_2 precipitates alone. Precipitates containing C as well as SiO_2 are postulated [12] and then the amount of oxygen in precipitates reduces to an acceptable level.

Table 3. Precipitate parameters after 600°C and 550°C treatments

Si Crystal Spec & Oxy Conc cm^{-3}	600°C 100h Nitrogen	600°C 300h Nitrogen	Parameters of precipitates
1e18[O] Undoped	No SANS observed	Isotropic 10.6 1.1 1.9e15 9.6e17	LxwL(Å) R_g(Å) C_g(mb sr^{-1} atom^{-1}) n_p(cm^{-3}) n_pN_p[O] (cm^{-3})
	550°C 100h Nitrogen	550°C 300h Nitrogen	
8.8e17[O] Undoped	No SANS observed	Very Weak 10 > R_g > 3 .03	R_g(Å) C_g(mb sr^{-1} atom^{-1}) n_p(cm^{-3}) n_pN_p[O] (cm^{-3})

the sensitivity limit of the D11 instrument and the data obtained so far do not have sufficient statistical accuracy to perform a full Guinier analysis. However if the observed scattering is due to SiO_2 precipitates, they must have a radius of gyration which is less than 10Å but greater than 3Å [18].

6. References

1. Stewart R.J. (1986) Neutron scattering from defects in materials, in A.V. Chadwick and M. Terenzi (eds.), *Defects in solids - Modern techniques*, Plenum Press, New York 95-130
2. Stewart R.J. (1988) Small angle neutron scattering from materials, in R.J. Newport, B.D. Rainford and R. Cywinski (eds.), *Neutron Scattering at a Pulsed Source*, Adam Hilger, Bristol 259-271
3. Ibel K. (1976) The neutron small angle camera D11 at the high flux reactor, Grenoble, *J. Appl. Crystallogr* **9** 296-309
4. Gupta S., Messoloras S., Schneider J.R. and Stewart R.J. (1992) Temperature dependent low-Q scattering from silicon single crystals, *J. Phys. Condens. Matter* **4** 1-8
5. Wada K. and Inoue N. (1980) Diffusion limited growth of oxide precipitates in Czochralski silicon, *J. Cryst. Growth* **49** 749-752
6. Bourret A., Thibauld-Desseaux J. and Seidman D.N. (1984) Early stages of oxygen segregation and precipitation in silicon, *J. Appl. Phys.* **55** 825-836
7. Livingston F.M., Messoloras S., Newman R.C., Pike B.C., Stewart R.J., Binns M.J., Brown W.P. and Wilkes J.G. (1984) An infrared and neutron scattering analysis of the precipitation of oxygen in dislocation-free silicon, *J. Phys. C: Solid St. Phys.* **17** 6253-6276
8. Messoloras S., Schneider J.R., Stewart R.J. and Zulehner W. (1988) Amorphous oxide precipitates in silicon single crystals, *Nature* **336** 364-365
9. Bergholz W., Binns M.J., Booker G.R., Hutchison J.C., Kinder S.H., Messoloras S., Newman R.C., Stewart R.J. and Wilkes J.G. (1989) A study of oxygen precipitation in silicon using high resolution transmission electron microscopy, small-angle neutron scattering and infrared absorption, *Phil. Mag.* **59** 499-522
10. Messoloras S., Schneider J.R., Stewart R.J. and Zulehner W. (1989) Anisotropic small-angle neutron scattering from oxide precipitates in silicon single crystals, *Semicond. Sci. Technol.* **4** 340-344
11. Gupta S., Messoloras S., Schneider J.R., Stewart R.J. and Zulehner W. (1990) Orientation of oxygen precipitates in silicon, *Semicond. Sci. Technol.* **5** 783-784
12. Gupta S., Messoloras S., Schneider J.R., Stewart R.J. and Zulehner W. (1992) Oxygen precipitation in carbon doped silicon, *Semicond. Sci. Technol.* **7** 6-11
13. Gupta S., Messoloras S., Schneider J.R., Stewart R.J. and Zulehner W. (1991) Oxygen precipitation in heavily boron doped silicon, *J. Appl. Cryst.* **24** 576-580
14. Gupta S., Messoloras S., Schneider J.R., Stewart R.J. and Zulehner W. (1992) Oxygen precipitation in antimony doped silicon, *Semicond. Sci. Technol.* **7** 443-451
15. Bouchard R., Schneider J.R., Gupta S., Messoloras S., Stewart R.J., Nagasawa H. and Zulehner W. (1995) Distribution of SiO_2 precipitates in large, oxygen rich Czochralski-grown silicon single crystals after annealing at 750°C, *J. Appl. Phys.* **77**, 553-562
16. Baghdadi A., Bullis W.M., Croakrin M.C., Yue-Zhen Li, Scace R.I., Series R.W., Stallhofer P. and Watanabe M. (1989) Interlaboratory determination of the calibration factor for the measurement of the interstitial oxygen content of silicon by infrared absorption, *J. Electrochem Soc.* **136** 2015-2024
17. Messoloras S., Rycroft S., Stewart R.J. and Zulehner W. (1996) Private Communication.
18. Messoloras S., Rycroft S., Stewart R.J. and Binns M.J. (1996) Private Communication.

ATOMIC COMPOSITION, STRUCTURE AND VIBRATIONAL EXCITATION OF SUBSTITUTIONAL CARBON-OXYGEN COMPLEXES IN SILICON

H. YAMADA-KANETA, Y. SHIRAKAWA, and C. KANETA*)
Process Development Div., Fujitsu Ltd.
1015 Kamikodanaka, Nakahara-ku, Kawasaki 211, Japan
*) *Fujitsu Laboratories Ltd.,*
10-1 Morinosato-Wakamiya, Atsugi 234-01, Japan

1. Introduction

The substitutional carbon (C_s) impurity in silicon serves as the nucleation site for the interstitial oxygen (O_i) aggregation [1,2]. The smallest aggregates in this case are the C_s-O_i complexes, which we simply refer to as the C-O complexes. Newman and co-workers [3,4] showed that the C-O complexes can be directly observed by means of the low-temperature infrared absorption. Figure 1 is a typical absorption spectrum [5,6], in which we find the peaks A, B, B', C, C' and D due to the bond-stretching local modes (**BSLMs**) of the O_i perturbed by the carbon. Bean and Newman [4] proposed that the C-O complex causing the peak A involves a C_s atom and an O_i atom occupying the second or third neighbor interstitial site to the C_s. This was supported by the *ab initio* calculation by C. Kaneta et al. [7] that consistently explained the structure and the absorption energies of the peak-A complex. However, identification of the C-O complexes causing the other peaks seems to be inconclusive. This article aims to clarify the atomic compositions of the C-O complexes causing the peaks in Fig. 1. We utilize the following mass-action law governing the equilibrium concentrations of the $(C_s)_m$-$(O_i)_n$ complex, the complex of m C_s atom(s) and n O_i atom(s):

$$\left[(C_s)_m\text{-}(O_i)_n \right]_{eq} = \kappa(T) \left[C_s \right]^m \left[O_i \right]^n , \qquad (1)$$

where $\kappa(T)$, the equilibrium constant of the chemical reaction $m\ C_s + n\ O_i \rightleftharpoons (C_s)_m$-$(O_i)_n$, depends only on the annealing temperature T. The symbol [ ] represents the concentrations of the impurities (complexes). We will show that, in the annealing process, there is a kind of thermal equilibrium where the above mass-action law holds.

The absorption peaks by the C-O complexes in Fig.1 are located near the 1136.4-cm^{-1} line of the isolated O_i. This suggests that the original BSLM of the isolated O_i (A_{2u} mode [9]) is not so strongly perturbed by the C_s in the complexes. However, the effect of the C_s on the 2-dimensional (**2D**) low-energy anharmonic excitation (**LEAE**) of the O_i [8,9] may be significant. This paper also aims to uncover how the 2D LEAE of the O_i in each $(C_s)_m$-$(O_i)_n$ complex is changed from that of the isolated O_i.

R. Jones (ed.), Early Stages of Oxygen Precipitation in Silicon, 389–396.
© *1996 Kluwer Academic Publishers.*

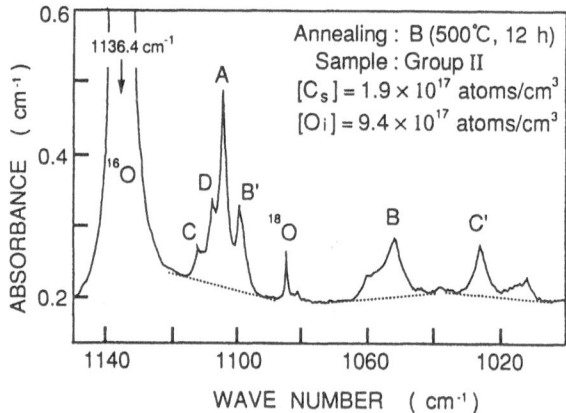

Figure 1. Infrared absorption peaks by the isolated O_i (^{16}O, ^{18}O) and the O_i in the C-O complexes (A - D) observed at 4.2 K. The wave numbers of the peaks are (A) 1104, (B) 1052, (B') 1099, (C) 1112, (C') 1026, and (D) 1108 cm^{-1}. The ordinate is the absorbance per sample thickness of 1 cm.

2. Mass-action Law and Atomic Compositions

We used seven groups of samples cut from seven different Czochralski-grown ingots. Specifications of the samples are given in Table I. We adopted three kinds of isothermal annealings A, B and C shown in Table II. The details of these annealings and the infrared absorption measurements have been presented elsewhere [5,6].

TABLE 1. The impurity concentrations of the samples. The $[O_i]$ and $[C_s]$ were determined by the room-temperature infrared absorption method with the conversion constants used in [5].

Samples	$[C_s]$ (10^{17} atoms/cm^3)	$[O_i]$ (10^{17} atoms/cm^3)	Dopants (10^{15} atoms/cm^3)
Group-I	2.5 - 3.0	7.5 - 8.0	1.5 (Boron)
Group-II	1.5 - 2.0	9.0 - 9.5	undoped
Group-III	3.3 - 3.6	7.6 - 8.0	undoped
Group-IV	0.6 - 0.7	5.4 - 5.5	undoped
Group-V	1.5 - 1.6	4.9 - 5.3	undoped
Group-VI	2.6 - 2.8	4.6 - 4.8	undoped
Group-VII	2.7 - 2.9	8.2 - 8.4	undoped

TABLE 2. Conditions of the one-step annealings A and B and the two-step annealing C.

Annealing	First step	Second step
A	700 °C, 0 -128 h	——
B	500 - 1330 °C, 0 -16 h	——
C	500 °C, 50 h	515 -575 °C, 0 -12 h

2.1. C-O$_A$ AND C-O$_D$ COMPLEXES

Figure 2 shows the annealing time dependencies of the relative concentrations $[C_s]$, $[O_i]$, and $[C\text{-}O_A]$. We refer to the C-O complex causing the peak A as the C-O$_A$ complex. Hereafter we adopt the similar manner of naming the C-O complexes.

Although both of $[C_s]$ and $[O_i]$ exhibit no detectable change for about 20 h, the $[C\text{-}O_A]$ rapidly reaches the quasi thermal-equilibrium (**QTE**) value $[C\text{-}O_A]_{eq}$ which continues until the $[C_s]$ and $[O_i]$ begin to decrease. Figure 3 details the QTE stages for the group-I and group-III samples. We also obtained similar plots for the group-II samples subjected to the annealing B (not shown). The QTE stages for the low-temperature annealings (Fig. 3(b)) were observed in the second step of the annealing C. [To clearly observe the QTE stages in the low-temperature annealings, we utilized the dissociation (decreasing) process rather than the formation (increasing) process of the C-O complex. We once increased the complex concentration in the first step of the annealing C, and then decreased it in the second step.] From these results for the three different sample groups, we found that $[C\text{-}O_A]_{eq}/[C_s][O_i]$ is a sample-independent function of the annealing temperature T, i.e., $[C\text{-}O_A]_{eq}/[C_s][O_i] = \kappa_A(T)$. This is shown in Fig. 4(a): The plots in the figure suggest a function of T common to three sample groups having different $[C_s]$ and $[O_i]$, and different thermal (crystal-growth) histories.

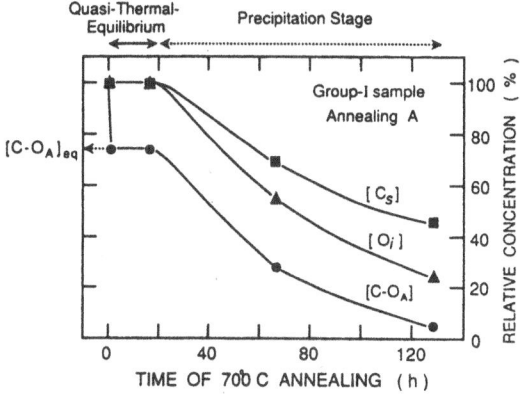

Figure 2. Overall annealing behavior of the $[C\text{-}O_A]$, $[C_s]$, and $[O_i]$ observed for the isothermal annealing at 700 °C. The concentrations (ordinate) are relative to those of the as-grown crystal.

We can interpret the above finding,

$$\left[C\text{-}O_A\right]_{eq} = \kappa_A(T) \left[C_s\right] \left[O_i\right] \quad , \tag{2}$$

as follows: This equation is similar to the mass-action law $[XY] = \kappa(T) [X] [Y]$ for the equilibrium chemical reaction $X + Y \rightleftarrows XY$ at temperature T. By analogy, we consider that, in the flat regions of the curves (plots) in Figs. 2 and 3, the impurity reaction $C_s + O_i \rightleftarrows CO$ is in thermal equilibrium. In the present experiment, the $[O_i]$ are higher than the solubility limits at the annealing temperatures, and thus, the flat regions of the curves are followed by the oxygen precipitation process (the precipitation stage in Fig. 2) going toward the true thermal equilibrium. In this meaning, we refer to these flat regions as the QTE state. If the $[O_i]$ is below the solubility limit, the flat region will last eternally as the true thermal equilibrium.

In the light of the general mass-action law $[X_m Y_n] = \kappa(T) [X]^m [Y]^n$ for the equilibrium chemical reaction $m X + n Y \rightleftarrows X_m Y_n$, Eq. (2) indicates that the C-O_A

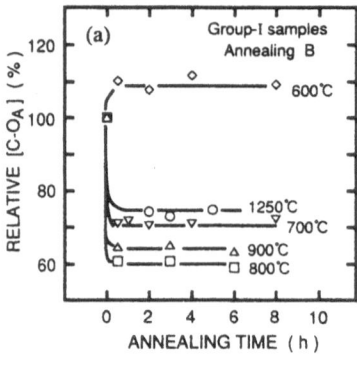 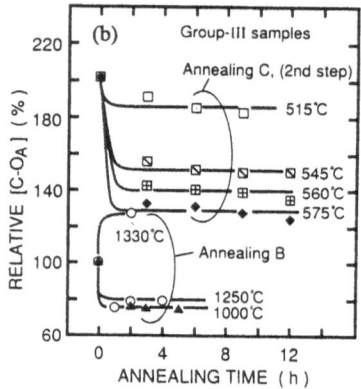

Figure 3. Isothermal annealing behavior of the [C-O$_A$] measured for (a) the group-I samples, and (b) the group-III samples. The ordinates are the [C-O$_A$] relative to those of the as-grown crystals, and the abscissa is the time of the annealing B or the time of the second step of the annealing C. In (b), the initial plots for the second step of the annealing C indicate [C-O$_A$] reached in the first step of the annealing C. The flat region of each curve corresponds to the quasi thermal-equilibrium state. The annealing temperature is given by each curve.

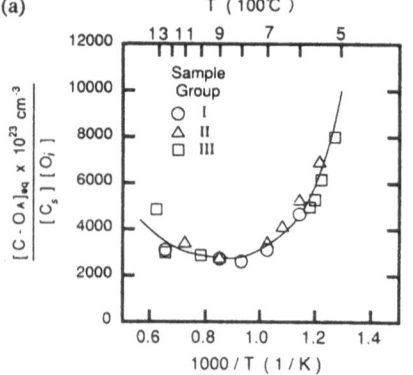 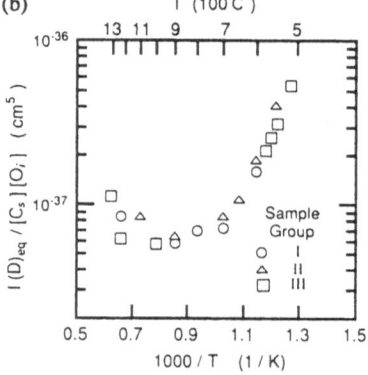

Figure 4. Annealing temperature dependence of (a) 10^{23} cm^{-3} × [C-O$_A$]$_{eq}$/[C$_s$][O$_i$], and (b) I(D)$_{eq}$/[C$_s$][O$_i$], measured for the sample groups I, II, and III. Each plot represents several measured values (plots) of (a) [C-O$_A$]$_{eq}$, and (b) the peak-D heights belonging to such the quasi thermal-equilibrium stage as shown in Fig. 3. The solid curve in (a) is the result of model calculation presented in [5].

complex is composed of a C_s atom and an O_i atom. Accordingly, the C-O$_A$ complex has only one BSLM of the O_i that causes the peak A. Thus, the peaks other than the A in Fig. 1 are not due to the C-O$_A$ complex. The present determination of the atomic composition of the C-O$_A$ complex confirms Bean and Newman's conclusion [4].

Since the peak D is small and mounting on the steep slope of the peak A (Fig. 1), we can not determine the concentration of the C-O$_D$ complex, [C-O$_D$]. Accordingly, we deal with the height of the peak D relative to the base line (dotted line) shown in Fig. 1. The observed annealing behavior of the peak-D height was similar to that of the peak

A: The peak-D height, $I(D)$, rapidly reached the constant value $I(D)_{eq}$ corresponding to the QTE state. As shown in Fig. 4(b), $I(D)_{eq}/[C_s][O_i] \blacksquare \kappa_D$ is again a sample-independent function of T. Here, the peak height $I(D)_{eq}$ involves the contribution from the C-O$_A$ complex, because of the overlapping of the peaks A and D. However, from Eq. (2) this contribution should also be proportional to the product $[C_s][O_i]$. Thus, the QTE concentration of the C-O$_D$ complex $[C\text{-}O_D]_{eq}$ is expressed as

$$\left[C\text{-}O_D\right]_{eq} = \kappa_D(T)\left[C_s\right]\left[O_i\right] . \tag{3}$$

This indicates that also the C-O$_D$ complex involves a C_s atom and an O_i atom. Accordingly, the peaks other than the D in Fig. 1 are not due to the C-O$_D$ complex.

The intensity ratio of the peak D to the peak A was seen to depend on the samples. We therefore conclude that the C-O$_D$ complex is different from the C-O$_A$ complex.

2.2. C-O$_B$ AND C-O$_C$ COMPLEXES.

Figure 5 shows that, for all measured samples, annealed or not annealed, the intensity ratio of the peak B to B', and that of the peak C to C' are always constant, respectively. Therefore, the peaks B and B' are due to two different BSLMs in the same C-O complex, namely, the C-O$_B$ complex; and the peaks C and C' are due to two different BSLMs in another kind of the C-O complex, namely, the C-O$_C$ complex. Accordingly, the C-O$_B$ and the C-O$_C$ complexes both contain at least two O$_i$ atoms.

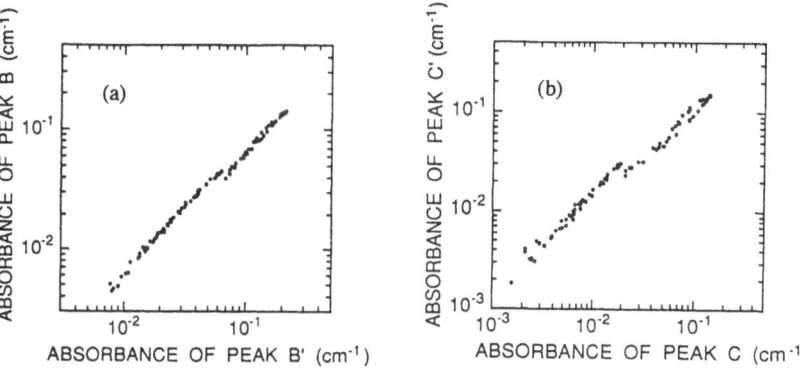

Figure 5. Correlations of the peak intensities between (a) the peaks B and B', and (b) the peaks C and C'.

Figure 6 shows the annealing behaviors of the peaks B and C' of the group-VII samples observed in the second step of the annealing C. Similar to the peaks A (Fig. 3) and D, the absorbances of the peaks B and C' also reach the constant values depending on the annealing temperature. We assume that these constant-value stages in Fig. 6 correspond to the QTE states where the mass-actin law (Eq. (1)) holds. Since the right-hand side of Eq. (1) is written as the product of the peak intensity $\alpha^{(eq)}$ of the $(C_s)_m$-$(O_i)_n$ complex in the QTE state and the proportionality constant λ, we obtain

$$\alpha^{(eq)}(T) / \left[C_s\right]^m \left[O_i\right]^n = \kappa(T)/\lambda . \tag{4}$$

This means that, for a given temperature, the right-hand side is sample-independent. We use this property to determine the numbers m and n for the C-O_B and C-O_C complexes.

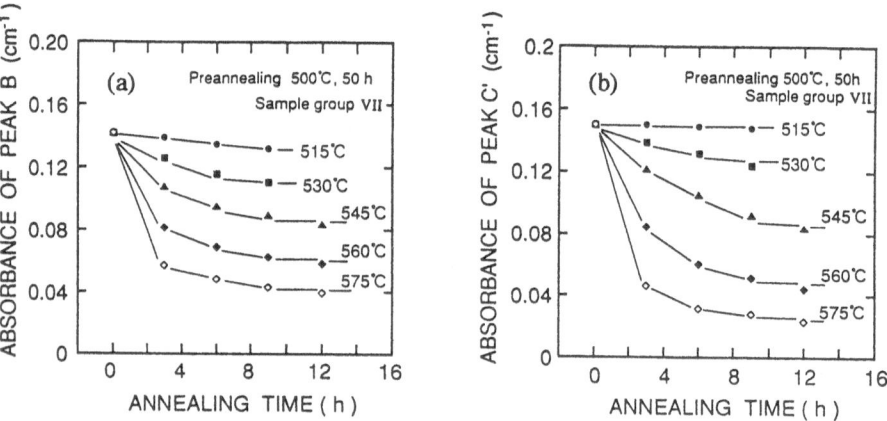

Figure 6. Annealing behaviors of the intensities of (a) the peak B, and (b) the peak C'. The quasi thermal-equilibrium states correspond to the flat regions of the curves appearing after annealing of about 8 h.

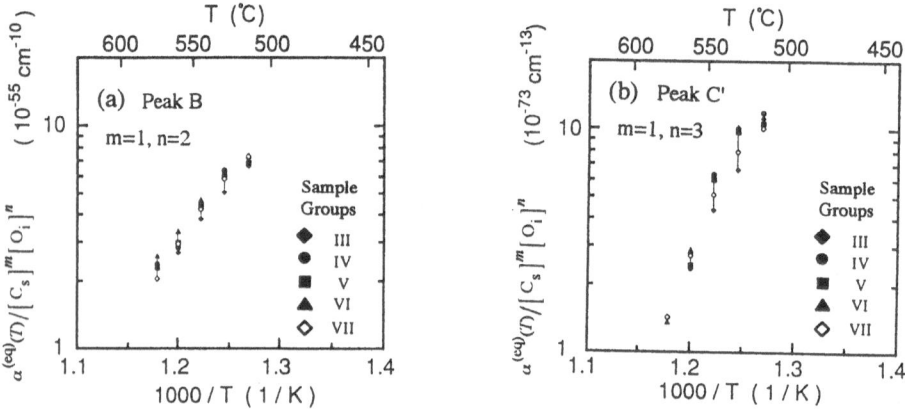

Figure 7. Plots of $\alpha^{(eq)}(T) / [C_s]^m[O_i]^n$ for (a) the peak B with the choice of $(m,n) = (1, 2)$, and for (b) the peak C' with the choice of $(m, n) = (1, 3)$.

Figure 7(a) shows the plots of the right-hand side of Eq. (4) calculated for the peak B with the assumption of $m=1$ and $n=2$. In this case, dispersion of the plots with respect to the sample group is negligibly small for all temperatures, suggesting a function of T common to all sample groups. We found that, for other choice of (m, n), the plots disperse significantly [6]. We therefore conclude that the complex B causing the peaks B and B' comprises one C_s atom and two O_i atoms. We found that the similar plots for the peak C' exhibit minimum dispersion for $m=1$ and $n=3$ (Fig. 7(b)), and for other choice of (m, n) the plots exhibit much larger dispersion. Thus we conclude that the C-O_C complex causing the peaks C and C' involves a C_s atom and three O_i atoms.

3. Effect of Carbon on Off-center Nature of Oxygen

Due to the low-lying states of the 2D LEAE coupled to the BSLM [8,9], thermal population of the ground state from which the 1136.4-cm^{-1} transition occurs rapidly changes in the low-temperature region. This is why the intensity of the BSLM peak at 1136.4 cm^{-1} exhibits the strong temperature dependence in the range 10 - 70 K (Fig. 2 in [10]). If the off-center (off-axis) nature [8,9] of the O_i becomes the stronger, i.e., if the area of the 2D motion of the O_i becomes the wider, the excitation energy of the 2D LEAE becomes the smaller. Therefore, strong temperature dependence of the intensity of the BSLM peak in the low-temperature region is an indication of strong off-center nature of the O_i. Temperature dependence of the peak D shown in Fig. 8 is comparable to that of the isolated O_i (Fig. 2 in [10]), although the peak D largely overlaps with the peak A.

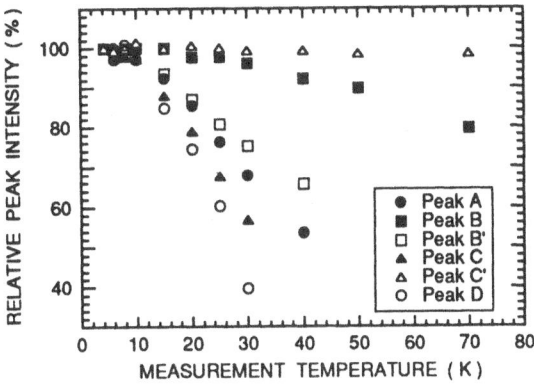

Figure 8. Temperature dependences of the normalized peak intensities (heights) measured from the baseline shown in Fig. 1. The peak intensities at 4.2 K are normalized to 100 %.

If we could properly determine (separate) the true height of the peak D, it may exhibit stronger temperature dependence than in Fig. 8. Thus the O_i in the C-O_D complex may have stronger off-center nature than the isolated O_i. The temperature dependence of the peak-A intensity shown in Fig. 8 appears to be somewhat weaker than that of the isolated O_i. This can be qualitatively explained by the *ab initio* calculation by C. Kaneta et al. [7,11]. They showed that the O_i causing the peak A has weaker off-center nature than the isolated O_i. In any case, for the group of the peaks D, C, A, and B' located near the isolated O_i peak at 1136.4 cm^{-1}, the temperature dependences are similar to that of the isolated O_i. Accordingly, the O_i in the C-O_A and C-O_D complexes, the O_i in the C-O_B complex causing the peak B', and the O_i in the C-O_C complex causing the peak C are considered to have the 2D LEAE (off-center nature) similar to that of the isolated O_i. In contrast with these peaks, temperature dependences of the peaks B and C' located far from the 1136.4-cm^{-1} peak are much weaker (Fig. 8). Thus, for the second O_i atoms in the C-O_B and the C-O_C complexes, which cause the peaks B and C' respectively, the off-center nature would be negligibly weak.

Lassmann [12] has reported the phonon absorption peak at 330 GHz (11.0 cm^{-1}) due to the C-O complex of unknown nature. For a group-VII sample subjected to the 450-°C annealing for 80 h in nitrogen ambient, we have observed, at 7K, the far-infrared absorption line at about 8 cm^{-1}. This absorption disappeared at about 15 K. This also implies that some kind of C-O complex has such the low-energy excitation as reported by Lassmann [12]. From the above argument, we are left possibility of ascribing this low energy excitation to the O_i in the C-O_D complex. Another candidate is the third (unidentified) oxygen in the C-O_C complex whose BSLM peak has not yet been observed. However, these are only speculations, and should be examined in future.

4. Conclusion

Table 3 summarizes the identification of the C-O complex peaks in Fig. 1, the atomic compositions, and the off-center nature of the oxygens in the complexes.

TABLE 3. Result of identification of the C-O complexes. The fourth column shows strength of off-center nature of the O_i in the complex compared with that of the isolated O_i. For example, the notation O_i (B') means one of the O_i atoms in the complex B that causes the peak B'.

C-O complex	Assigned peaks	Composition	Strength of off-center nature of O_i
A	A	C_1-O_1	weaker
B	B, B'	C_1-O_2	O_i (B): negligible, O_i (B'): weaker
C	C, C'	C_1-O_3	O_i (C): weaker or comparable, O_i (C'): negligible
D	D	C_1-O_1	comparable or stronger

5. References

1. Kishino, S., Matsushita, Y., and Kanamori, M. (1979) Carbon and oxygen role for thermally induced microdefect formation in silicon crystals, *Appl. Phys. Lett.* **35**, 213- 215.
2. Leroueille, J. (1981) Influence of carbon on oxygen behavior in silicon, *Phys. Status Solidi* A **67**, 177-181.
3. Newman, R.C. and Smith, R.S. (1969) Vibrational absorption of carbon and carbon-oxygen complexes in silicon, *J. Phys. Chem. Solids* **30**, 1493-1505.
4. Bean, A.R. and Newman, R.C. (1972) The effect of carbon on thermal donor formation in heat treated pulled silicon crystals, *J. Phys. Chem. Solids* **33**, 255-268.
5. Shirakawa, Y., Yamada-Kaneta, H., and Mori, H. (1995) Annealing behavior of carbon-oxygen complexes in silicon crystals observed by low-temperature infrared absorption, *J. Appl. Phys.* **77**, 41-46.
6. Shirakawa, Y., Yamada-Kaneta, H., and Mori, H. (1994) Annealing behavior of carbon-oxygen complexes in silicon crystals observed by low-temperature infrared absorption, *Extended Abstracts of the Electrochemical Society Meeting* **94-1**, 425-426.
7. Kaneta, C., Sasaki, T., and Katayama-Yoshida, H. (1992) Atomic configuration, stabilizing mechanism, and impurity vibrations of carbon-oxygen complexes in crystalline silicon, *Phys. Rev.* B **46**, 13179-13185.
8. Bosomworth, D.R., Hayes, W. Spray, A.R.L., and Watkins, G.D. (1970) Absorption of oxygen in silicon in the near and the far infrared, *Proc. Roy. Soc. Lond.* A **317**, 133-152.
9. Yamada-Kaneta, H., Kaneta, C., and Ogawa. T. (1990) Theory of local-phonon coupled low-energy anharmonic excitation of the interstitial oxygen in silicon, *Phys. Rev.* B **42**, 9650-9656.
10. Khirunenko, L.I., Shakhovtsov, V.I., and Shinkarenko, V.K. (1986) Investigation of vibrational absorption spectra of oxygen in Si:Ge solid solutions, *Sov. Phys. Semicond.* **20**, 1388-1399.
11. Kaneta, C., Sasaki, T., and Katayama-Yoshida, H. (1994) Effect of carbon on anharmonic vibration of oxygen in crystalline silicon, *Matter. Sci. Forum* **143-147**, 957-962.
12. Lassmann, K (1995) Phonon spectroscopy of low-energy excitations of defects in semiconductors, *Matter. Sci. Forum* **196-201**, 1563-1570.

INFLUENCE OF ISOVALENT DOPING ON THE PROCESSES OF THERMAL DONORS FORMATION IN SILICON

L. I. KHIRUNENKO, V. I. SHAKHOVTSOV,
V. V. SHUMOV AND V. I. YASHNIK
*Institute of Physics of the National
Academy of Sciences of Ukraine
Pr.Nauki, 46, 252650, Kiev-22,
UKRAINE*

1. Introduction

The doping of CZ silicon with the isovalent impurity germanium is known to lead to a decrease in the rate of thermal donor (TD) formation during heat treatment at 450–650 °C [1,2]. This effect is still not completely understood. Some authors suggested it was due internal elastic stresses caused by the Ge [1]. These stresses may increase the activation energy for oxygen diffusion during TD formation, though some authors have noted the enhanced oxygen diffusion in silicon doped with germanium [3]. In references 4 and 5 the influence of Ge doping was explained by a decrease in the diffusion coefficient of oxygen atoms and by the delayed decay of the supersaturated solid solution of oxygen. The change in the rate of TD formation in Si⟨Ge⟩ was explained in reference 6 by the decrease of the capture radius at defect nucleation sites. All these models may be correct but they cannot explain the concentration dependence of Ge in Si⟨Ge⟩ on the rate. The decrease of TD generation is only obseveved for Ge concentrations higher than $(2–3) \times 10^{19}$ cm^{-3}. However, even at lower concentrations than this, considerable elastic stresses in Si⟨Ge⟩ are found [7]. That leads, for example, to a nonuniform broadening of the IR absorption resonances and EPR lines, without affecting the TD formation rate.

Thus in our opinion there is an additional cause for the observed effects. In this paper we present some results of our experiments concerning this problem.

R. Jones (ed.), Early Stages of Oxygen Precipitation in Silicon, 397–402.
© *1996 Kluwer Academic Publishers.*

2. Experimental

Samples of n-type (phosphorous) CZ silicon undoped and doped with Ge at concentration $N_{Ge} = 5 \times 10^{18} - 2 \times 10^{20}$ cm^{-3}, and with resistivity 25–40 Ωcm were studied. The concentrations of oxygen and carbon measured by IR absorption were $(6-8) \times 10^{17}$ cm^{-3} and $(1-2) \times 10^{18}$ cm^{-3}, respectively. Heat treatment was performed, without pre-annealing, at $T \approx 450\,°C$ in air. The measurements were made by IR Fourier spectroscopy. This method enables a sequence of thermal donors to be studied in one experiment.

3. Results and Discussion

Germanium in silicon is known to influence essentially the absorption spectra of oxygen [8,9]. Two effects have been observed. The first is a strong broadening — more than 20 times — of the highly stress sensitive, low frequency vibrational ν_2 mode in Si$_2$O quasimolecules (at 29.3 cm^{-1}), and a weaker broadening of the ν_3 band (at 1136 cm^{-1}). The ν_2 mode is then essentially shifted to low frequences. This effect may be explained by chaotic deformation fields which occur in silicon doped with germanium. The second effect is the formation of new types of Si$_2$O quasimolecules leading to the appearance of new ν_3 vibrational modes at 1130.22 and 1118.65 cm^{-1} (see figure 1). Such an effect has not been observed for all known impurities in Si. This may be indicative of a correlated distribution for Si$_2$O quasimolecules and Ge-atoms in Si$\langle$Ge$\rangle$. The frequency of the ν_3 mode is known to depend weakly on the kind of neighbouring atoms bonded to Si, and is the same in many other compounds where the Si$_2$O quasimolecule exists. It depends strongly on the angle α between the bonds in the Si$_2$O [10]. For an isolated, bent triatomic molecule, X$_2$Y, the ν_3 vibrational mode is described by the following formula [11]:

$$\nu_3{}^2 = \left[1 + \frac{2\,M_x \sin^2\alpha}{M_y}\right] \cdot \frac{k}{4\,\pi\,M_x}, \tag{1}$$

where M_x, M_y are the masses of atoms of types X and Y (here atoms of Si and Ge, respectively); k is the force constant; and 2α is the angle between the bonds in X$_2$Y. For silicon $2\alpha \approx 162$–$164°$ [12,13]. At a concentration of Ge $\leq 10^{20}$ cm^{-3}, the main characteristics (energetic structure, phonon spectra, mobilities $etc.$) for Si$\langle$Ge$\rangle$ are the same as in undoped Si, and the force constant is believed to vary insignificantly for these crystals. In this case, in accordance with equation 1, the appearance of a new vibrational mode corresponds to a decrease of the angle α in the Si$_2$O quasimolecule

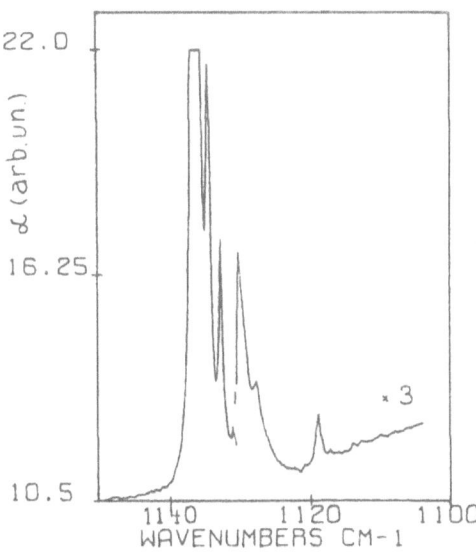

Figure 1. Absorption spectrum of oxygen in Si⟨Ge⟩ with $N_{Ge} \simeq 1 \times 10^{20}$ cm^{-3}.

with $\Delta\alpha \approx 2.4°$ and $\Delta\alpha \approx 5.9°$ for the 1130.22 and 1118.65 cm^{-1} bands, respectively.

It is therefore reasonable to expect that such correlated interactions between germanium and oxygen in silicon should effect the processes of TD formation.

We have studied thermal donors in the neutral charge state, TD^0, in Si⟨Ge⟩ at five hourly intervals during heat treatment. The effect of doping with Ge on the IR absorption spectrum is illustrated in figure 2. Up to a concentration of $N_{Ge} \leq 10^{18}$ cm^{-3} no changes were observed in spectra. At $N_{Ge} \geq 10^{18}$ cm^{-3} the intensity of some TD lines is reduced, and a weak broadening and structure is observed for individual lines. The structure becomes more complicated as the germanium content increases. An additional complex structure of individual lines is observed at a germanium concentration of more than 2×10^{19} cm^{-3}. This structure shows most clearly after about 40 hours annealing time. At that time a decrease of the intensity of all lines associated with undoped Si occurs. The energetic position of the additional lines in the absorption spectra is very close to that in undoped Si. The structure of bands is especialy strongly pronounced for the transitions 1s→2p$_\pm$ and 1s→2p$_0$ for TD$_3$ (about 470 cm^{-1} and 423 cm^{-1}, respectively), and 1s→2p$_\pm$ for TD$_4$ (about 452 cm^{-1}). The spectra for these transitions are shown in figure 3. For Si all these bands show a doublet structure due to the splitting of the oxygen donor 1s ground state, because the degeneracy is lifted by the interaction with the conduction band valleys. This effect is also observed for all hydrogen like centers in Si [14,15]. Unfortunately, TD structure may be studied only up to a concentration $N_{Ge} \leq (3\text{-}4) \times 10^{19}$ cm^{-3} because there is a strong broadening of the TD absorption lines at higher Ge concentrations.

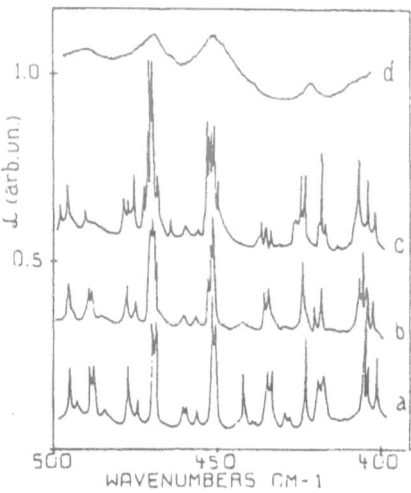

Figure 2. Influence of doping with Ge on the TD^0 absorption spectra in Si. N_{Ge}, cm^{-3}: (a) – 0; (b) – 3×10^{18}; (c) – 2.6×10^{19}; (d) – 1.5×10^{20}.

It should be noticed that at $N_{Ge} \approx 10^{19}$ cm^{-3} the area under above mentioned bands in Si⟨Ge⟩ is 1.5 times larger than that in Si. However, the total TD concentration is the same as in Si, and only when $N_{Ge} \geq 2 \times 10^{19}$ cm^{-3} is a decrease in TD concentration observed. By $N_{Ge} \approx 10^{20}$ cm^{-3} it is half that in Si. The efficiency of new TD formation is lower than that in undoped Si.

A weak additional illumination was used to investigate further the origin of the structure of the absorption lines. The illumination intensity was chosen so as to only alter the population of the 1s ground state components. In Si the bands associated with the transition from the excited 1s state to the 2p$_0$ and 2p$_\pm$ states grew in intensity at the expense of the corresponding transitions from the ground state (see figure 3). In the case of silicon doped with germanium the intensity of a few components grows only for one kind of transitions. Hence, these results seem to suggest the existence of new kinds of TDs in Si⟨Ge⟩ which arise from a correlative distribution of oxygen and germanium in the Si-lattice.

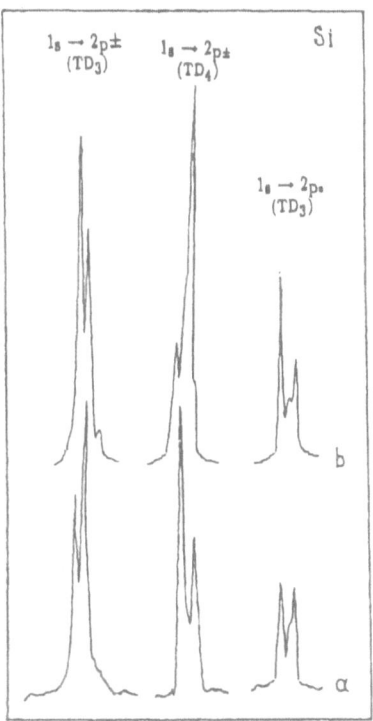 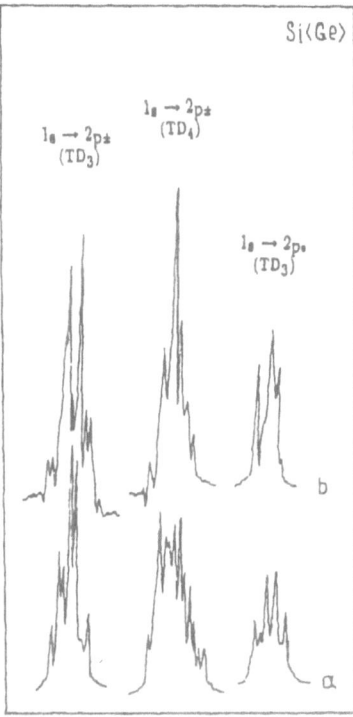

Figure 3. The effect of illumination on the absorption spectra of some transitions for TDs in Si and Si⟨Ge⟩ with $N_{Ge} \simeq 2.6 \times 10^{19}$ cm^{-3}. (a) – before illumination; (b) – with illumination.

Using the effective mass theory approximation, we identified several groups of lines as being related to these new TDs. They are marked on figure 2.

The results of our investigations suggest to us the following conclusions. Firstly, that due to internal elastic stresses in Ge doped Si, a decrease in the capture radius of oxygen by an initial assembly of atoms occurs. This leads to a decrease in the efficiency of TD formation from what is typical for undoped Si. Secondly, a correlated distribution of oxygen and germanium in the Si-lattice is causing the apperance of new kinds of TD. The formation rate of these TDs is lower than that in undoped Si. The competition between these two processes leads to the observed dependence of the TD formation rate on the germanium concentation in silicon.

References

1. Babitskii, Yu. M., Gorbacheva, N. I., Grinshtein, P. M., and Milvidskii, M. G.

402

(1984) The generation of thermal donors in silicon doped with germanium, *Fizika i Tekhnika Poluprovodnikov* **24**, 1129–1132.

2. Babitskii, Yu. M., Gorbacheva, N. I., Grinstein, P. M., Iliin, M. A., Kuznetsov, V. P., Milvidskii, M. G., and Turovskii, B. M. (1988) Generation kinetics of low temperature oxygen donors in silicon with isovalent impurities, *Fizika i Tekhnika Poluprovodnikov* **22**, 307–312.

3. Tipping, A. K., Newman, R. C., Newton, D. C., and Tucker, J. H. (1986) Enhanced oxygen diffusion in silicon at low temperatures, *Materials Science Forum* **10–12**, 887–892.

4. Korliakov, D. N. (1991) The influence of germanium on oxygen diffusion in single crystal silicon, *Neorganicheskie Materialy* **27**, 1333–1336.

5. Babitch, B. M., Baran, N. P., Zotov, K. I., Kiritsa, V. L., and Kovaltchuk, V. B. (1995) Low temperature diffusion of oxygen and formation of thermal donors in Si doped with isovalent impurity of Ge, *Fizika i Tekhnika Poluprovodnikov* **22**, 307–312.

6. Brinkevich, D. I., Markevich, V. P., Murin, L. I., and Petrov, V. V. (1992) Kinetics of thermal donor formation in Si⟨Ge,O⟩ crystals, *Fizika i Tekhnika Poluprovodnikov* **26**, 682–690.

7. Kustov, V. E., Krytskaia, T. V., Tripachko, N. A., and Shakhovtsov, V. I. (1988) Influence of Ge on innear elastic stresses in oxygen content silicon, *Fizika i Tekhnika Poluprovodnikov* **22**, 313–315.

8. Khirunenko, L. I., Shakhovtsov, V. I., and Shinkarenko, V. K. (1986) The investigation of vibration spectra of oxygen absorption in Si⟨Ge⟩ solid solutions, *Fizika i Tekhnika Poluprovodnikov* **20**, 2222–2225.

9. Yamada-Kaneta, H., Kaneta, C. and Ogawa, T., (1993) Infrared absorption by interstitial oxygen in germanium-doped silicon crystals, *Phys. Rev. B* **47**, 9338–9345.

10. Chumaevskii, N. A. (1971) *Vibration spectra of elementorganic compounds of IVB and VB group elements*, Nauka, Moscow.

11. Herzberg, G. (1945) *Infrared and Raman spectra of polyatomic molecules*, New York.

12. Bosomworth, D. R., Nayes N., Spray, A. R. L., and Watkins, G. D. (1970) Absorption of oxygen in the near and far infrared, *Proc. Roy. Soc. London* **317**, 133–152.

13. Pajot, B., and Cales, B. (1985) Infrared spectroscopy of interstitial oxygen in silicon, *Materials Research Society Symposium Proceedings* **59**, 39–44.

14. Stavola, M., and Lee, K. M. (1986) The electronic structure and atomic symmetry of the oxygen donor in silicon, *Materials Research Society Symposia Proceedings* **59**, 95–109.

15. Stoneham, A. M. (1975) *Theory of defects in solids* **1**, Clarendon Press, Oxford.

SOME PROPERTIES OF OXYGEN-RELATED RADIATION INDUCED DEFECTS IN SILICON AND GERMANIUM

L. I. KHIRUNENKO, V. I. SHAKHOVTSOV,
V. V. SHUMOV AND V. I. YASHNIK
*Institute of Physics of the National
Academy of Sciences of Ukraine
Pr.Nauki, 46, 252650, Kiev-22,
UKRAINE*

1. Introduction

The VO- or A-centre is known to be the main oxygen-related radiation defect CZ silicon and germanium. Its structure and annealing kinetics have been extensively investigated previously [1-7].

The VO-centre in Si has a 830 cm^{-1} local vibrational mode (for VO^0) at 300 K that shifts to 835 cm^{-1} at 4.2 K [8]. It anneals out at 300-350 °C through two stages: fast and slow [8]. The fast stage depends on the carbon content and leads to an increase of oxygen concentration upon annealing. At the slow stage some VO-centres transform into VO$_2$ complexes and others may be captured by unknown defects [9]. VO$_2$ is formed by the diffusion of VO to an interstitial oxygen atom O$_i$. At 450 °C the VO$_2$ complexes are transformed into VO$_3$ complexes by thermally activated diffusion of O$_i$ to VO$_2$-centres. The 889 cm^{-1} band corresponds to local vibrational modes of VO$_2$-centres, and three lines (905, 969 and 1001 cm^{-1}) correspond to VO$_3$ (at room temperature).

Assignment of the 889 cm^{-1} band to VO$_2$, however, is uncertain for the following reasons: there is no loss of O$_i$ when the 889 cm^{-1} band occurs; and for samples containing two isotopes (^{16}O and ^{18}O), only two vibrational modes were observed (not three) and the magnitude of isotope shift does not correspond to the presence of two atoms of oxygen [10,11].

The EPR investigations of reference 12 suggested the presence of a number of multivacancy-oxygen complexes in Si following irradiation and annealing. V$_2$O complexes form when irradiated and annealed at the

R. Jones (ed.), Early Stages of Oxygen Precipitation in Silicon, 403–409.
© *1996 Kluwer Academic Publishers.*

same temperature as for VO. The V_3O and V_2O_2 complexes appear after annealing of V_2O, and then at $T \geq 400\,°C$, V_2O_2 transforms into V_3O_2 or V_3O_3. As can be seen, the processes contributing to the annealing of VO-centres in Si are complex and ambiguous.

Natural germainuim has a complex isotopic composition which makes it difficult to use EPR methods for determining the electronic structure and properties of A-centres. However, is thought that these are similar to silicon [2,3,5].

In CZ Ge two unidentified oxygen-related centres (719 and 736 cm^{-1}) appear immediately following irradiation at 80 K. Both the 719 and 736 cm^{-1} bands disappear after annealing at 150 K, and a new 620 cm^{-1} band appears. This band is associated with VO-centres in Ge. Broad EPR lines correspond to this centre [13]. Annealing at $50\,°C$ results in the decay of the 620 cm^{-1} band and the growth of two new bands at 808 and 715 cm^{-1}. These bands then anneal at $100–125\,°C$ and another pair of bands at 802 and 731 cm^{-1} start to develop. At $200\,°C$ the appearance of two more bands at 772 and 780 cm^{-1} is observed. As all these lines appear only in oxygen containing Ge, they are thought to involve vacancy-oxygen defects.

In this paper the peculiarities of the VO-centre's transformation, which we observed after annealing of electron and neutron irradiated Si, are presented. These processes in Si and Ge are then compared.

2. Experimental

Both n- and p-type CZ Si with a resistivity of 19–30 Ωcm was studied. The concentration of oxygen and carbon, determined from IR measurements, in the Si was $(7–9) \times 10^{17}$ and $(1–2) \times 10^{16}$ cm^{-3}, respectively. The irradiation of the Si was performed either with thermal neutrons with fluence up to 5×10^{17} cm^{-2} (the temperature of irradiation did not exceed $100\,°C$) or with 3.5 Mev electrons at fluence up to 1.2×10^{18} cm^{-2} at 90 and 300 K. The samples of CZ natural Ge and isotopically enriched ^{74}Ge had a resistivity in the range 1–10 Ωcm and an oxygen concentration $2 \times 10^{16} - 1.2 \times 10^{17}$ cm^{-3}. These samples were irradiated at 90 K with 3.5 MeV electrons of fluence 5.7×10^{17} cm^{-2}.

An isochronal annealing of the irradiated samples up to $600\,°C$ was performed using 20-minute annealing times and in temperature increments of 25 K. The measurements were carried out by IR Fourier spectroscopy.

3. Results and Discussion

Figure 1 shows the effect of thermal annealing on the absorption band of A-centres for electron-irradiated Si. It is seen that with increasing annealing

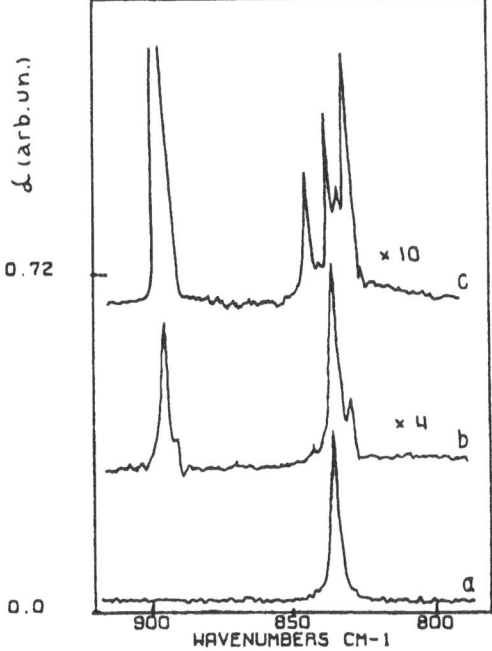

Figure 1. Annealing of A-centre in electron irradiated Si with fluence 1×10^{18} cm^{-2}. (a) After irradiation; (b) $T_{ann.} = 300\,°$C; (c) $T_{ann.} = 370\,°$C.

temperature, the band of VO-centres (830 cm^{-1}) broadens, and then a number of components split off from the main band. For clarity the scales on the ordinate axis for different spectra on figure 1 are shifted vertically. The fine structure appears simultaneously with appearence of the 889 cm^{-1} band. This structure exists up to 550 °C.

A similar effect is observed in silicon irradiated with neutrons [14] (fig. 2). But in contrast to electron-irradiated Si, before the appearance of the 889 cm^{-1} band two additional bands around 919.6 and 1008.9 cm^{-1} have been found. These bands appear at a temperature $\simeq 230\,°$C, when annealing of divacancies is observed. It is known that the annealing of divacancies is generally believed to take place by diffusion until they are trapped and removed by sinks. If one takes this into account then it is reasonable to assume that the additional lines correspond to new centres which form when divacancies are trapped by some defects (*e.g.* oxygen, impurity), because the A-centres are immobile at this temperature. The disordered regions arising in neutron irradiated Si are well known to be the multivacancy complexes surrounded with an impurity shell. They start to anneal at $T \geq 100\,°$C and are completely annealed out at 700–800 °C. An interaction between the divacancies liberated in this process,

406

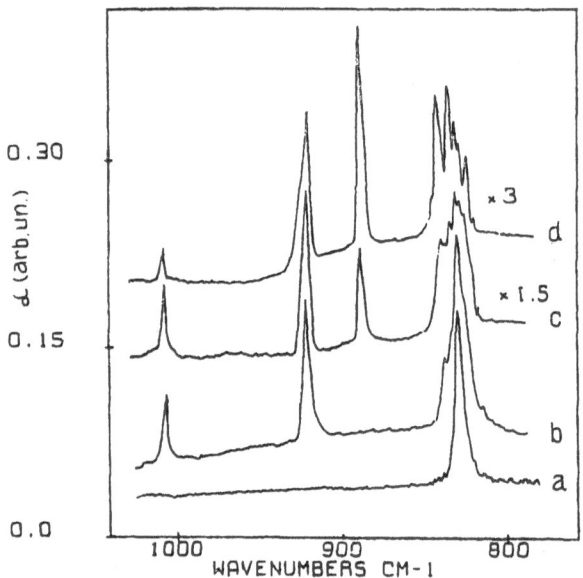

Figure 2. The effect of annealing on absorption spectra of the neutron irradiated Si. (a) Before annealing; (b) $T_{ann.} = 250\,°C$; (c) $T_{ann.} = 315\,°C$; (d) $T_{an} = 380\,°C$.

and impurities, therefore is possible. The 919.6 and 1008.9 cm^{-1} bands anneal at 450 °C. The 889 cm^{-1} band grow in intensity when A-centres anneal out.

Previously in reference 15 the splitting of one component with a frequency of 842 cm^{-1} from the A-centre band was seen, and splitting of three components was observed only at a single annealing temperature (about 300 °C) in Si irradiated with neutrons.

The presence of structure is not accidental and is associated with VO-centres. We investigated Si doped with germanium at a concentration $N_{Ge} = 2 \times 10^{20}$ cm^{-3}. Atoms of Ge in Si are well known to effectively interact with vacancies on irradiation with the formation of GeV-centres [16]. There are nearly no VO-centres present after irradiation at $T \leq 200$ K and they appear only after the GeV-centres have annealed (250–280 K) [17,18]. Thus we can investigate the processes of forming A-centres and their subsequent annealing. Irradiation of Si⟨Ge⟩ at the temperature $T \leq 90$ K was carried out using 3.5 Mev electrons. After irradiation no VO-centres were observed and no structure in the vicinity of 830 cm^{-1} was seen (fig. 3). After annealing the Si⟨Ge⟩ in the temperature range 200–300 K, A-centres appear. The structure in the region of the VO-centres' band appears only when divacancies and VO-centres anneal.

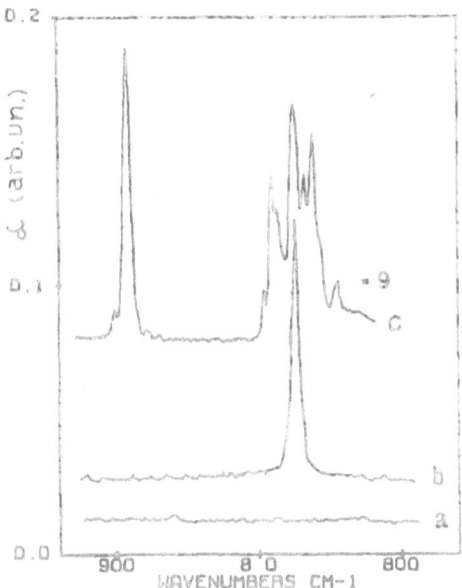

Figure 3. Absorption spectra of Si⟨Ge⟩ subjected to electron irradiation at $T \leq 90$ K with fluence 1×10^{18} cm^{-2} and annealing. (a) After irradiation; (b) $T_{ann.} = 300\,°\mathrm{C}$; (c) $T_{ann.} = 370\,°\mathrm{C}$.

The temperature at which a structure near the A-centre band appears depends on the presence of internal elastic stresses in samples. For undoped electron irradiated silicon this temperature is around $300\,°\mathrm{C}$. It is reduced when there are additional internal elastic stresses, *e.g.* with Ge-doping ($N_{Ge} = 2 \times 10^{20}$ cm^{-3}). The broadening of the band at 830 cm^{-1} can be detected at $230\,°\mathrm{C}$. A similar effect occurs in neutron-irradiated samples, where a high concentration of disordered regions leads to considerable internal stresses. In the later two cases (Ge-doping and neutron irradiation) the structure appears when the divacancy anneals out, whilst for electron irradiated – particularly undoped – Si, it appears when the A-centres anneal out. But in both cases the structure appears when vacancies are liberated.

It should be noted that the structure is not observed in FZ Si where the concentration of oxygen is very small. Taken together the above results seem to suggest that the appearance of the structure in the vicinity of A-centre band is caused by the formation of V_nO complexes upon annealing.

Isochronal annealing studies of irradiated crystals have thus demonstrated that all the components appearing in the region of the band related to A-centres can be divided into at least three groups, with annealing activation energies of; $E_1 = 2.35$ eV; $E_2 = 2.59$ eV, and $E_3 = 2.68$ eV.

In germanium the process of A-centres annealing is rather different from that in Si. We have carried out complex IR and EPR investigations of the A-centres in natural-abundance and isotopically enriched ^{74}Ge that has been electron irradiated at $T \leq 90$ K and then annealed for 15 minutes at 250 K. After the isochronal annealing we observed the A-centre band at 620 cm^{-1} both in natural and isotopically enriched ^{74}Ge. The EPR spectra of A-centres in the natural Ge consist of broad, overlapping lines that cannot be resolved into known centres. In isotopically enriched ^{74}Ge hyperfine structure was visible in the spectra. This is shown in figure 4, where a fluence of 5×10^{17} cm^{-2} was used for the irradiation treatment. The g-factors corresponding to the A-centre lines were calculated from the experimental data. They are were in the range 1.9912 to 2.0374. When A-centres were annealed at a higher temperature of about 50 °C there was no structure visible in the vicinity of the band at 620 cm^{-1}. Only two bands at 808 and 715 cm^{-1} — analogous to VO$_2$ complexes in Si — were seen. EPR spectra of these crystals also show a line with g = 1.95 (fig. 4).

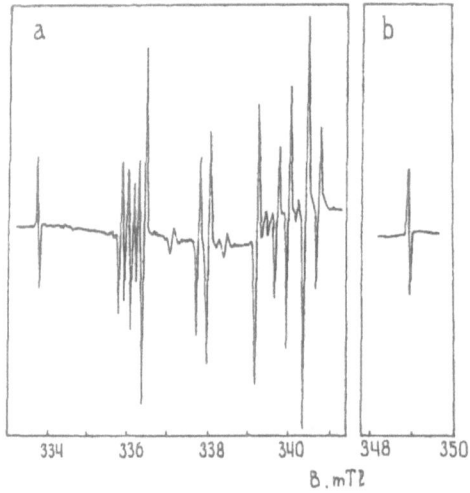

Figure 4. EPR spectra of the ^{74}Ge; (a) after irradiation with electron fluence 5×10^{17} cm^{-2} at $T \leq 90$ K; and (b) after annealing at 50 °C.

Thus, these results show that annealing behaviour of irradiated Si has a complex yet well-defined pattern, which depends on the kind of irradiation and consequent inner elastic stresses that are generated. Further studies by IR, EPR and DLTS on crystals containing different isotopes of oxygen are needed to elucidate further details.

409

Acknowledgments

This work was partly supported by the INTAS (project 93-0320).

References

1. Watkins, G. D., Corbett, J. W., and Walker, R. M., (1959) Spin Resonance in Electron Irradiated Silicon, *J. App. Phys.* **30**, 1198–1203.
2. Whan, R. E. (1965) Evidence for low-temperature motion of vacancies in germanium, *App. Phys. Lett.* **6**, 221–223.
3. Whan, R. E. (1965) Investigations of oxygen-defect interaction between 25 and 700 K in irradiated germanium, *Phys. Rev.* **140**, A690–A698.
4. Watkins, G. D., and Corbett, J. W. (1961) Defects in irradiated silicon. I. Electron spin resonance of the Si-A center, *Phys. Rev.* **121**, 1001–1013.
5. Dosthodjaev, T. N., Emtsev, V. V., Kortchajkina, R. L., and Mashovec, T. V. (1977) The nature of point defects arising in germanium at low temperature irradiation, *Fizika i Tekhnika Poluprovodnikov* **11**, 2128–2134.
6. Kemp, R., Sieverts, E. G., and Ammerlaan, C. A. J. (1986) The electronic structure of the oxygen-vacancy complex in silicon, *Materials Science Forum* **10–12**, 875–880.
7. Morrison, S. R., and Newman, R. C. (1973) The question of vacancies in germanium, *J. Phys. C: Solid State Physics* **6**, 1981–1988.
8. Lindström, J. L., and Svensson, B. G. (1986) Oxygen-related defects in silicon, *Materials Research Society Symposia Proceedings* **59**, 45–58.
9. Ewels, C. P., Jones, R., and Öberg, S. (1995) First principles investigation of vacancy oxygen defects in Si, *Materials Science Forum* **196–201**, 1297–1302.
10. Stein, H. J. (1986) Oxygen isotope effect on the 889 cm^{-1} band in silicon, *App. Phys. Lett.* **48**, 1540.
11. Pajot, B. (1994) Some Atomic Configurations of Oxygen, in F. Shimura (ed.), *Oxygen in Silicon*, in the series: Semiconductors and Semimetals, Academic Press **42**, pp. 191–249.
12. Lee, Y. H., and Corbett, J. W. (1976) EPR studies of defects in electron-irradiated silicon: A triplet state of vacancy-oxygen complexes, *Phys. Rev. B* **13**, 2653–2666.
13. Baldwin, J. A. (1964) Electron paramagnetic resonance in irradiated oxygen-doped germanium, *J. App. Phys.* **36**, 793–795.
14. Pomozov, Yu. V., Khirunenko, L. I., Shakhovtsov, V. I., and Yashnik, V. I. (1990) Transformation of point defects at annealing of Si and Si-Ge irradiated by neutrons, *Fizika i Tekhnika Poluprovodnikov* **24**, 993–996.
15. Ramdas, A. K., and Rao, M. G. (1966) Infrared Absorption Spectra of Oxygen-Defect Complexes in Irradiated Silicon, *Phys. Rev.* **142**, 451–456.
16. Watkins, G. D., Troxell, J. R., and Chatterjee, A. P. (1979) Vacancies and interstitials in silicon, *Institute Physics Conference* **46**: Chapter 1, 16–30.
17. Brelot, A., and Charlemagne, J. (1971) Infrared studies of low temperature electron irradiated silicon, containing germanium, oxygen and carbon, in J. W. Corbett and G. D. Watkins (eds.), *Proceedings International Conference on Radiation Effects in Semiconductors*, Gordon and Breach Science Publishers, London – New York – Paris, pp. 161–169.
18. Khirunenko, L. I., Shakhovtsov, V. I., Shumov, V. V., and Yashnik, V. I. (1995) Reactions between point defects in silicon doped with germanium, *Materials Science Forum* **196–201**, 1381–1384.

DEFECT PROFILING OF OXYGEN-RELATED DEFECTS USING A SLOW POSITRON BEAM

A. P. KNIGHTS, R. D. GOLDBERG, U. MYLER, AND P. J. SIMPSON
Department of Physics, The University of Western Ontario,
London, Ontario,
Canada, N6A 3K7

Abstract

Variable-energy positron annihilation spectroscopy (PAS) is a relatively new technique for probing subsurface defects, and has provided novel insights into defects associated with silicon-based systems such as SiO_2/Si, silicon nitrides, SIMOX and ion-irradiated Si. The technique entails measurement of Doppler broadening of the annihilation radiation from positrons implanted monoenergetically and subsequently thermalised in the sample, which is in turn dependent on the electronic environment. Positrons trapped by defects can thus be distinguished from those in undefected material. By controlling the incident positron energy, depth profiling of defects to several microns can be achieved. The technique is described, together with a study of defects induced by oxygen implantation for fluences of 5×10^{11}, 1×10^{13} and $1 \times 10^{14} cm^{-2}$. The positrons are trapped at open-volume type defects (probably divacancies) in the as-implanted samples. Changes in defect structure are observed and the sensitivity of the technique to the chemical environment of the defects is demonstrated.

1. Introduction

The study of defects in silicon is becoming ever more important with the continuing drive towards device miniaturisation. Despite a large effort being devoted to understanding the formation, evolution and structure of such defects, a complete understanding is still lacking [1].

Variable energy positron annihilation spectroscopy (PAS) is becoming a standard technique for investigating defects in silicon-based systems [2,3]. It is non-destructive, sensitive to defect concentrations of ~0.1ppm, and requires no special sample preparation. It is dependent on the availability of slow positron beams that allow mono-energetic positrons to be implanted at controlled depths up to a few microns.

Of particular interest to these proceedings are previous studies of the interaction of

411

R. Jones (ed.), Early Stages of Oxygen Precipitation in Silicon, 411–418.
© 1996 Kluwer Academic Publishers.

O with Si using PAS. Nielsen et al. [4] studied Si implanted with O at 200keV to doses of 1.4-1.7 x 10^{17}cm^{-2}. They detected open volume defects in the near-surface region which were assumed to be vacancy clusters resulting from the agglomeration of defects created during implantation. On annealing to 1300°C, the open volumes disappeared and positron trapping at oxygen inclusions was identified. A sensitivity to variations in dislocation density resulting from the implantation and subsequent annealing was also demonstrated.

A considerable effort has been devoted to the study of SiO$_2$ since the identification of a distinct PAS signal from the SiO$_2$/Si interface[5]. Leung et al. [6] showed that differences in the interface signal measured from samples oxidised in different ambients were related to the quality of the grown oxide. Several other studies have examined the effect of radiation damage and annealing on SiO$_2$ structures, yielding unique information on defect creation and evolution [3].

The first aim of this paper is to give an introduction to PAS, describing how a measurement is made and how information is extracted from the results. The second is to show its applicability to oxygen related defects in Si by describing a study of defects produced by O implantation for fluences ranging from 5x10^{11} to 1x10^{14}cm^{-2}, at an ion energy of 1MeV.

2. The PAS technique

2.1. SLOW POSITRON BEAMS

Using positrons to probe the near-surface region of a solid requires the ability to form mono-energetic beams in the energy range 0-60keV [2]. Early slow positron beams were designed to be multi-tasking, however recently apparatus has been constructed purely for application to the study of defects, as in the case of the slow positron beam at The University of Western Ontario (UWO) [7].

Positrons emitted from a ^{22}Na source (used in the majority of slow positron beams) have a continuous energy spectrum up to 544keV. A 1μm tungsten foil is placed in front of the source to moderate the emitted positrons. The W foil slows the positrons to thermal energies allowing a small fraction (~1x10^{-4}) to be reemitted from the side of the foil opposite the source. These thermal positrons are subsequently accelerated to energies ranging from 0-60keV by floating the entire source end of the apparatus to a positive potential. Unmoderated fast positrons and gamma rays are filtered from the beamline using an ExB velocity selector. The low efficiency of the moderation process results in sources of ~50-100mCi being required to obtain an acceptable beam current (~10^5 slow e+/second). The UWO beam operates under a vacuum of ~10^{-7} Torr.

2.2 POSITRON IMPLANTATION, THERMAL DIFFUSION AND TRAPPING

Upon implantation into a sample, positrons rapidly lose energy , thermalising in ~10ps. The mean depth z (Å) of implantation may be varied by changing the incident beam energy E (keV) such that $z = (400/\rho) E^{1.6}$, where ρ is the density of the solid (g cm^{-3}). The implantation profile also becomes broader with increasing energy and has a FWHM of ~ 1μm at 20keV (corresponding to a mean implantation depth of 2.7μm). The depth at which positrons annihilate depends not only on the implantation profile but also on the diffusion that occurs after the positron has thermalised. The positron diffusion coefficient (D$_+$) in Si at room temperature is taken to be in the range 2.1-2.7 cm^{2}s^{-1}, corresponding to a thermal diffusion length of ~240nm [2]. Thermal positrons diffuse through defect-free Si for ~200ps before annihilating. In the presence of defects the number of freely diffusing positrons decreases at a rate $\lambda_{eff} = \lambda_f + \nu C$ (where λ_f is the annihilation rate for positrons in the bulk solid (4.55 x 10^9 s^{-1} for Si [8]), ν is the trapping rate of the defects (3 x 10^{14} s^{-1} for point defects in Si [8]) and C is the fractional concentration of defects. The surface of a solid is also an efficient trap for positrons, particularly in the case of Si covered with a native oxide. Annihilation may thus occur from three distinct states: freely diffusing, trapped at a defect, or trapped at the surface of the solid.

A freely diffusing positron is extremely sensitive to changes in the potential it experiences in the solid. In a crystalline lattice containing missing ion cores (ie. vacancies), the positron is highly susceptible to trapping into a vacant lattice site which it sees as a strong potential well. Consequently, positrons are strongly attracted to open-volume-type defects, such as voids or vacancies, and in principal are sensitive to any defect producing a distortion in the lattice potential.

2.3 OBSERVABLES - THE 'S' PARAMETER

The annihilation of a positron-electron pair produces photons with a total energy of $2m_0c^2$ where m_0c^2 (=511keV) denotes the rest mass energy of an electron. Generally the process results in the emission of two photons allowing the conservation of both energy and momentum. Positrons participating in the annihilation process are predominantly thermalised, therefore the motion of the positron-electron pair immediately prior to annihilation will be dominated by the motion of the electron. The requirement to conserve momentum produces a Doppler shift in the annihilation gamma rays from 511keV. The annihilation photo-peak, measured using an intrinsic Ge detector, is broadened by the Doppler shift. Although in principle the photo-peak can be deconvoluted to extract the electron momentum distribution, a simple shape parameter is usually employed to characterize the annihilation line as this ensures

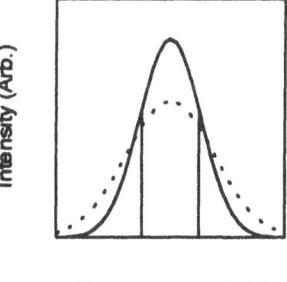

Figure 1. The S parameter is defined as the central area divided by the entire peak area

414

statistical robustness. The most common shape parameter, the S parameter, is defined graphically in figure 1. The area under a central region of the photo-peak is divided by the entire peak area to obtain the value of S (the dotted spectrum in figure 1 producing a relatively lower value of S). Positrons annihilating in states of differing local electronic environment produce different S parameters. The value of the S parameter can often reveal the type of defect in which the positron is trapped prior to annihilation. In the case of positrons trapped in open-volume defects, a high value of S is obtained due to the reduced probability of annihilation with higher momentum core electrons. By varying the incident energy (E) of the positron beam an S versus E curve is obtained, from which the depth distribution of defects may be extracted.

The S parameter versus positron energy data are analysed using computer fitting routines such as POSTRAP [9]. POSTRAP calculates the implantation profile and then solves the positron diffusion equation taking into account the effects of defects and electric fields. It calculates, for each incident positron energy E, the fractions $F_d(E)$ of positrons which annihilate in defects, from the freely diffusing state within the undefected bulk $F_f(E)$, and which annihilate at the surface $F_s(E)$, for a model of defects in the sample. The experimental lineshape can then be fitted using the equation

$$S(E) = S_s F_s(E) + S_f F_f(E) + S_d F_d(E) \tag{1}$$

where S_s, S_f and S_d are the characteristic line-shape parameters for annihilation at the surface, in the bulk, and trapped at a defect respectively.

2.4 EXTENDED ELECTRON MOMENTUM SPECTROSCOPY

The S parameter provides reliable but limited information on the defect species, however in general details of the structure and chemical environment of defects cannot be unambiguously determined. This is due to the considerable background associated with the tails of the annihilation photo-peak (the region containing chemical and structural information resulting from annihilations with the higher momentum core electrons).

By detecting *both* gamma rays from the annihilation process in coincidence using two detectors placed opposite one another, it is possible to significantly reduce the background. Subsequent analysis of the high energy tail of the photo-peak reveals structure clearly dependent on the momentum of the core electrons (the low-energy tail is masked by incomplete charge collection in the detector). Figure 3 shows an annihilation photo-peak from Si measured with two detectors, and the same photo-peak measured with a single detector. A successful application of this technique has

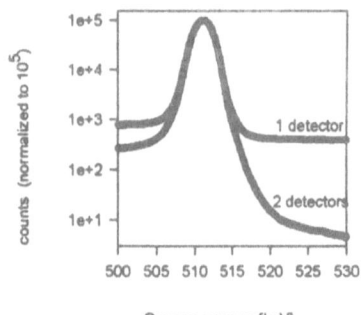

Figure 2. Annihilation photo-peaks measured using the 1 and 2 detector techniques

been the identification of As decorated vacancies in Si [10]. A disadvantage of the two detector technique is the severely increased counting time necessitated by the need to avoid pulse pile-up.

3. A Study of Defects Induced by Oxygen Implantation of Silicon

3.1 EXPERIMENTAL

Samples of p-type boron doped Cz-Si, with a resistivity of 2.5-4Ωcm were implantated using the 1.7MV Tandetron accelerator at UWO. Oxygen ions were implanted to fluences of 5×10^{11}, 1×10^{13} and 1×10^{14} cm^{-2} at an incident energy of 1MeV. All implantations were performed at room temperature at an angle of 7° to avoid channeling. Single-detector PAS spectra were measured for positron energies ranging from 0.5-40keV. The 1×10^{14}cm^{-2}, 1MeV sample was subsequently annealed to temperatures of 280, 500 and 750° C for 1000s in a nitrogen ambient. PAS measurements were made after each anneal using both the 1 detector and 2 detector techniques.

3.2. RESULTS AND DISCUSSION

3.2.1. *Depth profiling of implantation-induced defects*

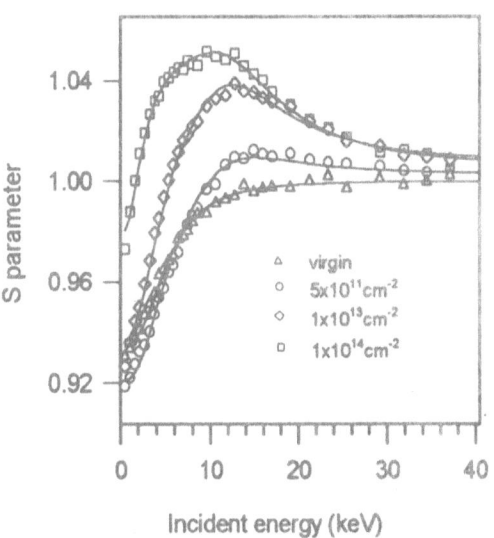

Figure 3. S parameter v incident positron energy for various implantation fluences

Tentative defect identification. Figure 3 shows the positron S parameter versus incident energy for the 1 MeV O implanted samples for fluences of 5×10^{11}, 1×10^{13} and 1×10^{14} cm^{-2}, and also a spectrum obtained from an unimplanted virgin sample. The values of S have been normalised to the value obtained from undefected bulk silicon. The data have been modelled using the program POSTRAP5 [9], the results of which are shown as solid lines.

The S parameter curve for the virgin sample exhibits a smooth rise from the surface S value of 0.93 (typical for a thin oxide layer [3]), to the normalised bulk value at an incident energy of ~20keV. For higher incident energies the value of S remains unchanged as a negligible

416

fraction of the positrons are able to diffuse back to the surface. Modelling of these data yields a thermal positron diffusion length of 240 ± 10nm in good agreement with previously measured values [11]. Upon irradiation, the S value for incident energies up to ~30keV is seen to increase (the reduction of S at very low energies for the $5 \times 10^{11} \text{cm}^{-2}$

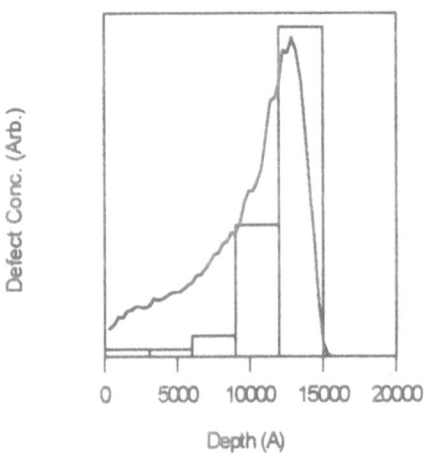

Figure 4. Model used in fitting PAS data (blocks),and vacancy distribution from TRIM

irradiated sample is an artifact of a change in the surface S value only). The maximum value of S increases with increasing ion fluence due to the trapping of positrons in the increasing number of implantation-induced defects. The broad spectrum obtained from the $1 \times 10^{14} \text{ cm}^{-2}$ sample suggests that the positron trapping is approaching saturation (ie. the introduction of further defects does not increase the maximum value of S further). Hence, the maximum value of S for this curve is indicative of the value of S that would be obtained if all of the positrons were trapped into the implantation induced defects. This value is close to that obtained (at UWO) from proton-irradiated Si samples in which the positron trapping was saturated and was assumed to be predominantly at divacancies. It is therefore possible that the defect at which the positrons are trapped in the implanted samples is the divacancy or structurally similar defect.

3.2.2. *Fluence dependence of the defect profile.* Details of the distribution and concentration of the defects can be obtained from the POSTRAP fits. The defect model required to successfully fit the data is shown in figure 4 together with the vacancy distribution obtained from the program TRIM[12]. The shape and depth of the model defect distribution are similar to those from TRIM. It is noted that the PAS model requires fewer defects in the near surface region than predicted by TRIM, perhaps an indication of defect mobility subsequent to implantation. The same defect distribution (with scaled concentrations) was used to fit all of the data plotted in figure 3.

The defect concentrations for the various implantations, also deduced from POSTRAP, are 1.6×10^{13}, 3.2×10^{14} and $3.2 \times 10^{15} \text{cm}^{-2}$ for increasing irradiation fluence. The concentrations of defects calculated by TRIM were approximately an order of magnitude greater than those deduced from PAS. This is an indication of defect recombination, which is not accounted for by TRIM.

Figure 5 shows S parameter versus incident energy for the sample implanted to a fluence of $1 \times 10^{14} \text{cm}^{-2}$ after annealing at various temperatures for 1000s. The spectrum for the as-implanted sample has been shifted up by 0.005 to maintain clarity. The S parameter spectrum after an anneal at 280°C is essentially identical to that for the as-implanted sample. This suggests that the annealing stage of the positron-trapping defects is above this temperature. After annealing to 500°C, a reduction of the S parameter corresponding to the implanted region is observed, together with an increase in the positron diffusion length, indicating partial annealing of the original defects. A further increase in annealing temperature to 750°C reduces the level of the maximum S parameter to that of the virgin Si. Although this apparently indicates the complete annealing of the vacancy-type defects, the presence of residual defects can be identified by a short positron diffusion length of ~75nm.

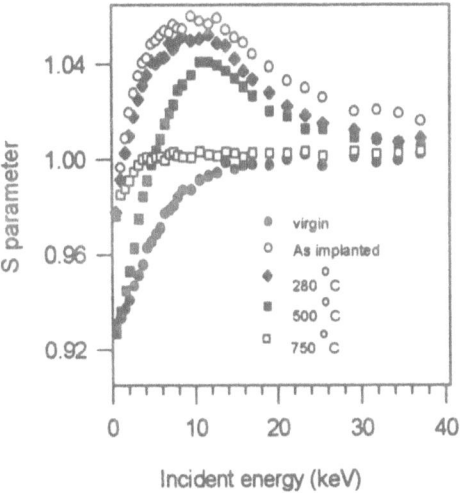

Figure 5. S parameter v incident positron energy for annealed sample

3.2.3. Extended electron momentum spectroscopy - chemical environment identification

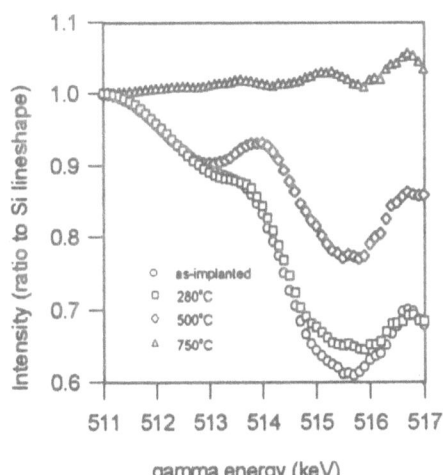

Figure 6. Two detector measurement of the annihilation photo-peak after various anneals

The two-detector technique described in section 2.4 was used to measure the detailed structure of the high energy slope of the photo-peak for the $1 \times 10^{14} cm^{-2}$ sample after the various anneals. All spectra were measured at an incident energy of 12keV. The data shown in figure 6 are obtained by dividing the spectra by the spectrum of bulk crystalline Si (ie. a Si spectrum would be represented by a horizontal line of intensity 1). The as-implanted, 280°C and 500°C annealed samples produce valleys (at 515.5 keV), consistent with a narrowing of the photo-peak. After the 500°C anneal a significant reduction in the depth of the valley is observed with a peak situated at ~514keV. This peak

418

has been positively identified as resulting from positron annihilations with oxygen and its intensity is at least twice that which would be obtained from positron annihilations in the surface oxide at this incident energy. Therefore, after the 500°C anneal, vacancy-type defects are still present, however one might expect a considerable number of these defects to have O attached to them. An alternative explanation may be that in addition to the vacancy-type defects, the 500°C produces a new O-based positron trapping defect. Theoretical studies are currently underway to predict lineshape structure for different defect types. It is hoped that these, together with a more comprehensive experimental study, may shed further light on the positron trapping defects in samples of this type.

4. References

[1] Corbett J. W., Karins J. P., and Tan T. Y., (1981) Ion-induced defects in semiconductors, *Nucl. Inst. and Methods*, **182/183**, 457-476
[2] Schultz P. J. and Lynn K. G. (1988) Interaction of positron beams with surfaces, thin films, and interfaces, *Rev. Mod. Phys.* , **60**, 701-779
[3] Asoka-Kumar P., Lynn K. G., and Welch D. O. (1994) Characterization of defects in Si and SiO$_2$-Si using positrons, *J. Appl. Phys.*, **76**, 4935-4981
[4] Nielsen B., Lynn K. G., Leung T. C., Cordts B. F., and Seraphin S. (1991) Defects in oxygen-implanted silicon-on-insulator structures probed with positrons, *Phys. Rev. B.*, **44**, 1812-1816
[5] Nielsen B., Lynn K. G., Chen Y-C., and Welch D. O. (1987), SiO$_2$/Si interface probed with a variable energy positron beam, *Appl. Phys. Lett.*, **51**, 1022-1023
[6] Leung T. C., Asoka-Kumar P., Nielsen B., and Lynn K. G. (1993), Study of SiO$_2$-Si and metal-oxide-semiconductor structures using positrons, *J. Appl. Phys.*, **73**, 168-177
[7] Leung T. C., Knights A. P., Perquin P., Schmidt I., Schultz P. J., and Simpson P. J. (1996) A variable-energy positron-beam apparatus for defect profiling, *Meas. Sci. and Tech.*, (submitted for publication)
[8] Dannefaer S (1987) Defects and oxygen in silicon studied by positrons *Phys Status Solidi A* **102** 481-491
[9] Aers G. C. (1990) An algorithm for defect profiling using variable energy positrons, in *Positron beams for solids and surfaces* P.J. Schultz, G. R. Massoumi, and P. J. Simpson, AIP conf. proc. #218, New York, pp.162-170
[10] Lawther D. W., Myler U., Simpson P. J., Rousseau P. M., Griffin P. B., and Plummer J. D. (1995), Vacancy generation resulting from electrical deactivation of arsenic, *Appl. Phys. Lett.*, **67**, 3573-3577
[11] Nielsen B., Lynn K. G., and Schultz P. J. (1985) Positron diffusion in Si, *Phys. Rev. B*, **32**, 2296-2301
[12] Ziegler J. F., Biersack J. P., and Littmark U. (1985) *The stopping and range of ions in solids* Pergamon, New York

SHALLOW N-O DONORS IN SILICON

Adam Gali, József Miro and Peter Deák
Department of Atomic Physics, Technical University Budapest,
Budafoki út 8., H-1111 Budapest, Hungary

Abstract. Shallow thermal donors (STD) are electrically active, oxygen-related, thermally formed defects observed in silicon after annealing between 300 and 600 °C. There are experimental indications that STD's might be related to nitrogen. EPR studies find C_{2v} symmetry for such defects. Recently, theoretical investigation on an NO complex has shown that it is a shallow donor. We discuss the possibility of adding an oxygen interstitial to the NO complex. By using semi-empirical cyclic cluster model calculations, two metastable structures have been found. One of these has C_{2v} symmetry. The approximate spin distributions are reported for both of them.

1. Introduction

The inert atmosphere for processing of Si is commonly nitrogen. Indeed, the equilibrium solid solubility of nitrogen is low in silicon and the nitrogen impurity is known to be electrically inactive but it may interact with other defects, especially oxygen. Recently, it has been shown that nitrogen doping of Czochralski-silicon in the melt can be used to suppress vacancy and interstitial defects and to increase gate oxide integrity [1].

Nitrogen occurs as substitutional impurity in irradiated silicon [2] but the majority of the nitrogen concentration is in interstitial pairs to which infrared (IR) bands at 962 and 766 cm^{-1} have been assigned [3]. In samples implanted with a ^{14}N/^{15}N mixture, it could be shown that the members of the pair occupy equivalent positions [4]. Channeling studies by the same authors in ^{15}N-implanted Si has shown that the nitrogen atoms are at 1.1±0.1 Å from lattice positions along <001> directions, with a maximum deviation of 0.2 Å perpendicular to the <001> axis. Calculations have predicted a structure, where the N atoms form a four member ring with two lattice silicon atoms [5]. The displacement of the nitrogen atoms from lattice sites were calculated to be 1.1 Å along <001> and 0.2 Å along <110>. Taking into account the inaccuracy in calculating frequencies, the computed vibration modes, 919 and 689 cm^{-1}, fit the experimentally observed frequencies well and the isotope shifts are also correctly reproduced. So the (NN) ring structure of nitrogen pairs in silicon can be regarded as established.

It is well known that shallow, single donor complexes, called shallow thermal donors (STD), are formed in oxygen-rich silicon after annealing between 300 and 600 °C [6,7]. It has been a long controversy whether these donor centers contain nitrogen or not.

419

R. Jones (ed.), Early Stages of Oxygen Precipitation in Silicon, 419–425.

420

No direct spectroscopic evidence exists for the involvement of N in these defects, but a number of studies [7-13] have shown that their appearance is related to the presence of nitrogen. Griffin et al. [9] has measured nitrogen dependency of forming STD. In the case of low N concentration the STD's concentration increases with increasing of N concentration, but in the case of high N concentration the STD-s disappear. On the other hand, the STD-s show up after longer anneals in oxygen rich samples without intentional nitrogen doping [6,14-17]. It appears likely that the term STD is used for different families of defects, as has been shown by Hara et al. [12]. Some of them appear to be related to hydrogen passivated thermal double donors (TDD) [15], probably giving rise to the low symmetry NL-10 signal in ESR [18], while others may contain nitrogen [12]. One of the STD-s found in Al doped samples has proven bistable [17].

Hara et al. [12] attributed the C_{2v} component of the NL-10 signal to a nitrogen containing defect. Based on the analysis of annealing kinetics, Suezawa and coworkers [8] postulated that the STD-s consist of a core of one oxygen atom and an NN pair, with additional NN pairs contributing to the bigger members of the family. A C_{2v} defect with two nitrogens and one oxygen at the core would imply an oxygen on the C_2 axis, contradicting ENDOR results [19]. Also, the stable configuration of two nitrogens and an oxygen has lower symmetry and was calculated to be electrically inactive [20]. Deák and coworkers [21] has proposed an NO ring acting as the core of shallow donors in a theoretical study (see Fig. 1). This structure has C_{1s} symmetry. The donor state is due to the extra electron of the oxygen. It has been shown that the binding energy of nitrogen in this complex is lower than in an NN pair: this explains, why a high concentration of N does not provide for high concentration of STD-s. On the other hand, the calculations resulted in a stronger bonding of O in NO than in a di-oxigen complex. That explains, why STD formation retards the TDD formation [21]. The question is wether this NO ring can capture another oxygen to form a C_{2v} core for STD-s.

The aim of this paper is to study this possibility by self-consistent semi-empirical quantum chemical calculations (briefly described in Section 2).

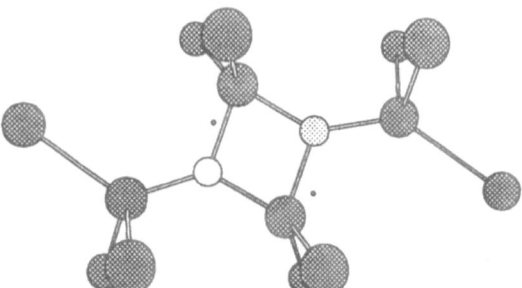

Figure 1.The NO complex
(dark shaded circles: Si, light shaded circle: N, empty circle: O, the small circles reperesent the original lattice positions).

2. Model and computational method

The cyclic cluster model (CCM) [22] has been proven to be an economic and fruitful way of modeling defect complexes in silicon [23,24]. A 32-atom unit cell with cyclic boundary conditions is used here to compare the stabilities of different complexes. The center of the reduced Brillouin zone for this unit cell ($K=0$) represents a special k vector set (Γ, three X and twelve $\Sigma<1/3,1/3,0>$ points) of the primitive Brillouin zone, allowing for a reasonable simulation of the crystal in a $\{K=0\}$ approximation. (In addition, only interaction within the first Wigner-Seitz cell are taken into account in the CCM. This reduces the interaction between repeated defects).

For the investigation of relative stabilities of complicated defect complexes with many possible configurations, semi-empirical quantum chemical methods offer an economic solution. Even though they are semi-quantitative in nature, they are capable of predicting trends reliably. The applicability of these basically "chemical" approximations to solid state studies has been tested in crystalline silicon [22] and α-quartz [25]. They have been successfully applied to predict the behavior of hydrogen [24] and oxygen complexes [23], to investigate the phenomenon of quantum confinement [26], as well as surface reconstruction and adsorption phenomena [27,28] in Si.

The total energy of the Si_{32} CCM have been minimized subject to T_d symmetry within the framework of the self-consistent PM3 semi-empirical approximation [29]. The results of a CCM-PM3 calculation for crystalline silicon are shown in Table I. The PM3 method was also shown to be successful in calculating oxygen defects in silicon [30]. Even though its performance in Si-N systems is less convincing (the relative electronegativity of N is underestimated), it was favored over the AM1 method [31], since the latter gives very poor results for Si-O bond distances. (PM3 and AM1 are different parametrizations of the Modified Neglect of Diatomic Overlap approximation introduced to Hartree-Fock molecular orbital theory by the Dewar group; for further details see refs. [29,30].) The stability of different defect complexes were compared by calculating their total energy, minimized with respect to all atomic coordinates except the 3rd and 4th neighbor silicon shell in CCM.

Table I. Properties of crystalline silicon.

Properties of c-Si	PM3-CCM	Exptl.
Bond length (Å)	2.36	2.35
Lattice energy (eV/atom)	5.20	4.70
Valence bandwidth (eV)	10.20	12.30
Gap (eV)	0.99	1.17
Ionization treshold (eV)	7.64	5.35
Raman frequency (cm^{-1})	484	518

In calculating properties of a defect wave function with energy in the gap, it is extremely important that the approximation for the Hamiltonian allows a correct description of states at both the valence and conduction band edges [22]. Therefore, the CNDO/S approximation (as described in refs. [22,23]) has been used for that purpose at defect ge-

422

ometries obtained from the PM3 calculation. Parameters for nitrogen have been worked out by fitting to ionization potentials of trisililamine, $N(SiH_3)_3$.

3. Results

The addition of an oxygen atom to the $(NO)^+$ complex has been studied in a Si_{32} CCM. Regarding the EMT nature of the donor orbital and the relatively small size of the cluster, the geometry optimization has been done in the positive charge state. Three stable complexes have been found. The binding energy (ΔH) is calculated from the total energies of the separate clusters:

$$\Delta H = E[Si_{32}(NOO)^+]-E[Si_{32}(NO)^+]-E[Si_{32}O]+E[Si_{32}] \qquad (1)$$

where NOO is the triatomic complex, NO is the ring in Figure 1. and O is a single interstitial oxygen in the Si_{32} CCM. The results can be seen in Table II. The most stable configuration is an (NO)O complex with monoclinic (C_s) symmetry as shown in Figure 2a. A CNDO/S calculation for the neutral charge state at this geometry gives a shallow EMT donor orbital with spin distribution shown in Table III.

Table II. Binding energies of the NOO configurations

Complex	Binding energy [eV]	Symmetry
(NO)O (see Figure 2a)	-1.43	C_{1s}
O(NO)	-1.05	C_{1s}
(ONO) (see Figure 2b)	-0.73	C_{2v}

It is basically an asymmetric combination of orbitals from the conduction band edge in the [001] direction. The spin density is small on the nitrogen and oxygen atoms, especially taking into account the fact that the finite cluster limits the delocalization possibility for the orbital. An (ONO) configuration with C_{2v} symmetry (Figure 2b) has been found to be metastable with respect to a monoclinic O(NO) configuration. (The latter is analogous to the complex in Fig. 2a, with N and Or changing position.) The energy difference between the two is only 0.32 eV, so it is conceivable that further oxygens in the (110) plane will stabilize the (ONO) structure, because the latter can relieve the stress caused by an adjacent interstitial oxygen (c.f. the inward "bending" of atoms 5 in Figure 2b). In this interesting structure, the central silicon is overcoordinated. The extra bond to the nitrogen accommodates the sixth electron of one of the oxygens (with energy in the valence band), so only the other is given up for an EMT orbital. The spin distribution on the donor orbital from a CNDO/S calculation for the neutral charge state is shown in

Table IV. The orbital is a symmetric combination of conduction band edge orbitals along the [001] axis.

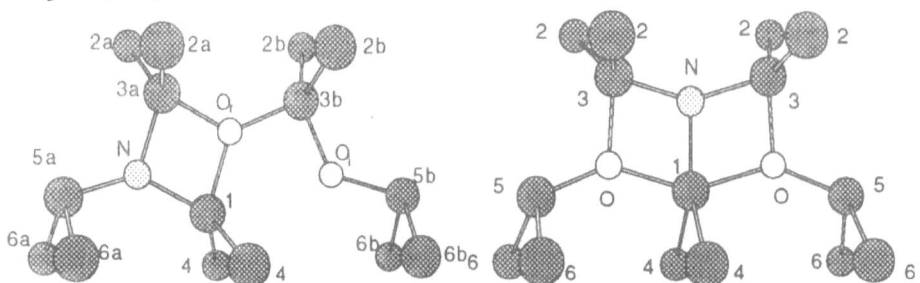

Fig 2a
The (NO)O complex with C_s symmetry.

Fig 2b
The(ONO) complex with C_{2v} symmetry.

(Numbers denote atomic positions used in Table III-IV.)

Table III.: Spin distribution of the donor orbital in the (NO)O complex

Position (see Figure 2a)	s (%)	$p_x + p_y$ (%)	p_z (%)
N	0.03	0.07	0.29
O_r	0.00	0.00	0.08
O_i	0.00	0.01	0.07
Si(1)	2.35	0.00	0.81
Si(2a)	0.87	0.97	2.43
Si(2b)	0.85	0.34	1.92
Si(3a)	1.90	0.11	0.50
Si(3b)	1.15	0.01	0.85
Si(4)	1.44	0.59	1.61
Si(5a)	0.52	0.00	0.57
Si(5b)	0.71	0.02	0.89
Si(6a)	0.49	0.05	1.18
Si(6b)	0.82	0.07	1.18

424

Table IV.: Spin distribution of the donor orbital in the (ONO) complex

Position (see Figure 2b)	s (%)	$p_x + p_y$ (%)	p_z (%)
N	0.01	0	1.25
O	0.00	0.17	0.11
Si(1)	0.27	0	0.05
Si(2)	0.22	0.88	2.10
Si(3)	1.78	0.00	0.74
Si(4)	0.15	0.40	0.04
Si(5)	0.01	0.22	3.47
Si(6)	1.89	0.11	0.08

4. Conclusions

It was found that the electrically active (NO) ring may capture another interstitial oxygen in the (110) plane.

The most stable NO^+O systems have C_{1s} symmetry. They give rise to an EMT donor orbital constructed from a pair of states at the conduction band edge in the [001] direction. The spin distribution is very low on the N and O atoms of the core. One metastable structure with C_{2v} symmetry has been found. This (ONO) double-ring structure might be expected to be stabilized by adjacent oxygens in the (110) plane. If that is the case, the spin density on the nitrogen atom of this center seems to be bigger than to escape ENDOR detection, provided a sufficient concentration of this complex is available.

Acknowledgment

The authors wish to thank the support of the British-Hungarian Intergovernmental S&T Cooperation administered jointly by the British Council and the Hungarian National Committee for Technological Development (Project Nr. 11). This work was also supported by the Hungarian OTKA grant Nr. 2744 and the British SRC grant GR/H1404. Helpful discussions with J. Weber and L. C. Snyder are appreciated.

References

1. Ammon, W. v., Dreier, P., Hensel, W. and Lambert, U. (1995) *Mater. Sci. & Eng. B.* to be published
2. Brower, K. L., (1982) *Phys. Rev. B* **26**, 6040
3. Stein, H. J., (1986) *MRS Symp. Proc. Vol.* **59** [Mater. Res. Soc., Pittsburg, 1986] p. 523.
4. Rasmussen, F. Berg, Nielsen, B. Bech, Jones, R. and Öberg, S., (1994) *Mater. Sci. Forum* **143-147**, 1221
5. Jones, R., Öberg, S., Rasmussen, F. Berg and Nielsen, B. Bech, (1994) *Phys. Rev. Lett.* **72**, 1882
6. Navarro, H., Griffin, J., Weber, J. and Genzel, L., (1986) *Sol. State Commun.* **58**, 151
7. Suezawa, M., Sumino, K., Harada, H. and Abe, T., (1986) *Jpn. J. Appl. Phys.* **25**, L859
8. Suezawa, M., Sumino, K., Harada, H. and Abe, T., (1988) *Jpn. J. Appl. Phys.* **27**, 62
9. Griffin, J. A., Hartung, J., Weber, J., Navarro, H. and Genzel, L., (1989) *Appl. Phys. A* **48**, 41
10. Steele, A. G., Lenchyshyn, L. C. and Thewalt, M. L. W., (1990) *Appl. Phys. Lett.* B**56**, 148
11. Hara, A., Fukuda, T., Miyabo, T. and Hirai, I., (1989) *Jpn. J. Appl. Phys.* **28**, 142
12. Hara, A., Masaki, A., Koizuka, M. and Fukuda, T., (1994) *J. Appl. Phys.* **75**, 2929
13. Chen, C. S., Li, C. F., Ye, H. J., Shen, S. C. and Yang, D. R., (1994) *J. Appl. Phys.* **76**, 3347
14. Clayburn, M. and Newman, R. C., (1989) *Mater. Sci. Forum Vols.* **38-41** [Trans Tech Publications, Switzerland 1989] p. 613.
15. McQuaid, S. A., Newman, R. C. and Lightowlers, E. C., (1994) *Semicond. Sci. Techn.* **9**, 1736
16. Heijmink Liesert, B. J., Gregorkiewicz, T. and Ammerlaan, C. A. J., (1993) *Phys. Rev. B* **47**, 7005
17. Kaczor, P., Godlewski, M. and Gregorkiewicz, T., (1994) *Mater. Sci. Forum Vols.* **143-147** [Trans Tech Publications, Switzerland 1994] p. 1185.
18. Martynov, Yu. V., Gregorkiewicz, T. and Ammerlaan, C. A. J., (1995) *Phys. Rev. Lett.* **74**, 2030
19. Gregorkiewicz, T., Bekman, H. H. P. Th. and Ammerlaan, C. A. J., (1990) *Phys. Rev. B* **41**, 12628
20. Jones, R., Ewels, C., Goss, J., Deák, P., Miro, J., Öberg, S. and Rasmussen, F. Berg, (1994) *Semicond. Sci. Techn.* **9**, 2145
21. Gali, A., Miro, J., Deák, P. and Jones, R. (1996) submitted to *Phys. Rev. B*
22. Deák, P. and Synder, L. C., (1987) *Phys. Rev. B.* **36**, 9619
23. Deák, P., Synder, L. C. and Corbett, J.W., (1992) *Phys. Rev. B* **45**, 11612
24. Deák, P., Snyder, L.C. and Corbett, J. W., (1988) *Phys. Rev. B* **37**, 6887-6892
25. Deák, P. and Giber, J., (1982) *Phys. Lett.* **88A**, 237
26. Baierle, R. J., Caldas, M. J., Rodrigues, C. W., Ossicini, S. and Molinari, E., (1995) *Thin Solid Films*, to be published
27. Badziag, P., Verwoerd, W. S. and Van Hove, M. A., (1991) *Phys. Rev. B* **43**, 2058
28. Craig, B. I. and Smith, P. V., (1991) *Surf. Sci.* **239**, 36
29. Sewart, J. J. P., (1991) *J. Comput. Chem.* **12**, 320
30. Verwoerd, W. S. and Weimer, K., (1991) *J. Comput. Chem.* **12**, 417
31. Dewar, M. J. S., Zoebisch, E. G., Healy, E. F. and Stewart, J. J. P., (1985) *J. Am. Chem. Soc.* **107**, 3902
32. Yang, D., Que, D. and Sumino, K., (1995) *J. Appl. Phys.* **77**, 943

THE C·Si·O·Si (·C) FOUR-MEMBER RING AND THE Si-G15 CENTRE

L. C. SNYDER AND R. WU
The University at Albany
1400 Washington Avenue, Albany, New York 12222

AND

P. DEAK
Physical Institute of the Technical University
Budapest H-111, Budafoki ut 8, Hungary

Abstract

Based on geometry optimisation molecular cluster calculations employing the MINDO/3 method, we have concluded that the interstitial-carbon interstitial-oxygen complex in silicon has a C·Si·O·Si (·C) four-member ring structure as the most stable state. Our conclusion is similar to that of Jones and Oberg who carried out a LDF·(local density functional) pseudopotential calculation on a similar molecular cluster. The computed properties of our four-member ring structure are compared with those of the Si-G15 defect.

1. Introduction

The Si-G15 defect, first observed in its electron spin resonance spectrum and reported in 1964 by Watkins [1], is one of the dominant spectroscopic features produced by the room-temperature radiation of Czochralski silicon. Later an independent observation by Almeleh and Goldstein [2] was labelled as the K centre, which is now recognised as the same species as Si-G15. Trombetta and Watkins [3] utilised stress-induced alignment as measured by both EPR (electron paramagnetic resonance) and PL (photoluminescence) to provide for the first time, an unambiguous demonstration that the extensively studied photoluminescence C-line [4] and the Si-G15 do indeed arise from the same defect. Davies *et al.* [5] examined the relationship between the C-line and an infrared local mode absorption band labelled C(3) [6,7], making the comparisons on the same sample, and presented a range of results that showed that C(3) and the C-line probably occurred at the same centre.

The EPR spectrum of Si-G15 is not produced in float-zone refined silicon, demonstrating its dependence upon oxygen for formation [1]. Observations of the Zero-phonon PL C-line energy shift due to ^{13}C confirmed the presence of a carbon atom [8].

427

R. Jones (ed.), Early Stages of Oxygen Precipitation in Silicon, 427–432.

Trombetta and Watkins proposed a [C_i , O_i] model [3]. This model contains a divalent interstitial oxygen atom and a trivalent interstitial carbon atom bonded to a common silicon atom.

We will suggest [9] that a C·Si·O·Si(·C) four-member ring is the structure of the Si-G15 defect. Our computed structure is close to the structure computed by Jones and Oberg [10] using LDF pseudopotential calculations.

2. Calculations

Our calculations have employed the MINDO/3 semi-empirical electronic structure method and the $Si_{40}H_{54}$ molecular cluster depicted in Figure 1.

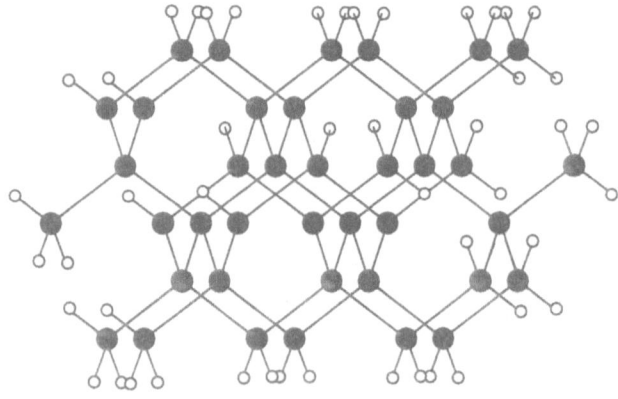

Figure 1. The basic $Si_{40}H_{54}$ molecular cluster.

In computing the equilibrium geometry of the [C_i , O_i] complex, we have varied in three dimensions the positions of fourteen nearest neighbour and next nearest neighbour silicon atoms, one carbon atom and one oxygen atom. The computed equilibrium structure of the triplet state of the uncharged [C_i , O_i] complex is depicted in Figure 2. The computed structure corresponding to the positive ion of the four-member ring model is depicted in Figure 3. Vibrational frequency calculations have been made for both the [C_i , O_i] structure and the four-member ring structure. In the calculations of vibrations we have allowed the oxygen atom, the carbon atom and four first neighbour silicon atoms to move. The remaining atoms retain the positions they have in the basic $Si_{40}H_{54}$ cluster and simply provide the potential for that motion. The revised MINDO/3 program [11] was employed for all the calculations of vibrations.

3. Calculated Geometries and Total Energy Values

For both the [C_i , O_i] and the four-member ring structures geometry optimisation calculations were made for the singlet and triplet spin state neutral systems, and for the positive ion doublet spin state. In Figure 2 we show the computed geometry for the triplet state of the neutral Trombetta and Watkins [C_i , O_i] complex. This complex was computed to be unstable in the mono-positive charge state with respect to the four-member ring structure.

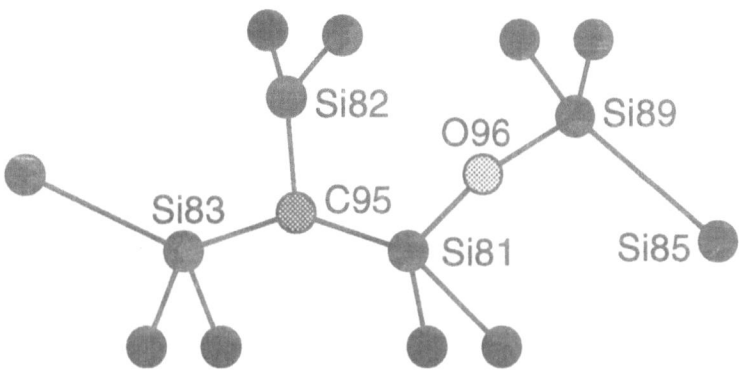

Figure 2. Computed core of the [C_i , O_i] complex

Our computed structure for the positive ion of the four-member ring complex is depicted in Figure 3. This structure is quite similar to that computed by Jones and Oberg [10]. They differ in that the Si84-O96 bond in the Jones and Oberg calculation is about 0.7 Angstroms longer.

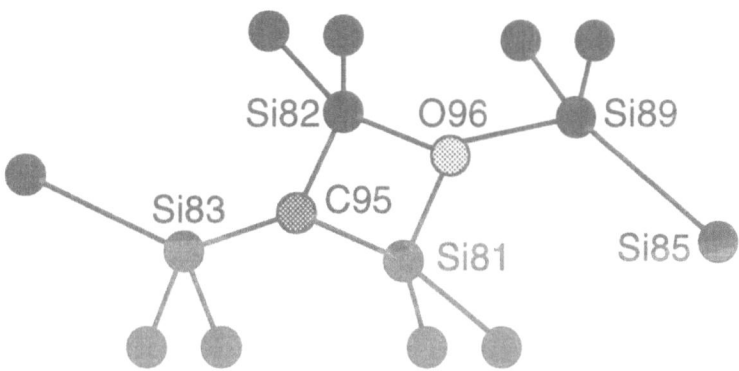

Figure 3. Computed core of the four-member ring structure

We note that the four member ring structure is isoelectronic with the dominant N pair defect in silicon and has a structure very close to that found in LDF pseudopotential calculations by Jones and Oberg [12].

In Table 1 are given the total energy computed for geometry of minimum energy for each of several electronic states of the [C_i , O_i] complex and the four-member ring.

Table 1. Computed Total Energies

Structure	State	Total Energy (eV)
Four-member ring	Singlet	-5025.60
	Triplet	-5026.32
	Mono-positive ion	-5021.15
[C_i , O_i]	Singlet	-5025.06
	Triplet	-5025.87
	Mono-positive ion	goes to 4-ring

4. Interpretation of EPR Spectra

Using a ^{13}C enriched sample (up to 60%), Trombetta and Watkins [3] found that satellites due to ^{13}C (nuclear spin I=1/2) showed the presence of a single carbon in the Si-G15 defect. They noted that the hyperfine tensor could be interpreted as an unpaired spin on a carbon atom in a p-orbital pointing in the [110] direction. This result is very similar to the previously determined property of the EPR of the Si-G12 centre, which is well known as the isolated split-interstitial carbon structure [1,7].

Table 2. Computed Diagonal Elements of the Spin Density

Structure	Atoms				
	C95		O96	Si81	Si82
four-member	0.054	s	-0.001	-0.016	-0.016
ring positive	0.060	x	-0.000	-0.007	-0.042
ion	0.048	y	-0.001	-0.039	-0.010
	0.801	z	-0.005	-0.039	-0.042
[C_i , O_i]	0.050		-0.000	-0.001	0.080
triplet	0.034		-0.002	-0.015	-0.017
	0.042		-0.002	-0.005	0.643
	0.798		-0.005	-0.037	-0.029
(C_i)$^+$	0.056			-0.017	-0.010
	0.058			-0.044	-0.015
	0.046			-0.010	-0.037
	0.822			-0.046	-0.033

In Table 2 calculated values of spin density of several systems are listed. Calculations for systems with unpaired spins were made with the unrestricted Hartree-

Fock version of the MINDO/3 program. The calculations on the mono-positive charge state of the four-member ring and the isolated split interstitial carbon will be discussed in some detail. The calculated unpaired p_z spin density of the isolated split-interstitial carbon positive ion, the carbon of the four-member ring positive ion, and the carbon of the triplet state of the [C_i , O_i] complex are 0.82, 0.80, and 0.80 respectively. The calculated values of the spin density in the p_z orbital for the four-member ring and the $[C_i$, O_i] model are very close to each other and close to that for the isolated interstitial carbon positive ion. This suggests that both four-member ring and the [C_i , O_i] complex are good candidate structures for the Si-G15 defect.

5. Interpretation of Vibrational Frequencies

The observed and computed vibrational frequencies of C-O complexes are summarised in Table 3. The IR (infrared) spectrum of C(3) exhibited two mainly carbon vibrational lines at 1115.5 cm^{-1} and 865.2 cm^{-1} as determined from their carbon isotope effects [6,13-15]. The isotope effect of the PL sidebands of the C-line assign the 861.4 cm^{-1} to carbon but could not assign a vibrational 1116.3 cm^{-1} line to the vibration of carbon or oxygen. Davies *et al.* [5] have reported two oxygen lines at 742 cm^{-1} and 550 cm^{-1} which exhibited no carbon isotope effect and thus were reported to be due to oxygen. The oxygen isotope effect in the PL experiment was large for the line at 738.0 cm^{-1} and small for the line at 548.8 cm^{-1}.

Comparison of calculated and observed vibrational frequencies of oxygen defects described earlier [11] lead to the conclusion that our calculated frequency is usually about 11% higher than the observed frequency for oxygen localised mode vibration To obtain the correction factor for a carbon vibration, the vibrational frequency calculation was done for substitutional carbon. The computed frequency is 590 cm^{-1}, which is 97% of the well-known 607 cm^{-1} C_s frequency [6]. Counting on these relations of theory to experiment, we conclude that the computed frequencies of the four member ring fit the experimental results qualitatively well. The computed pattern of vibrations for the [Ci , Oi] complex is quite different from those observed as shown in Table 3.

The calculated ^{13}C shifts of the ^{12}C vibrations of the four-member ring structure are 32.1 cm^{-1} and 22.0 cm^{-1} for the 1115.5 cm^{-1} and 865.2 cm^{-1} bands respectively, while the observed shifts are 37.2 cm^{-1} and 23.4 cm^{-1} [6].

Table 3. Computed and Observed Vibrational Frequencies (cm^{-1}). The values in parentheses are corrected using the scale factors.

Source		Carbon				Oxygen			
					Computed				
(4-ring)⁺	$^{12}C^{16}O$	1020.4	(1052.0)	779.5	(803.6)	873.2	(786.7)	673.2	606.5
	$^{13}C^{16}O$	989.3	(1019.9)	758.2	(781.6)	871.2	(784.9)	672.1	605.5
	$^{12}C^{18}O$	1018.3	(1049.9)	777.9	(802.0)	836.9	(754.0)	648.5	584.2
[C_i , O_i] triplet	$^{12}C^{16}O$	926.6		861.5	(888.1)	1141.7	(1027.9)		
					Observed				
C(3)	$^{12}C^{16}O$	1115.5		865.2		742		550	
	$^{13}C^{16}O$	1078.3		841.8					
C-line	$^{12}C^{16}O$			861.4		738.0		584.8	

The vibrational frequency calculations for the four-member ring and the [C_i , O_i] structure suggest that the four-member ring is the best candidate for the structure of Si-G15, as was concluded by Jones and Oberg [10].

References.

1. Watkins, G. D. (1965) *Radiation Damage in Semiconductors*, edited by P. Baruch, Dunod, Paris, 97.
2. Almeleh, N. and Goldstein, B. (1966) *Phys. Rev.*, **149**, 687.
3. Trombetta, J. M. and Watkins, G. D. (1987) *Appl. Phys. Lett.*, **51**, 1103.
4. Yukhnevich, A. V., and Tkachev, V. D. (1966) *Sov. Phys. Solid State*, **8**, 1004.
5. Davies, G., Oates, A. S., Newman, R. C., Woolley, R., Lightowers, E. C., Binns, M. J., and Wilkes, J. G. (1986) *J. Phys. C: Solid State Phys.*, **19**, 841.
6. Newman, R. C. (1973) *Infrared Studies of Crystal Defects*, Barnes & Noble, New York.
7. Svennson, B. G., and Lindstrom, J. L. (1986) *Phys. Stat. Sol. (a)*, **95**, 537.
8. Davies, G., Lightowlers, E. C., Woolley, R., Newman, R. C., and Oates, A. S. (1984) J. Phys. C: Solid State Phys., **17**, 499.
9. Wu, R.: Ph.D. Thesis "The Modification and Use of the MINDO/3 Method on Oxygen-Related Defects in Silicon", The University at Albany, Albany, New York (1989).
10. Jones, R., and Oberg, S. (1992) Oxygen Frustration and the Interstitial Carbon-Oxygen Complex in Si, *Phys. Rev. Letts.* **68**, 86-89.
11. Snyder, L., Wu, R., and Deak, P. (1989) Initial Applications of the Molecular Model to Compute Defect Vibrations of Oxygen in Silicon, *Radiation Effects and Defects in Solids*, **111-112**, 393-398.
12. Jones, R., and Oberg, S. (1994) Identification of the Dominant Nitrogen Defect in Silicon, *Phys. Rev. Letts.* **72**, 1882-1885.
13. Ramdas, A. K., and Rao, M. G., (1966) *Phys. Rev.*, **142**, 451.
14. Bean, A. R., Newman, R. C., and Smith, R. S., (1970) *J. Phys. Chem. Solids*, **31**, 739
15. Newman, R. C., and Bean, A. R., (1971) *Radiat. Eff.*, **8**, 189.

KINETICS OF OXYGEN LOSS AND THERMAL DONOR FORMATION IN SILICON: THE RAPID DIFFUSION OF OXYGEN CLUSTERS

S.A. MCQUAID
Laboratorio de Microelectrónica, Departamento de Física Aplicada,
Universidad Autónoma de Madrid, Cantoblanco, 28049 Madrid, Spain
R.C. NEWMAN
Interdisciplinary Research Centre for Semiconductor Materials, Imperial
College, Prince Consort Road, London SW7 2BZ, United Kingdom

Abstract. The kinetics of the loss of interstitial oxygen from solution are second order at low temperatures ($\leq 400°C$) and the rate is within a factor of 10 of that expected for loss of $2O_i$ per O_i-O_i interaction by normal oxygen diffusion. At higher temperatures the rates are reduced while the order tends to increase. This effect can be explained if dimerization is the rate limiting step during the formation of relatively large clusters. An explanation is proposed in which no complex is required to diffuse spontaneously.

1. Introduction.

Oxygen dissolved in molten silicon from the quartz crucible remains in supersaturated solution in as-grown material because the diffusion barrier (2.53eV) for isolated atoms (O_i) is larger than that of any other interstitial defect. During anneals at 450°C thermal donors (TD) form at rates which increase with the fourth power of $[O_i]$ so it was proposed [1] that (i) O-clustering was responsible, (ii) mainly O_4 were formed and (iii) these and smaller unstable clusters O_3 and O_2 were donors. Re-examination of this model revealed inconsistent requirements for $[O_3]$ and $[O_2]$ which could be avoided by assuming that O_4 formed by the reaction of two rapidly diffusing O_2 *molecules* [2], the first suggestion that O-clusters might diffuse faster than O_i. Although subsequent calculations demonstrated that such molecules would not be stable in silicon, an alternative mechanism was proposed for rapid diffusion of O_2 [3]. In any case, the O_i diffusivity, D_{oxy}, required by the original model [1] was orders of magnitude larger than that expected by extrapolation of values subsequently measured over a wide temperature range. A model of O-clustering which could account for the fact that the TD series consisted of many more than 3 centres [4] required even larger enhancements of D_{oxy}, and it was suggested that a complex with a self interstitial, O_i-Si_i, might diffuse rapidly. On the other hand, the rate of loss of $[O_i]$ during TD formation coincided with that expected for O_i-O_i interaction by normal D_{oxy} [5,6]. New models were proposed which assumed that this process followed second order kinetics while the rate of TD formation depended on $[O_i]^4$. Suppression of TD formation in material containing substitutional

433

R. Jones (ed.), Early Stages of Oxygen Precipitation in Silicon, 433–440.
© 1996 *Kluwer Academic Publishers.*

carbon ($[C_s] > 10^{17} cm^{-3}$) was argued to provide evidence for involvement of Si_i [5] so if O_2 formation caused ejection of Si_i, then TDs could be considered to be Si_i-clusters nucleated on O_3 [7]. Alternatively, if O_2 diffused rapidly and were trapped predominantly by reaction with Si_i, TDs could correspond to oxygen clusters formed in a parallel reaction, while the effect of a high $[C_s]$ might be to trap mobile O_2 [8]. However, recent measurements established that rates of *both* $[O_i]$-loss and TD-formation depend on $[O_i]^4$, but only for anneals at $\sim 450°C$ [9].

2. Measured [O_i]-loss and TD formation.

The dependence of $d[O_i]/dt$ and $d[TD]/dt$ on $[O_i]^n$, where n is the order of the formation processes, changes with anneal temperature such that n increases to ~ 8 for anneals at $500°C$ and tends to a value of 2 for those at low temperatures ($T < 400°C$) while the ratio $\Delta[O_i]/\Delta[TD] \sim 10$ remains independent of temperature in this range [9]. Measured $d[O_i]/dt$ at the low temperatures were only marginally faster (by factors of 3-10) than the rate expected due to loss of $2O_i$ by O_i-O_i interaction for normal D_{oxy} (Fig. 1). Tendency to higher order kinetics at higher temperatures was shown to coincide with increasing *reductions* in the rate of the

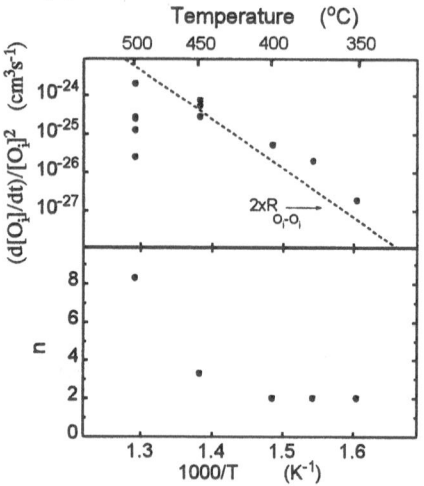

Figure 1. Measured rates of [O_i]-loss during isothermal anneals with a line representing the rate expected for O_i-O_i by D_{oxy} (above) and the dependence on $[O_i]^n$ in different samples (below).

process and the uniformity of these reductions with anneal time indicated that the concentration of the unstable O-cluster was in equilibrium with $[O_i]$ during the anneals.

3. Oxygen clustering: a serial reaction.

Clustering of randomly diffusing O_i is described by $O_i + O_m \rightleftarrows O_{m+1}$, where O_m represents a cluster of m O-atoms. During anneals of an initially dispersed solution of $[O_i] = 10^{18} cm^{-3}$ using a value of $k_{fl} = 2 \times 10^{-25} cm^3 s^{-1}$ (corresponding to $k_{fl} = 4\pi r_c D_{oxy}$ for normal D_{oxy} at $450°C$ and $r_c = 5Å$) most of the O_i atoms lost from solution were incorporated as O_2 dimers so that $[O_2]$ tended to $[O_i]$ and the concentration of each successively larger cluster was smaller by more than an order of magnitude (Fig. 2a). In this immobile cluster model, reduction of $d[O_i]/dt$ could only occur by dimer dissociation, a time-independent rate reduced by the factor $[O_2]/[O_i]$ would be observed when this ratio had reached a steady-state value (Fig. 2b) and the reduced rate would depend on $[O_i]^3$ [9]. The magnitude of the reduction derived from measurements (less than by a factor of 2, Fig. 1) at a temperature at which such kinetics are expected would imply $[O_2]_{eq} \geq ([O_i]/2)$. Absence of significant $[O_i]$-loss before the reductions became effective would then require that as-grown material contained as many O-atoms in dimers as in isolated form. Similarly, dissociation of all clusters smaller than O_8 could account for n=8 but, in the same way, as-grown material would then have to contain as many O-atoms in all clusters up to O_7 as in isolated form. This is an unrealistic description of as-grown material and, besides, the observed increases in order do not coincide with changes of $\Delta[O_i]/\Delta[TD]$, suggesting that the clusters formed (whatever the structure of TD) are not very different over the whole temperatures range.

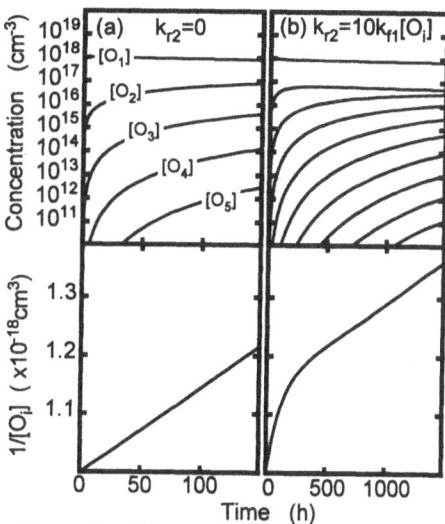

Figure 2. $[O_i]$-loss and formation of stable immobile clusters (a) and the effect of dimer dissociation (b).

3.1. Effect of O_2 reconfiguration.

If O_2 quickly relaxed into a more stable configuration (e.g. by $O_2 \rightarrow O_2V + Si_i$ [5]) then prior dissociation could account for uniform reductions in $d[O_i]/dt$. However, models demonstrated that the reduced rate would remain second order, irrespective of whether Si_i formed clusters and whether reverse reactions were due to $Si_i + O_2V$ recombination. We conclude that the observed changes in n (Fig. 1) cannot be explained by models which assume that oxygen clusters predominantly in O_2-type defects. Note that if O_2V was assumed to diffuse rapidly to O_i then a transition to third order kinetics was induced by its dissociation ($O_2V \rightarrow 2O_i + V$) at a rate which was much greater than that of its capture. Correspondingly, if an O_2Si_i complex were to diffuse rapidly to O_i, its relatively fast dissociation would induce increases of order up to n=3. In effect, these models are elaborations of the proposal that O_2 diffuses rapidly.

3.2. Effect of spontaneous rapid diffusion of O_2.

If O_2 dimers diffused much more rapidly than O_i atoms ($D_{O2} \gg D_{oxy}$) a steady-state dimer concentration would be established even if they were stable. The $[O_i]$-loss process would remain second order but lead mainly to increases in $[O_3]$. Dimer dissociation at a rate which was increasingly competitive with that of its diffusion to O_i would reduce $d[O_i]/dt$ with insignificant incubation time and n would tend to increase reaching a maximum of n=3 when the overall rate had been reduced by more than a factor of 10 [9]. In general, the maximum order obtainable by such a serial reaction would correspond to the number of O-atoms in the cluster whose formation results from the irreversible capture of the diffusing dimers. In other words, if trimers dissociated quickly at all temperatures ($O_3 \rightarrow O_2 + O_1$) and larger clusters (m > 3) were stable, then formation of the smallest of these (O_M) would establish steady-state values of $[O_3]$ and $[O_2]$. Addition of further O_2 to O_M would cause a family of stable clusters to evolve rapidly. Dimer dissociation would then reduce $d[O_i]/dt$ and cause n to increase up to a maximum value of M, after the overall rate had been reduced by a factor of 10 and $n \approx m/2$ for reductions by a factor of ~ 2 [10]. Thus $n \sim 4$ for anneals at $\sim 450°C$ and $n \sim 8$ for anneals at $\sim 500°C$ and the corresponding reductions of rate (Fig. 1) would suggest that $M \sim 8$ at *both* temperatures. Dissociation of smaller clusters would have had to occur by emission of O_2 rather than O_i to comply with the measured second order kinetics at low temperatures ($T \leq 400°C$), possibly a further consequence of a relatively low barrier for dimer diffusion.

3.3. Thermal donor formation by dimer agglomeration.

The series of stable oxygen clusters could then be associated with TDs and a more reasonable evolution of dissociation rates with cluster size would lead to the predominance of an odd numbered series [11]. Clusters O_m (m$\rangle$2) were assumed to dissociate by O_2-emission at a rate which decreased by a factor of 5 for each additional atom contained and trimer dissociation rate was chosen so that ($[O_3]/[O_2])_{eq}$ was large so the dominant clusters were all members of the same series (O_{2N+1}). Clusters larger than O_7 trapped O_2 much faster than they dissociated and a value of $D_{O2}/D_{oxy}=3 \times 10^7$ (comparable to that deduced if near-surface out-diffusion was due to D_{O2} [8]) ensured that the incubation time was short. Weak infrared (IR) absorption bands at 1012 and 1006cm^{-1} detected in as-grown material [12] were supposed to correspond to the smaller unstable clusters (O_3 and O_5) since they relate to two different O-clusters (distinct from those of TDs) whose concentrations quickly reach equilibrium during anneals. Obviously the model was designed to account for measured changes in rate and order of d$[O_i]$/dt and d[TD]/dt but modifications were necessary to reproduce the well-known evolution of TDs.

Stable O-clusters reached the same maximum concentrations in the initial simulation (Fig. 3a), whereas measured maximum concentrations of TDN increase by a factor of around 30 from N=1 to 3 [13,14]. For a serial reaction, this implies that the corresponding forward rates *decrease* by these factors [13], contrary to the expectation for trapping by clusters of increasing size. Applying empirical factors reproduced the increasing maximum concentrations but in addition caused $\Delta[O_i]/\Delta[O_m]$ to

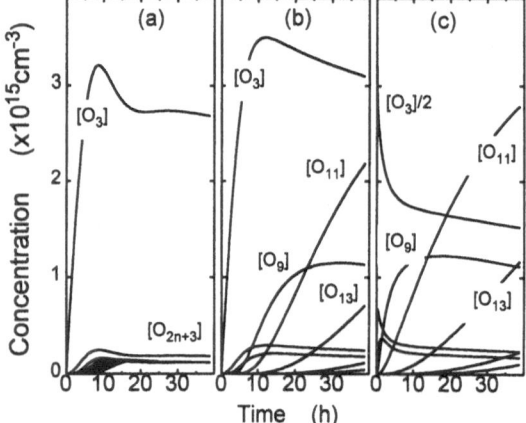

Figure 3. O-Clustering by dimer agglomeration where clusters larger than O_7 are stable (a) effect of reductions in capture rates for clusters associated with TDs (b) and that of initial $[O_3]$ and $[O_5]$ (c).

tend to a constant value of 11, as required, because d$[O_{11}]$/dt (d[TD3]/dt) dominated (Fig. 3b). Such reductions might be physically justified if O_2 alternated between neutral and donor configurations in its diffusion path [3] so that Coulombic repulsion from TDN (N$\geq$3) occurred, while this effect might be smaller for TDs 1 and 2 since they can be present in a configuration X without electrical activity [14]. However, the relative stability of X at room temperature appears to require the Fermi level to be high in the gap, which is not expected at the anneal temperatures. The observed decrease by a factor of $\sim$2 of the 1012cm^{-1} IR-band, at a rate similar to that at which [TD2] reached it's maximum value during anneals of as-grown material [12] was reproduced by setting an appropriate initial $[O_3]$ relative to its equilibrium value (Fig. 3c), but $[O_3]/[O_3]$ reached a steady-state value relatively quickly unlike the observed behaviour for 1006cm^{-1} IR band. The defect giving rise to IR absorption at 1006cm^{-1} does not appear to be an earlier member of the same series which include TDs. While the interaction $O_2 + O_1 \rightleftarrows O_3$ might lead to catalysed diffusion jumps of O_i and thereby account for the apparent enhancement of D_{oxy} by a factor ~ 3 at low temperatures (Fig. 1), the fact that $[O_2] \propto [O_i]^{0.5}$ would imply that true second order kinetics would not be observed. For these reasons, this model seems to be unrealistic. The most obvious problem is that cluster calculations [3] for pure O-agglomorates did not reproduce TD properties.

4. Involvement of self-interstitials.

A cluster of two O and one Si_i has been proposed to account uniquely for TD activity [3] and the different centres in the series would then arise from additional O or Si_i added to this core. Thus Si_i would have to be generated at a rate not smaller than $d[O_i]/dt$ by more than a factor of 10. The rate of $Si_i + V$ generation by Frankel pair dissociation estimated using the same prefactor as that for O_i hopping and an activation energy of 7eV (the sum of typically calculated formation energies of Si_i and V) was ~ 22 orders of magnitude too small. The activation energy would have had to have been only ~ 4eV. So if TDs contain at least one Si_i then we are led to suppose that Si_i are ejected by O-clusters. Since the strain in O_2 has been argued to be insufficient to cause such ejection [3], some larger cluster must be responsible and its formation at the required rate would imply that dimerization was the rate limiting step in its formation, reinforcing the implication that O_2 diffuses rapidly.

Supposing O_3 relaxes by Si_i emission at a rate much faster than that at which another O_i diffuses to it, then the initial rate of Si_i introduction would be close to $d[O_3]/dt$, i.e., more than an order of magnitude smaller than the rate of O_i-O_i interaction (Fig. 4a). However, the Si_i-diffusivity, D_{Si}, appears to be so large that it has never been unambiguously evaluated. Using the same prefactor as that for D_{oxy} but with an activation energy of 0.2eV leads to $D_{Si}/D_{oxy} \sim 1.7 \times 10^{16}$ at 450°C. Rapidly diffusing Si_i would have to travel large distances before recombining with a centre from which a Si_i had previously been ejected $(O_3 V)$. A value of $D_{Si} = 1.7 \times 10^{12} D_{oxy}$ was sufficient to cause the flux of Si_i ($k_{fl}[Si_i]$) to exceed that of O_i ($k_{fl}[O_i]$) by a factor of 10^5. Accounting for the formation of complexes (below) in reasonable computation times restricts the value of D_{Si} which can be used to $\sim 1.7 \times 10^6 D_{oxy}$ but even so the flux of Si_i tended to exceed that of O_i by a factor of 10^3 (Fig. 4a). Obviously Si_i could interact with the other defects present in higher concentrations. Complexes attributed to O_i-Si_i have been detected following low temperature irradiation but they dissociate before the samples reach room temperature [15]. Enhanced oxygen diffusion during room temperature irradiation has been adequately accounted for by sequential trapping of V and Si_i by O_i [16]. Since there is no evidence that $Si_i \rightleftarrows O_i$ causes catalysed diffusion jumps of O_i then the absence of large enhancements of D_{oxy} during anneals does not exclude the possibility that the Si_i flux is relatively large. If Si_i were trapped efficiently by O_2 then such a large flux would be prevented, irrespective of how fast it diffused (Fig. 4b). On the other hand, if O_2Si_i were unstable then this constraint would not apply. Formation and dissociation of what is assumed to be a TD-core could cause O_2 to make a diffusion jump. Limitation of the value of D_{Si} was compensated by assuming that the diffusion jump distance (d_J) was artificially large ($d_J = 100d_{nn}$, the nearest neighbour distance) (Fig. 4c). If the real value of D_{Si} was 10^4 larger than that used in these calculations then $d_J = d_{nn}$ would lead to a similar evolution.

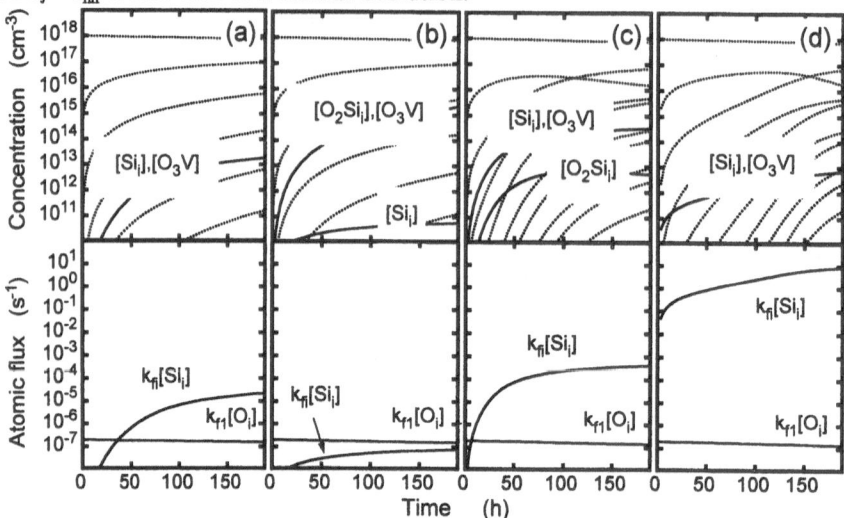

Figure 4. $[O_i]$-loss and cluster formation with $O_n \rightleftarrows O_nV + Si_i$ for $n \geq 3$ (upper) and atomic fluxes of Si_i and O_i (bottom). The effects of (b) stable trapping of Si_i by O_2, (c) O_2Si_i dissociation with a catalysed diffusion jump and (d) a similar process but where a diffusion jump occurred for $Si_i \rightleftarrows O_2Si_i$.

If O_2 undergoes such catalysed diffusion jumps, the measured increases of n (Fig. 1) can be explained in a new way. Increasing the rate of dimer dissociation reduced $d[O_i]/dt$ after an incubation period of $\sim 100h$ and the order tended to a maximum value of 5 for reductions by more than a factor of 10. Where a complex had been assumed to diffuse spontaneously leading to O_3-formation, the maximum value of n was 3 (see above). Modifications of the present model ignoring O_2Si_i formation so that a larger value of D_{Si}/D_{oxy} (1.7×10^{13}) and $d_J = d_{nn}$ were permitted, demonstrated that if Si_i was released by O_4 or O_5, maximum orders of 6 or 7 respectively were achieved in the same way. In general then, n tended to $M + 2$ where M is the number of O-atoms in the first cluster which releases Si_i faster than another O_i diffuses to it. Although it would thus be possible to explain the kinetics of $[O_i]$-loss by assuming that Si_i were ejected quickly after formation of O_6, the generation of Si_i at a rate not less than 1 order of magnitude smaller than $d[O_i]/dt$ would not then be possible.

Nucleation of $O_2(Si_i)_n$ (one possible identification of the TD series) would not prevent catalysis provided that the formation rate was smaller than the rate of $[O_i]$-loss by many orders of magnitude. This condition aparently precludes such association with TDs and, in any case, their formation did not increase the maximum order realisable. Considering instead reactions between Si_i and larger O-clusters, then if Si_i formed a stable complex with O_3 the enhanced diffusion of O_2 would be prevented, but if O_3Si_i were also unstable (assumed to dissociate at the same rate as O_2Si_i) then catalysis would still occur but again the maximum order was not changed. It is conceivable that O_3 may also undergo catalysed diffusion jumps, but dimer dissociation under these circumstances led to only fractionally higher orders than $n=5$, irrespective of whether Si_i were released by O_3 or O_4 (since O_3 formation remained the rate limiting step in O_4-formation and therefore Si_i emission). Thus subsequent forward reactions do not affect the changes of order, as was the case for the spontaneously diffusing dimer model (Section 3.2). Trimer dissociaton served only to counteract the catalysis of O_2 diffusion and thus could not account for reductions with increasing n (Section 3). To increase the maximum value of n in this model, it is necessary to increase the order of the Si_i-O_2 interaction. This can be realised by assuming that the diffusion jump occurred by the interaction $Si_i \rightleftarrows O_2Si_i$. Simulation of this process proved prone to instability and the problem was avoided by not explicitly calculating $[O_2Si_i]$ so that $D_{Si}/D_{oxy} = 1.7 \times 10^{12}$ could be used (Fig. 4d). Note that since $[O_2Si_i]$ would be similar to $[O_3V]$ then catalysis of D_{O2} would be of the same order as $Si_i + O_3V$ recombination, thus catalysis of O_3 diffusion might occur and be of comparable magnitude. The Si_i-emission rate was set larger than that at which another O_i would diffuse to O_n by the factor 4×10^3. For an assumed jump distance $d_J = d_{nn}$ then $[O_2Si_i] \sim [Si_i][O_2]/10^{15} cm^{-3}$ so that results (Fig. 4d) would lead to $[O_2]/[O_2Si_i] \sim 100$. The essential test was that increasing dimer dissociation in this case led to reduced $d[O_i]/dt$ so that n tended to a maximum of ~ 9, i.e., $\sim 2(M+2)$, sufficient to account for experimental observations.

5. Discussion and Conclusions.

Second order kinetics measured at low temperatures establish that O_i-O_i is the rate limiting step in $[O_i]$-loss and TD-formation. Variations in order with dissociation as the temperature is increased provides information about the next rate limitation. If a complex diffuses spontaneously then this would correspond to the formation of a stable agglomorate which would have to contain ~ 8 O-atoms, so the problem of how smaller clusters dissociate would arise. If small clusters diffuse by catalytic interaction with Si_i, then the control of their diffusivity by $[Si_i]$ would introduce feedback into the serial reaction of O-clustering. The second rate limitation would then involve the ejection of Si_i, and quantitatively the cluster responsible during the initial stages of anneals could be as small as O_3. This kinetic model would be consistent with the proposed identity of the TD core [3]. The requirement that Si_i is not trapped by any stable cluster would require that different members of the TD series arise from incorporation of additional O-atoms rather than additional Si_i. Dimer diffusion to O_i rather than O_M would not automatically account for the formation of a series of O-clusters implied by the TD evolution. It would be necessary to suppose that larger O-clusters also diffuse by fast emission of Si_i and recombination. Since [TD] would merely be the fraction of $[O_x]$ to which an Si_i is weakly bound, the problem of reduced forward rates for the early members in the TD series may relate to increased binding of Si_i and/or decreasing probability that catalytic diffusion jumps of the corresponding O_m occur. Such cluster diffusion would lead to the loss of more than 2 O_i per O_i-O_i interaction, so the apparent enhancement of D_{oxy} by a factor of ~ 3 at low temperatures (Fig. 1) may not be real, avoiding any contradiction with the observed $n=2$. Note that we have not demonstrated

440

$\Delta[O_i]/\Delta[O_mSi_j] \sim 10$, but since the clusters predominantly formed are all assumed to emit Si_i then in principle this result seems achievable. The correlation of annealing behaviour of the 1012 and 1006cm^{-1} IR bands and [TD] [12] could be explained if they related to O_xV_y clusters. During extended anneals nucleation of stable Si_i aggregates would be expected to occur (possibly on O-clusters). Sinking of the catalyst would be consistent with a) loss of TD activity, b) reduction of $d[O_i]/dt$ by factor of ~ 3 without change of n [9], c) black-dot contrast in TEM images (O-clusters) and d) Ribbon-like defects $(O_x(Si_i)_n)$. Suppression of TD formation in carbon-doped material presumably occurs as originally suggested [5], i.e., $Si_i + C_s \rightleftarrows C_i$, with the added implication that C_i also catalyses dimer diffusion. Since some large cluster should release a second Si_i, then the observed loss of $2O_i$s per C_s [5] would suggest that this occured for O_4 or O_5.

We thank J. Piqueras for help and encouragement, T. Hallberg for providing a copy of his thesis and R. Jones for very useful comments. Work was funded by EC HCM programme P-24 0920 176.

References
1. Kaiser, W., Frisch, H.L., and Reiss, H., (1956) Mechanism of the formation of donor states in heat-treated silicon, *Phys. Rev.* **112**, 1546-1554.
2. Gösele, U. and Tan, T.Y., (1982) Oxygen diffusion and thermal donor formation in silicon, *Appl. Phys. A* **28**, 79-92.
3. Deák, P., Snyder, L.C. and Corbett, J.W., (1992) Theoretical studies on the core structure of the 450°C oxygen thermal donors in silicon, *Phys. Rev. B* **45**, 11612-11626.
4. Ourmazd, A., Schröter, W. and Bourret, A., (1984) Oxygen-related thermal donors in silicon: A new structural and kinetic model, *J. Appl. Phys.* **56**, 1670-1681.
5. Newman, R.C., (1985) Thermal donors in silicon: oxygen clusters or self-interstitial aggregates, *J. Phys. C* **18**, L967-L972.
6. Tan, T.Y., Kleinhenz, R. and Schneider C.P., (1986) On the kinetics of oxygen clustering and thermal donor formation in Czochralski silicon, *Mater. Res. Soc. Symp. Proc.* **59**, 195-204.
7. Mathiot, D., (1987) Thermal donor formation in silicon: A new kinetic model based on self-interstitial aggregation, *Appl. Phys. Lett.* **51**, 904-906.
8. Gösele, U., Ahn, K.-Y., Marioton, B.P.R., Tan, T.Y. and Lee, S.T., (1989) Do oxygen molecules contribute to oxygen diffusion and thermal donor formation in silicon?, *Appl. Phys. A* **48**, 219-228.
9. McQuaid, S.A., Binns, M.J., Londos, C.A., Tucker, J.H., Brown, A.R. and Newman, R.C., (1995) Oxygen loss during thermal donor formation in Czochralski silicon: New insights into oxygen diffusion mechanisms, *J. Appl. Phys.* **77**, 1427-1442.
10. McQuaid, S.A., Newman, R.C. and Muñoz, E., (in press) The role of rapidly diffusing dimers in oxygen loss and the association of thermal donors with small oxygen clusters, *Mat. Sci. Eng. B*
11. McQuaid, S.A., Newman, R.C. and Muñoz, E., (in press) Models of oxygen loss and thermal donor formation in silicon by the clustering of rapidly diffusing oxygen dimers, Proc. of 18th Int. Conf. on Defects in Semiconductors, Sendai, Japan.
12. Lindström, J.L. and Hallberg, T., (1995) Vibrational infrared-absorption bands related to thermal donors in silicon, *J. Appl. Phys.* **77**, 2684-2690.
13. Borenstein, J.T., Peak, D. and Corbett, J.W., (1986) Formation kinetics of thermal donors in silicon, *Mater. Res. Soc. Symp. Proc.* **59**, 173-179.
14. Wagner, P. and Hage, J., (1989) Thermal double donors in silicon, *Appl. Phys. A* **49**, 123-138.
15. Brelot, A. and Charlemagne, J., (1971) Infrared studies of low temperature electron irradiated silicon containing high concentrations of Germanium and Carbon, in J.W. Corbett and G.D. Watkins (eds.), *Radiation effects in semiconductors*, Gordon and Breach, London, pp. 161-169.
16. Oates, A.S., Binns, M.J., Newman, R.C., Tucker, J.H., Wilkes, J.G. and Wilkinson, A., (1984) The mechanism of radiation-enhanced diffusion of oxygen in silicon at room temperature, *J. Phys. C: Sol. St. Phys.* **17**, 5695-57051.

MOLECULAR DYNAMICS STUDY OF OXYGEN DEFECTS IN SILICON

P. J. GRÖNBERG[1], J. VON BOEHM[2] AND R. M. NIEMINEN[1]

[1] *Laboratory of Physics, Helsinki University of Technology, 02150 Espoo, Finland*
[2] *Laboratory of Computational Dynamics, Helsinki University of Technology, 02150 Espoo, Finland*

The molecular dynamics method in conjunction with the classical Jiang-Brown interaction model has been adopted for investigating oxygen-related defects in silicon. In addition to reproducing the results by Jiang and Brown we find that the diffusion barrier for vacancy-oxygen diffusion is 1.84 eV, which is substantially lower than the value of 2.5 eV for intrinsic oxygen diffusion.

1. Introduction

Most silicon wafers are grown by the Czochralski process, in which a single-crystal seed is first dipped into the silicon melt and then slowly withdrawn. The crucible containing the melt is typically made of quartz (SiO_2). Because molten silicon reacts to a large extent with every crucible material, oxygen enters the melt as the silica dissolves in the liquid silicon. Part of the dissolved oxygen is incorporated into the growing crystal with concentrations up to 10^{18} cm^{-3} [1]. It is now believed that an oxygen atom breaks a Si$-$Si bond in the $< 111 >$ direction and forms two Si$-$O bonds to occupy a puckered, bond centered, interstitial position [1]-[9].

Different techniques have been used to study oxygen diffusivity in silicon at high temperatures and the agreement among the results is excellent [1]. In this intrinsic oxygen diffusion the oxygen atoms are believed to jump from one interstitial site to another with the activation energy of 2.5 eV [8], [10]. Heck *et al.* [11] found differences concerning diffusivity at 1100^{0}C and suggested that oxygen diffuses via a vacancy-dominant mechanism. Lee and Nichols [12], however, found the independence of oxygen diffusivity on point-defect concentrations in the sample at temperatures 700-1160^{0}C.

R. Jones (ed.), Early Stages of Oxygen Precipitation in Silicon, 441–446.
© *1996 Kluwer Academic Publishers.*

Enhanced oxygen diffusivity in silicon has been observed at temperatures below 700^0C [13], [14]. This anomalous oxygen diffusivity has been explained with the formation of vacancy-oxygen complexes (A centers), which capture Si-interstitial and emit interstitial oxygen atoms [14], [15]. Such a process is assumed to occur more easily than the intrinsic oxygen jump from one interstitial site to another. On the other hand, no evidence of interaction between silicon interstitials and oxygen atoms in the temperature range 25-500^{0}C has been given [1].

We use molecular dynamics (MD) simulations to study oxygen-related defects and oxygen diffusion in silicon. The interactions between atoms are modelled with the empirical potential constructed by Jiang and Brown [16].

2. Methods

The Jiang-Brown (JB) potential is constructed by combining the Stillinger-Weber (SW) potential [17] originally proposed for pure silicon with the potential developed by van Beest et al. [18], [19] for simulating the polymorphs of silica. This composite interatomic potential can be written as follows [16]:

$$\mathcal{V} = \sum_i e_i(q_i) + \sum_{i<j} [\phi_{ij} + \nu_2(i,j)] + \sum_{i<j<k} \nu_3(i,j,k) . \qquad (1)$$

where (i,j,k) are taken over all atoms. In this model effective charges q_i have been given for the atoms to take into account the polarization that an electronegative oxygen atom induces on the surrounding silicon atoms.

The first term in (1) consists of the ionization energies $e_i(q_i)$:

$$e_i(q_i) = \begin{cases} e_0 \exp[\frac{-1}{(q_i-q^0_{Si})}] , & \text{if } q_i > q^0_{Si} \text{ and } i \subset \text{Si} \\ e_0 \exp[\frac{1}{(q_i-q^0_O)}] , & \text{if } q_i < q^0_O \text{ and } i \subset \text{O} , \\ 0 & \text{otherwise} \end{cases} \qquad (2)$$

where e_0, q^0_{Si} and q^0_O are adjustable. The ionization energies restrict the values of the effective charges on the atoms and thus prevent an oxygen atom from making too many bonds with the neighboring silicon atoms.

The second sum in (1) consists of the pairwise interactions, where the first term ϕ_{ij} is originally the two-body term from the van Beest potential. It contains the long-range Coulombic interactions between the atoms having an effective charge. The pair potential term $\nu_2(i,j)$ in (1) contains the pair term from the SW potential multiplied by a so-called bond-softening function g_{ij} defined as:

$$g_{ij} = \begin{cases} \exp\left[\frac{1}{q_s}\right] \exp\left[\frac{1}{q_i+q_j-q_s}\right] , & \text{if } q_i + q_j < q_s \\ 0 & \text{if } q_i + q_j \geq q_s \end{cases} , \qquad (3)$$

where q_s is a parameter. The bond-softening function weakens the interaction between the two Si-atoms adjacent to an oxygen atom.

The last sum in (1) contains the three-body terms from the SW potential multiplied by the corresponding bond-softening functions g_{ij} (3). The three-body terms maintain the diamond structure of the silicon crystal.

A silicon lattice of density $\rho = 4.9942 \cdot 10^{28}$ m^{-3} was constructed by repeating periodically an MD box containing 512 atoms. An additional cutoff distance r_{cut} was introduced for the van Beest pair potential term ϕ_{ij}. The cutoff distance was chosen to be larger than the longest distance between two charged atoms inside the MD supercell so that all the long-range Coulombic interactions were included.

Properties of the system at zero temperature were computed using a quasidynamic method where a large positive friction coefficient was used to damp out kinetic energy. At every 20th time step the volume of the MD box was changed to lower the internal pressure of the system towards 0 Pa.

The diffusion path and the diffusion barrier for the intrinsic oxygen diffusion were determined by a series of energy minimization calculations performed with a constraint of a constant cone angle θ between the Si-O bond and the axis connecting the two silicon atoms bonded to the oxygen atom in the initial configuration [20]-[21]. By increasing the angle θ gradually after every relaxation the oxygen atom was driven from the initial position to the next equilibrium interstitial site in the (110) plane.

Intuitively we assumed that a vacancy-oxygen complex might diffuse with a following mechanism: The silicon atom Si1 jumps into the vacant site, while the oxygen atom O moves toward the new vacant site previously occupied by the atom Si1 (see Fig. 1). The bond between the oxygen atom O and the silicon atom Si2 breaks up and a new bond is formed between the atoms O and Si3. The bond between the atoms O and Si1 is preserved. The resulting configuration corresponds to the vacancy-oxygen complex, which is migrated by a single jump from the initial configuration. Let the displacement vectors from the initial to the final positions for the atoms O and Si1 be $\vec{d}_O$ and $\vec{d}_{Si1}$, respectively. The diffusion of the vacancy-oxygen complex was investigated by a series of energy minimization calculations, where the atoms O and Si1 were forced to move on the separate planes perpendicular to the displacement vectors $\vec{d}_O$ and $\vec{d}_{Si1}$, respectively. By moving these planes simultaneously step by step in the direction of the respective displacement vectors the atoms O and Si1 were driven from their initial positions toward the next equilibrium configuration of the vacancy-oxygen complex. The relative displacement x changing from 0 to 1 describes

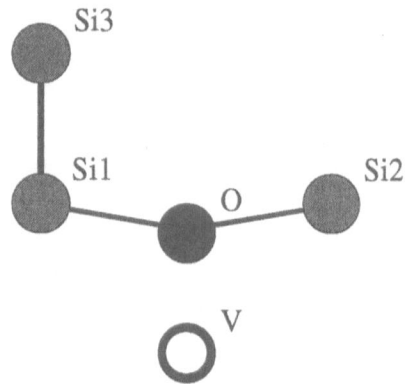

Figure 1. The initial configuration of the vacancy-oxygen complex.

the simultaneous shifts of the planes. The energy minimization calculation was performed at each step.

3. Results and discussion

We find that the equilibrium site of an oxygen interstitial is located in the middle of a Si-Si bond with the Si-O bond length of 1.78 Å in agreement with [16], [20], [22]. *Ab initio* calculations [4]-[9] give a slightly different, puckered configuration. The formation energy for the oxygen interstitial was calculated to be $E_f(O) = 1.13$ eV in fair agreement with [16], [21].

Our calculations reproduce the same configuration for the O_2-cluster as obtained by Jiang and Brown [21]. This configuration differs from the staggered O_2-cluster obtained by Needels *et al.* [7], because the oxygen atoms are not bonded to a common silicon atom. Our calculated binding energy for the O_2-cluster $E_b(O_2) = 2E_f(O) - E_f(O_2) = 1.30$ eV differs slightly from the Jiang-Brown value of 1.18 eV [21]. This deviation might stem from the fact that we have used a larger MD box and apparently a larger cutoff than in [21]. Further studies of larger oxygen clusters were not performed, because the JB potential [16] unfortunately contains an error and will not go smoothly to the correct SiO_2 limit [22].

Our calculated equilibrium configuration of a vacancy-oxygen complex is similar to that proposed by Watkins and Corbett [23]. The oxygen atom lies above the vacant site and is bonded to the two neighboring silicon atoms in the (110) plane. Our calculated angle between the Si-O bonds and the Si-O bond length are 169^0 and 1.84 Å, respectively, close to the values 168^0 and 1.82 Å of [16]. We find the oxygen atom to be displaced from the vacant site by 1.15 Å in reasonable agreement with the calculated displacement of 1.0 - 1.1 Å by DeLeo *et al.* [24] and the displacement of 0.95 Å obtained by van Oosten *et al.* [25] from an *ab initio* cluster calculation. Our calculated binding energy for the vacancy-oxygen complex is $E_b(VO) = E_f(O) + E_f(V) - E_f(VO) = 2.90$ eV.

During the intrinsic oxygen diffusion simulation the total energy of the system E was calculated at each step. E as a function of the angle θ agrees well with the calculation by Jiang and Brown [20]-[21]. The energy barrier of 2.4 eV is obtained at $\theta = 78^0$. The method used may not give the global minimum of the energy barrier from one interstitial site to another but it gives one possible diffusion path over the coupled-barrier described by Ramamoorthy and Pantelides [26].

Our results of the MD simulation for the vacancy-oxygen complex diffusion are presented in Fig. 2. The top of the barrier is reached at x=0.53. However, our method does not produce the expected final equilibrium configuration (the expected path is indicated by the dotted line in Fig: 2) but a non-equilibrium one in a local minimum at 1.3 eV. Nevertheless, we expect our calculated value of 1.84 eV to faithfully represent the global minimum of the energy barrier. The diffusion barrier for the vacancy-oxygen complex thus appears to be about 0.7 eV lower than the intrinsic diffusion barrier of 2.5 eV, which indicates that the oxygen atom diffuses more easily as a part of the vacancy-oxygen complex than via jumps from one interstitial site to another.

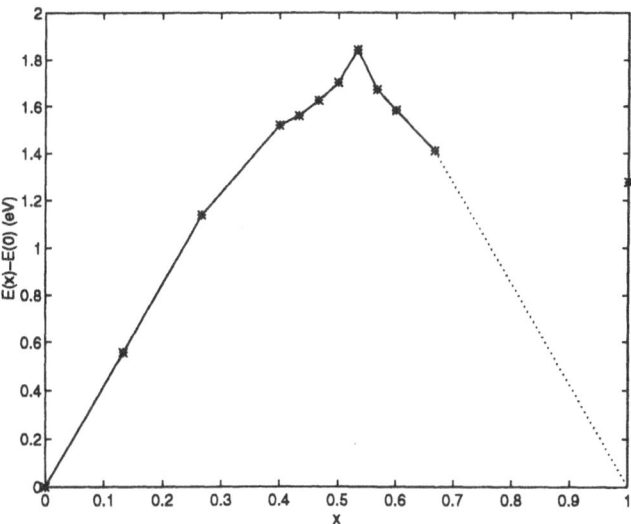

Figure 2. The energy difference between the intermediate and the initial configuration is plotted as a function of the relative displacement x in the simulation of the vacancy-oxygen complex diffusion.

Acknowledgements

This work has been supported by a grant from VTT Electronics, Espoo, Finland.

References

1. A. Borghesi, B. Pivac, A. Sassella and A. Stella, J. Appl. Phys. **77**, 4169 (1995).
2. J. W. Corbett, R. S. McDonald and G. D. Watkins, J. Phys. Chem. Solids **25**, 873 (1964).
3. D. R. Bosomworth, W. Hayes, A. R. L. Spray and G. D. Watkins, Proc. R. Soc. London Ser. A **317**, 133 (1970).
4. J. Plans, G. Diaz, E. Martinez and F. Yndurain, Phys. Rev. B **35**, 788 (1987).
5. E. Martinez, J. Plans and F. Yndurain, Phys. Rev. B **36**, 8043 (1987).
6. M. Saito and A. Oshiyama, Phys. Rev. B **38**, 10711 (1988).
7. M. Needels, J. D. Joannopoulos, Y. Bar-Yam and S. T. Pantelides, Phys. Rev. B **43**, 4208 (1991).
8. P. J. Kelly and R. Car, Phys. Rev. B **45**, 6543 (1992).
9. R. Jones, A. Umerski and S. Öberg, Phys. Rev. B **45**, 11321 (1992).
10. J. C. Mikkelsen, Jr., in *Oxygen, Carbon, Hydrogen and Nitrogen in Silicon* edited by J. C. Mikkelsen, Jr., S. J. Pearton, J. W. Corbett and S. J. Pennycook, (Materials Research Society, Princeton, NJ, 1986), p. 19.
11. D. Heck, R. E. Tressler and J. Monkowski, J. Appl. Phys. **54**, 5739 (1983).
12. S. T. Lee and D. Nichols, in *Oxygen, Carbon, Hydrogen and Nitrogen in Silicon* edited by J. C. Mikkelsen, Jr., S. J. Pearton, J. W. Corbett and S. J. Pennycook, (Materials Research Society, Princeton, NJ, 1986), p. 31.
13. M. Stavola, J. R. Patel, L. C. Kimerling and P. E. Freeland, Appl. Phys. Lett. **42**, 73 (1983).
14. A. S. Oates, M. J. Binns, R. C. Newman, J. H. Tucker, J. G. Wilkes and A. Wilkinson, J. Phys. C. **17**, 5695 (1984).
15. R. C. Newman, in *Defects in Electronic Materials*, edited by M. Stavola, S. J. Pearton and G. Davies (Materials Research Society, Pittsburgh, PA, 1988), p. 25.
16. Z. Jiang and R. A. Brown, Chem. Eng. Sci. **49**, 2991 (1994).
17. F. H. Stillinger and T. A. Weber, Phys. Rev. B **31**, 5262 (1985).
18. B. W. H. van Beest, G. J. Kramer and R. A. van Santen, Phys. Rev. Lett. **64**, 1955 (1990).
19. G. J. Kramer, N. P. Farragher and B. W. H. van Beest, Phys. Rev. B **43**, 5068 (1991).
20. Z. Jiang and R. A. Brown, Phys. Rev. Lett. **74**, 2046 (1995).
21. Z. Jiang and R. A. Brown, in *Defect- and Impurity- Engineerred Semiconductors and Devices*, (MRS Proceeding, in press).
22. Private communications from Z. Jiang.
23. G. D. Watkins and J, W. Corbett, Phys. Rev. **121**, 1001 (1961).
24. G. G. DeLeo, W. B. Fowler and G. D. Watkins, Phys. Rev. B **29**, 3193 (1984).
25. A. B. van Oosten, A. M. Frens and J. Schmidt, Phys. Rev. B **50**, 5239 (1994).
26. M. Ramamoorthy and S. T. Pantelides, Phys. Rev. Lett. **76**, 267 (1996).

A KINETIC MODEL FOR PRECIPITATION OF OXYGEN IN SILICON

S. SENKADER AND G. HOBLER

Institut für Festkörperelektronik, TU Wien
Gußhausstr. 27-29/E362, A-1040 Wien, Austria

1. Introduction

Oxygen is an inevitably incorporated impurity in Czohralski-grown silicon crystals. As the Si material is treated at temperatures less than 1200°C, oxygen precipitates as SiO_x. The oxygen induced defects are effective gettering sites for unwanted metallic contaminants. Excessive precipitation, however, causes warpage of silicon wafers. Therefore, the understanding and control of the behavior of oxygen during heat treatments is essential.

Most of the precipitation models reported during the last two decades are based mainly on the classical theory of nucleation and/or the theory of diffusion limited precipitation. These models suffer from several shortcomings. For example they do not account for different sizes of precipitates occuring in a given sample. Nucleation and growth cannot be treated consistently within one model. Furthermore, only the growth of precipitates and not their dissolution can be described by these models. Our recently developed model [1, 2] based on the work of Schrems [3] treats precipitation nucleation and growth/dissolution simultaneously. The model describes the kinetics of oxygen precipitates by rate equations combined with a Fokker-Planck equation, and is able to calculate the size distribution of precipitates.

In our paper we present a model based on our recent model development. After describing the model we show simulation results and their comparision with the experimental ones. With the same parameter set we can simulate different experiments. In particular, we discuss the simulations of 2-step, 3-step, and multi-step annealings. Furhermore, an attempt is made to model the influence of the hydrogen annealing on the formation of precipitate free zone (Denuded Zone; DZ).

R. Jones (ed.), Early Stages of Oxygen Precipitation in Silicon, 447–454.
© *1996 Kluwer Academic Publishers.*

2. The Model

2.1. KINETICS OF OXYGEN PRECIPITATION

We describe the oxygen precipitation process as unimolecular because experimental observations indicate that the rate of precipitation is directly proportional to the oxygen concentration, C_O, rather than proportional to C_O^2. Thus, the governing reaction at the precipitate-matrix interface is $\frac{1}{2}Si + O \rightleftharpoons \frac{1}{2}SiO_2$. The precipitate containing n oxygen atoms can build up or decay by interactions with single oxygen atoms resulting in

$$(n) + (1) \xrightarrow{g(n,x,t)} (n+1) \quad \text{or} \quad (n) \xrightarrow{d(n,x,t)} (n-1) + (1). \quad (1)$$

$g(n,x,t)$ and $d(n,x,t)$ are the growth and dissolution rates, respectively, where x denotes the spatial coordinate and t is time. The evolution of the precipitate density thus can be described by a rate equation for each precipitate size as

$$\frac{\partial f(n,x,t)}{\partial t} = J(n,x,t) - J(n+1,x,t) \qquad n \geq 2, \quad (2)$$

$$J(n,x,t) = g(n-1,x,t) \cdot f(n-1,x,t) - d(n,x,t) \cdot f(n,x,t), \quad (3)$$

where $f(n,x,t)$ denotes the concentration of precipitates with n oxygen atoms.

In practice, precipitates with a size of several hundred nm can occur. The number of rate equations required, therefore, will exceed 10^9. An approximation based on a Taylor series expansion is made to transform Eqs.(2),(3) into a partial differantial equation called the Fokker-Planck equation (FPE):

$$\frac{\partial f(n,x,t)}{\partial t} = -\frac{\partial I(n,x,t)}{\partial n}, \quad I(n,x,t) = -B \cdot \frac{\partial f(n,x,t)}{\partial n} + A \cdot f(n,x,t) \quad (4)$$

$$A = \frac{g(n,x,t) + d(n,x,t)}{2}, \qquad B = g(n,x,t) - d(n,x,t) - \frac{\partial A}{\partial n}. \quad (5)$$

The expansion of the rate equations into a single FPE is not valid for small precipitates [4]. The use of the rate equations for small precipitates ($n \leq 20$) together with the FPE for larger ones allows us to deal with all precipitate sizes in a unified model.

In our formulation we assume that precipitates of size two and larger are immobile. For oxygen atoms, $C_O = f(1,x,t)$, an additional equation is considered:

$$\frac{\partial C_O}{\partial t} = \frac{\partial^2 C_O}{\partial x^2} - \frac{\partial}{\partial t} \sum_{n \geq 2} f(n,x,t) \cdot n. \quad (6)$$

On the right hand side of Eq.(6) the first term describes diffusion of oxygen while the second term corresponds to the change in the amount of precipitated oxygen.

2.2. GROWTH AND DISSOLUTION RATES

Based on the theory of reaction rates [5] the growth rate of a spherical precipitate containing n oxygen atoms can be written as

$$g(n, x, t) = 4\pi r^2 \delta \, C_O^{if} \, \nu \, \exp\left(\frac{-\Delta G_{n \to n+1}}{kT}\right), \qquad (7)$$

where r is the precipitate radius, δ is the thickness of the precipitate-matrix interface assumed to be roughly equal to the elementary jump length of oxygen atoms ($\approx 0.2\,nm$). C_O^{if} is the concentration of oxygen atoms at the precipitate-matrix interface, and ν is the jump frequency of oxygen atoms for incorporation into the precipitate. ν is assumed to be equal to the jump frequency of oxygen atoms during diffusion and is given by [6, 7]

$$\nu = \frac{6 \cdot D_{Oc}}{g \cdot \delta^2}, \qquad (8)$$

where g is the number of equivalent diffusion paths ($g = 36$), and D_{Oc} is the prefactor in the diffuson coefficient of oxygen, D_O [8]. The energy barrier, $\Delta G_{n \to n+1}$, is given by [9]

$$\Delta G_{n \to n+1} = G_{ac} + \frac{1}{2}\frac{\partial G_n}{\partial n}, \qquad (9)$$

where G_{ac} is the free energy of activation. G_n denotes the Gibbs free energy of a precipitate containing n oxygen atoms and is written as the sum of volume energy and interfacial energy

$$G_n = -nkT \ln \frac{C_O}{C_O^{eq}} + 4\pi r^2 \gamma, \qquad (10)$$

where k is the Boltzman constant, T is the temperature, C_O^{eq} is the equilibrium concentration of oxygen [10], and γ is the interfacial energy density.

The dissolution rate $d(n, x, t)$ can be obtained by means of the growth law [11]

$$\frac{dn}{dt} = g(n, x, t) - d(n, x, t) = \kappa_r(C_O^{if} - C_O^{if,eq}), \qquad (11)$$

$$d(n, x, t) = \kappa_r \cdot C_O^{if,eq}, \qquad \kappa_r = 4\pi r^2 \delta \, \nu \, \exp\left(\frac{-\Delta G_{n \to n+1}}{kT}\right), \qquad (12)$$

where $C_O^{if,eq}$ is the equilibrium concentration of oxygen at the precipitate-matrix interface calculated by setting $\partial G_n/\partial n = 0$ and solving $C_O = C_O^{if,eq}$.

Another formulation of the growth law, $\partial n/\partial t$, can be found in steady-state by solving the diffusion equation in spherical coordinates

$$\frac{\partial n}{\partial t} = \kappa_d(C_O - C_O^{if}) \qquad \kappa_d = \frac{\partial n}{\partial r} \cdot \frac{\Omega_{SiO_2}}{2} \cdot \frac{D_O}{r}, \qquad (13)$$

where Ω_{SiO_2} is the molecular volume of SiO_2 [12]. Using Eqs.(11),(12),(13) C_O^i can be written as

$$C_O^{if} = \frac{\kappa_r \cdot C_O^{if,eq} + \kappa_d \cdot C_O}{\kappa_r + \kappa_d}. \tag{14}$$

3. Results and Discussion

We have used above model to simulate the precipitation behavior of oxygen during 2-step, 3-step, and multi-step annealings. The experimental data have been taken from the literature. The values of $G_{ac} = 2.53\,\mathrm{eV}$ and $\gamma = 0.59\,\mathrm{J/m^2}$ are found by fitting the simulation results to the experimental data. These parameter values are used throughout all simulations. It is noted that all reported values of initial oxygen concentration are recalculated according to the conversion factor reported in [13].

3.1. 2-STEP AND MULTI-STEP ANNEALINGS

For routine testing of wafer quality the wafer manufacturers usually carry out a short 2-step low temperature-high temperature annealing known as "precipitation test", which is intended to show a similar amount of precipitation as a multi-step device manufacturing cycle. Figure 1 shows the dependence of oxygen loss on initial oxygen concentration for three different 2-step annealings. Simulations agree very well with the experimental results (triangles [14], squares [15], circles [15]).

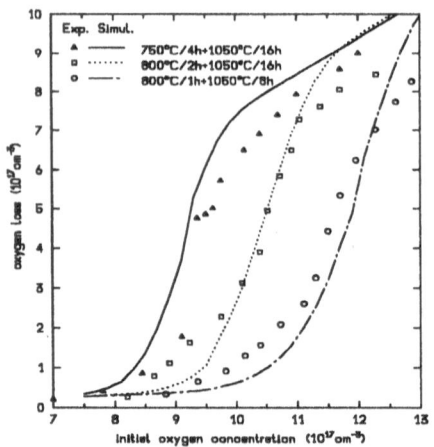

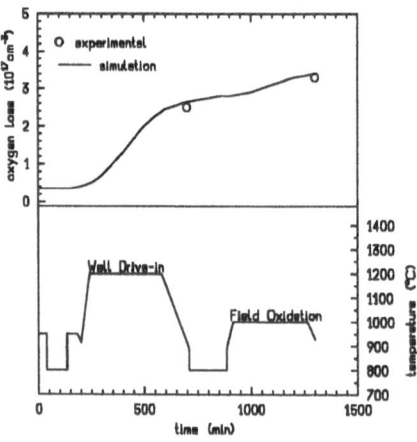

Figure 1. The dependence of oxygen loss on the initial oxygen concentration.

Figure 2. The change of the oxygen loss during a CMOS type process.

Figure 2 shows the change of oxygen loss during a CMOS type multi step annealing [16]. The initial oxygen concentration is $9.7 \times 10^{17}\,\mathrm{cm^{-3}}$ and

451

samples are annealed for 2h at 725°C before the CMOS cycle. The simulation suggests that major oxygen loss occurs during well drive-in at 1200°C and during field oxidation at 1000°C. After the well drive-in step an intermediate heat treatment at 800°C promotes the nucleation of precipitates, which can grow during the subsequent field oxidation step.

3.2. 3-STEP ANNEALINGS

In device processing technology, internal gettering of metallic contaminants via oxygen precipitates is considered as an important step. The subsurface region, where active devices are placed, however, must be free of defects. Processing cycles, therefore, include a 3-step annealing to create a defect free subsurface region (Denuded Zone; DZ) and a defect region in the bulk. The first step is at high temperatures (1000°C–1200°C) and is used to achive an oxygen-free region via outdiffusion. During the second step (at moderate temperatures, 600°C–800°C) precipitate nucleation in the oxygen rich bulk region takes place. Another high temperature step is subsequently used to grow the nucleated precipitates.

Figue 3 shows the change of precipitate density with the wafer depth for different denudation times. Simulation results agree excellently with the experimental ones [17] obtained by IR tomography.

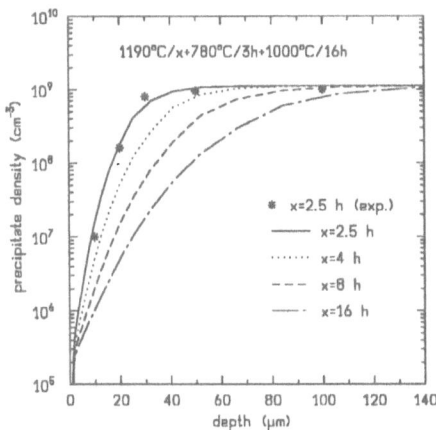
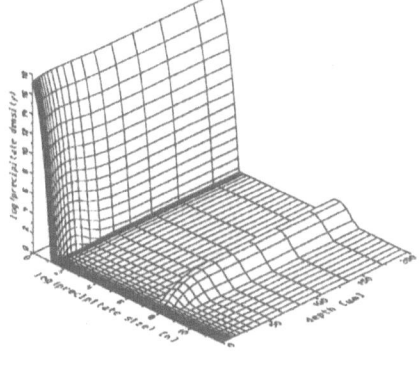

Figure 3. Precipitate density versus wafer depth after a 3-step annealing in neutral ambient.

Figure 4. Precipitate size distribution after 3. step

Figure 4 shows the size distribution of precipitates as function of the depth and the precipitate size after the 3rd step. From the figure one can see easily the near-surface precipitate free zone. The bulk, however, is rich of big precipitates. In Figure 5 the size distribution functions of the precipitates in the bulk are depicted after each annealing step. Moreover, the effect of the duration of 3rd step on the precipitation behavior is investigated.

The results shown in Fig. 5 suggest that a longer annealing during 3rd step (1000°/64 h) causes the growth of large precipitates at the expense of small ones (coarsening).

Because the formation of the DZ is closely related to the outdiffusion of the oxygen, DZ width varies with the oxygen concentration. This is plotted in Figure 5 for different denudation times. The simulations reproduce the experimental results [18] well except for low initial oxygen concentrations and long denudation times where simulation underestimates the DZ width.

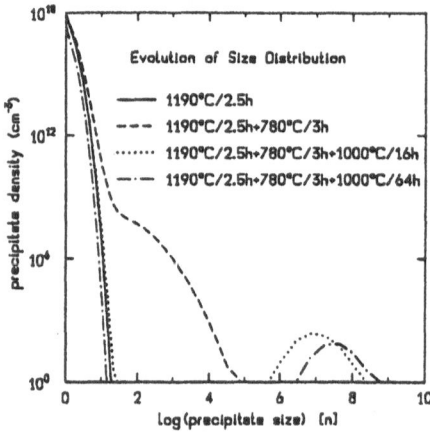 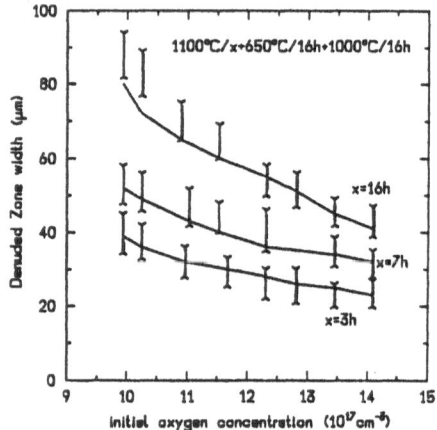

Figure 5. Change of the size distribution during 3-step annealing

Figure 6. Experimental (bars) and simulated (lines) dependence of DZ width on the oxygen concentration.

3.3. THE EFFECT OF HYDROGEN ANNEALING

Since the crystal quality of the wafer surface plays an important role for the performance of the devices placed in the subsurface region, many attemps have recently been made to improve surface and subsurface characteristics of wafers. One of the methods is to use a high temperature hydrogen annealing. It is a common observation that hydrogen annealed wafers have better gate oxide integrity and deeper DZ. One can think that deeper DZ results if hydrogen annealing enhances oxygen outdiffusion. We have investigated the influence of hydrogen annealing assuming that hydrogen affects the oxygen diffusion coefficient and/or the oxygen concentration at the surface. Simulations suggest that the experimental results may be reproduced only using a new value of the oxygen concentration at the surface. This value has been obtained by using oxygen outdiffusion profiles published previously [19, 20]. Using the relationship $C_O^{surf} = 6.492 \times 10^{25} \exp(-2.62/kT)\,\mathrm{cm}^{-3}$ we could find a very good agreement between simulations and experiments.

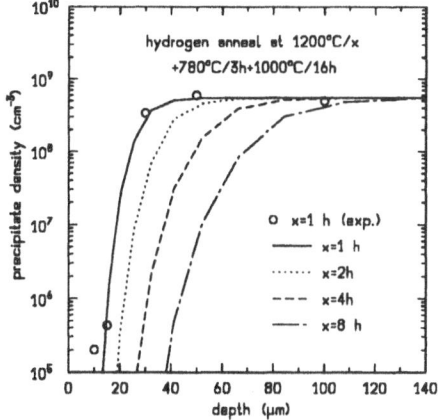 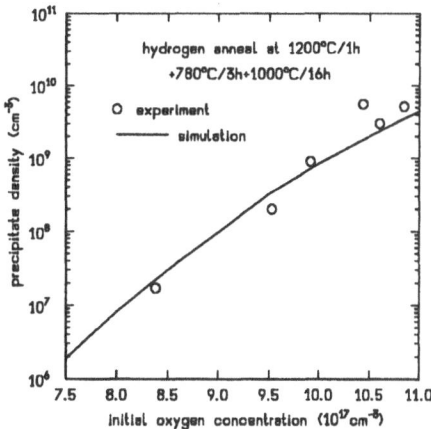

Figure 7. Change of the precipitate density with the wafer depth after 3-step annealing. The first step is carried out in hydrogen ambient.

Figure 8. Experimental and simulated dependence of the precipitate density on the oxygen concentration for a 3-step annealing. The first step is carried out in hydrogen ambient.

In Figure 7 experimental [17] and simulated results (circles and lines, respectively) of the precipitate density are shown as a function of the wafer depth for a 3-step annealing. The first high temperature annealing at 1200°C is carried out in hydrogen ambient. Comparision with the 3-step annealing in neutral ambient (Fig. 3) indicates that though the precipitate densities in the bulk are in the same order, hydrogen annealing leads to less precipitation in the near surface region. Finally, in Figure 8 the dependence of the precipitate density on the initial oxygen concentration is plotted.

4. Summary

We have presented a model, which simulates the precipitation behavior of oxygen in silicon. Using the same parameters we are able to reproduce different kinds of experimental observations. In particular, we have simulated the oxygen precipitation during 2-step, 3-step, and multi-step annealings. Moreover, investigations of hydrogen annealing on precipitation suggest that using only a new value for the oxygen concentration at the wafer surface is sufficient to simulate the experiments.

Acknowledgments: This work is supported by the "Jubiläums Fonds der Österreichischen Nationalbank (Project No. 5473)". We are grateful to Dr. J. Esfandyari (Wacker-Siltronic Gmbh, Germany) for his help.

454

References

1. Senkader, S., Esfandyari, J., and Hobler, G. (1995) A model for oxygen precipitation in silicon including bulk stacking fault growth, *J. Appl. Phys.* **78**, 6469.
2. Esfandyari, J., Schmeiser, C., Senkader, S., Hobler, G., and Murphy, B. (1996) Computer simulation of oxygen precipitation in Cz-grown silicon during HI-LO-HI anneals, *J. Electrochem. Soc.* **143**, 995.
3. Schrems, M. (1994) Simulation of oxygen precipitation, in Oxygen in Silicon, F. Shimura (ed.), *Semiconductors and Semimetals* **42**, Academic, New York, pp. 391.
4. Frenkel, J. (1946) *Kinetic theory of liquids*, Oxford University, Oxford.
5. Christian, C., W. (1975) *The theory of transformation in metals and alloys, Part I: Equilibrium and general kinetic theory*, Pergamon, Oxford.
6. Vineyard, G., H. (1957) Frequency factors and isotrope effects in solid state rate processes, *J. Phys. Chem. Solids* **3**, 121.
7. Weiser, K. (1962) Theory of diffusion and equilibrium position of interstitial impurities in the diamond lattice, *Phys. Rev.* ,**126**, 1427.
8. Mikkelsen, Jr., J.C. (1986) Diffusivity and solubility of oxygen in silicon, in Mikkelsen, Jr., J.C, Pearton, S.J., Corbett, J.W., and Pennycook, S.J. (eds.), *Oxygen, carbon, hydrogen and nitrogen in crystalline silicon*, Mater. Res. Soc. Proc. **59**, Mater. Res. Soc., Pittsburg.
9. Turnbull D. and Fisher, J.C. (1949) Rate of Nucleation in Condensed Systems, *J. Chem. Phys.* **17**, 71.
10. Craven, R.A. (1981) Oxygen precipitation in Cz-silicon, in Huff, H.R, Kriegler, R.T., Takeishi, Y. (eds.), *Semiconductor Silicon/1981*, Electrochem. Soc., Inc., Pennington, NJ, p. 245.
11. Burke, J. (1965) *The kinetics of phase transformations in metals*, Pergamon, Oxford.
12. Tiller, W.A., Oh, S (1988) The effect of frenkel defect formation on spherical SiO_2 precipitate growth in silicon wafers, *J. Appl. Phys.* **64**, 375.
13. Baghdadi, A., Bullis, W.M., Croarkin, M.C., Li, Y., Scace, R.I., Series, R.W., Stallhofer, P., and Watanabe, M. (1989) Interlaboratory determination of the calibration factor for the measurement of the interstitial oxygen content of silicon by infrared absorption, *J. Electrochem. Soc* **136**, 2015.
14. Schrems, M., Brabec, T., Budil, Pötzl, H., M., Guerrero, E., Huber, D., and Pongratz, P. (1990) Computer aided investigation of oxygen precipitation, in *Proc. IC-STDCS'89*, 245.
15. Chiou, H., D. (1987) Oxygen precipitation behavior and control in silicon, *Solid State Technol.* **30**, 77.
16. Isomae, S. (1991) Computer aided simulation for oxygen precipitation in silicon, *J. Appl. Phys.* **70**, 4217.
17. Kubota, H., Numano, M., Amai, T., Miyashita, M., Samata, S., and Matsushita, Y. (1994) Perfect silicon surface by hydrogen annealing, in Huff, H.R., Bergholz, W., Sumino, K. (eds.), *Semiconductor Silicon/1994*, The Electrochem. Soc., Inc., Pennington, NJ.
18. Isomae, S., Aoki, S., and Watanabe, K. (1984) Depth profiles of interstitial oxygen concentrations in silicon subjected to three-step annealing, *J. Appl. Phys.* **55**, 817.
19. Zhong, L. and Shimura, F. (1993) Hydrogen enhanced outdiffusion of oxygen in Czochralski silicon, *J. Appl. Phys.* **73**, 707.
20. Maddalon-Vinante, C., Barbier, D., Erramli, H., and Blondiaux, G. (1993) Charged particle activation analysis study of the oxygen outdiffusion from Czochralski-grown silicon during classical and rapid thermal annealing in various gas ambient, *J. Appl. Phys.* **74**, 6115.
21. Samata, S., Numano, M., Amai, T., Matsushita, T., Kobayashi, H., Yamamoto, A., Kawaguchi, T., Nadahara, S., and Yamabe, K. (1993) Hydrogen annealing of silicon wafer, in *Ext. Abst. of Electrochem. Soc.* **93-2**, The Electrochem. Soc., Inc., Pennington, NJ, pp. 426.

OXYGEN GETTERING AND THERMAL DONOR FORMATION AT POST-IMPLANTATION ANNEALING OF SILICON

A.G. ULYASHIN, Yu.A. BUMAI, A.I. IVANOV
Belarussian State Polytechnical Academy
F.Skorina 65 Ave. Minsk 220027, BELARUS

V.S. VARICHENKO, N.M. KAZYCHITS
Belarussian State University
F.Skorina 4 Ave. Minsk 220080, BELARUS

A.M. ZAITSEV, W.R. FAHRNER
FernUniversity of Hagen
Haldener Str. 182, Hagen 588084, GERMANY

1. Introduction

Behavior of residual oxygen in Czochralski-grown (Cz) silicon after post-implantation annealing is of a great interest due to the formation of oxygen-related electrically active centers - thermal donors as well as oxygen precipitates. Implantation of light ions with the energies of 100-200 keV or heavy ions with MeV energies in silicon introduces two quite well separated damage zones - one around the peak of the impurity concentration profile and one in front of it. During the post-implantation heat treatment the mobile oxygen, metallic, hydrogen or other atoms are gettered into the regions with higher defect concentration, so that there is a difference in the rates of secondary defect formation, oxygen precipitation processes and impurity-defect reactions between the different damage zones [1-3]. The aim of this work is to investigate the thermal donor formation effect for different damage zones in Cz-silicon implanted with 210 Mev krypton and 160 keV hydrogen ions under post-implantation annealing and low energy hydrogen plasma treatment conditions. Moreover, correlation between the positions of the regions with an enhanced thermal donor formation and oxygen gettering regions has been considered.

2. Experimental

Samples used in this study were p-type, 10 Ω cm or n-type, 0.5 Ω cm Cz-silicon with an oxygen concentration of about 10^{18} cm^{-3}. The p-type, 1 Ω cm float zone (FZ) silicon

R. Jones (ed.), Early Stages of Oxygen Precipitation in Silicon, 455–462.
© 1996 *Kluwer Academic Publishers.*

samples were used in some cases to investigate the effect of oxygen on defect formation during the annealing. Hydrogen was implanted into the samples with a dose of 3×10^{16} cm^{-2} at the energy of 160 keV at room temperature. Ion implantation with 210 Mev krypton ions with doses of 2.56×10^{13} and 1×10^{14} cm^{-2} was carried out at a U-400 accelerator in the Joint Institute for Nuclear Research (Dubna, Russia). The temperature of the samples during the irradiation did not exceed 80°C. After implantation, the samples were annealed in N$_2$ ambient at temperatures ranging from 400 to 1000°C in a conventional furnace tube. The ion implanted samples were beveled mechanically for the depth resolved spreading resistance measurements. A two-point probe instrument with tungsten carbide tips was used. Secondary ion mass spectrometry (SIMS) measurements were carried out using a CAMECA IMS-4F ion microprobe for hydrogen implanted samples. The beam probe was formed by Cs$^+$ ions with an energy of 10 keV. To minimize crater-edge effects, secondary ions were collected only from a spot of 10 μm in the central region inside the total rastered area of 250x250 μm^2 for reduction of the typical problems of hydrogen depth profiling related to the contamination of the residual gas species [4,5]. The SIMS raw data were converted into hydrogen or oxygen concentration by employing Si standards: a silicon sample implanted by hydrogen ions at 160 keV to a dose of 3×10^{16} cm^{-2} and an oxygen containing one characterized by IR measurements. Exposure for some Kr implanted Si samples to a DC hydrogen plasma (<0.1 W/cm^2) has been carried out in a reactor for reactive ion etching with a plate voltage of 300 V for 4 hours at 400°C. The type of conductivity in the different damage regions after annealing was determined by the thermal microprobe method.

3. Results

Fig.1 shows the changes of the spreading resistance profiles for Kr-implanted p-type Cz-silicon with a dose and annealing temperature up to 600°C. The spreading resistance curves follow to a certain extent the radiation damage depth distribution. Accumulation of radiation defects at a low dose provides the increased spreading resistance in a depth interval extending from the surface to a depth of about 30 μm for the as-implanted sample (Fig.1a). For the higher implantation dose, a widening of spreading resistance profile up to 36 μm is observed. This may be explained by both the accumulation of the radiation defects at the end of the ion implanted layer and the diffusion of radiation defects or by ion channeling effect [6]. It is clearly visible from Fig.1b that no recovery of sample conductivity occurs at the annealing temperatures below 600°C.

The changes of the spreading resistance profiles in the samples annealed at the temperatures above 600°C are shown in Fig 2. The formation of buried n-type layers with high enough conductivities around the projected range (R$_p$) for the samples implanted with different Kr doses is clearly seen from Fig.2. The n-type region is wider for the sample implanted with a higher dose. This correlates with the deeper penetration of the radiation damages with dose increasing, as shown in Fig.1a. But no n-type layer formation has been observed for the annealing temperatures below 600°C (Fig.1b) in contrast to high temperature annealing (Fig.2).

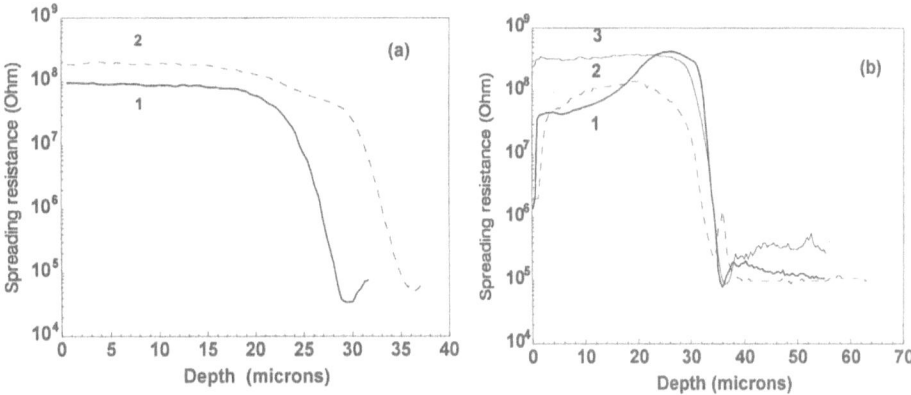

Figure 1. Spreading resistance
versus depth in p-type as-implanted Cz-silicon samples for Kr doses: 1—2.56×10^{13} cm^{-2}; 2—1×10^{14} cm^{-2} (a) and in the samples implanted with a dose 1×10^{14} cm^{-2} and annealed at the temperatures: 1—500°C; 2—600°C; 3—400°C. (b). The annealing time was 15 min.

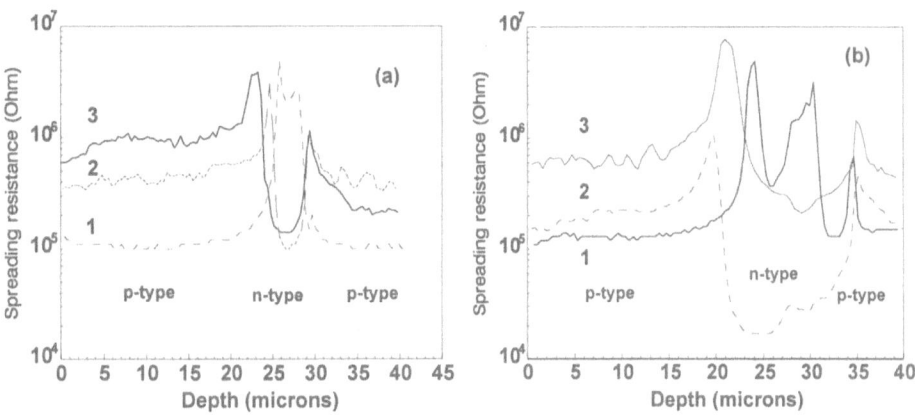

Figure 2. Depth profiles of spreading resistance in p-type Cz-silicon samples implanted by Kr ions with doses 2.56×10^{13} cm^{-2} (a) and 1×10^{14} cm^{-2} (b) and annealed at the temperatures: 1—900°C; 2—800°C; 3—700°C. The annealing time was 15 min.

Fig.3 shows the analogous spreading resistance profiles in n-type Cz-silicon samples after Kr implantation and annealing. The behavior of the spreading resistance profiles with annealing temperature for these samples indicates that the complete recovery of conductivity at the R_p position occurs only at 700°C. At the same time the region in front of R_p has been annealed at considerably lower temperatures.

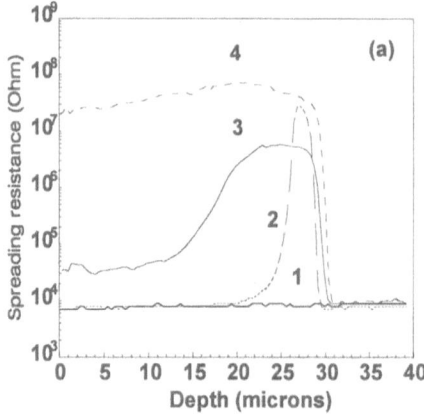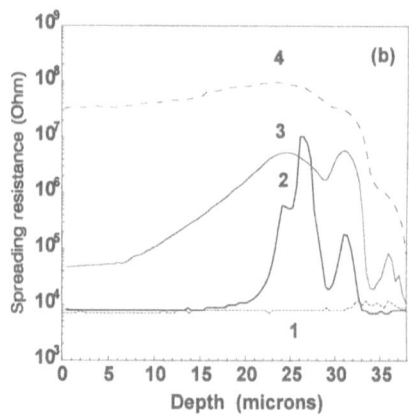

Figure 3. Depth profiles of spreading resistance in n-type Cz-silicon samples implanted by Kr ions with doses 2.56×10^{13} cm^{-2} (a) and 1×10^{14} cm^{-2} (b) and annealed at the temperatures: 1—700°C; 2—600°C; 3—500°C; 4—400°C. The annealing time was 15 min.

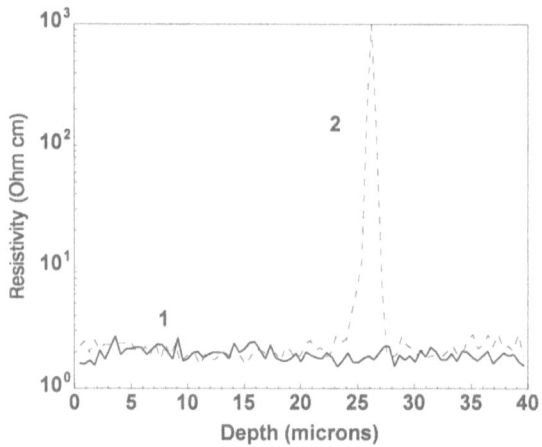

Figure 4. Depth profiles of resistivity in p-type FZ-silicon samples implanted by Kr ions with dose of 1×10^{14} cm^{-2} and annealed at the temperatures: 1 — 900°C; 2 — 800° C.
The annealing time was 15 min.

For FZ-silicon samples implanted by Kr ions, the complete recovery of conductivity and the annealing of damages occur at 900°C (Fig.4). It is necessary to note that the peak of spreading resistance for p-type FZ-silicon is located in the same damage region around R_p position as for the n-type regions in p-type Cz-silicon samples as shown in Fig.2.

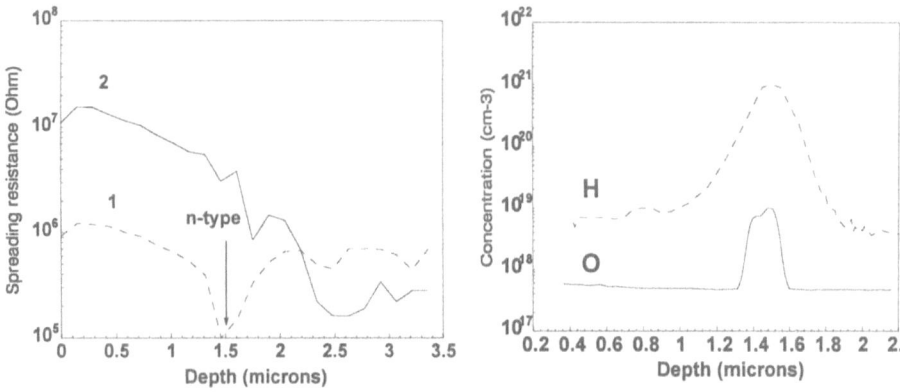

Figure 5. Depth profiles of spreading resistance in p-type Cz-silicon samples implanted by hydrogen: 1 — as-implanted, with dose 3×10^{16} cm^{-2}, 2 — annealed at 600°C for 20 min, and SIMS depth profiles of hydrogen in as-implanted sample and of oxygen in the one annealed at 1000°C.

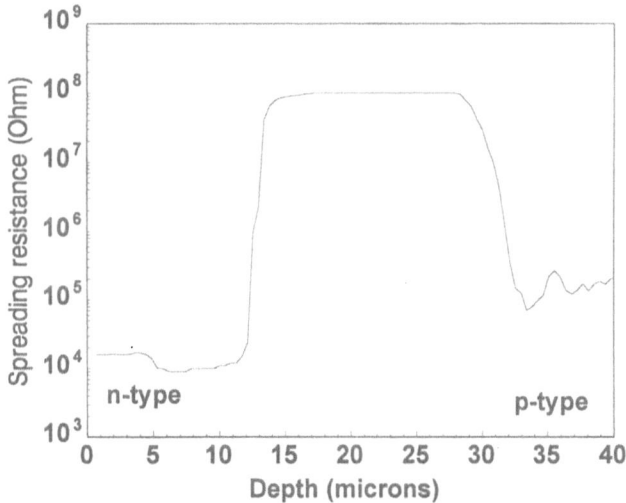

Figure 6. Depth profile of spreading resistance in p-type Cz-silicon sample implanted by Kr ions with a dose of 1×10^{14} cm^{-2} and then annealed in a hydrogen plasma at 400°C. The annealing time was 4 h.

The formation of n-type region around R_p position is observed for hydrogen implanted p-type Cz-silicon samples after the annealing at 600°C (Fig. 5).

Fig.6 shows spreading resistance profile in the p-type Cz-silicon sample implanted with Kr-ions and subsequently annealed in hydrogen plasma at 400°C. The conversion of conductivity from p- to n-type in the first 13 μm of the sample followed by the high resistivity damage region around R_p position has been observed.

4. Discussion

The annealing of the high energy implanted silicon is accompanied by the modification of radiation defects and their disappearance which is detected by the corresponding changes of the spreading resistance profiles. For 210 MeV Kr-ion doses used in this study, defect reconstruction and annealing finish at the temperatures higher than 600°C. The different rates of resistance recovery in the subsurface region where the point radiation defects dominate and a region around R_p where the large lattice displacements occur are observed. It is necessary to note that both annealing and donor doping in the subsurface region ceased after the low temperature (400°C) hydrogen plasma treatment without any modification to the spreading resistance for the region around R_p. The annealing is believed to occur due to the lower annealing temperature of radiation defects in the samples with hydrogen contamination [7,8].

For oxygen containing Cz-silicon, the further increase of the annealing temperature causes the formation of n-type layer located at R_p. The p-n-p structures have been formed for p-Si (10 Ω cm) substrate. These structures are stable to at least 1000°C.

It is necessary to note that no appearance of an n-type conductivity layer has been observed for FZ-silicon with low oxygen contamination. The sharp resistance peak at the projected range position for Fz-silicon annealed at 800°C (Fig.4) disappears after annealing at 900°C. The existence of this peak in silicon annealed at 800°C is explained by the incomplete electrical activation of boron dopant at this temperature.

Comparing the results on the annealing behavior of implanted Cz- and FZ-silicon, we can conclude that the conduction properties of the silicon region around the ion projected range for noble gas ion implantation are controlled by the concentration of the residual oxygen atoms. It is well known that oxygen gettering occurs in this region under high temperature annealing [1].

The enhanced thermal donor generation in the region around R_p is supposed to be responsible for the n-type conductivity range creation in oxygen containing material. For the samples annealed up to 600°C, the spreading resistance profiles are defined by unannealed radiation defects rather than thermal donors appearing. After annealing of these radiation defects, the contribution from thermal donors entirely defines a modification of spreading resistance profiles. The concentration of thermal donors in local regions reached the value of 10^{16} cm^{-3} after annealing at 700-800°C for 15 min. The enhancement of the generation rate of oxygen-related thermal donors in this local region is possibly due to an increase of concentration of hydrogen atoms, self- and oxygen interstitials at a buried damage region and by an increase of local pressure in this region [9].

Hydrogen implantation and subsequent annealing above 300°C is known to produce shallow donors in both n- and p-type silicon [10,11]. These donors anneals out at a temperature of about 600°C. In our case, the n-type buried region formation in hydrogen implanted p-type silicon after annealing at 600°C (Fig.5) is attributed to oxygen-related thermal donors. The donors appearing in Kr-implanted Cz-silicon after hydrogen plasma treatment at 400°C (Fig.6) may be both hydrogen-related shallow donors and oxygen-related thermal double donors.

5. Conclusions

Electrical activity has been observed for local Cz-silicon regions in front of and around projected ranges for krypton and hydrogen ions after an appropriate annealing. It is established that these regions exhibit n-type conductivity and are dependent on the residual oxygen contamination in silicon. The results obtained in each annealing stage can be summarized as follows:

— The annealing of buried damage layers produced by high energy Kr implantation begins at 600°C and leads to the creation of buried n-type regions in p-type Cz-silicon (p-n-p structure) due to the enhanced formation of thermal donors which are stable up to 1000°C;

— The low temperature annealing (400°C) of Cz-silicon implanted with Kr ions in a hydrogen plasma leads to the creation of an n-type layer in the subsurface region with individual ion tracks (n-i-p structure) due to the formation of hydrogen- and oxygen-related thermal donors and hydrogen passivation of radiation defects. The buried defect layers are barriers for hydrogen diffusion preventing the thermal donor formation behind n-type region;

— The positions of local buried regions with an enhanced rate of thermal donor formation around the projected range in ion implanted Cz-silicon annealed at 600-800°C coincides with the oxygen precipitation regions formed at higher annealing temperatures.

6. References

1. La Ferla, A., Galvagno, G., Rainery, V., Priolo, F., Carnera, A., Gasparotto, A. and Rimini, E. (1995) Dopant, defects and oxygen interaction in MeV implanted Czochralski silicon, *Nucl. Instr. and Meth. B* **96**, 232-235.
2. Tamura, M. and Suzuki, T. (1989) Damage formation and annealing of high energy ion implantation in Si, *Nucl. Instr. and Meth. B* **39**, 318-329.
3. Mohadjery, B., Williams, J.S. and Wong-Leung J. (1995) Gettering of nickel to cavities in silicon introduced by hydrogen implantation, *Appl. Phys. Lett.* **66**, 1889-1891.
4. Magee, C.W. and Botnick, E.M. (1981) Hydrogen depth profiling using SIMS - Problems and their solutions, *J.Vac.Sci.Technol* **19**, 47-52.

5. Cerofolini, G.F., Meda, L., Volpones, C., Ottaviany, G., DeFayette, J., Dierckx, R., Donelli, D., Orlandini, M., Anderle, M., Canteri, R., Claeyes, C. and Vanhellemont, J. (1990) Structure and evolution of the displacement field in hydrogen-implanted silicon, *Phys. Rev.B* **41**, 12607-12618.

6. Varichenko, V.S., Zaitsev, A.M., Melnikov, A.A., Fahrner ,W.R., Kasytchits, N.M., Penina, N.M. and Erchak, D.P. (1994) Defect production in silicon implanted with 13.6 MeV boron ions, *Nucl. Instr. and Meth. B* **94,** 259-265.

7. Shlopak, N.V., Bumai, Yu.A. and Ulyashin , A.G. (1993) Hydrogen passivation of γ-induced radiation defects in n-type Si epilayers. *Phys. Stat. Sol. (a)* **137**, 165-171.

8. Shlopak, N.V., Ulyashin, A.G. and Bumai, Yu.A. (1994) Hydrogen effects on the electrical properties of electron -irradiated n-type Si epilayers, *Nucl. Instr. and Meth. B* **86**, 298-302.

9. Emtsev, V.V., Andreev, B.A., Misiuk, A. and Schmalz K. (1996) *Formation of thermal donors in Czochralski grown silicon under hydrostatic pressure up to 1 GPa,* (in this book).

10. Pearton, S.J, Corbett, J.M. and Stavola M. (1992) *Hydrogen in Crystalline Semiconductors*, Springer-Verlag New York Berlin Heidelberg.

11. Ohmura, Y., Iwaki, M., Shiokawa, T., Toyoda, K., Namba, S. and Okabe, Y. (1989) A comparative study of D and H implantation-induced shallow donors in Si, *Nucl. Instr. and Meth. B* **39**, 377-380.

CARBON-HYDROGEN-OXYGEN RELATED CENTRE RESPONSIBLE FOR THE I-LINE LUMINESCENCE SYSTEM

J.E.GOWER, E.C.LIGHTOWLERS, G.DAVIES AND
A.N.SAFONOV
Physics Department, King's College London,
Strand, London, WC2R 2LS, U.K.

Abstract. The I-line luminescence system with its zero-phonon transition at 0.965 eV is created when Czochralski silicon is annealed at temperatures greater than 400 °C. The intensity of this luminescence system has been shown to increase with carbon and oxygen concentrations implying that both of these impurities are responsible for the luminescence system. Investigations of silicon doped with hydrogen and carbon isotope mixtures have shown that this centre contains at least one carbon and one hydrogen atom. In this paper we report a more detailed study. Silicon has been deliberately doped with an oxygen isotope mixture proving unequivocally that at least one oxygen atom is present. Temperature controlled measurements have revealed a second zero-phonon transition at an energy 2.38 meV higher than the I-line, consistent with an excited electronic state. Uniaxial stress perturbations couple the two excited states producing a non-linear splitting of the luminescence lines. Analysis of this splitting shows that the symmetry of the centre is monoclinic I.

1. Introduction

When Czochralski (CZ) silicon is heated at temperatures above 450 °C, oxygen becomes mobile allowing the formation of oxygen dimers. The oxygen dimers release silicon self-interstitials, which in silicon with a small carbon concentration leads to the formation of the thermal donors[1] . In silicon with a high carbon concentration, however, the thermal donors are not formed and the silicon self-interstitial releases carbon from its substitutional site. Substitutional carbon and interstitial oxygen tend to be spatially correlated in the crystal, probably because the carbon contraction of the lattice tends to offset the oxygen induced expansion, so that spatial correlation allows the strain energy to be minimised. Consequently it is not surprising that heat treatment of high-carbon CZ silicon allows the formation of carbon-oxygen complexes.

In this paper we examine one of the complexes produced by heat treatment. The complex produces the so-called "I" luminescence line. The I-line was first observed in

R. Jones (ed.), Early Stages of Oxygen Precipitation in Silicon, 463–468.
© 1996 *Kluwer Academic Publishers.*

1981 by Minaev and Mudyri[2]. It is readily observed in the luminescence spectra of high carbon Czochralski silicon annealed at temperatures greater than 400 °C. The optical centre is known to contain carbon[3], and it is assumed from its production to contain oxygen. Also, unexpectedly, the centre contains hydrogen[3].

In sec.3 we confirm the presence of carbon and hydrogen in the centre and, for the first time, we establish unambiguously the presence of oxygen. In sec.5 we derive the symmetry of the centre. These results on the chemistry and symmetry of the centre place considerable constraints on its possible structure. We also show in sec.4 that the centre has two excited states separated by 2.38 meV. Both the excited state structure and the response to stress are similar to those of the carbon-hydrogen "T" line which can also be generated by heating carbon containing silicon[4].

2. Experimental

All the samples studied were Czochralski silicon with high concentrations of carbon, oxygen and hydrogen. Typically $[O] \sim 10^{18}$ cm^{-3}, $[C] \sim 2*10^{17}$ cm^{-3} and $[H] \sim 10^{16}$ cm^{-3}. Various annealing studies were done using irradiated and non-irradiated samples to maximise the luminescence efficiency of the centre. It was found that maximum luminescence occurred when the samples were hydrogenated at 1300 °C, neutron irradiated at room temperature and then annealed at 600 °C for 30 mins.

The photoluminescence (PL) was excited by a 488 nm argon-ion laser, with an incident power of typically 200 mW, and a beam half width of ~1 mm. The spectra were taken with the samples either immersed in liquid helium at 4.2 K or bathed in a low temperature helium atmosphere, where the temperature was varied between 8 K and 60 K. The spectra were recorded using either a Nicolet 60SX or a Bomem DA8 Fourier transform spectrometer. Both spectrometers were fitted with a North Coast cooled germanium diode detector.

3. Isotope structure

To verify the presence of carbon, oxygen and hydrogen in the centre, silicon was doped with mixed isotopes of these 3 elements. The carbon and hydrogen data were obtained by Lightowlers et al[3], but this paper shows the first splitting associated with oxygen isotopes. Silicon was doped with an ^{16}O-^{18}O isotope mixture and the spectrum is shown in figure 1b. Although there is no third line corresponding to a centre with one ^{18}O and one ^{16}O, it should not be assumed that the I-line contains only one oxygen atom. Similarly, assumptions on the number of carbon and hydrogen atoms present in the centre should not be inferred from the spectra. Fig. 1a shows the I-line when natural isotopes are used, Fig. 1c shows the carbon isotope splitting, and Fig. 1d shows the hydrogen-deuterium splitting.

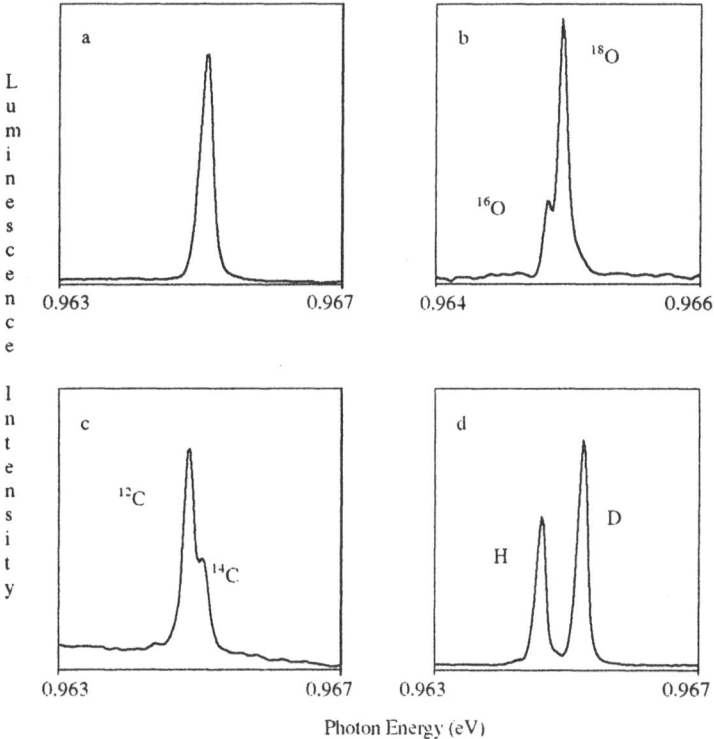

Fig 1. Isotope splittings of the I-line. (a) Natural isotope material (b) $^{18}O:^{16}O \sim 3:1$ (c) $^{12}C:^{14}C \sim 2.2:1$ (d) $D:H \sim 1.3:1$

4. Temperature dependence

Temperature controlled measurements of several samples revealed the presence of a second zero-phonon line at an energy 2.38 meV higher than the I line. Luminescence from this excited state is shown in figure 2a. An Arrhenius plot (Fig. 2b) of the relative intensities of the two excited states delivered an energy difference between the two as ~ 2 meV, close to the spectroscopic value obtained. It also indicated that the ratio of the transition probabilities of the two excited states $t_2/t_1 = 1/2$, again similar to that of the T-line[5].

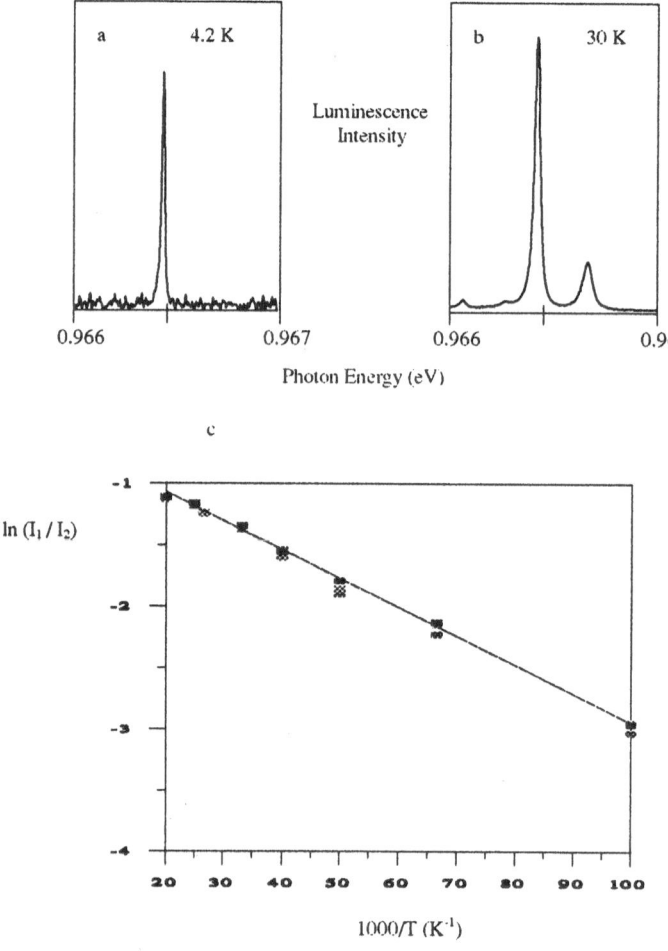

Figure 2. Spectra of the I-line at (a) 4.2 K and (b) 30 K. (c) shows a plot of log of the ratio of intensities against the reciprocal temperature

5. Uniaxial stress

Uniaxial stress measurements were carried out with compressions applied along the three major crystallographic orientations (<100>, <110>, <111>). The splittings of the two excited states are shown in figure 3. The number of components in the splitting is consistent with the centre possessing monoclinic I (C1h) symmetry[6]. There is no thermalisation between the stress split states, consistent with a lack of electronic degeneracy. To confirm this assignment we have fitted the data using perturbation theory for a monoclinic I centre. It is clear from the non-linear nature of the splitting that there is coupling induced between the states.

According to perturbation theory the following symmetric interaction matrix is derived by group theory[7].

$$\begin{bmatrix} a & b \\ b & c \end{bmatrix}$$

$a = A_1s_{zz} + A_2(s_{xx} + s_{yy}) + 2A_3s_{xy} + 4A_4(s_{yz} - s_{zx})$

$b = W_1s_{zz} + W_2(s_{xx} + s_{yy}) + 2W_3s_{xy} + 4W_4(s_{yz} - s_{zx})$

$c = B_1s_{zz} + B_2(s_{xx} + s_{yy}) + 2B_3s_{xy} + 4B_4(s_{yz} - s_{zx}) + D$

A_i, B_i are the splitting parameters of the I line and the excited state, W_i are the interaction parameters, s_{ij} are the stress tensor components of the applied stress, and D is the energy difference between the two excited states.

Using a least squared method a fit to the experimental data was obtained using this matrix. The fit is shown in figure 3 below with the solid lines representing the theoretical fit. This gives us values for A1 etc., in units of meV/GPa as:-

A1 = -8.6	B1 = -21.2	W1 = -17.1
A2 = 4.1	B2 = 14.5	W2 = 16.9
A3 = 17.8	B3 = 4.0	W3 = 15.8
A4 = 6.7	B4 = -11.4	W4 = -8.8

The fit to the I-line is excellent, confirming the assignment to a monoclinic I centre. Agreement for the higher energy transition is less good, possibly indicating further interactions with higher lying states which are not included in our model.

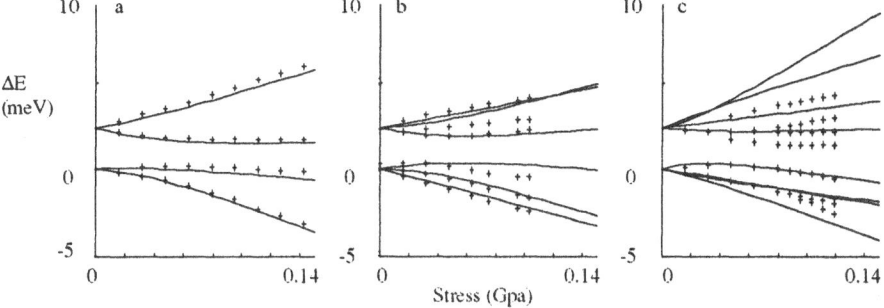

Figure 3 Uniaxial stress splittings of the I-line for compressions along
(a)<001>,(b)<111>,(c)<110>

6. Conclusions

These preliminary investigations on the I centre have shown us that it contains at least one carbon and one hydrogen atom. We have established unambiguously that oxygen is also a constituent of the same centre.

We have also shown that uniaxial stress reveals the symmetry of the centre to be monoclinic I, and the interaction under stress of the two excited states can be quantitatively understood.

7. Acknowledgments

This work was supported by EPSRC, and JEG thanks EPSRC for the provision of a research studentship.

8. References

[1] Newman, R.C., (1985) Thermal donors in silicon: oxygen centres or self-interstitial aggregates, *J.Phys.C:Solid State Phys.* 18, L967-L972.

[2] Minaev, N.S. & Mudyri, A.V. (1981) Thermally-induced defects in silicon containing oxygen and carbon, *Phys Stat Solidi* A 68, 561-566

[3] Lightowlers, E.C., Newman, R.C., Tucker, J.H. (1994) Hydrogen-related luminescence centres in thermally treated Czochralski silicon, *Semicond. Sci. Technol.* 9, 1370-1374

[4] Safonov, A.N., Lightowlers, E.C. & Davies, G. (1995) Carbon-hydrogen deep level luminescence centre in silicon responsible for the T-line, *Materials Science Forum*, Vols 196-201, 909-914

[5] Irion,E., Bürger,N., Thonke,K. ,Sauer,R., (1985) The defect luminescence spectrum at 0.9351 eV in carbon-doped heat-treated or irradiated silicon, J.Phys. C: Solid State Phys. 18, 5069-5082

[6] Kaplyanskii, A.A. (1964) Noncubic centres in cubic crystals and their piezospectroscopic investigation, *Opt. Spectrosc.*, 16 329-337

[7] Thonke, K., Bürger, N, and Sauer, R., (1985) Gallium related 0.875 eV photoluminescence defect spectrum in irradiated silicon, *Phys. Rev. B* Vol32, 10 6720-6730

LUMINESCENCE INVESTIGATIONS OF THE INTERACTION OF OXYGEN WITH DISLOCATIONS IN CZ Si

V. HIGGS[*]

Physics Department, King's College London, The Strand,

London WC2R 2LS, UK.

[*] *Present Address: Bio-Rad Micromeasurements,*

Hemel Hempstead, HP2 7TD, UK.

Abstract

Photoluminescence (PL) spectroscopy and cathodoluminescence (CL) imaging measurements have been carried out to characterize the interaction of oxygen and transition metals with dislocations in CZ Si. The CL and PL spectra recorded from CZ Si samples ($[O]=10^{17}$-10^{19} cm^{-3}) deformed at 700^0C for 15 minutes contained the characteristic D-band features (D1-D4). As the deformation time increased (>2 hours) the D-band luminescence features were quenched. The quenching of the D-band features was more rapid for the CZ Si samples containing more oxygen. IR absorption measurements showed that there was no detectable changes in the bulk oxygen levels for all the deformation treatments carried out at $T=700^0$C, TEM investigations did not reveal any changes in the dislocation structure or precipitation. It is suggested that the presence of the D-band features in the as-deformed CZ samples is due to the presence of residual transition metal impurities in the starting material. At higher deformation temperatures ($T=800^0$C) no D-bands are observed in the as-deformed state due to oxygen segregation. Following a longer deformation time (>1 hour) the loss of oxygen from solution is observed by IR measurements, subsequent intentional contamination with Fe produces an increase in oxygen loss from solution and an increase in dislocation donor formation.

1. Introduction

Czochralski (CZ) Si single crystals continues to be the low cost Si source for the next generation of Si semiconductor devices. During the CZ growth process oxygen is incorporated directly in Si from the silica crucible, and is an ubiquitous impurity in Si. The presence of oxygen in Si has a beneficial effect mainly related to intrinsic gettering effects, however if oxygen precipitation is not controlled, oxygen related defects can

R. Jones (ed.), Early Stages of Oxygen Precipitation in Silicon, 469–476.
© *1996 Kluwer Academic Publishers.*

form in the device active region and degrade device performance [1]. To optimise and control oxygen precipitation and oxygen related defects it is necessary to understand the properties of oxygen interacting with other impurities present in Si, such as transition metals. In this investigation photoluminescence (PL) spectroscopy and cathodoluminescence (CL) imaging were used to characterize the interaction of oxygen with dislocations in CZ Si.

2. Experimental

CZ Si samples were obtained which contained different bulk oxygen levels ([O]=10^{17}-10^{19} cm^{-3}) and FZ samples were prepared as control samples to check the possibility of contamination occurring during deformation. Dislocation sources were nucleated by scratching the surface parallel to the [011] with a diamond on which a constant load was applied (0.15 N). Then the sample was deformed elastically at room temperature by cantilever bending and annealed under stress at 700^{0}C or 800^{0}C for different times (15 minutes-14 hours). All the samples were RCA surface cleaned before deformation or post-annealing to avoid transition metal contamination.

Photoluminescence (PL) spectroscopy was excited using an Ar$^+$ laser with the samples immersed in liquid helium at 4.2 K. The spectra were recorded using a Nicolet 60 SX Fourier transform spectrometer using a germanium photodiode.

CL measurements were made at T$\approx$5 K using the CL mode of a Jeol 35 C SEM. Custom designed optics were used to collect the luminescence and direct the luminescence into a Bruker IFS 66 Fourier transform spectrometer or focus the luminescence on a Ge photodiode through a narrow band pass filter. IR absorption measurements were carried out to determine the interstitial oxygen remaining after thermal treatments.

3. Results and Discussion

PL spectra recorded from the CZ Si samples ([O]=10^{17}-10^{19} cm^{-3}) deformed at 700 ^{0}C for 15 minutes contained the characteristic dislocation related D-band features (D1-D4). As the deformation time increased (> 2 hours) the D-band luminescence features were quenched, this was a common trend irrespective of the starting [O]. The quenching of the D-band features was more rapid for the CZ Si samples containing more oxygen. Fig. 1a shows the PL spectrum recorded from a CZ Si sample deformed at T=700 ^{0}C for 15 minutes containing a dislocation density of 5 x 10^7cm^{-2}, and [O] = 5 x 10^{17} atoms.cm^{-3} . After a deformation time of 2 hours the D-band features are quenched (see Fig. 1b). The samples that were deformed at T= 800^{0}C for a short deformation time (15 minutes) with a low oxygen concentration [O]=2 x 10^{17} atoms.cm^{-3} contained no D-band features in the PL spectra, this is believed to be due to oxygen segregation. In contrast, the FZ control samples deformed at 700 ^{0}C for 15 minutes contained no D-band features. D-band features were only observed in the FZ deformed samples after deliberate transition metal contamination. In accord with previous deformation experiments [2] these control experiments imply that no transition metal contamination is occurring during deformation, however we cannot

eliminate the possibility that there is some trace metallic impurities already present in the CZ samples. It is known that there is always a higher concentration of metal impurities in CZ Si as compared to FZ Si, arising from contamination during growth or following annealing to destroy thermal donors. Electrical characterisation methods have shown that Fe is present typically with a concentration of 10^9-10^{10} atoms.cm^{-3} in CZ Si [3]. Therefore it is possible that the D-band features in CZ Si are produced by residual transition metal impurities already present in the starting material, and the interaction of metal impurities with oxygen or non-equilibrium point defects during deformation may be an important factor.

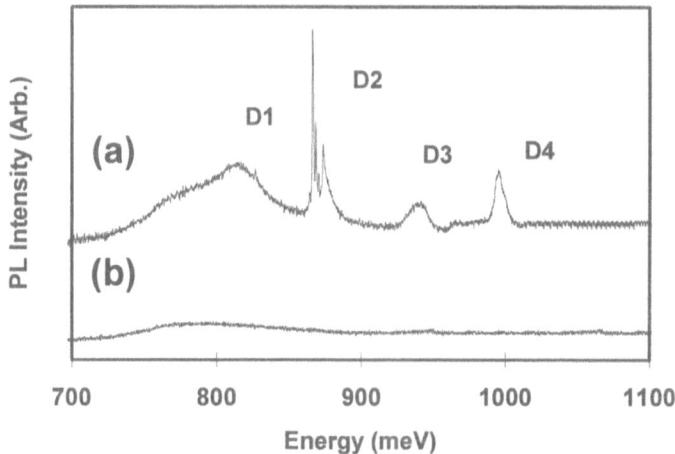

Figure 1. PL spectra recorded from CZ Si deformed for a) 15 minutes, b) 2 hours.

In a previous investigation analysing the effect of impurities on D-band formation in FZ Si it has been shown that the production of the D-bands is independent of type of transition metal used [2]. This effect could be due to the metals diffusing to the dislocation core causing some structural changes, such as reconstruction defects [4], kinks or jogs, independent of the nature of the metal. In addition, the formation of complexes between reconstruction defects and other defects such as vacancies and kinks have been investigated using theoretical calculations [5], such defect complexes could also react with metal impurity atoms.

Similarly, the metal impurities maybe indirectly involved in the formation of the D1 and D2 centres by acting as a nucleation centre for point defect aggregation. The tendency for point defects to agglomerate in semiconductors is strongly correlated to their charge state [6], the metal atoms could be involved in this type of mechanism.

As mentioned above it is difficult to differentiate between the effect due to metal atoms and intrinsic point defects in the production of the D-bands, the following experiments were carried out to try to elucidate this complex problem.

472

To try to understand the effects observed in CZ Si it is instructive to examine the different properties of the D-band features. D1 and D2 have in the past been attributed to various electronic transitions, point defects trapped in the strain field of the dislocation, the stacking faults between the dislocation or dislocation kinks. Whereas. D3 and D4 are commonly associated with transitions close or at the dislocation core. further work analysing dislocations in SiGe alloys has shown that D3 is likely to be the phonon replica of D4 [7]. CL investigations have shown that the distribution of D1 and D2 depends on the electrical activity of the dislocation, in lightly contaminated dislocations the D1 and D2 features can be seen in the regions between the dislocations and also at the position of the dislocation [7]. This suggests that the D1 and D2 features have point defect like characteristics.

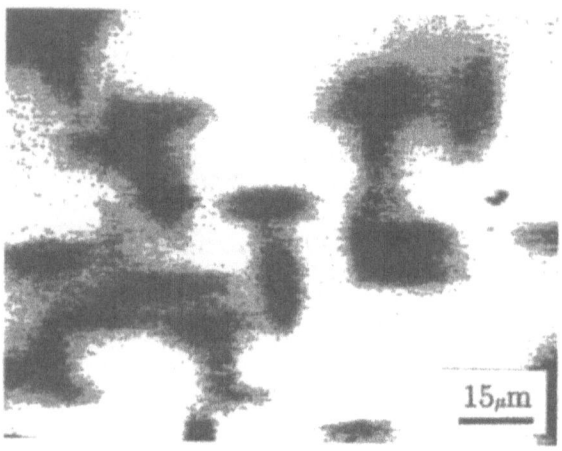

Figure 2. Monochromatic CL image using Si substrate feature.

The depth dependence of the D-band features present was examined using chemical etching to remove the surface of the deformed sample and then the PL spectrum recorded after each etch step. To obtain a more surface sensitive measurement. PL spectra were recorded using a excitation wavelength to give high depth discrimination (459.7 nm). The D1 and D2 features decreased in intensity by a factor of 8 times after the top 1500 nm of the sample surface was removed and then remained approximately constant following the removal of the next 2-3 μm. The observed distribution could be explained on the basis that the centres responsible have point defect properties and are preferentially formed in the surface region because the Si surface is acting as a sink for the annihilation of point defects. The surface region will also act as a sink for any residual transition metals and will aggregate both transition metals and intrinsic point defects.

To investigate this type of surface point defect interaction oxygen induced stacking fault (OISF) samples were prepared under clean conditions at high temperature (1100 ^{0}C) in a commercial oxidation furnace (GEC Plessey UK). Low temperature CL spectra contain only the D1 and D2 bands and the Si band edge features. CL images were recorded using the Si intrinsic luminescence, the defects are observed due to the reduction in the luminescence intensity and the dark contrast [7] is used to identify the position of the stacking faults, The monochromatic image recorded using the Si band edge feature showing the stacking faults is shown in Fig. 2. The corresponding D2 monochromatic image is shown in Fig. 3. The intensity distribution of the D2 band can be seen to be observed not only close to the position of the OISFs, but also in the region in between. the same behaviour was observed for the D1 feature. The distribution of the D1 and D2 features is similar to that observed for dislocations in Si and SiGe [7]. These results again support the suggestion that the spatial distribution of the D1 and D2 bands is characteristic of point defects. The oxidation of Si occurs by transport of oxygen atoms through the growing oxide and the conversion of Si to SiO_2 at the Si-oxide interface. The oxidation process is also associated with a volume expansion of a factor of two, which is accommodated by the viscoelastic flow of the oxide and partly by the injection of Si self interstitials into the Si crystal. The excess concentration of Si interstitials produced during oxidation can nucleate oxidation induced stacking faults. The exact mechanism for the nucleation of OISFs is still a contentious issue, and many possible causes have been postulated, mechanical damage, SiO_2 precipitates and impurity atoms. During the oxidation process the Si-oxide interface is a sink for point defects including transition metals and again it is possible that the origin of the D1 and D2 features is related to the presence of both point defects and transition metals produced during oxidation. The reaction of Si interstitials with metals atoms is known to play a role in metal silicide and oxide formation and dissolution, the preliminary stages of these reactions maybe generating the centres responsible for D1 and D2.

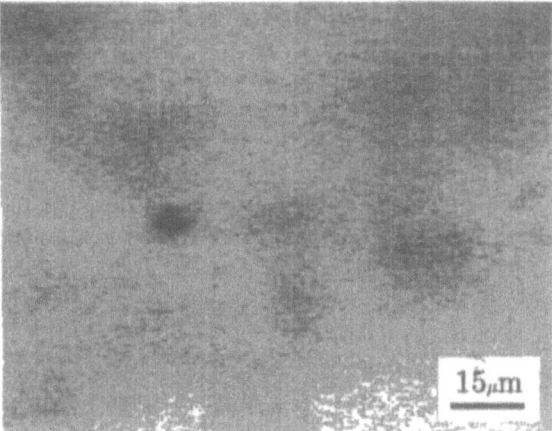

Figure 3. Monochromatic CL image using D2 feature.

These results raise the question, is it possible to produce the D1 and D2 centres without the presence of extended defects ?, to examine this further a series of oxidation treatments were carried out at low temperature to avoid OISF formation but to produce point defect interactions. After oxidation at 700 °C for 1 hour on n-type FZ Si (resistivity 100 Ohm.cm), PL spectra contain only the luminescence features from the phosphorous bound exciton, however if the sample is intentionally Cu contaminated (4×10^{10} atoms.cm^{-2}) before oxidation then very weak D1 and D2 features can be observed (see Fig. 4a). Similar results can produced for p-type FZ Si as shown in Fig. b. In contrast, FZ samples only Cu contaminated at 700°C for 1 hour give rise to the

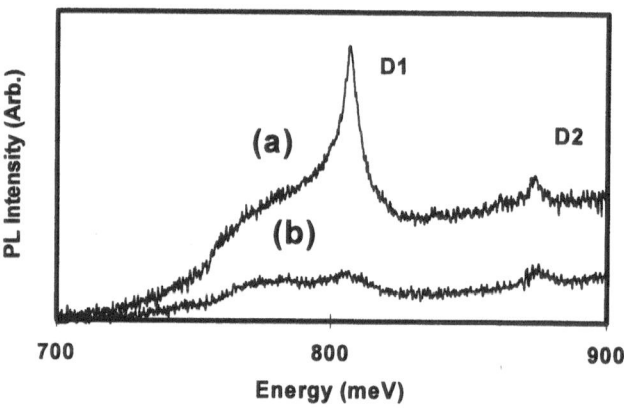

Figure 4. PL spectra recorded from FZ Si after oxidation/ Cu contamination, a) n-type type and b) p-type.

characteristic luminescence spectrum of Cu pairs in Si. The D-bands can also be produced by Fe contamination instead of Cu. These last two experiments suggest that transition metal contamination is required to produce the point defect reactions necessary for the formation D1 and D2 centres, and extended defects are not necessary. Therefore the oxidation process and the production of dislocations act as a preferential sink to getter both the transition metals and intrinsic point defects. As mentioned earlier the D1 and D2 bands are independent of the nature of the transition metal used, but seem catalyse the formation of the centres necessary to from D1 and D2. Assuming that during the oxidation process excess Si interstitials are produced, the transition metals maybe facilitating the nucleation of Si interstitial clusters by forming complexes, the formation of Si interstitial and impurity complexes are well known (e.g. C, Al and B). Rod-like defects formed in Si are special type of agglomeration of Si interstitials, high resolution TEM investigations have shown that B, P Sb and even Si interstitials produced during Si implantation can act as nucleation centres [8]. In addition, it has been shown that a strong luminescence line is correlated with the formation and destruction of the rod-like defects [9], D1 and D2 features were observed during the creation of the rod-like defects, supporting the idea D1 and D2 are related to that Si interstitials agglomerates.

Low temperature oxidation and metal contamination treatment was also tried on CZ samples, yet no D1 and D2 features were observed. Also it was found that after Cu contamination the characteristic spectrum of Cu pairs was not found. This result could be explained by the fact that non-equilibrium point defects present in CZ Si capture the

Cu atoms so they are not available for nucleation of the centres responsible for D1 and D2. By increasing the Cu contamination level (5 x10^{11} atoms.cm^{-2}) prior to oxidation the D1 and D2 bands were observed. Further investigation of Cu contamination on heat treated CZ Si containing oxygen precipitates is complicated by the various possible reactions (metal silicate, silicide and oxide formation) [10].

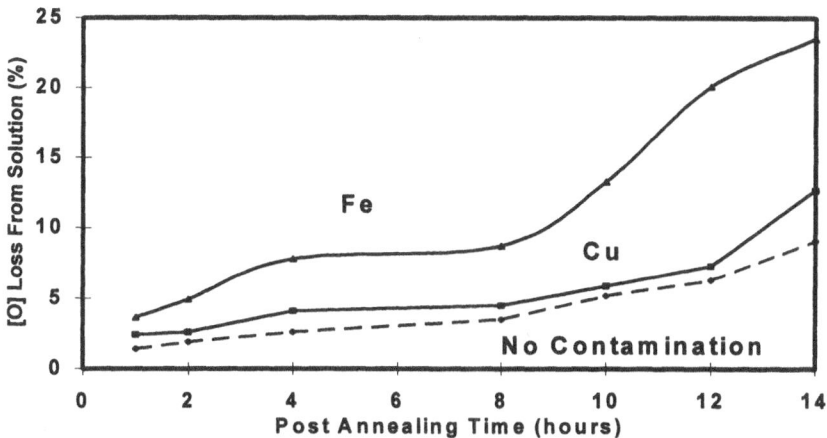

Figure 5. Effect of metal contamination on loss of interstitial oxygen concentration.

The influence of transition metals on dislocations in CZ Si was further analysed by carrying out deformation treatments at a higher temperature (T=800^{0}C for 15-840 minutes). No D-bands were observed from the sample deformed for 15 minutes, this is believed to be due to oxygen segregation close to dislocation core, as confirmed by deep level formation [7]. Only after the deformation time increased (> 15 minutes) the D1 and D2 features were observed which corresponded to a marked loss of interstitial oxygen (5-10%). This is related to oxygen segregation producing Si interstitials which can interact with residual transition metal impurities.

Ageing the sample deformed for 15 minutes by post-annealing to the same time period as produced by longer deformation times produced a similar measured loss of interstitial oxygen. Fig. 5 shows the effect of annealing the short deformation time sample for 1-14 hours. Intentional Cu (4 x10^{12} atoms.cm^{-3}) contamination prior to annealing the sample has no effect the rate of loss of interstitial oxygen, however in contrast Fe contamination (5 x10^{12} atoms.cm^{-3}) has a more pronounced effect, increasing the loss of interstitial oxygen from solution. After this treatment it is not clear if Fe contamination is creating sites for oxide nucleation and inducing nucleation in the bulk or the loss of oxygen is caused by enhanced segregation at the dislocations. Electrical donors related to oxygen atoms segregated on dislocations is known to be produced during deformation [11], to understand if Fe is affecting the segregation of oxygen at the dislocation core, resistivity and Hall measurements were used to measure the concentration of dislocation donors before and after Fe contamination. Preliminary results indicate an increase in donor concentration, which could be interpreted as enhanced oxygen-dislocation interaction caused by Fe contamination. A more detailed

476

investigation of the influence of transition metal contamination on oxygen precipitation was presented at this conference [12], the enhancement of oxygen loss from solution was found to be significant after Fe contamination at high temperatures $T= 900\text{-}1000^0C$ and high Fe concentrations (2×10^{14} atoms.cm^{-3}), and Cu did not affect oxygen precipitation. In the present investigation much lower levels of deliberate contamination were used, it is suggested that the effect of Fe is to have a catalytic effect on oxygen segregation at or near the dislocation core.

4. Conclusions

The production of the D-bands in CZ Si is possibly due to residual transition metal contamination. The D-bands were quenched as the interaction of oxygen and dislocations is increased by longer deformation times and or higher bulk oxygen concentration, before there is any significant precipitation as detected by IR absorption measurements. At a higher deformation temperature ($T= 800^0C$) deep levels are formed and dislocation donors. Post-contamination of dislocated samples with Fe influences the interaction of oxygen at the dislocation while Cu has no influence.

5. Acknowledgements

The author would like to thank Prof. E. C. Lightowlers (King's College London) and Prof. B. Pichaud (MATOP) for interesting discussions and Bio-Rad Micromeasurements UK for facilitating my participation in this conference.

6. References

1. Boughesi, A., Pivac, B., Sassella, A., and Stella, A. (1995) J. Appl. Phys. **77**, 4169.
2. Higgs, V., Norman, C. E., Lightowlers, E. C., and Kigthley, P. (1992) Mat. Res. Symp. Proc. **163**, 57.
3. Falster, R., private communication.
4. Hirsch, P. (1980) J. Microsoc. **118**, 3.
5. Teichler, H. (1990) Proc. Polyse 90, Werner., J.H., and Strunk, H. P., (eds.), Heidelberg, Springer Press.
6. Lagowski, J., Gatos, H. C., Aoyama, T., and Lin, D. G. (1984) Appl. Phys. Lett. **45**, 680.
7. Higgs, V., Zhou, Q., and Rozgonyi, G. A. (1994) Mat. Sci. and Eng. **B24**, 48.
8. Werner, P., Reiche, M., and Heydenreich, J. (1993) Phys. Stat. Sol. **137**, 533.
9. Lightowlers, E. C., Jeyanathan, L., Safnov, A. N., Higgs, V., and Davies, G. (1994) **B24**, 144.
10. Colas, E., and Weber, E.R (1986) Appl. Phys. Lett. **48**,1371.
11. Koguchi, M., Yonenaga, I., and Sumino, K. (1982) **21**, L411.
12. Sumino, K., this conference.

AN ISOCHRONAL ANNEALING STUDY OF THE KINETICS
OF VO AND VO₂ DEFECTS IN NEUTRON IRRADIATED SI

C. A. LONDOS, N. V. SARLIS AND L. G. FYTROS

University of Athens, Physics Department, Solid State Section
Panepistimiopolis, Zografos, Athens 157 84 Greece

Abstract. The infrared spectra of room-temperature, neutron-irradiated, Czochralski-grown Si were investigated. During the anneal, the 827 cm^{-1} VO defect band decreases, and another band at 885 cm^{-1} of VO$_2$ centres increases. The kinetics of the evolution of these two defects was investigated. The decay of VO is dominated by a second order reaction (VO + Si$_i$ → O$_i$) with an activation energy 1.70 eV. The growth of VO$_2$ exhibits two stages. Below 360 °C, a first order reaction (VO + O$_i$ → VO$_2$) with an activation energy of 1.46 eV is more likely to occur. Above 360 °C, the growth of VO$_2$ seems to be correlated with mechanisms which do not involve VO centres. This behaviour may become explicable, at least qualitatively, in terms of a mechanism that involves the formation of oxygen dimers.

1. Introduction

Radiation-induced defects in Si have been the subject of many investigations, and numerous experimental and theoretical works have been reported so far. The A-centre (VO) is one of the most common defects encountered in irradiated Si. It is registered in all electrical ($E_c - 0.17$ eV level) and/or optical (827 cm^{-1} LVM) studies irrespective of the kind of irradiation (*e.g.* electrons, protons, neutrons, α-particles, γ-rays, *etc.*) Although a lot of information has been gathered about the structure and the electronic and optical properties of the centre, its exact annealing behaviour has not been completely clarified so far. The main reason for this is that the annealing of A-centres can follow simultaneously more than one reaction channel. Among them are the following:

477

R. Jones (ed.), Early Stages of Oxygen Precipitation in Silicon, 477–484.

i) Dissociation of A-centres

$$VO \rightarrow V + O_i, \tag{1}$$

ii) Destruction of A-centres by silicon self-interstitials emitted from defect clusters

$$VO + Si_i \rightarrow O_i, \tag{2}$$

iii) Association of VO with O_i atoms

$$VO + O_i \rightarrow VO_2, \tag{3}$$

and/or with another VO centre

$$VO + VO \rightarrow V_2O_2 \rightarrow VO_2 + V. \tag{4}$$

In principle, all these processes could occur in parallel. However it depends on the dose and the kind of irradiation, the oxygen content of the material, the temperature, *etc.*, as to which mechanism will prevail. It is therefore evident that the annealing process of the A-centre is compounded and complicated. Consequently, experimental data should be treated very carefully in order to distinguish the effect of the various processes and avoid misleading conclusions. Activation energies for example, reported from Arrhenius plots where only one process has been considered, may not represent true values if one or more competing processes with different activation energies occur in parallel. The scatter of reported values of activation energies cited in the literature reflects the above complexities as a result of the fact that a consensus among researchers has not been reached concerning the exact kinetics describing the phenomenon [1, 2, 3, 4].

In IR studies the decay of the 827 cm^{-1} signal from A-centres is accompanied by the growth of an 885 cm^{-1} signal attributed to VO$_2$ defects [2]. In this paper we shall examine the annealing kinetics of A-centres and the correlated growth kinetics of VO$_2$ defects using data taken from isochronal annealing studies of fast neutron irradiated Cz-grown Si. We shall consider mainly two scenaria of combining processes for the decay of VO and the growth of VO$_2$ defects in an attempt to establish which are the most suitable, but also physically meaningful mechanisms, that explain the whole phenomenon.

2. Experimental

Cz-grown Si samples with undetactable levels of substitutional carbon C_s ($[C_s] < 10^{16}$ cm^{-3}) and $[O_i] \approx 10^{18}$ cm^{-3} were irradiated ($T_{irr} < 50\,°\text{C}$) by fast neutrons at a dose equal to 10^{17} cm^{-2}. The samples were wrapped in

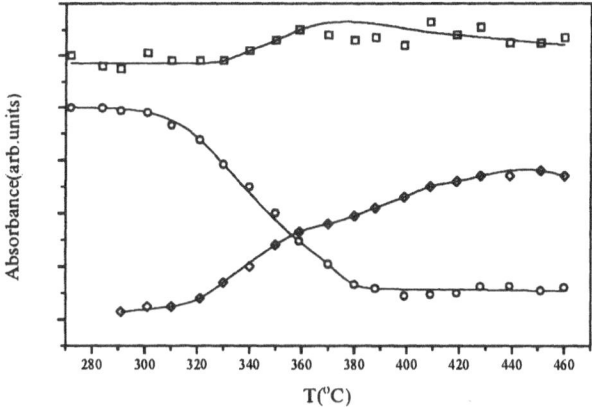

Figure 1. The evolution of VO (circles) and VO_2 (rhombs) defects as a function of temperature (15 min. isochronal annealing). The variation of O_i (squares) is also shown.

Cd foils to eliminate the effect of thermal neutrons. The heat treatments for annealing studies were carried out in open furnaces. The infrared absorption spectra were recorded at room temperature by using a JASCO-IR 700 dual beam spectrometer. Control samples of equal thickness from Float-Zone material were used during the measurements in order to subtract the two-phonon inherent absorption of the Si lattice.

3. Experimental Results and Discussion

As expected, the A-centre band at 827 cm^{-1} appears in our IR spectra, immediately after irradiation. On increasing the temperature as the isochronal anneal is carried out, the shape of the band is distorted by the emergence of satellite bands previously reported in the literature [5]. To extract the individual contribution of each component, the data were analyzed by computer deconvolution using Lorentzian profiles for each band [6]. Figure 1 shows the annealing behaviour of the A-centre band after subtracting the contributions of the satellite bands. The growth of the 885 cm^{-1} band attributed to VO_2 centre is also shown. The study of the evolution of these two defects will be the subject of this investigation.

3.1. THE DECAY OF VO

As it was already mentioned there are many reactions, Equations (1)–(4), that could take place in this temperature range. However, the irradiation is rather heavy and the isochronal annealing time $\tau = 15$ min. is rather small.

One would expect therefore, that the contribution of some of the reactions will be negligible in comparison with some others [3].

Since in neutron-irradiated Si large defect clusters are formed which liberate Si_i, in the temperature range under consideration, the most likely process for the decay of VO is that corresponding to equation (2), in agreement with the increase of the O_i concentration shown in figure 1. The process is generally described by a rate equation

$$\frac{d[VO]}{dt} = -k_1 [VO] [Si_i],\tag{5}$$

which may be approximated for the purposes of the present analysis, by

$$\frac{d[VO]}{dt} = -k_1 [VO]^2.\tag{6}$$

By integration, one gets

$$[VO](t) = \frac{[VO]_0}{1 + [VO]_0 k_1 t},\tag{7}$$

and

$$k_1 \tau = \frac{1}{[VO]'} - \frac{1}{[VO]_0},\tag{8}$$

where k_1 has a temperature dependence of the form $k_1 \propto e^{-E_a/kT}$, E_a being the activation energy, $[VO]_0$ is the initial concentration of A-centres, $[VO]'$ the concentration of VO after annealing for time τ at temperature T, and $[VO](t)$ the concentration of VO at any time during the isochronal annealing at T. By fitting $\ln k_1$ versus $1/T$ in the range 310–380 °C (figure 2), we find an activation energy $E_a = 1.70 \pm 0.05$ eV with a correlation coefficient r = 0.997. Since the value inferred by the fitting differs from the true mean value of $\ln k_1$ that would have been obtained by an infinite number of experiments, we plot in figure 2 the 80% confidence interval curves [7]. This means, that according to the fitting, the true mean value of $\ln k_1$ has an 80% probability of lying inside the area enclosed by the 80% confidence interval curves.

3.2. THE GROWTH OF VO_2

Evidently, from figure 1 the growth of VO_2 is related to the decay of A-centres. This relation may be modelled by combining equation (2) either with equation (3) or with equation (4).

3.2.1. *Scenario I*
In the first alternative the production of VO_2 is a first order process

$$\frac{d[VO_2]}{dt} = -k_2 [VO],\tag{9}$$

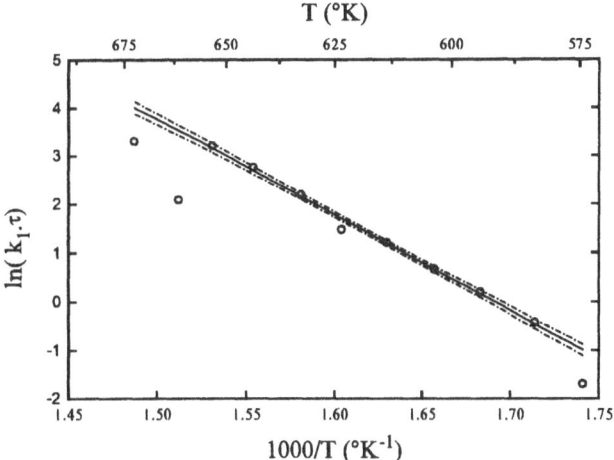

Figure 2. The Arrhenius plot for the decay of VO; the 80% confidence interval is enclosed by the two lateral curves.

where $[VO](t)$ is given by equation (7). By integration, one gets

$$k_2 = k_1 \frac{[VO_2]' - [VO_2]_0}{\ln([VO]_0 / [VO]')}, \tag{10}$$

where $[VO_2]_0$ and $[VO_2]'$ are defined in the same way as $[VO]_0$ and $[VO]'$ respectively. By subsituting equation (8) into (10) the following expression for k_2 is obtained

$$k_2 \tau = \left(\frac{1}{[VO]'} - \frac{1}{[VO]_0}\right) \frac{[VO_2]' - [VO_2]_0}{\ln([VO]_0 / [VO]')}. \tag{11}$$

The Arrhenius plot for the rate constant k_2 is shown in figure 3. As it is seen from this graph two stages possibly exist. In the first stage, below 360 °C, the logarithm of the rate constant can be fitted to a linear expression of $1/T$, and an activation energy of $E_a = (1.46 \pm 0.29)$ eV is obtained, with a correlation coefficient r = 0.962. In figure 3, the 80% confidence interval curves are also shown [7].

3.2.2. *Scenario II*
In the second alternative, the production of VO_2 is a second order process,

$$\frac{d[VO_2]}{dt} = -k_2'[VO]^2, \tag{12}$$

where $[VO](t)$ is given by equation (7). By integration one gets

$$k_2' = k_1 \frac{[VO_2]' - [VO_2]_0}{[VO]_0 - [VO]'}. \tag{13}$$

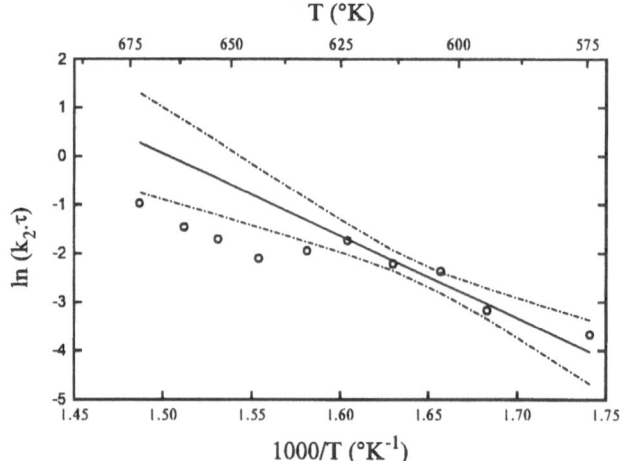

Figure 3. The Arrhenius plot for the growth of VO_2 (first order kinetics); the 80% confidence interval is enclosed by the two lateral curves.

By subsituting equation (8) into (14) the following expression for k_2' is obtained

$$k_2'\tau = \left(\frac{1}{[VO]'} - \frac{1}{[VO]_0}\right) \frac{[VO_2]' - [VO_2]_0}{[VO]_0 - [VO]'}. \qquad (14)$$

The Arrhenius plot for the rate constant k_2' is shown in figure 4. We observe again the existence of two characteristic stages below and above 360 °C. The activation energy, extracted from fitting the data of the first stage, has a value $E_a = (1.96 \pm 0.27)$ eV, with a correlation coefficient r = 0.981. In figure 4, the 80% confidence interval curves are also shown [7].

Let us consider now the second stage, above 360 °C, of the growth of VO_2. As it is seen from both figures 3 and 4, experimental points lie outside the 80% confidence interval curves for temperatures in this range. In other words, the experimental values of $\ln k_2$ or $\ln k_2'$ are not described adequately by the regression lines. Moreover, for temperatures above 380 °C, we notice that VO_2 continues to grow (figure 1), although the VO signal has almost stabilized at low amplitudes. These observations imply that another mechanism for the formation of VO_2 is involved. One possibility is through oxygen dimers. The notion of oxygen dimers that would migrate more easily than oxygen interstitials atoms through the Si lattice was firstly proposed by Gösele and Tan in order to explain experimental data concerning thermal donor behaviour [8]. The idea was also used to suggest an alternative process for VO_2 formation as a result of lattice vacancy trapping by oxygen dimers [9]. Such an explanation could account – at least qualitatively – for our experimental observations too. Notably, the oxygen concentration in figure 1 shows a tendency to decrease

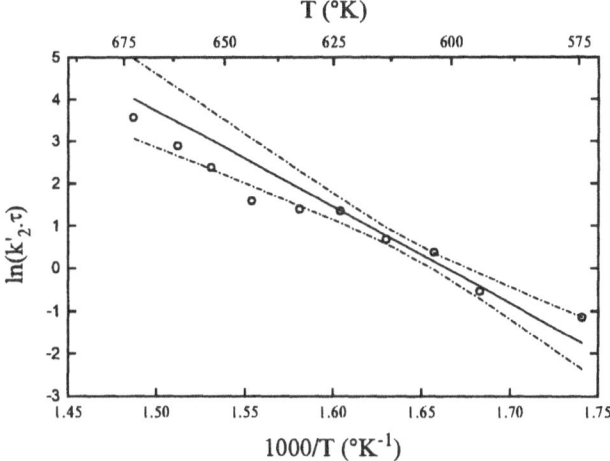

Figure 4. The Arrhenius plot for the growth of VO$_2$ (second order kinetics); the 80% confidence interval is enclosed by the two lateral curves.

in this temperature range, a fact which is in accord with processes involving oxygen dimerization. Quantitatively, however, the picture is not so clear. A more detailed report on the variation of O$_i$ concentration and its relation to the processes of formation of the various VO$_n$ defects in Si will be published shortly [10].

3.3. RECAPITULATION

The results are summarized in Table 1. Although scenario II exhibits a

TABLE 1. The two scenaria for the decay of VO and VO$_2$.

	Scenario I	Scenario II
Reactions	VO + Si$_i$ → O$_i$	VO + Si$_i$ → O$_i$
	VO + O$_i$ → VO$_2$	VO + VO → VO$_2$ + V
Rate Equations	$d[\text{VO}]/dt = -k_1\,[\text{VO}]^2$	$d[\text{VO}]/dt = -k_1\,[\text{VO}]^2$
	$d[\text{VO}_2]/dt = -k_2\,[\text{VO}]$	$d[\text{VO}_2]/dt = -k_2'\,[\text{VO}]^2$
Activation Energies	$E_{a1} = (1.70 \pm 0.05)$ eV	$E_{a1} = (1.70 \pm 0.05)$ eV
	$E_{a2} = (1.46 \pm 0.29)$ eV	$E_{a2'} = (1.96 \pm 0.27)$ eV
Correlation Coefficients	$r_1 = 0.997$	$r_1 = 0.997$
	$r_2 = 0.962$	$r_{2'} = 0.981$

higher correlation coefficient, scenario I is more likely to occur in our opinion, since $[O_i]$ is about two orders of magnitude higher than [VO] and therefore reaction (3) is expected to dominate over reaction (4).

4. Conclusions

Two scenaria have been employed for the study of the decay of VO and the correlated growth of VO_2 defects. In both cases the disappearance of A-centres was studied in terms of annihilation of VO by Si self-interstitials liberated from defect clusters that are expected to be present in neutron-irradiated material. The growth of VO_2 exhibits two stages. In the first stage, the growth was discussed in terms of either O_i being trapped by VO (equation 3), or VO defects pairing up to form V_2O_2 defects which in turn decompose into VO_2 and V (equation 4). In our opinion, a defect reaction model postulating the processes $VO + Si_i \rightarrow O_i$ and $VO + O_i \rightarrow VO_2$ is more physically meaningful due to the high oxygen content of the Si material. In the second stage, the growth is likely to be related to oxygen dimer formation. Fast moving oxygen dimers provide an alternative channel for the creation of VO_2 defects through capturing of free lattice vacancies.

Acknowledgments

N.V.S. was supported by a grant of the National Scholarship Foundation of Greece, IKY.

References

1. Benski, G., and Augestiniak, W. M. (1957) Annealing of electron bombardment damage in silicon crystals, *Phys. Rev.* **108**, 645–648.
2. Corbett, J. W., Watkins, G. D., and McDonald, R. C. (1964) New oxygen infrared bands in annealed irradiated silicon, *Phys. Rev.* **135**, A1381–A1385.
3. Svensson, B. G., and Lindström, J. L. (1986) Kinetic study of the 830 and 889 cm^{-1} infrared bands during annealing of irradiated silicon, *Phys. Rev. B* **34**, 8709–8717.
4. Tipping, A. K., Newmann, R. C., Newton, D. C., and Tucker, J. H. (1986) Enhanced oxygen diffussion in silicon at low temperatures, in J. von Bardeleben (ed.), *Defects in Semiconductors Materials Science Forum* **10–12**, pp. 887–892.
5. Ramdas, A. K., and Rao, M. G. (1966) Infrared Absorption spectra of oxygen-defect complexes in irradiated Silicon, *Phys. Rev.* **142**, 451–456.
6. Londos, C. A., Sarlis, N. V., Georgiou, G., and Fytros, L. G. (unpublished data).
7. Crow, E. L., Davis, A. F., and Maxfield, M. W. (1960) *Statistics Manual*, Dover Publications, Inc., New York.
8. Gösele, U., and Tan, T. Y. (1982) Oxygen diffusion and thermal donor formation in silicon, *Appl. Phys. A* **28**, 79–92.
9. Svenson, B. G., and Lindström, J. L. (1985) Growth of the 889 cm^{-1} infrared band in annealed electron-irradiated Si, *Appl. Phys. Lett.* **47**, 841–843.
10. Sarlis, N. V., Londos, C. A., and Fytros, L. G. (to be published).

UNIFORM STRESS EFFECT ON NUCLEATION OF OXYGEN PRECIPITATES IN CZOCHRALSKI GROWN SILICON

A. MISIUK
Institute of Electron Technology
Al.Lotnikow 32/46, 02-668 Warszawa, POLAND

1. Introduction

Precipitation of oxygen in Czochralski grown silicon, Cz-Si, at higher temperatures is concomitant with stress [1,2]. This follows mostly from the difference in volume between the clustered (precipitated) oxygen atoms and that of the Si atoms originally present in the same region of the lattice. Stress-related effects in Cz-Si are present even at the earliest stages of clustering, that is at the "thermal donor, TD, temperature range" (TDs exert a compressive stress on the surrounding Si lattice [3]). The same takes place in the case of new donors, NDs, and of nucleation centres for oxygen precipitation, NCs, created at about 1000K.

Investigation of Cz-Si samples annealed under hydrostatic pressure, HP, can help in understanding stress-related effects during oxygen precipitate nucleation. Such an approach proved its efficacy for the SiO_{2-x}/Si system [4,5].

This paper describes investigations of HP induced phenomena in Cz-Si annealed at the temperature range within which TDs, NDs and NCs are created.

2. Experimental

Silicon samples of about $10 \times 10 \times 0.6$ mm^3 dimension were cut from commercially available Cz-Si wafers of (100) orientation.

The oxygen interstitial concentration in the n- and p-type Cz-Si samples, c_o, determined by Fourier Transform Infrared Spectrometry, FTIR, using ASTM F 121-83 standard of 2.45×10^{17} cm^{-3}, was within the $(6.3-11) \times 10^{17}$ cm^{-3} range. The carbon concentration, c_c, was below 3×10^{16} cm^{-3}. The properties of the crystals investigated are summarised in Table 1.

Samples B and C were preannealed at 723K under atmospheric pressure (10^5Pa) for up to 96 hours to create TDs.

Samples were placed into silica or alumina crucibles and annealed under hydrostatic pressure (HT-HP treatment) exerted by purified argon gas (HP up to 1.1×10^9 Pa) at 723-1000K (1230K) for up to 10 hours. After the HT-HP treatment, some samples were additionally annealed under 10^5 Pa at 1230 and 1400K for 5 hours to investigate the effect of HP during the creation of NCs on oxygen precipitation.

After removal of the (5-50)µm thick surface layer, sample properties were investigated by FTIR, selective chemical etching, X-ray and electrical methods.

R. Jones (ed.), Early Stages of Oxygen Precipitation in Silicon, 485–492.
© *1996 Kluwer Academic Publishers.*

486

TABLE 1. Preannealing conditions (at 10^5Pa), initial interstitial oxygen concentration, type of conductivity, carrier concentration of studied samples.

Sample	Preannealing (sample, K, hours)	$c_o \times 10^{-17} cm^{-3}$	Type	$N_n, N_p \times 10^{-14}$ cm^{-3}
A	--	6.4	n	9
B	--	6.5	n	7.5
BI	(B, at: 723, 10)	6.3	n	10
C	--	8.0	n	7.5
CI	(C, at: 723, 96)	7.5	n	60
D	--	7.0	p	14
E	--	10-11	p	11

3. Results

3.1. STRESS - RELATED EFFECTS AT 723K

Annealing the Cz-Si samples at 723K under HP up to 1.1 GPa markedly influences their electrical properties, namely resistivity and carrier concentration (Fig. 1). The effects observed were greater for samples with higher c_o. Annealing of sample BI at HP-873K was performed to check the effect HP has on "donor killing" for the sample containing TDs.

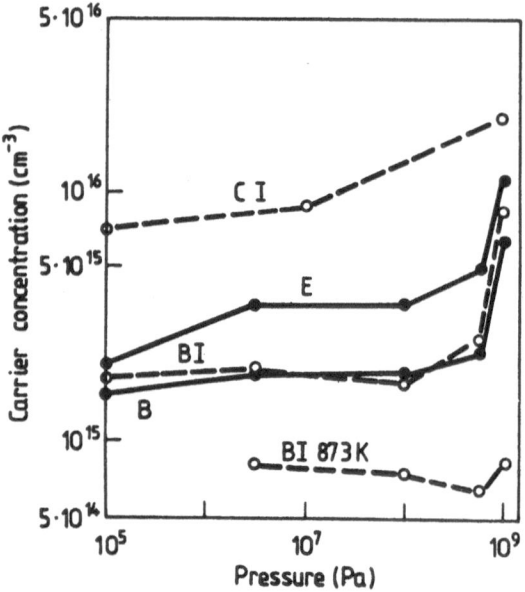

Figure 1. Carrier concentration dependence on HP for Cz-Si samples annealed at 723K, except sample BI 873K - annealed at 873K. Time of HP treatment: 10 hours except sample CI - 6 hours.

The "added" carrier (electron) concentration, ΔN_n, increases with c_o accordingly to the power law dependency [6]:

$$\Delta N_n \sim c_o{}^n \tag{1}$$

A value of $n=4$ was estimated for anneals at 10^5 Pa (annealing time - 10 hours). This value diminished with pressure, reaching $n \sim 1$ for samples annealed at HP=1.1×10^9 Pa. The HT-HP treatment also caused a pronounced rise in N_n for samples BI and CI where TDs are present initially.

After the HT-HP treatment at 0.6×10^9 - 1.1×10^9 Pa, the samples exhibited a slight (up to 10%) diminishment in c_o.

3.2. STRESS - RELATED EFFECTS AT 873K

As it concerned a change of carrier concentration, the effect of HP during annealing at 873-920K was similar but less pronounced than that for the samples annealed at 723K. For example, whereas ΔN_n reached the value above 1×10^{16} cm^{-3} for sample E (Table 1, Fig. 1) treated at 1.1×10^9Pa - 723K for 10 hours, the same treatment but at 873K produced $\Delta N_p \sim 5\times10^{14}$ cm^{-3}. Typical examples of resistivity changes are presented in Fig. 2.

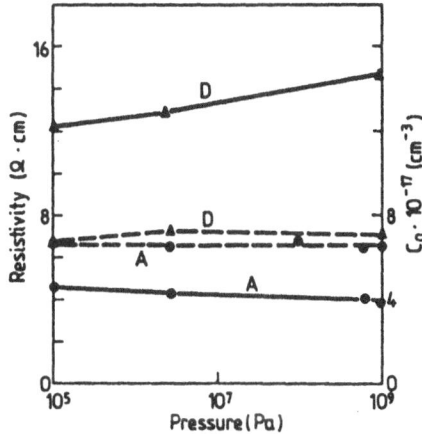

Figure 2. Resistivity (solid line) and c_o (dashed line) dependence on HP
for samples A and D subjected to annealing at 873K for 10 hours.

The resistivity changes can be interpreted as the electron concentration rising with HP. c_o was almost pressure independent for samples with $c_o \leq 8\times10^{17}$ cm^{-3}, but diminished with HP for sample E where $c_o \sim 1\times10^{18}$ cm^3.

The rise of N_n with HP was stronger for the samples with higher c_o. For example the "added concentration" of electrons for sample E treated at 6×10^8Pa - 873K for 10 hours was 5×10^{14}cm^{-3} but about 5 times lower for sample A. ΔN changed markedly between the samples of different origin, being distinctly dependent also on sample characteristics other than c_o.

3.3 STRESS - RELATED EFFECTS AT 1000K (NUCLEATION STAGE)

The density, d, of saucer pit defects, SPDs, and of precipitate defect complexes, PDCs, revealed after etching in the Yang solution, rose with HP for the samples with $c_o \sim 6.5\times10^{17}$ cm^{-3} annealed at 1000K - HP (Table 2).

TABLE 2. Total density of oxygen-related defects for sample C annealed under HP for 5 hours at 1000K

HP (Pa)	10^5	10^7	10^8	6×10^8	10^9
d (cm^{-2})	6×10^3	1.4×10^4	6×10^3	1.8×10^4	3×10^5

For the samples with $c_o \geq 8 \times 10^{17}$cm^{-3}, d was about 1×10^4 cm^{-2}, practically independent of HP.

The effect of HP upon c_o for samples C and E annealed under HP at 1000K for 5 hours is presented in Fig. 3. HP markedly influenced c_o only in the case of the samples with the high initial concentration of interstitial oxygen.

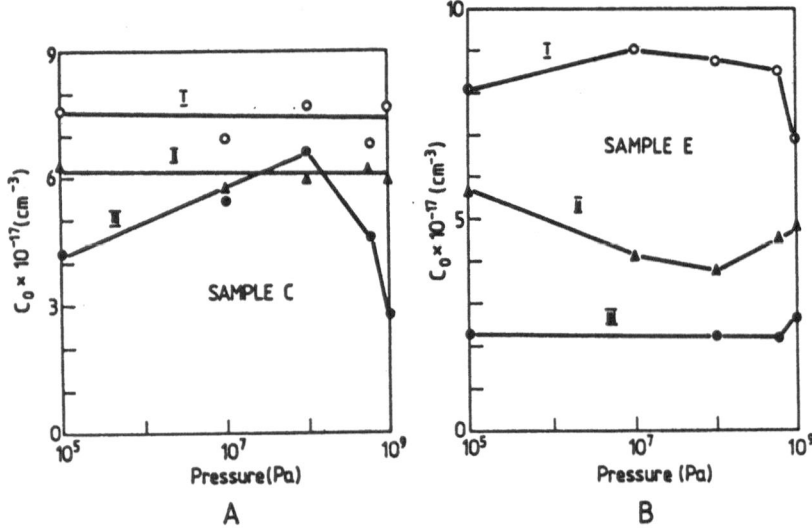

Figure 3. A. Data for C sample with initial $c_o = 8 \times 10^{17}$cm^{-3}.

B. Data for E sample ($c_o = 1 \times 10^{18}$ cm^{-3}).

I. Dependence of c_o on HP for samples annealed at 1000K for 5 hours.

II & III. Oxygen precipitation at 10^5 Pa, 5 hours.
Dependence of c_o on HP during nucleation stage.

Triangles (II):c_o after annealing at 1400K.

Filled circles (III): c_o after annealing at 1230K.

3.4. EFFECT OF HP APPLIED DURING NUCLEATION STAGE ON OXYGEN PRECIPITATION AT 1230-1400K

Annealing Cz-Si samples under HP at 1230K can promote massive precipitation of oxygen even in the case of samples initially without NCs (Fig. 4).

This effect was dependent on c_o and the initial defect structure of Cz-Si samples as well as on the annealing temperature [7].

Subsequent annealing for 5 hours at 1230K-10^5 Pa or at 1400K-10^5 Pa of samples containing NCs created under HP caused further oxygen precipitation at NCs. The dependence of c_o on HP applied during the nucleation stage differs for samples C and E (Fig. 3: II, III).

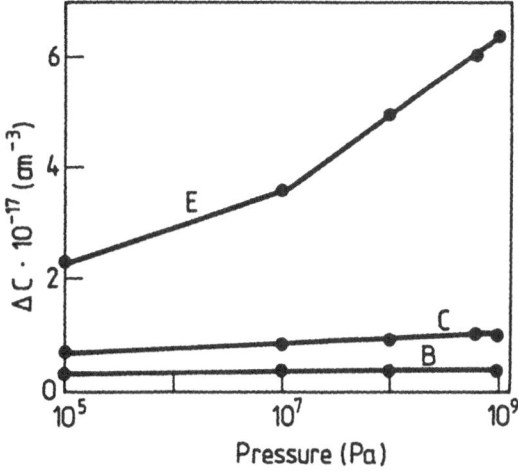

Figure 4. Precipitated oxygen, ΔC, dependence on HP for samples
annealed at 1230K for 5 hours .

The total defect density, d (of PDCs, stacking faults, SFs, and dislocations, Ds), for sample E annealed at 1230K-10^5 Pa for 5 hours was about 2×10^6cm^{-2} being independent of HP applied during the nucleation stage. In contrast, d rose with HP applied during the nucleation stage when annealing at 1400K-10^5 Pa for 5 hours. It increased from 6×10^5 cm^{-2} for the sample with NCs created at 1000K-10^5 Pa (5 hours) to 2×10^6 cm^{-2} for the sample with NCs created at 1000K-10^9 Pa - 5 hours.

The structural distortion defined as the value of root-mean-square atom displacement, U [8] varied with HP. The lowest value of 0.0018nm was attained for sample E with nucleation performed under 10^7 Pa and annealed at 1400K for 5 hours, but reached the value of 0.0054nm for the same sample with nucleation performed at 10^9 Pa.

4. Discussion

A variety of effects were observed in Cz-Si annealed under hydrostatic pressure. The common features were the enhanced generation of donor states (in the TD and ND temperature ranges) and the stress-stimulated precipitation of oxygen. The observed effects are strongly dependent on c_o (Eq. 1). In most cases, the defect density rose with HP, i.e. the HP treatment promoted effects related to the rise of the concentration of structural irregularities.

There is experimental evidence [9] that HT-HP treatment results in a reduction of the diffusion coefficient of the interstitial oxygen.

Because of the different compressibilities, γ, of Si and SiO$_{2-x}$ ($\gamma=1/K$, where K is the bulk modulus: $K_{Si} = 9.8\times10^{10}$ Pa, $K_{SiO_2} = 4\times10^{10}$ Pa) the value of misfit, ε, at the oxygen cluster / Si lattice boundary is dependent on HP [5,10]. There is an equivalence[4] between the effect of annealing (temperature rise ΔT) and the effect induced by HP. The HP and ΔT changes causing the same change of ε are related by Eq. 2:

$$\Delta T \approx HP \frac{\left(\dfrac{1}{K_m} - \dfrac{1}{K_p} \right)}{\beta_p - \beta_m} \tag{2}$$

where the subscripts m and p concern lattice and cluster / precipitate material, and β are the respective volume thermal expansion coefficients ($\beta_{Si}=1.3 \times 10^{-5}$ K^{-1}, $\beta_{SiO_2}=1.6 \times 10^{-6}$ K^{-1}). In the case of Cz-Si with precipitates composed of amorphous SiO_2, the effect of HP=1GPa would be the same (in terms of $\Delta\varepsilon$) as that caused by annealing at about 1300K [4]. In the case of clusters of SiO_{2-x} composition with x far from zero, ΔT would be much smaller. Nevertheless it is clear that misfit can be created even at the initially misfit‑ free structural inhomogeneities.

Various defects are known to be present in as-grown Cz-Si [11]. It is my opinion that even very small oxygen clusters are activated under the HT-HP conditions, to serve as the nucleation centres for NC growth. This means that after the HT-HP treatment, more NCs might be present in the lattice, and their concentration would be dependent on the homogeneity of the as-grown Cz-Si.

So one can expect that under the HP conditions there will be more NCs created (when compared to the "ambient pressure conditions"). A considerable part of them will be of small dimensions, resulting from mentioned HP-retarded oxygen diffusivity [9], some of them with donor activity.

Activation of structural irregularities under HP was observed even for the non-preannealed Cz-Si samples with the high c_o value (Fig. 4, Ref. [7]). The rise of N_n with HP for the samples annealed at 723K and at 823K is also consistent with proposed explanation.

In the case of annealing at 1000 K, more defects were created at high HP (Table 2, Fig. 3, I), especially for the samples with considerable inhomogeneity (large c_o - sample E). Further annealing at 1230 K (Fig. 3, III) caused promoted oxygen precipitation for HP-preannealed samples. The non-monotonic c_o dependence on HP is a result of the wide spectrum of oxygen cluster radii and their different composition. Part of the small oxygen clusters were dissolving into the lattice at 1230 K and more at 1400 K, and subsequently not acting as the NCs for further oxygen precipitation.

The above model for HP-induced effects is, of course, oversimplified (e.g. the effect of HP on the precipitate radius is not taken into account). However, the model is consistent with experimental observations.

The nature of HP-induced donor centres also needs to be discussed.

Annealing Cz-Si at 723 and 873 K under HP for 6-10 hours caused a lot of effects related to a change in carrier concentration. Two of them are very significant: (1) a pronounced rise of the N_n concentration with HP and the corresponding changes of sample resistivity and (2) a complicated dependence of the donor killing effect on HP (sample BI 873K - Fig. 1). The rise of electron concentration is distinctly dependent on c_o and is more pronounced for the samples annealed at 723K than at 873K.

The first effect can be attributed to enhanced HP-induced creation of TDs and NDs.

However, such an explanation is not sufficient. The change in n (Eq. 1) with HP and the enhanced HP effect on N_n for samples subjected to prolonged preannealing at 10^5 Pa (Fig. 1) were observed. This suggests that part of the HP-induced donors are created somewhat independently of the typical TD generation process and seems to speak in favour of the different nature of the HP-induced donor centres. The same was suggested by Emtsev [12].

As it concerns the donor-killing effect, N_n (HP) changes can be interpreted as a result of the "723K donor states" being removed at 873K with simultaneous generation of "873K donor states", both processes being dependent on HP (Fig. 2).

Next, possible artifacts following from unintentional sample contamination are to be discussed.

The carbon concentration did not change after processing. The most probable other contaminant would be aluminium. It is known, however, that Al contamination would promote creation of the p-type conductivity, the opposite of what is observed.

Other probable contaminants would be transition metals, influencing the kinetics of TD creation [13]. However, at low temperatures of treatment (723-823 K) the diffusion rate of most transition metals into Si would be low and hence the equilibrium concentrations would also be low, i.e. not much above 10^{12} cm^{-3}.

Another possibility follows from "internal contamination". High pressure can promote dissociation of impurity precipitates present in semiconductors materials [14]. This effect is not expected, however, to be of significance because of the standard high purity of Cz-Si. DLTS investigation of the HT-HP processed samples proved that the concentration of deep levels which could be attributed to metallic impurities was below 1×10^{14} cm^{-3}.

Summarising, there exists strong evidence that the observed effects are primarily of intrinsic nature. Of course, a partial influence of the above mentioned or other artifacts can not be excluded.

5. Conclusions

A number of stress-induced effects were observed in Cz-Si samples annealed at 723-1000K (1230K) under hydrostatic pressure. The common feature observed is the enhanced generation of the donor states (in the TD and ND temperature ranges) and stimulated precipitation of oxygen. In most cases the sample defect density rose with HP. The effects were strongly dependent on the oxygen concentration.

The above phenomena can be explained qualitatively in terms of the HP-induced misfit at the oxygen cluster/lattice boundary and also of the activation of "additional" nucleation centres for oxygen precipitation.

6. Acknowledgements

I thank Dr W.Jung, Mr M.Rozental (Institute of Electron Technology, Warszawa), M.sc. B.Surma (Institute of Electronic Materials Technology, Warszawa) and Prof. V.Khrupa (Institute of Semiconductor Physics, Kiev, Ukraine) for the sample measurements. A stimulating discussion with Prof. V.V.Emtsev from A.F.Ioffe Institute, St.Petersburg, Russia and Prof. J.Vanhellemont from IMEC, Leuven, Belgium is gratefully acknowledged.

This work was supported in part by the Polish Committee for Scientific Research and the Foundation for Polish-German Collaboration.

7. References

1. Reiche, M., Nitzsche, W. (1992) The influence of stresses on the surface-near defect structure, *Mat.Res.Soc.Symp.Proc.* **262**,621-626.
2. S.Kimura, T.Ishikawa (1995) Relation between lattice strain and anomalous oxygen precipitation in a Czochralski-grown silicon, *J.Appl.Phys.* **77**, 528-532.
3. Wagner, P. and Hage, J. (1989) Thermal double donors in silicon, *Appl.Phys.A* **49**, 123-138.
4. Misiuk, A., Datsenko, L.I., Surma, B. and Popov, V.P (1994) Oxygen precipitation in Cz-Si under uniform stress, ed. C.Hill, P.Ashburn *Proceed. ESSDERC'94 Conference, Edinburgh 1994*, Editions Frontiers, 243-246.
5. Misiuk, A. (1995) Microstress in high pressure-temperature treated Czochralski grown silicon with oxygen concentration up to $1.2 \times 10^{18} cm^{-3}$, in ed. J. Zmija, A. Rogalski, J. Zielinski, *Solid State Crystals: Materials Science and Applications, Proc. SPIE* **2373**, 335-340.
6. Londos, C.A., Binns, M.J., Brown, A.R., McQuaid, S.A., Newman, R.C. (1993) Effect of oxygen concentration on the kinetics of thermal donor formation in silicon at temperatures between 350 and 500 C, *Appl.Phys.Lett.* **62**, 1525-1526.
7. Misiuk, A., Surma, B., Hartwig, J. (1996) Stress-induced oxygen precipitation in Cz-Si, *Materials Sci.Eng.B*, **36**,30-32..
8. Datsenko, L.I., Misiuk, A., Hartwig, J., Briginets, A., Khrupa, V.I.(1994) Influence of preannealing on perfection of Czochralski grown silicon crystals subjected to high pressure treatment, *Acta Phys. Polon. A* **86**, 585-590.
9. Antonova, I.V., Fedina, L.I., Misiuk, A., Popov, V.P., Shaimeev, S.S. (1995) DLTS study of oxygen precipitates evolution in Czochralski grown silicon at high pressure -high temperature, in ed. H.Morawiec & D.Stroz, *Applied Crystallography*, World Scientific, 324-327.
10. Misiuk, A., Adamczewska, J., Bak-Misiuk, J., Hartwig, J., Morawski, A., Witczak, Z. (1991) Hydrostatic pressure effect on oxygen precipitates in silicon single crystal, *Acta Phys. Polon. A* **80**, 317-320.
11. Ikari, A., Kawakami, K., Haga, H., Uedono, A., Wei, L., Kawano, T., Tanigawa, S. (1994) Defects in Czochralski grown silicon crystals investigated by positron annihilation, *Jpn. J.Appl. Phys.***33**, 5585-5589.
12. Emtsev, V.V. (1995) *private communication.*
13. Bakhadyrkhanov, M.K., Askarov, Sh.I., Narkulov, N., Srazhev, S.N., Toshboev, T.U. (1995) The influence of fast-diffusing impurities on the kinetics of thermodonor generation in silicon at 900-500 C, *Phys.Tekhn.Polupr.*, **29**, 1396-1402.
14. Fistul, V.I., Turaev, A.R., Zainabidinov, S.Z. (1993) Baric impurity effects in silicon, *phys.stat.sol. (a)* **136**, 337-349.

ON THE IMPACT OF GROWN-IN SILICON OXIDE PRECIPITATE NUCLEI ON SILICON GATE OXIDE INTEGRITY

J. VANHELLEMONT[a], G. KISSINGER[b], K. KENIS[a], M. DEPAS[a], D. GRÄF[c], U. LAMBERT[c] and P. WAGNER[c]

[a]IMEC, Kapeldreef 75, B-3001 Leuven, Belgium
[b]Institute for Semiconductor Physics, D-15204 Frankfurt (Oder), Germany
[c]Wacker Siltronic AG, P.O. Box 1140, D-84479 Burghausen, Germany

1. Introduction

Small point defect clusters grown-in during Czochralski (Cz) crystal pulling can have an important impact on gate oxide integrity (GOI) for present day high quality silicon substrates. A pronounced effect of the crystal pulling conditions on GOI is indeed observed for gate oxide thicknesses between 10 and 100 nm as was already reported more than 20 years ago [1]. Beginning of the 1980's the link with microdefects grown-in in the silicon substrate was already made [2,3] but it was only the last five years that this correlation has been studied extensively [e.g. 4-9]. Due to the low density and small size of these grown-in defects, few techniques are available to study their nature which is therefore not yet well understood. Recently, powerful tools based on light scattering using visible [10], near [11] and mid infra red [12] light have become available and allow to reveal the minute lattice defects present in as-grown silicon substrates.

Visible light scattering reveals defects on the polished wafer surface as localised light scatterers (LLS). When the wafers are been cleaned using Standard Cleaning (SC) procedures, e.g. the SC1 clean in $NH_4OH/H_2O_2/H_2O$ mixtures, small etch pits develop on the wafer surface which are visible as LLS. As these LLS originate from the substrate they are also commonly called Crystal Originated Particles (COP's) as they were first observed by particle detection tools. The size of the scatterers can be estimated by comparison with scattering by calibrated latex spheres and is expressed as latex sphere equivalent (LSE) size. Smallest scatterer sizes which can be detected are of the order of 0.08 μm. The COP density and size distribution depends strongly of the crystal pulling conditions and thus also on the wafer diameter. A recent review on the possibilities of visible light scattering for silicon wafer surface inspection can be found in reference 10.

The use of infra red light has the advantage that the silicon substrate becomes transparent and a volume density of light scatterers can be obtained. Large volumes can be probed allowing to detect defect densities in the range 10^5-10^{10} cm^{-3}. The upper defect density detection limit is determined by the overlap of the scatterers in the image. The lower size detection limit for near IR light and assuming SiO_2 inclusions, was determined to be about 0.04 μm [13].

A complementary technique to reveal grown-in point defect clusters is preferential etching. Flow pattern defect (FPD) etching [14], reveals differences between crystals

R. Jones (ed.), Early Stages of Oxygen Precipitation in Silicon, 493–500.

grown under different pulling conditions. Immersing the wafers vertical in Secco etch, V shaped so called flow patterns develop on the wafer surface as well as small etch pits, Secco Etch Pits (SEP's), which can be easily imaged with an optical microscope.

Although no clear identification of the defects revealed by the different techniques has been obtained so far - it is even not 100% clear that the same defects are detected -, the defect densities correlate quite well with gate oxide defect densities obtained from dielectric breakdown studies.

In the present paper results are presented of an ongoing study [15,16] of the evolution of the grown-in defect density and size as function of crystal pulling conditions and additional thermal treatments and of the correlation with gate oxide breakdown characteristics. Light scattering results are compared with gate oxide breakdown characteristics. The impact of crystal cooling rate and vacancy supersaturation on silicon oxide precipitate nucleation and critical size is briefly discussed and theoretical predictions are compared with the defect observations.

2. Experimental

150 mm and 200 mm, p-type Cz substrates are studied obtained from crystals grown with different cooling ramps. In all cases the ratio of pulling rate over temperature gradient was above the critical value of 0.13 $mm^2K^{-1}min^{-1}$ for stacking fault ring formation [17], i.e. yielding crystals which are vacancy-rich. The wafers are labelled A, B, C and D with decreasing cooling rate.

The COP density is determined using Tencor SS6200 and Censor ANS-100 instruments both on the as-polished wafers and after a COP test consisting of a 4h SC1 treatment in $NH_4OH/H_2O_2/H_2O = 1/1/5$ at 85°C. By this treatment about 150 nm silicon is etched off and the integrated number of grown-in defects in the removed volume is observed as COP's with different sizes [18].

The crystal defect density before and after gate oxidation is studied with infrared laser scattering tomography (LST) using a MILSA IRHQ-2 instrument. Observations are performed after cleaving the wafers using a Nd-YAG beam incident perpendicular on the cleavage plane. The scattered IR light is detected in the direction perpendicular to the wafer surface and the incident beam (= cross-section observation).

Flow pattern defects and SEP's are counted using optical microscopy after vertical immersion of the wafers in unstirred Secco etch for 30 min.

6.5 and 15 nm gate oxides are grown at 900°C on 150 mm wafers. The wafers are ramped-up in a 5% O_2/95% N_2 atmosphere which is also used to grow the 6.5 nm oxides. The 15 nm oxides are grown in dry oxygen. 1, 4 and 16 mm^2 capacitors are fabricated using 500 nm phosphorus solid source doped polycrystalline silicon contacts patterned by wet etching. Backside contacts are made by Al/Si sputtering and 30 min sintering at 435°C.

Gate oxide breakdown characteristics are measured using voltage to breakdown (E_{BD}) characteristics by ramping up the voltage in steps of 0.1V with a holding time of 10 ms between each step (= 10 Vs^{-1}). Gate oxide defect densities are calculated assuming an intrinsic breakdown field of 11 $MVcm^{-1}$.

3. Observations and Discussion

Crystal originated particles

The COP density and size distribution depends strongly on the cooling rate as illustrated in Fig. 1 for 200 mm wafers both as polished and after COP test [19]. The lines drawn through the experimental data are obtained using cubic spline fitting. Polishing of the wafers reveals only the largest grown-in defects which can be observed as LLS. Therefore at first sight there seems not to be a clear correlation between cooling rate and defect density on the as polished wafers. After the COP test which enlarges the average size of etch pits the expected trend is however observed: the faster the cooling rate the larger the defect density and the smaller their size.

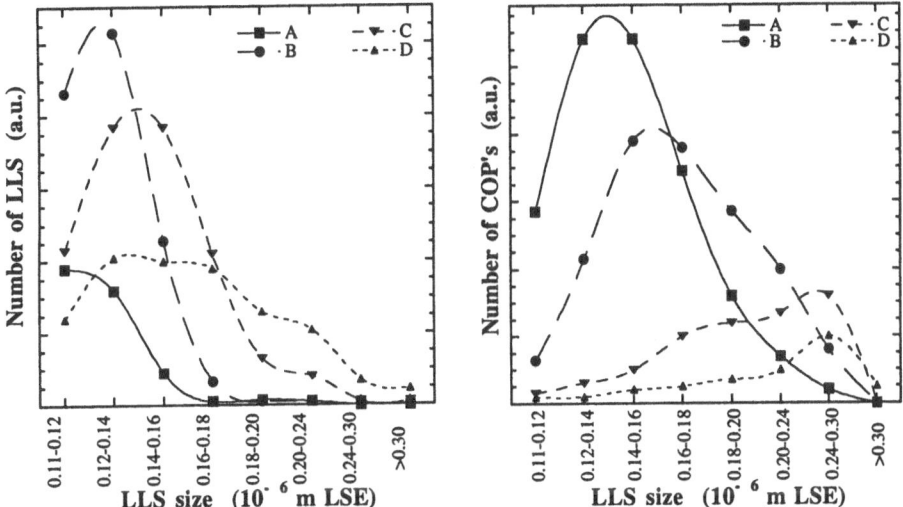

Fig. 1: LLS and COP size distribution before (left) and after (right) SC1-treatment at 85°C for 4h of 200 mm wafers [19].

LST results

IRLST is one of the few techniques which allows to observe the low density of defects in as-grown silicon material. Defects are made visible when their scattering size is larger than a critical value. Assuming that the scatterer is a silicon oxide particle, the critical diameter is close to 40 nm. Smaller SiO_x particles will thus not be observed.

Figure 2 shows the radial distribution of the LST defect density in 150 mm wafers with different cooling rates. The observed defect distribution is quite uniform but drops off quite abrupt in the last 10 mm near the rim of the wafer. A clear dependence of LST defect density on crystal cooling rate is observed: increasing cooling rate leads to increasing density of defects. Average N_{LST} values for 150 mm wafers both as-grown and after gate oxidation, are given in Table I showing that the defect density apparently increases by the heat treatment.

One can obtain an idea of the size of the scattering centres by calibrating the instrument using samples with silicon oxide precipitates with known sizes. Using this approach the average size of ten LST defects in different as-grown 150 and 200 mm wafers was obtained as illustrated in Fig. 3. It is clear that with increasing cooling rate, the LST defect density increases while the average size of the defects decreases. After gate oxidation (70 min at 900°C), the average size and apparent density of the LST defects increases but the size distribution keeps the same overall shape. The larger size of the LST defects in the 200 mm wafers illustrates the increasing thermal budget with increasing crystal diameter. The large diameter crystal will spend a longer time in a temperature window where cluster growth by diffusion of point defects is possible. That this indeed occurs is illustrated by the observations after a 70 min treatment at 900°C of the 150 mm wafers (= 15 nm gate oxidation).

One can wonder if there exists a correlation between the surface and bulk LLS's observed by visible and near IR light, respectively. Taking for the typical size of the COP's the value at the maximum density in Fig.1(right), a close correlation between the COP density and N_{LST} is observed as shown in Fig. 4 suggesting that both techniques reveal the same defect.

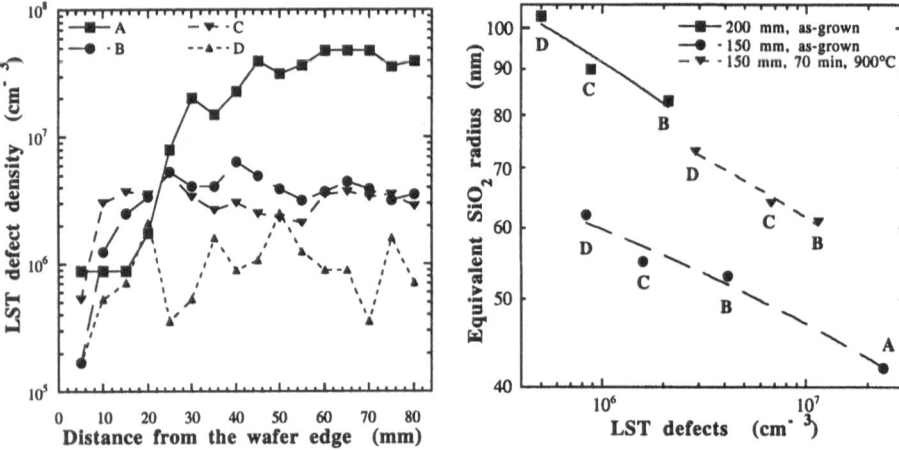

Fig. 2: LST defect density for 150 mm wafers with different cooling ramps.

Fig. 3: Exponential dependence of equivalent SiO_2 radius on N_{LST}.

TABLE I. Average LST defect and FPD densities in 150 mm wafers . (- = not measured)

Substrate type	N_{LST} before	N_{LST} 6.4 nm	N_{LST} 15 nm	N_{FPD} before	N_{FPD} 6.4 nm	N_{FPD} 15 nm
	10^5 cm^{-3}	10^5 cm^{-3}	10^5 cm^{-3}	10^5 cm^{-3}	10^5 cm^{-3}	10^5 cm^{-3}
A1	238	-	374	6.5	-	-
B1	41.2	115	119	5.9	5.0	12.2
C1	15.8	67.0	47.8	5.7	4.0	6.3
D1	8.35	28.7	20.2	2.9	2.2	5.5

Flow pattern defects

FPD's are often associated in literature with large vacancy clusters as they are not observed in slowly pulled crystals which are known to be self-interstitial-rich. Large vacancy clusters are also efficient scattering centres and therefore can not be differentiated from SiO_x particles by a single LST measurement. It can however be expected that the growth and anneal kinetics of both types of scattering particles are strongly different during thermal treatments. Figure 5 shows the radial distribution of FPD's in 150 and 200 mm wafers. The lowest density is observed in the slow cooled wafers. As is illustrated in Table I, gate oxidation has little influence of the FPD density although in general there is also some increase after gate oxidation.

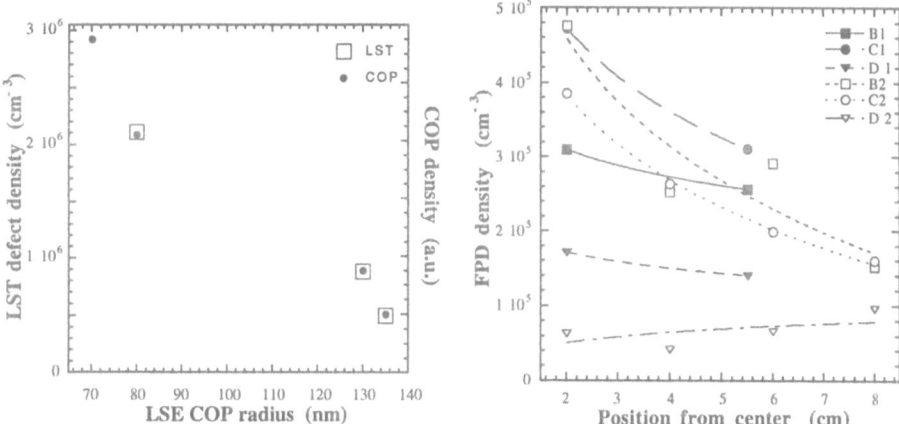

Fig. 4: N_{LST} and N_{COP} vs. COP radius for 200 mm wafers.

Fig. 5: FPD defect density before gate oxidation.

Gate oxide integrity and link with grown-in substrate defects

When growing a gate oxide by thermal oxidation grown-in defects present in the silicon substrate are gradually incorporated into the oxide layer. The number of gate oxide defects therefore increases initially with increasing oxide thickness. It is assumed that a substrate defect which is incorporated in the oxide causes a weak spot with poorer dielectric properties leading to a smaller effective oxide thickness [20]. An increase of the number of gate oxide defects, i.e. spots where the oxide breaks down for electrical fields smaller than the intrinsic breakdown field, is thus observed with increasing oxide thickness reaching a maximum defect density between 30 and 100 nm. This maximum is determined by the density of substrate defects and their average size. When the oxide thickness becomes larger than the typical defect size the midfield breakdown field will shift towards higher values closer to the intrinsic level and the number of gate oxide defects will thus go down.

Figure 6 illustrates the strong impact of the substrate on the 15 nm gate oxide breakdown characteristics. The lines in the figure are cubic spline fits through the measurement points which are averages taken over two wafers. The breakdown frequency has a maximum for electrical fields between 6 and 8 MVcm^{-1}. The maximum decreases with decreasing cooling rate while the corresponding breakdown field increases. For comparison also results for epitaxial wafers are shown revealing no substrate effect and only intrinsic breakdown close to 12 MVcm^{-1}. One can expect that the electrical field at the maxima of the midfield breakdown frequency would correlate with the average size of the substrate defects which are incorporated in the oxide [4]. This is illustrated in Fig. 7 for the 150 mm wafers used in the present study.

In Figure 8 the calculated gate oxide defect density (per unit volume of consumed silicon) is given as function of the LST defect density measured on the same 150 mm wafer. Average values are given for two wafers each. Results are also given for the case where 10^{12} Fe cm^{-2} was intentionally introduced before the gate oxidation [16]. The odd point for the highest defect density (A-type wafer) is probably due to the small size of the scatterers in the substrate so that the majority of them is not detected by LST.

COP's, LST and gate oxide defects: oxide precipitates?

Cooling of the silicon crystal from the melt temperature leads to a supersaturation of self-interstitials, vacancies and interstitial oxygen [21]. The rate of increase of supersaturation will increase with increasing cooling rate. It is well known that the critical size of a silicon oxide precipitate is strongly influenced by the supersaturation of oxygen and intrinsic point defects [22] and can be written as

$$r_c = \frac{2\sigma}{\dfrac{xkT}{\Omega_{SiOx}} \ln \dfrac{C_{OI}}{C_{OI}^*} (\dfrac{C_V}{C_V^*})^\beta (\dfrac{C_I}{C_I^*})^\gamma}$$

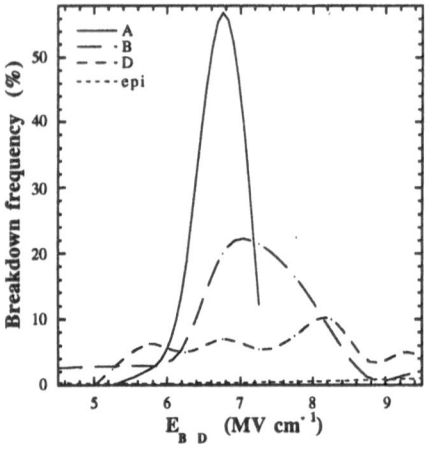

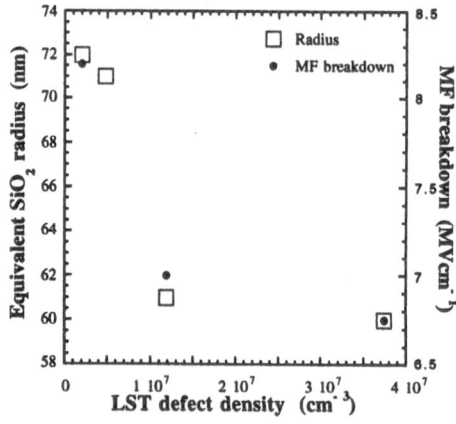

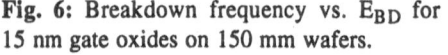

Fig. 6: Breakdown frequency vs. E_{BD} for 15 nm gate oxides on 150 mm wafers.

Fig. 7: LST defect size and E_{BD} vs. LST defect density on 150 mm wafers.

The silicon oxide precipitate is assumed to have a composition SiO_x. The interstitial oxygen, the vacancy and the self-interstitial concentration in the matrix are respectively C_{OI}, C_V and C_I while their thermal equilibrium values are C_{OI}^*, C_V^* and C_I^*. The molecular volume Ω_{SiOx} of all known silicon oxide phases is larger than the one of silicon (Ω_{Si}). Due to this the growth of the precipitate can cause an increase of its strain energy. It is assumed that the strain is relieved by the absorption of γ (< 0) self-interstitials and/or of β vacancies from the matrix per precipitated oxygen atom.

Assuming that the strain free critical precipitate radius is dominated only by the oxygen and vacancy supersaturation one obtains typical curves as shown in Fig. 9. It is clear that silicon oxide precipitates with sizes above 20 nm can only nucleate at temperatures above 1400K. Those precipitates will thus also be stable during treatments at temperatures below 1400 K, in good agreement with the reported stability of LST defects, FPD's and COP's. The similarities between the sizes of LST defects, COP's and r_c values suggest that (at least part of) the defects are silicon oxide precipitates and that they are causing the reduced gate oxide quality when being incorporated into the growing oxide.

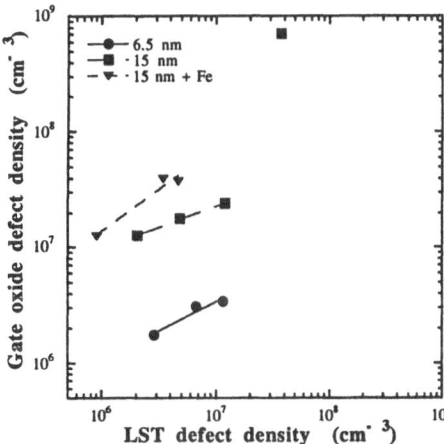 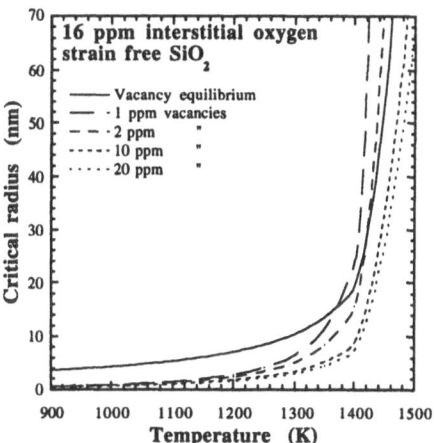

Fig. 8: Gate oxide defect density vs. LST defect density for 150 mm wafers with 6.5 and 15 nm gate oxides.

Fig. 9: Predicted temperature dependence of the critical size of spherical strain free silicon oxide precipitates with the vacancy concentration as parameter.

4. Conclusions

It is shown that pulling and cooling rate during Czochralski growth of silicon single crystals strongly influence the number and size of grown-in microdefects. This distribution of microdefects not only determines the interstitial oxygen precipitation in subsequent thermal treatments but also has a prominent effect on thermal oxide break-down characteristics. A close correlation between the density of lattice defects observed by light scattering tools and gate oxide defect density is observed.

Most of the observed defects are assumed to be silicon oxide precipitates with sizes between 40 and 160 nm and which are formed during crystal cooling. In some cases

500

(fast cooling) most of the grown-in precipitates are however not observed as their size is smaller than the detection limit of the scattering tools (LST 40 nm, COP 80 nm). After prolonged heat treatments these precipitate nuclei grow however to larger size silicon oxide precipitates resulting in LST defect densities in the 10^8 cm^{-3} range.

5. References

1. Li, S.P. and Maserjian, J. (1976) Effective defect density for MOS breakdown: dependence on oxide thickness, *IEEE Trans. Electron Devices* **ED-23**, 525-527.
2. Itsumi, M. and Kiyosumi, F. (1982) Origin and elimination of defects in SiO$_2$ thermally grown on Czochralski silicon substrate, *Appl. Phys. Lett.* **40**, 496-498.
3. Yamabe, K., Taniguchi, K. and Matsushita, Y. (1983) Thickness dependence of dielectric breakdown failure of thermal SiO$_2$ films, *Proceedings of IRPS*, IEEE New York, 184-190.
4. Itsumi, M., Nakajima, O. and Shiono, N. (1992) Oxide defects originating from Czochralski silicon substrates, J. Appl. Phys. **72**, 2185-2191.
5. Zemke, D., Gerlach, P., Zulehner, W. and Jacobs, K. (1994) Investigations on the correlation between growth rate and gate oxide integrity of Czochralski-grown silicon, *Journal of Crystal Growth* **139**, 37-46.
6. Satoh, Y., Murakami, Y., Furuya, H. and Shingyouji, T. (1994) Effect of bulk microdefects induced in heat-treated Czochralski silicon on dielectric breakdown of thermal SiO$_2$ films, *Appl. Phys. Lett.* **64**, 303-305.
7. Park, J.-G., Ushio, S., Takeno, H., Cho, K.-C., Kim, J.-K. and Rozgonyi, G.A. (1994) Comparison of oxide breakdown mechanisms due to D-defects and oxygen precipitates, *The Electrochemical Society Proceedings Volume* **94-33**, 53-64.
8. v. Ammon, W., Ehlert, A., Lambert, U., Gräf, D., Brohl, M. and Wagner, P. (1994) Gate oxide related bulk properties of oxygen doped floating zone and Czochralski silicon, *The Electrochemical Society Proceedings Volume* **94-10**, 136-147.
9. Fusegawa, I., Takano, K., Kimura, M. and Fujimaki, N. (1995) Review of the influence of micro crystal defects in silicon single crystals on gate oxide integrity, *Material Science Forum Vols.* **196-201**, 1683-1690.
10. Wagner, P. (1995) Automated surface inspection of bare polished silicon surfaces with light scattering techniques, *Electrochemical Society Proceedings Volume* **95-30**, 236-251.
11. Moriya, K. (1989) Observation of micro-defects in as-grown and heat treated Si crystals by infrared laser scattering tomography, *Journal of Crystal Growth* **94**, 182-196..
12. Voronkov, V.V., Zabolotskiy, S.E., Kalinushkin, V.P., Murin, D.I., Ploppa, M.G. and Yuryev, V.A. (1990) Application of elastic IR light scattering for investigation of large-scale electrically active defects in semiconductors, *Journal of Crystal Growth* **103**, 126-130.
13. Kissinger, G., Vanhellemont, J., Gräf,, D., Claeys, C. and H. Richter (1996) IR-LST a powerful tool to observe crystal defects in as-grown silicon, after device processing, and in heteroepitaxial layers, Inst.. Phys. Conf. Ser., in press.
14. Yamagashi, H., Fusegawa, I., Fujimaki, N. and Katayama, M. (1992) Recognition of D defects in silicon single crystals by preferential etching and effect on gate oxide integrity, *Semicond. Sci. Technol.* **7**, A135-A140.
15. Vanhellemont, J., Kissinger, G., Gräf, D., Kenis, K., Depas, M., Mertens, P., Lambert, U., Heyns, M., Claeys, C., Richter, H. and Wagner, P. (1995) Lattice defects in high quality as-grown Cz silicon, studied with light scattering and preferential etching techniques, *Materials Science Forum Vols.* **196-201**, 1755-1760.
16. Vanhellemont, J., Kissinger, G., Gräf, D., Kenis, K., Depas, M., Mertens, P., Lambert, U., Heyns, M., Claeys, C., Richter, H. and Wagner, P. (1996) Light scattering tomography study of lattice defects in high quality as-grown Cz silicon wafers and their evolution during gate oxidation, *Inst. Phys. Conf. Ser.*, in press.
17. Dornberger, E. and von Ammon, W. (1996) Dependence of ring like distributed stacking faults on the axial temperature gradient of growing Czochralski silicon crystals, *J. Electrochem. Soc.*, in press.
18. Ryuta, J., Morita, E., Tanaka, T. and Shimanuki, Y. (1992) Effect of crystal pulling rate on formation of crystal-originated particles on Si wafers, *Jap. J. Appl. Phys.* **31**, L293-L295.
19. Brohl, M., Gräf, D., Wagner, P., Lambert, U., Gerber, H.A. and Piontek,H. (1994) Monitoring of crystal bulk defects on polished Si (100)-surfaces, *The Electrochem. Soc. Extended Abstracts* 94-2, 619-620.
20. Lee, J.C., Chen, I.C. and Hu, C. (1988) Modelling and characterisation of gate oxide reliability, *IEEE Trans. Electron Devices* **35**, 2268-2278.
21. Iwasaki, T., Harada, H. and Haga, H. (1995) Influence of point defect concentration in growing Cz-Si on the formation temperature of the defects affecting gate oxide integrity, *Materials Science Forum Vols.* **196-201**, 1731-1736.
22. Vanhellemont, J. (1996) On the impact of interface energy and vacancy concentration on morphology changes and nucleation of silicon oxide precipitates in silicon, *Appl. Phys. Lett.*, in press.

LOW TEMPERATURE ANNEALING STUDIES OF THE DIVACANCY IN P-TYPE SILICON

M.-A. TRAUWAERT[1,2], J. VANHELLEMONT[1], A.-M. VAN BAVEL[2], P. CLAUWS[3], A. STESMANS[2], H.E. MAES[1] AND G. LANGOUCHE[2]
1. IMEC, Kapeldreef 75, B-3001 Leuven, Belgium
2. KUL, Celestijnenlaan 200D, B-3001 Leuven, Belgium
3. RUG, Krijgslaan 281-S1, B-9000 Gent, Belgium

Abstract

In the present study an $E_V+0.19$ eV deep level is observed in boron doped Si after electron irradiation at room temperature. After prolonged anneals at temperatures between 200 and 300°C a gradual shift of the deep level with annealing time occurs however towards $E_V+0.24$ eV. The observed transition is much faster in Cz than in FZ silicon suggesting the involvement of oxygen. It is shown that this apparent gradual shift of the deep level in the bandgap can be explained by a gradual transformation of the divacancy with trap parameters ($E_V+0.17$ eV, capture cross-section $\sigma \sim 10^{-17}$ cm^2) towards another defect with trap parameters ($E_V+0.24$ eV, $\sigma \sim 10^{-15}$ cm^2). The final defect is assumed to be a multi-vacancy/oxygen complex. First results are presented of an electron paramagnetic resonance (EPR) and infrared (IR) spectroscopy study before and after the transition to confirm this hypothesis.

1. Introduction

The divacancy is easily created in silicon by irradiation with high energy electrons and has therefore been studied more than thirty years by a wide variety of techniques. Deep level transient spectroscopy analyses of irradiated p-type silicon in literature make the link between the divacancy and a majority carrier trap with reported energy levels between $E_V+0.19$ and $E_V+0.24$ eV. Londos measured an activation energy of 0.19 eV for the divacancy, which was generated by 1.5 MeV electron irradiation at 80 K, and assumed that this low value was due to a Si self-interstitial, loosely bond to the divacancy [1]. During warming-up to room temperature, an increase of the activation energy was observed till 0.21 eV, which was interpreted as a dissociation of the complex, by which the Si self-interstitial is liberated.

The divacancy is stable at room temperature and starts to anneal out at temperatures between 270 and 300°C depending on the oxygen content of the substrate [2]. Most of the anneal-in defects, appearing when the divacancy anneals out, are preliminary associated with multi-vacancy/multi-oxygen or vacancy-doping atom complexes. However, those assignments are somewhat speculative and to our knowledge no one has performed a detailed study to unambiguously identify these anneal-in defects.

R. Jones (ed.), Early Stages of Oxygen Precipitation in Silicon, 501–508.
© *1996 Kluwer Academic Publishers.*

502

The annealing of vacancy- and oxygen-related irradiation induced defects has been extensively studied by Lee and Corbett [3] and Svensson et al. [4] using EPR and the IR (infrared) spectroscopy technique, respectively. Lee and Corbett claimed that the divacancy, the A-centre (VO) and the divacancy-oxygen complex (only observed in Cz material) can all be present below 350°C, the temperature at which the first two start to dissociate. The mobile vacancy or oxygen atom will be trapped by the divacancy-oxygen complex, resulting in V_xO_y structures. The treatments performed by Svensson et al. at temperatures ranging from typically 300°C up to 800°C resulted in the formation of V_xO_y complexes. In both studies it was suggested that the VO centre acted as nucleation centre, initiating the cluster formation. Models have been proposed where VO diffuses to the immobile O_i or where vacancies are released from dissociating VO or V_2 and subsequently trapped by VO. However, only the structural configuration of the different defects was established, and no electrical parameters was given.

In the present paper a detailed DLTS investigation of the thermal behaviour of the divacancy in silicon is presented. We observe a conversion in boron doped silicon of a defect level at $E_v+0.19$ eV, to a level at $E_v+0.24$ eV after anneal at 200°C, both levels associated in literature with the donor level of the divacancy. The conversion kinetics depend strongly on the oxygen content. Preliminary results from correlated spectroscopic techniques (EPR, IR) will be presented.

2. Experimental observations and discussion

2.1 DLTS MEASUREMENTS

Figure 1 shows a typical DLTS spectrum obtained on a 2 MeV electron irradiated (fluence: 1×10^{16} cm^{-2}) p-type Cz diode before and after a low temperature anneal at 200°C for 10 days [5]. In the as-irradiated sample two DLTS bands are observed. The H(0.19) with an activation energy of 0.192 eV and a capture cross-section $\sigma_h=3.8\times10^{-16}$ cm^2, is believed to be the donor level of the divacancy. The H(0.36) level is mostly associated with the interstitial carbon/interstitial oxygen (C_iO_i) complex since this dominant defect in Cz substrates is hardly observable in FZ Si [6].

An anneal of 10 days at 200°C causes the appearance of a new level, H(0.29), correlated by Kimerling et al. with the interstitial boron/substitutional carbon (B_iC_s) complex [7]. This complex is formed at 150°C, following the dissociation of B_iO_i, which acts as a minority carrier trap in irradiated p-type silicon. The most surprising observation is however the shift of the H(0.19) level to a new band H(0.24) positioned at higher temperatures. An activation energy of (0.236 ± 0.005) eV and a capture cross-section $\sigma_h=(1.1\pm0.5)\times10^{-14}$ cm^2 has been determined for this latter level.

This conversion has been studied in more detail by performing extensive isothermal anneal experiments at temperatures between 200 and 290°C on Cz ($[O_i]\sim9\times10^{17}$ cm^{-3}) and FZ ($[O_i]\sim10^{16}$ cm^{-3}) p-type Si samples. After each anneal step an Arrhenius analysis was performed in order to accurately follow the change in activation energy of the intermediate bands. From Figure 2 it is clear that the conversion does not happen at once. It appears that the H(0.19) level shifts gradually with annealing time. It is unlikely that this evolution from H(0.19) to H(0.24) is due to a continuous shift of the

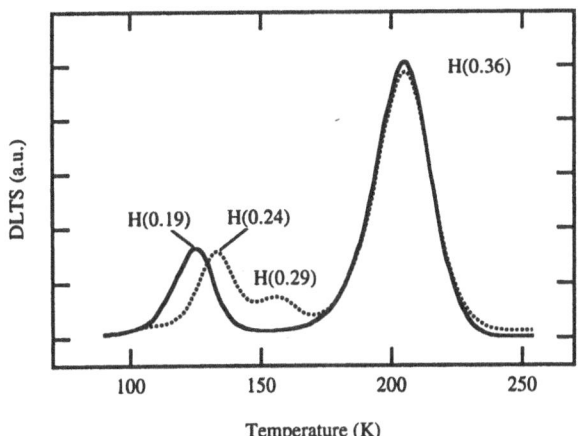

Figure 1.   DLTS spectrum (emission rate = 223 s-1) of a 2 MeV electron
              irradiated p-type Cz diode before (full line) and after (dotted line)
              anneal at 200°C for 10 days.

defect level in the bandgap. In [8] an alternative explanation whereby two different defect
complexes are involved is described. During thermal treatments an initial structure
anneals out, accompanied by the gradual appearance of a final one. The observed
intermediate DLTS spectra result then from the superposition of a fraction of the initial
and final DLTS bands. Due to the insufficient resolution of the DLTS system combined
with the specific defect signatures of the initial and final state, it is, however, not
possible to observe two different well-resolved DLTS bands using standard measurement
parameters.

By comparing the experimental observations with calculations the remaining fraction of
the initial state can be determined as a function of the anneal time for different anneal
times. This results in first order kinetics characterised by a thermally activated annealing
rate of $4.5 \times 10^4 \exp(-0.89 eV/kT) s^{-1}$ [9]. The deduced activation energy is similar to the
value determined by Wang for the partial annealing of the divacancy in n-type Si in the
temperature range from 240 and 300°C [10].

In Figure 2 also the influence of the oxygen content in the substrate on the conversion
is illustrated by comparing oxygen rich Cz and oxygen poor FZ diodes. The
experimental observations indicate that oxygen plays a role in the transformation: in Cz
substrates much shorter anneal times are required to reach the $E_V + 0.24 eV$ level,
compared with in the FZ ones. Therefor we suggest a transformation of the divacancy
into a divacancy-oxygen complex. As oxygen is present in much higher concentration in
the Czochralski wafers the conversion is thus indeed faster. This is in contrast with
Londos' assumption of an interaction with a Si self-interstitial to explain the low
energy value he observed [1]. The loosely bound Si self-interstitial induces a shift of the
activation energy towards lower values, which will recover following the thermally

504

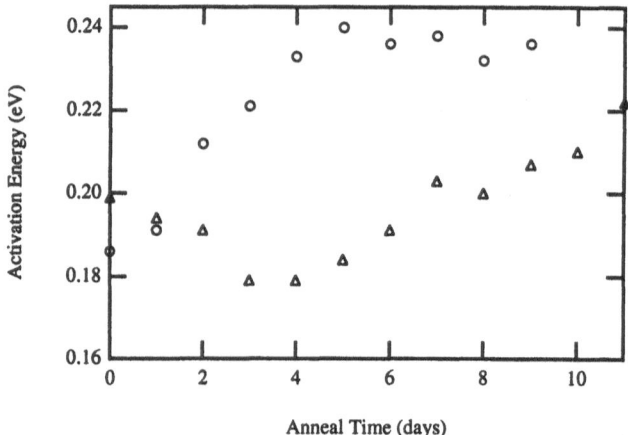

Figure 2.   Changes in the apparent activation energy of the H(0.19) level in p-type
Cz (circles) and in p-type FZ (triangles) Si after thermal treatments at
200°C.

activated dissociation of the V_2-Si self-interstitial complex. In his opinion the final
level thus corresponds with the divacancy.

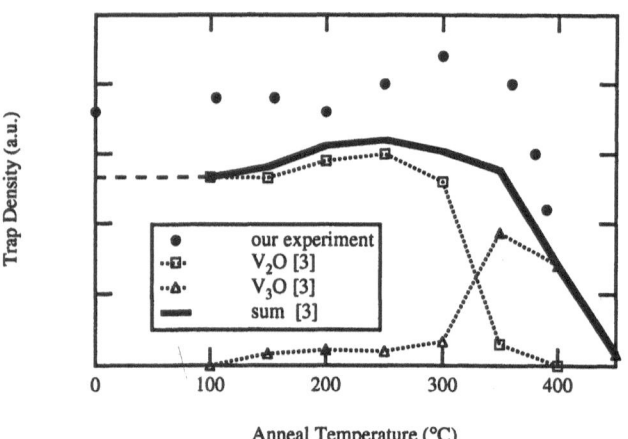

Figure 3:   Isochronal annealing (15 min) of the final H(0.24) state compared with the isochronal
annealing (30 min) of V_xO_y complexes reported by Lee and Corbett [3].

Furthermore, the high temperature isochronal annealing behaviour of the final H(0.24) level agrees considerably well with anneal results reported for the V_2O defect [3]. Lee and Corbett investigated, using EPR, the anneal stages, which occurred in Cz Si after the divacancy disappeared (>350°C). In Figure 3, the experimentally determined trap densities of the H(0.24) level are shown together with the behaviour of the V_2O and V_3O defect as determined by Lee and Corbett during isochronal anneal experiments of 15 and 30 min, respectively [3]. Qualitatively, our experimental observations correspond rather well with the sum (full line in Figure 3) of the concentrations of V_2O, V_3O and small contributions of higher degrees of V_xO_y complexes. This may indicate that at temperatures above 300°C, the measured DLTS band originates from a combination of these different vacancy-oxygen complexes. Determination of the activation energy from conventional Arrhenius analyses of the DLTS band during the isochronal anneal at higher temperatures resulted indeed in 0.26 eV and 0.29 eV after anneal at 300 and 360°C, respectively [11].

2.2 CORRELATION WITH IR AND EPR

To consolidate the proposed model for the low temperature anneal of the divacancy in p-type silicon observed with DLTS, correlation experiments have been performed. Since the IR and EPR signal for the divacancy have been well documented in literature from both techniques genuine information should be obtained.

The main and first problem encountered here, was the search for appropriate samples, since the DLTS samples as such could not be used. Different defect concentrations and specific sample preparations are required. Also the position of the Fermi level plays an important role, e.g. the divacancy being only EPR observable in singly positive or singly negative charge state. These differences have to be taken into account when comparing and interpreting the obtained results.

2.2.1 *FT-IR measurements*
The substrates and irradiation conditions have been selected in order to have comparable initial properties and a minimum of divacancy concentration to be detectable with FT-IR. Bare p-type Cz Si pieces were used with an oxygen content of 8.1×10^{17} cm^{-3}. The background doping was obtained from resistivity measurements before irradiation and varied between 5.5 and 7×10^{14} B cm^{-3}. The substrates were irradiated with 2 MeV electrons to a fluence of 2×10^{18} cm^{-2}. One set of samples has been annealed at 250°C for two hours. Under these thermal conditions the H(0.19) DLTS level was observed to be fully converted into the H(0.24) level in p-type Cz substrates.

In Figure 4 FT-IR spectra recorded at 5K are shown for different specimens: one has been irradiated at a fluence equal to 2×10^{18} cm^{-2} (upper curves); and the second was irradiated at a fluence of 2×10^{18} cm^{-2} and subsequently annealed at 250°C for two hours. The effect of the irradiation is clearly observed: vacancies are created, which are subsequently trapped by interstitial oxygen forming the A-centre (localised vibrational mode (LVM) at 835.5 cm^{-1}). The spectra were recorded after cooling with continuous illumination of the sample, in order to observe the 3.3 μm absorption band, arising from electronic transitions at the negatively charged divacancy [12]. The 3.3 μm peak, clearly present in the as-irradiated sample, is no longer observed after annealing. Similar

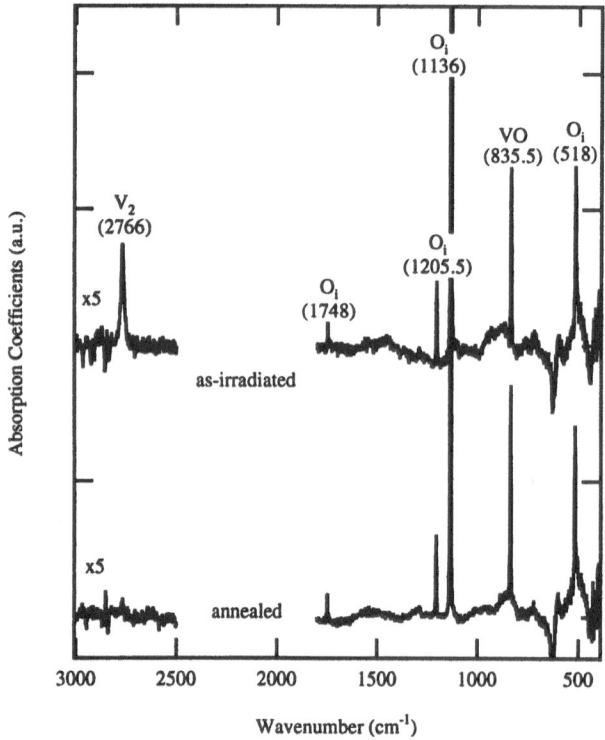

Figure 4:   5K recording of the FT-IR spectrum of irradiated p-type Cz silicon. upper curves: 2×10^{18} cm^{-2}, 2 MeV electron irradiated; lower curves: 2×10^{18} cm^{-2}, 2 MeV electron irradiated and subsequent anneal at 250°C for 2 h. The left hand side signal has been enlarged by a factor 5 (resolution = 4 cm^{-1}), compared to the right hand side signal (resolution = 2 cm^{-1}). The amplitude of the 1136 cm^{-1} O_i signal does not fit within the scales.

results, although less distinct due to the large band width, were obtained from room temperature recordings in the range of the 1.8 μm of the divacancy [8].

The DLTS experiments have been tentatively interpreted as a conversion from the divacancy to a divacancy-oxygen complex. Recent performed calculations predict a LVM for V_2O at almost the same frequency as for the VO defect, but still resolvable with FT-IR [9]. Annealing at conditions which induces the observed divacancy shift in the DLTS spectra, results in the disappearance of the divacancy related absorption bands. No new signals appeared, suggesting that, if new complexes have been formed, their concentration is below the detection limit. An increase of the VO has been observed between the as-irradiated and the annealed sample. This might result from the interaction between oxygen and vacancies released from the dissociation of divacancies. However, two different specimens were involved, probably misleading the interpretations.

2.2.2 *EPR measurements*

A major general conclusion of the preliminary EPR analysis is that, for the applied electron irradiation fluences, the divacancy, if observed at all, is not the major part of the observed spectra, neither before nor after annealing. Other centres dominate the spectra, different before and after annealing. Hence, it can be stated that the annealing conditions which induced the $E_V+0.19$ eV level shift towards $E_V+0.24$ eV, as observed by DLTS, resulted in an important change of the EPR spectra of irradiated substrates. And it is almost sure that part of the microscopic answer as to what constitutes the level shift, is buried in these spectral changes. However, as this could not be correlated with a vacancy pattern, the proposed identification is not yet confirmed. At this point it is however important to remark that no genuine information is available for the position of the Fermi level. Since higher irradiation fluences were required, it is not clear if the Fermi level remains close enough to the shallow acceptor levels, e.g. if the divacancy is still paramagnetic. The FT-IR measurements indicate that for a $2x10^{18}$ cm^{-2} radiation fluence (x10!) in the 5.5 to $7x10^{14}$ cm^{-3} doped samples the divacancy is neutral and as such not paramagnetic.

3. Conclusions

In this paper a thermally activated shift is reported of the H(0.19) DLTS band arising in boron doped irradiated Si substrates. This shift, observed after anneal at temperatures between 200 and 290°C in different substrates, is related to a slight increase of defect activation energy and an increase of the capture cross-section of the defect of two orders of magnitude. However, the origin of both defect structures is not as well established. Associations reported in literature and the observed oxygen influence strongly suggest a conversion from the divacancy to the divacancy-oxygen complexes.

FT-IR measurements performed on comparable structures confirm the presence of the divacancy in as-irradiated substrates. No divacancy related absorption bands, nor new arising complexes were observed after the thermal treatments. The thermal treatment thus induces the anneal out of the divacancy, but no information is available about the final structure.

EPR confirms that the thermal treatment highly influenced the defect structures in irradiated substrates. The g-map of the divacancy neither of V_xO_y related complexes was observed. However, it is not known if the Fermi level is pinned low enough in the bandgap for the divacancy to be paramagnetic. Therefore, further analysis should strongly depends on the availability of suitable starting samples, which establish clear divacancy signals.

Since the observed shift results in a new defect structure with a much higher capture cross-section, important effects on carrier lifetime can be expected [10]. Therefore the divacancy conversion could have considerable implications on the long term stability of electronic devices. Investigating the carrier lifetime in diodes after low temperature anneal should be useful for linking the fundamental defect reactions presented in this paper with macroscopic device aspects.

508

Acknowledgements

Part of this work was performed under IIKW contract 4.0007.93. W. Mondelaers is acknowledged for the 2 MeV electron irradiations which were performed in the linear accelerator (LINAC) of the University of Gent.

References

[1] Londos, C. A. (1992) The production and the evolution of A-centers and divacancies in silicon, *Phys. Stat. Sol. (a)* **132**, 43-50.

[2] Cheng, L. J. and Corelli, J. C. (1966) 1.8-, 3.3-, and 3.9- μm bands in irradiated silicon: correlations with the divacancy, *Phys. Rev.* **152**, 761-774.

[3] Lee, Y.-H. and Corbett, J. W. (1976) EPR studies of defects in electron-irradiated silicon: a triplet state of vacancy-oxygen complexes, *Phys. Rev. B* **13**, 2653-2666.

[4] Svensson, B. G. and Lindstrom, J. L. (1986) Kinetic study of the 830- and 889-cm^{-1} infrared bands during annealing of irradiated silicon, *Phys. Rev. B* **34**, 8709-8717.

[5] Trauwaert, M.-A., Vanhellemont, J., Maes, H. E., Van Bavel, A.-M., Langouche, G. and Clauws, P. (1995) Low temperature anneal of the divacancy in p-type silicon: a transformation from V(2) to V(x)O(y) complexes?, *Appl. Phys. Lett.* **66**, 3057-3058.

[6] Drevinsky, P. J., Caefer, C. E., Kimerling, L. C. and Benton, J. L. (1990) *Defect control in Semiconductors (341-345)*, Elsevier Science Publishers, Amsterdam.

[7] Kimerling, L. C., Asom, M. T., Benton, J. L., Drevinsky, P. J., and Caefer, C. E. (1989) Interstitial defect reactions in silicon, *Mat. Sci. Forum* **38-41**, 141-150.

[8] Trauwaert, M.-A. (1995) Radiation and Impurity related deep levels in Si: A Deep Level Transient Spectroscopy study correlated with other spectroscopic techniques, PhD, Leuven.

[9] Trauwaert, M.-A., Vanhellemont, J., Maes, H. E., Van Bavel, A.-M., Langouche, G. and Clauws, P. (1995) Influence of oxygen and carbon on the generation and annihilation of radiation defects in silicon, *Mat. Sci. & Eng. B.* **36**, 196-199.

[10] Wang, F. P., Sun, H. H. and Lu, F. (1990) Novel electrical and annealing properties of defects in electron irradiated silicon p^{+}-n junctions, *J. Appl. Phys.* **68**, 1535-1540.

[11] Trauwaert, M.-A., Vanhellemont, J., Maes, H. E., Van Bavel, A.-M., Langouche, G., Stesmans, A. and Clauws, P. (1995) On the behaviour of the divacancy in silicon during anneals between 200 and 350°C., *Mat. Sci. Forum* **196-201**, 1147-1152.

[12] Svensson, J. H., Svensson, B. G. and Monemar, B. (1988) Infrared absorption studies of the divacancy in silicon: new properties of the single negative charge state, *Phys. Rev. B* **38**, 4192-4197.

ANOMALOUS DISTRIBUTION OF OXYGEN PRECIPITATES IN A SILICON WAFER AFTER ANNEALING

H. ONO, T. IKARASHI, S. KIMURA, A. TANIKAWA,
and K. TERASHIMA
Microelectronics Research Laboratories, NEC Corporation,
34 Miyukigaoka, Tsukuba, Ibaraki 305, JAPAN

1. Introduction

Oxygen precipitation after thermal annealing is generally controlled by interstitial oxygen atoms and other unknown defects. This often results in wafers with an inhomogeneous distribution of oxygen precipitates, even though they have a homogeneous distribution of interstitial oxygen atoms.

It is known that anomalous ring-shaped distributions of oxidation-induced stacking faults (OSFs) appear in wafers which are cut from ingots grown at a specific growth rate [1,2]. These wafers also exhibit anomalous distributions of oxygen precipitates after annealing [3-5]. This is a typical example of wafers in which precipitation varies among different regions of the wafer, perhaps due to inhomogeneous distributions of unknown grown-in defects.

In this paper, we study anomalous ring-shaped distributions of oxygen precipitates in wafers using infrared absorption spectroscopy and X-ray topography. We also discuss how grown-in defects affect oxygen precipitation after annealing.

2. Experimental

Samples were cut from a 5-inch p-type <100> Si crystal grown at a constant growth rate of 0.8 mm/min. The initial concentration of interstitial oxygen was 1.4×10^{18} cm^{-3}. Mirror-polished wafers were annealed at 1000-1150°C for 16h in an N_2 or O_2 atmosphere (single-step annealing), or preannealed at 500-900°C for 16h in N_2 preceding postannealing at 1000°C for 16h in N_2 (two-step annealing). The thickness of the wafers was 0.65 mm or 5 mm.

The distribution of interstitial oxygen, oxygen precipitates and thermal donors was studied by using Fourier-transform infrared absorption spectroscopy (FTIR). Room-temperature FTIR profiles were obtained using a conventional micro-FTIR system with a HgCdTe detector. The aperture was 100 μm and the instrument resolution was 4 cm^{-1}. The profiles were also obtained at liquid-He temperature using a cryostat, in order to

R. Jones (ed.), Early Stages of Oxygen Precipitation in Silicon, 509–516.

510

detect thermal donors. In this case, the detector was Si-bolometer, the aperture 1mm and the resolution 0.5 cm⁻¹.

Lang X-ray topographs (XRT) were taken to show the distribution of strains due to precipitates and/or OSFs.

3. Results and Discussion

3.1. TWO-STEP ANNEALING

Most dramatic variation of the anomalous distribution of oxygen precipitates was found in two-step annealed wafers. Two-step annealing enhances oxygen precipitation, because the preannealing forms precipitation nuclei, and the postannealing grows the

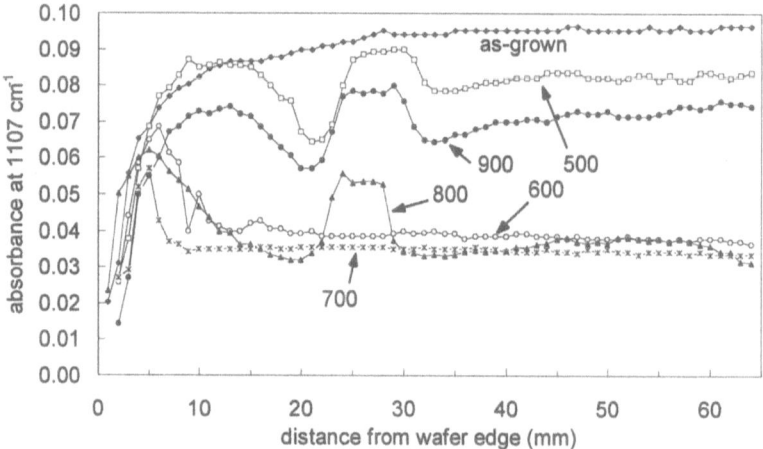

Figure 1. Preannealing-temperature dependence of FTIR profiles of interstitial oxygen after two-step annealing. The preannealing time was for 16h, and the postannealing was at 1000°C for 16h in N₂.

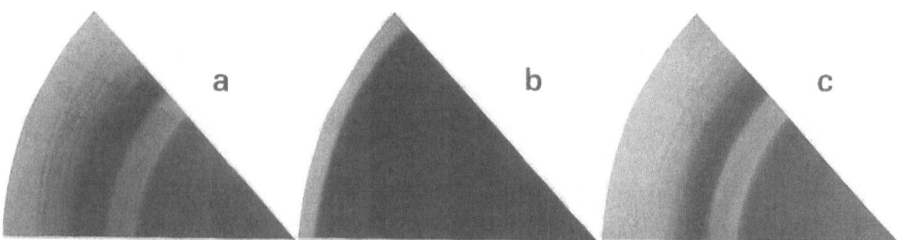

Figure 2. Preannealing-temperature dependence of XRT images after two-step annealing. (a)500°C, (b) 700°C, and (c) 900°C. The annealing conditions are the same as Fig. 1.

nuclei into large precipitates. Figures 1 and 2 show FTIR profiles and XRT images, respectively, of the wafers after two-step annealing, as a function of the preannealing temperature. In the as-grown wafer, interstitial oxygen atoms are homogeneously distributed except the peripheral region of the wafer. The peripheral region has less interstitial oxygen because of out diffusion during crystal growth. Anomalous precipitation behavior was observed depending on the preannealing temperature. Preannealing at a temperature of 600-700°C strongly nucleates all portions of the wafer. The nuclei density seems to be so high that it hides the inhomogeneity in the original crystal. Thus, the resulting precipitation distribution after post annealing is homogeneous; this is in contrast with the case at lower (500°C) or higher (800 and 900°C) temperatures.

Figures 3 and 4 show FTIR profiles and XRT images, respectively, of the wafers after two-step annealing as a function of the preannealing time at 700°C. Although the

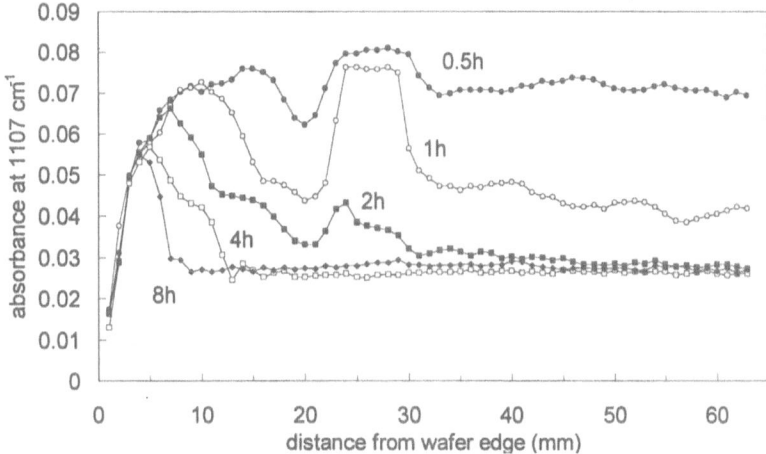

Figure 3. Preannealing-time dependence of FTIR profiles of interstitial oxygen after two-step annealing. The preannealing temperature was at 700°C and the postannealing was at 1000°C for 16h in N_2.

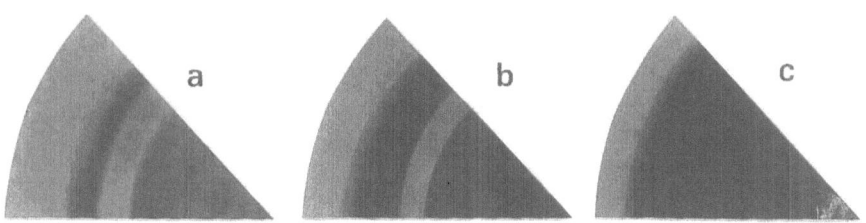

Figure 4. Preannealing-time dependence of XRT images after two-step annealing. (a) 0.5h, (b) 1h, and (c) 2h. Annealing conditions are the same as Fig. 3.

precipitation was homogeneous for the wafer preannealed for 16h at 700°C (Figs. 1 and 2), the precipitation was inhomogeneous for a shorter preannealing time at 700°C. This fact supports the above idea about precipitate nucleation. That is, too many nuclei formed at 700°C for a longer time more than 4h hide the preexisting inhomogenuity due to grown-in defects. On the other hand, preannealing at 800°C for 16h shows inhomogeneous precipitation (Fig. 1). In this case, we observed that even a long annealing time of 64h resulted in maintenance of inhomogeneous features.

3.2. STARTING MATERIAL

The initial concentration of interstitial oxygen and thermal donors often strongly affect the precipitation. We precisely investigated the distribution of interstitial oxygen (1107 cm^{-1}), oxygen precipitates (1125 cm^{-1}) and thermal donors using FTIR across the anomalous area in a 5mm-thick as-grown wafer. However, the interstitial oxygen concentration did not abruptly change across the area. The concentration variation should be less than 10^{16} cm^{-3}. The as-grown crystal already contains oxygen precipitates exhibiting a peak at 1125 cm^{-1}. These precipitates which are known as disk-shaped silica [6] were also homogeneously distributed within 4%.

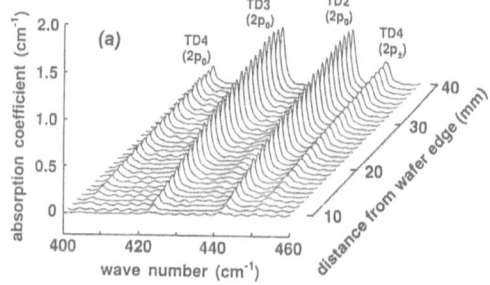

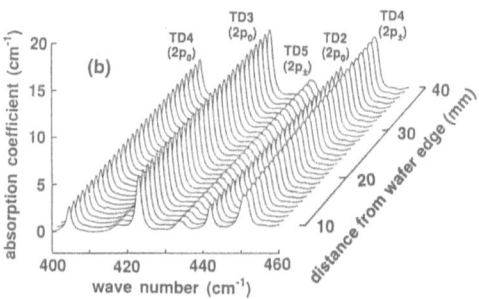

Figure 5. Low-temperature FTIR profiles of thermal donors in (a) an as-grown and (b) an annealed wafer (450°C, 8h)

The distribution of thermal donors observed by low-temperature FTIR is shown in Fig. 5. They are also homogeneous across the anomalous area in the wafer annealed at 450 °C as well as in the as-grown wafer. Our previous experiments [4] have shown that thermal donors do not affect the anomalous behavior of the oxygen precipitation.

3.3. SINGLE-STEP ANNEALING

A ring-shaped region can often be observed by XRT after single-step annealing [2]. We examined the oxygen precipitation after single-step annealing at a high temperature both in an O_2 atmosphere and in an N_2 atmosphere.

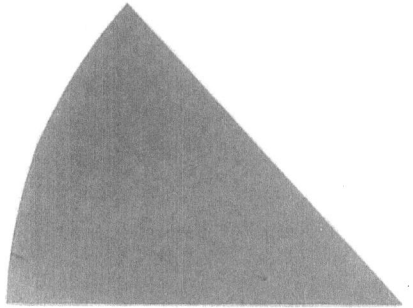

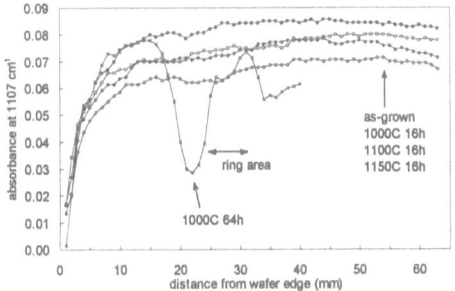
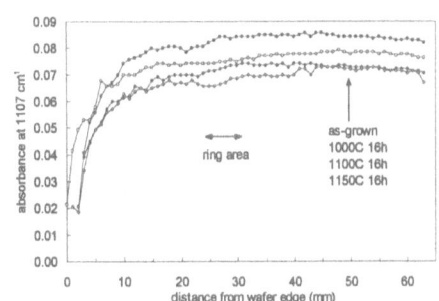

Figure 6. XRT and FTIR profiles of wafers annealed at a single temperature in an N_2 atmosphere.

Figure 7. XRT and FTIR profiles of wafers annealed at a single temperature in an O_2 atmosphere.

Figures 6 and 7 compare XRT images and FTIR profiles between the two different atmospheres. The XRT image in the O_2 atmosphere shows a very clear contrast with a ring-shape. However, FTIR shows that the oxygen profiles are independent of the atmospheres, that is, precipitation is not especially strong at the ring area for both atmospheres.

In the O_2 atmosphere, well known ring-shaped distribution of OSFs appears. Figure 8 shows the distribution of OSFs, which were revealed by Secco etching for 4 min. The OSF density did not significantly

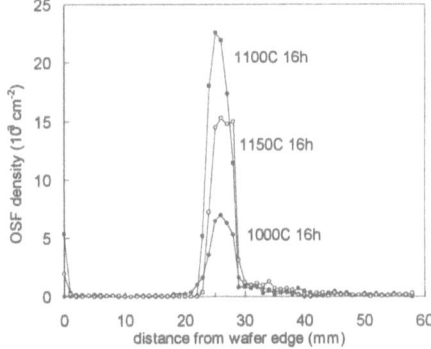

Figure 8. OSF distribution revealed by Secco etching

depend on the etching time, implying that the OSFs are located only in the region close to the wafer surface. A "ring area" is the region where OSFs are distributed with ring-shaped and anomalously rich in density [1]. The wafers in our study have a ring area 25-30 mm from the wafer edge. These results imply that the strong contrast observed at

514

the ring area by XRT was due to strong OSF strain, and that the precipitation was another problem which should be separately discussed with OSFs. These results also indicate that interstitial silicon atoms, which should be induced by surface oxidation, do not affect oxygen precipitation.

Furthermore, besides the ring area, we observed strong precipitation at the outer region of the ring area after long annealing of 64h in N_2 (Fig. 6). This area coincides with the area where small strain has been detected in an as-grown wafer by grazing-angle XRT using Synchrotron Radiation [7]. This implies that precipitation nuclei already exist inhomogeneously in an as-grown crystal, since the single-step annealing does not produce any renewal nuclei in the wafer.

3.4. PRECIPITATION MODEL

When discussing the process responsible for the anomalous distribution of oxygen precipitates after two-step annealing, we accept the following rule [8-11]. It has been reported that at the nucleation stage, a high density of small nuclei are formed at low temperatures, while a low density of large nuclei are formed at high temperatures. Figure 9 shows the nucleus density as a function of nucleus radii after preannealing at various temperatures. When the nuclei exist at the postannealing stage, smaller nuclei shrink and larger ones grow at a critical temperature. The critical radius at the postannealing temperature (1000°C) is shown as r_c in Fig. 9. Postannealing enhances only the precipitates which have a radius greater than r_c at postannealing temperature.

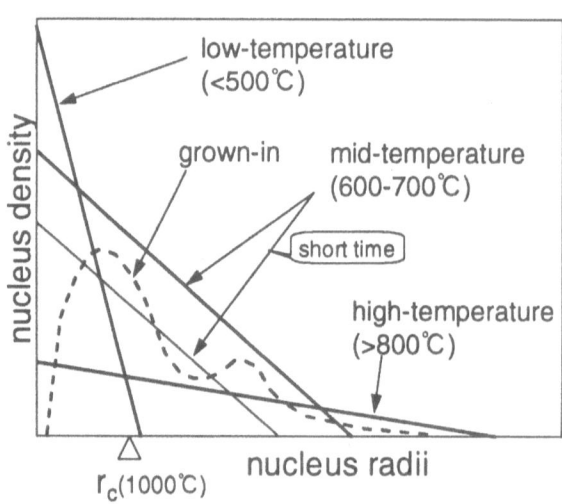

Figure 9. A model for precipitation nucleus after preannealing

Under these conditions, we propose the following model to explain the anomalous precipitation behavior of two-step annealing.

In our model, we assume that the grown-in defects, which should also be small precipitates of oxygen, distribute inhomogeneously in the wafer (Fig.9). They nucleated heterogeneously at a very high temperature during the cooling process after crystal growth so that their features (density vs. radii) are not uniform. Since Fig.9 shows a situation after preannealing, the grown-in defects with smaller radii have already shrunk. Thus, the curve for the grown-in defects in Fig.9 might have a maximum around r_c.

At a low temperature (<500°C), a high density of small nuclei are newly formed during the preannealing. Almost all of the nuclei shrink during the postannealing stage, because their radius is less than r_c at 1000°C. Oxygen atoms are also consumed in growing the preexisting grown-in precipitates into precipitates with a radius larger than r_c at 1000°C. This results in inhomogeneous precipitation after two-step annealing.

In the higher temperature case (>800°C), very few nuclei are formed. Instead of forming new nuclei, interstitial oxygen atoms are consumed in growing the preexisting grown-in precipitates into precipitates that have a radius larger than r_c at 1000°C. The precipitate nuclei should already be inhomogeneously distributed in the as-grown crystal. This again results in inhomogeneous precipitation after two-step annealing.

At mid-temperatures (600-700°C), there are so many newly formed nuclei with a radius larger than r_c that they hide the as-grown inhomogeneity. However, Figure 9 indicates that a short preannealing time at 600-700°C can not hide the inhomogeneity. On the other hand, even for a long annealing time the inhomogeneity can not be hidden by higher or lower temperature preannealing.

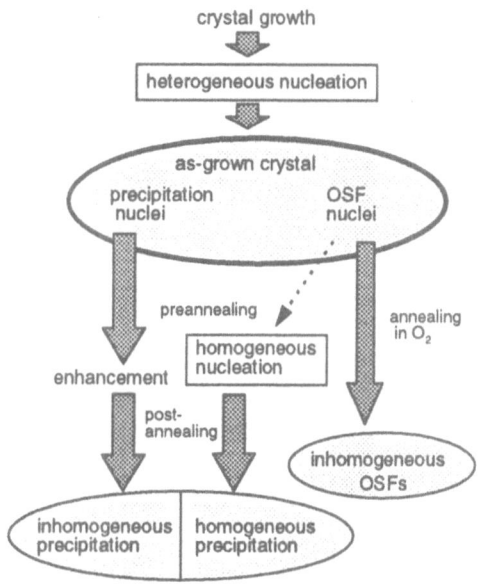

Figure 10. Schematic of the inhomogeneous distribution of oxygen precipitates after anneal processes

4. Conclusion

We studied the anomalous ring-shaped distributions of oxygen precipitates appearing after thermal annealing. We found that two-step annealing strongly

516

enhances the inhomogeneity of the oxygen precipitation. This feature can be explained by a simple model assuming the inhomogeneous distribution of the grown-in defects and homogeneous nucleation during the preannealing. The grown-in defects were heterogeneously nucleated by an interaction of oxygen with native point defects or impurity atoms at an elevated temperature during cooling process after crystal growth. In the model, homogeneous precipitation takes place if the nucleation during preannealing overwhelms the inhomogeneous precipitation due to the grown-in defects (Fig.10).

However, our FTIR measurements showed homogeneous distributions of oxygen precipitates, interstitial oxygen and thermal donors. The results also showed that the thermal donors and interstitial silicon atoms in the wafer do not affect oxygen precipitation. Therefore, we conclude that native point defects, impurities or their clusters in as-grown wafers are not responsible for inhomogeneous precipitation after thermal annealing. The anomalous distribution should be strongly affected by already-existing precipitates except OSF nuclei. The density of OSF nuclei, which were also heterogeneously nucleated at high temperatures during crystal growth, is too low to directly affect the inhomogeneous precipitation.

5. References

1. Harada, H., Abe, T., and Chikawa, J. (1986) Oxygen precipitation enhanced with vacancies in silicon, in H. R. Huff, T. Abe, and B. Kolbesen (eds.), *Semiconductor Silicon*, The Electrochemical Society, Pennington, NJ, pp. 76-85.
2. Hasebe, M., Takeoka, Y., Shinoyama, S., and Naito, S. (1989) Formation process of stacking faults with ringlike distribution in CZ-Si wafers, *Jpn. J. Appl. Phys.* **28**, L1999-L2002.
3. Ono, H. and Ikarashi, T. (1993) Ring-distribution of oxygen precipitates in Czochralski silicon revealed by low-temperature infrared absorption spectroscopy, *Appl. Phys. Lett.* **63**, 3303-3305.
4. Ono, H., Ikarashi, T., Kimura, S., and Tanikawa, A. (1994) Preanneal effect on the ring-shaped distribution of oxygen precipitates in Czochralski-grown silicon, *J. Appl. Phys.* **76**, 621-623.
5. Ono, H., Ikarashi, T., Kimura, S., and Tanikawa, A. (1995) Anomalous ring-shaped distribution of oxygen precipitates in a Czochralski-grown silicon crystal, *J. Appl. Phys.* **78**, 4395-4400.
6. Hu, S. M. (1980) Infrared absorption spectra of SiO₂ precipitates of various shapes in silicon: calculated and experimental, *J. Appl .Phys.* **51**, 5945-5948.
7. Kimura, S. and Ishikawa, T. (1995) Relation between lattice strain and anomalous oxygen precipitation in a Czochralski-grown silicon, *J. Appl. Phys.* **77**, 528
8. Hu, S. M. (1981) Precipitation of oxygen in silicon: Some phenomena and a nucleation model, *J. Appl. Phys.* **52**, 3974-3984.
9. Wilkes, J. G.(1983) The precipitation of oxygen in silicon, *J. Cryst. Growth* **65**, 214-230.
10. Livingston, F. M., Messoloras, S., Newman, R. C., Pike, B. C., Stewart, R. J., Binns, M. J., Brown, W. P., and Wilkes, J. G. (1984) An infrared and neutron scattering analysis of the precipitation of oxygen in dislocation-free silicon, *J. Phys. C : Solid State Phys.* **17**, 6253-6276.
11. Borghesi, A., Pivac, B., Sassella, A., and Stella, A. (1995) Oxygen precipitation in silicon, *J. Appl. Phys.* **77**, 4169-4244.

OXYGEN PRECIPITATION IN MCZ SILICON: BEHAVIOUR AND DEPENDENCE ON THE ORIGIN OF RAW MATERIAL AND GROWTH CONDITIONS

T. M. TKACHEVA
Institute of Metallurgy of the Russian Academy of Sciences
49, Leninsky prospekt, Moscow, 117333, Russia

G. N. PETROV AND K. L. ENISHERLOVA
Research and Production Company ELLINA-NT
38, Vavilov street, Moscow, 117942, Russia

AND

N. A. IASAMANOV
Museum of Earth of the Moscow State University
Vorobyovy Gory, Moscow, Russia

1. Introduction

The demands of device producers for very low impurity content in Si, gives rise to a necessity to pay attention to the impurity concentration in the raw material. The analysis of impurity distribution in quartz of various genotypes shows that the most chemically pure quartz is the quartz from the veins formed as a result of hydrothermal processes and metamorphic treatments.

The most common imperfections of the large-diameter dislocation-free silicon single crystals are impurity striations and grown in microdefects that are the consequence of fluctuations in the growth conditions. It is widely accepted that the use of an applied magnetic field ensures obtaining silicon ingots that are highly homogeneous both in the electrophysical properties, distribution of oxygen and grown-in micro defects [1].

It is well known that "oxygen in silicon" is one of the very important and complex problems. Oxygen in a silicon crystal stably occupies the bond centred interstitial site, forming a slightly bent Si-O-Si bond. It causes an

517

R. Jones (ed.), Early Stages of Oxygen Precipitation in Silicon, 517–525.
© *1996 Kluwer Academic Publishers.*

infrared absorption by a vibration parallel to the original Si-Si axis, and a group of lines by the two-dimensional anharmonic vibration perpendicular to the Si-Si axis [2]. It was established that oxygen has specific features which are revealed in peculiarities of its solubility, diffusion and its behaviour at defect formation. Oxygen in silicon can give rise to optically and electrically active stages only, and both stages can be revealed at the same time. The various heat treatments can give rise to the mutual transformation of these stages of oxygen. There are hundreds of papers that have discussed this problem. The appearance of "new" silicon – MCZ silicon, with its new and specific peculiarities in its properties, adds further questions about oxygens behaviour.

The aim of the present paper is to study oxygen behaviour in MCZ silicon under heat treatments, and illustrate the possibilities of MCZ Si in device manufacturing.

2. Experimental Procedure and Samples

The basic research method employed was IR spectroscopy with a calibration factor of 2.45×10^{17} cm^{-2} for measuring optically active oxygen concentration (OAOC), and 1.1×10^{17} cm^{-2} for the measurement of optically active carbon concentration. The total oxygen concentration (TOC) was measured using a special gas analysis method. The oxygen redistribution under heat treatment was investigated in 2.27 mm thick samples by IR measurements using a double channel scheme at liquid nitrogen temperature before and after the heat treatments.

The radial variation of resistivity was estimated by the four point probe method, using thick wafers cut from both the ingot top, and bottom, after post growth annealing.

The effect of the heat treatment is studied by using the following annealing conditions:
- 450°C for 2-18 hours in air;
- 1050°C for 16 hours in nitrogen;
- two-step gettering annealing – (750°C 4h + 1050°C 16h) in nitrogen with 5% wet oxygen (ASTM, test B);
- two step gettering annealing at 700°C 16h + 1000°C 4h in nitrogen.

There are three groups of samples studied, cut along the (100) plane from:
- MCZ ingots grown under an axial magnetic field (AMF) ranged from 0.07 to 0.32 T;
- MCZ ingots grown under a rotating magnetic field (RMF) of 0.005 T;
- MCZ ingots grown under combined magnetic fields (CMF).
The main parameters of the studied samples are shown in Tables 1,2.

The crystal structure was controlled by X-ray topography using Lang's technique with MoKα₁ irradiation in 220, 400, and 400 reflections.

The wafers of ingot number 3.3K were taken to produce high voltage test diodes and switch MOS-transistors. The wafers from ingot 3.2K were

TABLE 1. Parameters of studied samples.

Ingot no.	MF type, T	Cond. type	ϱ_{top}, Ω cm	Radial $\delta\varrho_{top}$, %	ϱ_{bot}, Ω cm	Radial $\delta\varrho_{bot}$, %
1.1A	AMF, 0.07	P	12	±3	12.8	±5
1.2A	AMF 0.085	P	14	±3.7	15	±5
1.3A	AMF 0.185	P	11	±5.2	10	±6.8
1.4A	AMF 0.2	P	7	±8	7.5	±8
1.5A	AMF 0.32	P	10	±13	10.8	±15
2.1R	RMF 0.005	N	7.5	±5.1	7.2	±5
3.1T	CMF	P	12	±4.5	13.8	±4.9
3.2T	CMF	N	12	±3	15	±3
3.3T	CMF	P	14	±2.8	14.9	±4
3.1K	CMF	P	25	±5.2	28.5	±8
3.2K	CMF	P	22.5	±1.2	23.5	±9
3.3K	CMF	N	70	±8	34.5	±7
3.4K	CMF	P	38	±3.2	31.3	±1.4
3.5K	CMF	P	45	±8.5	38	±5.6
3.6K	CMF	N	49	±5	48	±6

taken to technological cycle of large CCD matrix production. The samples cut from ingots grown by the conventional Czochralski technique are used as control ones.

3. Results and Discussion

Results of the resistivity, and of its radial variation measurements are shown in Tables 1 and 2. The longitudinal variations in the resistivity are rather small for all of the samples kinds investigated. Radial variations of resistivity are largest in samples grown under AMF, and smallest in the samples grown under CMF. The carbon concentration is only very high in the case where AMF was used, with additional changes in growth conditions.

TABLE 2. Parameters of studied samples (continued).

Ingot no.	MF type, T	$[O]_{top}$, $\times 10^{17}$ cm^{-3}	$[O]_{bot}$, $\times 10^{17}$ cm^{-3}	$[C]_{top}$, $\times 10^{16}$ cm^{-3}	$[C]_{bot}$, $\times 10^{16}$ cm^{-3}
1.1A	AMF, 0.07	18	19	3.0	3.0
1.2A	AMF 0.085	11	13	3.0	3.6
1.3A	AMF 0.185	7.9	8.2	4.8	6.0
1.4A	AMF 0.2	11.8	12	4.7	6.0
1.5A	AMF 0.32	26	30	15	17
2.1R	RMF 0.005	4.4	4.7	5.0	5.0
3.1T	CMF	3.3	3.5	2.5	2.7
3.2T	CMF	4.7	4.8	2.2	2.7
3.3T	CMF	7.6	7.7	2.0	2.2
3.1K	CMF	0.6	1.8	1.8	4.6
3.2K	CMF	9.6	7.1	2.0	4.1
3.3K	CMF	6.6	5.5	5.0	5.0
3.4K	CMF	10	8.8	2.0	3.8
3.5K	CMF	7.9	4.0	2.8	2.9
3.6K	CMF	3.9	3.3	2.0	1.5

The basic microdefects in these crystals are the perfect dislocation loops of interstitial type that have Burgers vectors $a/2\langle 110 \rangle$ and lie on $\{111\}$ planes. Under certain conditions, it is possible to obtain crystals that contain minute dislocation loops (of 0.1 to 0.3 μm in diameter), particles of α-quartz and α-cristobalite [3]. For the AMF-grown ingots, the increase in concentration of optically active oxygen and carbon is accompanied by an increasing microdefect density. As well as this, elastic strains are revealed by X-ray topography without any decoration, which exhibit a striated pattern similar to the striated distribution of microdefects. The most clear pattern has been observed when the ingot was grown under a high axial magnetic field of 0.32T. For RMF-growth, it was difficult to notice any relationship between the oxygen content and defect structure, except for the fact that in "defect free" regions the OAOC was, as a rule, 20 to 25% lower than in regions containing microdefects detected by X-ray topography. For the CMF-grown crystals, no connection between defect structure of an ingot

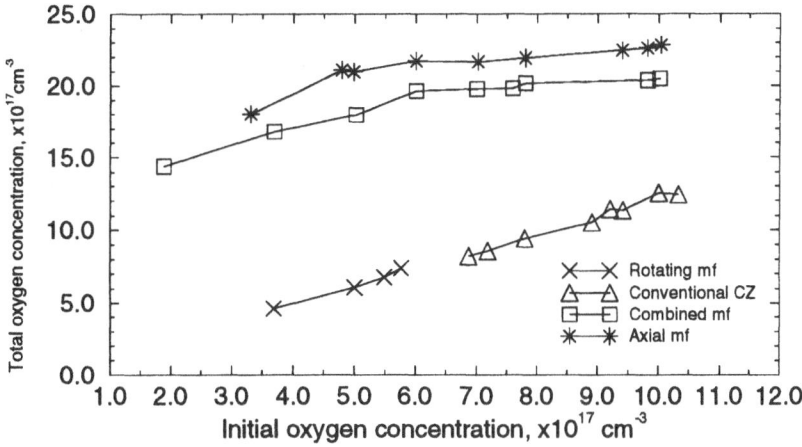

Figure 1. The relationship between TOC and the Initial OAOC in the Si samples grown by different growth technique

and the content of OAOC could be found.

The total oxygen and carbon concentrations were measured using a special gas analysis technique. In Fig 1. the relationship between the TOC and initial OAOC, measured by IR spectroscopy is shown.

The presence of the AMF, leads to a sharp increase of the TOC to the magnitude of the solution limit. For Si samples grown by conventional CZ, and under RMF, the difference between the OAOC and the TOC are practically constant, all in the range of the measured OAOC being equal to 15-20% of the TOC.

It is well known that the effective segregation coefficient increases with increasing magnetic field strength (in the case of AMF) [4]. Our result confirms this conclusion. It is necessary to take this increase into account when considering such a big increase in the TOC, for the samples grown in the presence of AMF only, and in the case of CMF (AMF + RMF).

The annealing of Si samples grown under CMF at 450°C shows that the generation of the traditional thermal donors in MCZ Si samples only begins after prolonged heat treatment (no less than 16-18 hours). To notice this effect, it is necessary to have the initial OAOC no less than 7×10^{17} cm^{-3}. It is shown that the generation of thermal donors is weaker in MCZ samples than the CZ samples. In MCZ samples, the concentration of thermal donors is $\sim$ 10-11 times less than in the CZ samples, after annealing at 450°C for

18 hours. The OAOC in MCZ and CZ samples was approximately equal in the initial state before annealing. At temperatures higher than 850°C, a rather slow precipitation of oxygen is observed.

After heat treatment at 1050°C for 16 hours in several MCZ samples with an initial OAOC in the range less than $(7\text{-}8)\times10^{17}$ cm^{-3} grown under CMF, it is possible to notice an increase of OAOC by 20-25% instead of the traditionally waited decrease of OAOC.

The very weak precipitation of oxygen at low temperature annealing, and the observed increase of OAOC are phenomenon of the same origin: growth under a magnetic field allows the oxygen atoms to relax into the silicon lattice, and occupy a stable position in such a manner that: (1) there is no vibration in the "as-grown" state, and as a result, there is no IR absorption, but, the other method allows us to reveal the TOC; (2) the low temperature heat treatment can not destroy these stable positions of the oxygen in the silicon lattice, so there is only a weak generation of thermal donors; (3) high temperature annealing can make the oxygen atom vibrate leading to IR absorption, and these atoms can participate in the precipitation processes. In our previous paper it was shown that the radiation defects connected with interstitial oxygen begin to form in MCZ Si at temperatures, and integral doses more than that of the samples grown by conventional CZ [5].

Fig 2. illustrates the relative decrease in OAOC after annealing in the regime of internal gettering (IG). The studied MCZ Si samples were grown under CMF. It is shown that it is necessary to have a higher OAOC for the beginning of the phase changing for samples grown under CMF, in comparison to the samples grown by conventional CZ. When the OAOC is more than $(6\text{-}6.5)\times10^{17}$ cm^{-3}, the phase changing processes are more intensive, and more rapid in the samples of MCZ than in the samples grown by conventional CZ. The annealing regime with the first stage at 750°C gives rise to the precipitation in MCZ Si samples with approximately the same rate as for the CZ samples. The data shown in Fig 2. confirms our conclusion about the existence of a so-called "incubation" period for oxygen atoms in the MCZ Si samples lattice, before they begin to precipitate into the oxygen solid solution decomposition.

The IG annealing allows the creation of a denuded zone near the surface of the silicon wafer. We observed this in several kinds of samples grown under an applied magnetic field after the IG heat treatments annealing at 750°C for 4 h + 1050°C for 16 h in nitrogen with 5% wet oxygen. Both the width of the defect-free, and the defect density in the getter zone differ depending on the wafer type, being 20μm and $\sim5\times10^8$ cm^{-3} for 3.3T, and 30μm and $\sim3.5\times10^8$ cm^{-3} for 3.2T (see Table 1). In the getter zone of the control sample grown by the conventional CZ method, the width of the

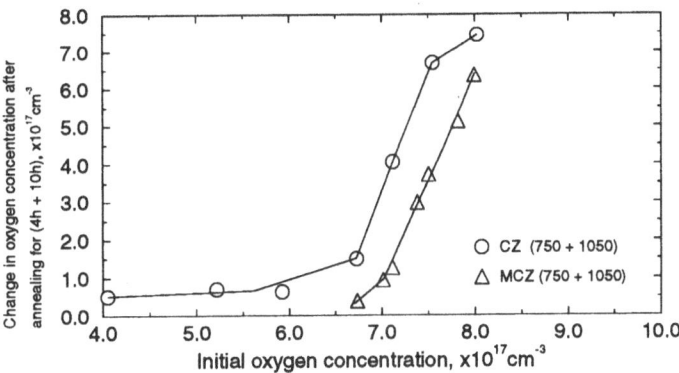

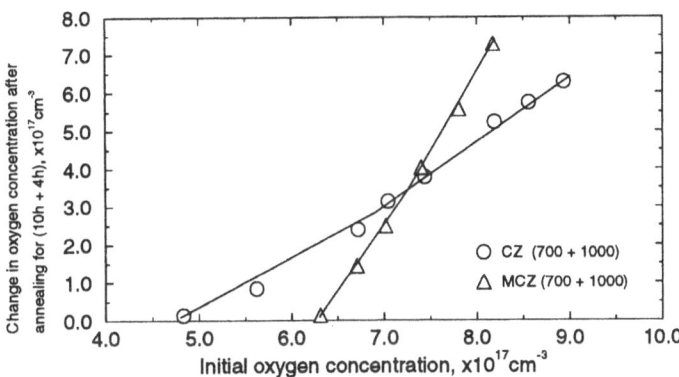

Figure 2. The decrease of OAOC in dependence on the initial OAOC as a result of the oxygen solid solution decomposition under IG annealing

getter zone is almost the same, and is equal to 27μm. However, the defect density is two orders of magnitudes smaller and is approximately equal to 5×10^6 cm^{-3}. It is necessary to note that the denuded zone in the 3.2T sample was only observed with the additional annealing stage at 450°C for 2 hours in air. We can notice again that for MCZ Si, it is necessary "to swing" the lattice with the oxygen atoms, before the precipitation begins. When OAOC in the MCZ samples is less than $(6\text{-}6.5)\times10^{17}$ cm^{-3} the phase changing processes do not begin without preliminary low temperature heat treatment (for example, 400°C for 2 hours). In this case, one can obtain a very clean and sharp denuded zone in MCZ samples, even for the samples

524

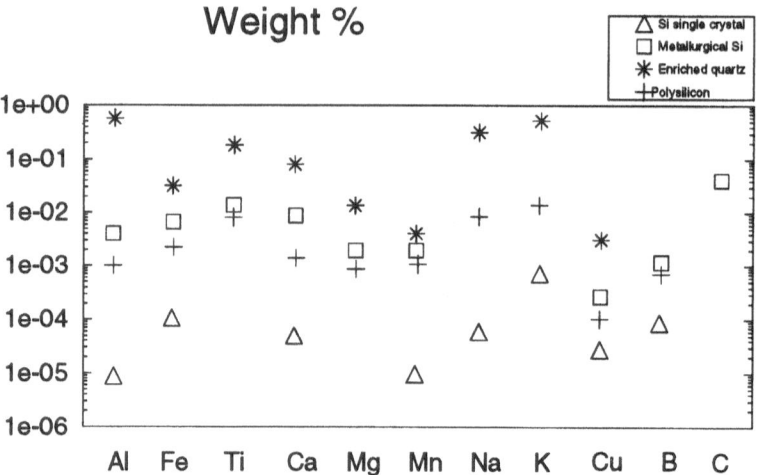

Figure 3. Elements content at every step of raw material treatment on the way from raw material mining to single crystal growth

with the initial OAOC of about 3×10^{17} cm^{-3} [4]. We observe the lower oxygen precipitation in MCZ samples without preliminary heat treatment in comparison with that in the CZ samples and the very high TOC in MCZ samples.

The very sharp denuded zone is observed in practically all of the MCZ samples investigated (1.3 A, 3.3 T, 3.2 K), with an initial OAOC of more than 7×10^{17} cm^{-3} after very simple and standard heat treatments at 1150-1200°C for 3 hours in an atmosphere containing chlorine. After such heat treatment, one can observe the denuded zone of 20-25μm in width with the sharp boundary getter zone. When using these samples for CCD matrix production, there is a decrease in the number of videodefects in the active part of the vide image from 100 or more (for CZ Si) to only 10 or less.

MOS-transistors made by using MCZ samples as substrates have characteristics close to the ideal, ie there is no "soft" CVC, and the reverse current increases quickly and abruptly. In these devices the residual pre-avalanche breakdown currents are of 3-4 orders lower than those in diodes made by using epitaxial structures on the substrates grown by the conventional CZ method.

So, the growth conditions play the most important role for the further behaviour of impurities introduced into the ingot during growth. Naturally, the quality and impurity content of the raw material is the first problem.

In Fig. 3 there is a comparison between the content of some of the important impurities, measured by a neutron activation method, after every step of single crystal growth starting from the raw material. It is clear that the impurity content can be reduced in the final single crystal not only by the cleaning process during the single crystal growth, but also by choosing the most clean raw material in the first place.

4. Conclusions

The main conclusions are as follows:

1. The precipitation and thermal donor generation in MCZ Si samples begins at temperatures and initial oxygen concentrations more than in the samples grown by CZ, in the case where each sample has the same magnitude of OAOC.

2. The IG annealing allows one to obtain a very clear denuded zone, in the MCZ Si, which has a high density of defects; using these MCZ Si samples, it is possible to produce large area CCD-imagers for the TV-chamber with the very low density of videodefects.

3. Using the MCZ Si samples it is possible to prepare switch MOS-transistors with close to "ideal" reverse current characteristics.

4. The quality and impurity content of the raw material (as a rule granular transport in natural quartz) is the decisive characteristic for the final impurity content in the single crystal.

References

1. Series, R. W. and Hurle, D. T. J. (1991) The Use of Magnetic Fields in Semiconductor Crystal Growth, *J. Cryst. Growth*, **113**, pp. 305–328
2. Kaneta, C., Sasaki, T. and Katayama-Yoshida, H. (1994) Effect of Carbon on Anharmonic Vibration of Oxygen in Crystalline Silicon *Material Science Forum*, **143–147, part 2** pp.957–962
3. Bochkarev, E. P., Gorin, S. N., Petrov, G. N. and Tkacheva, T. M. (1992) Defects in MCZ Silicon with Various Oxygen and Carbon Content *Material Science Forum*, **83–87, part 2** pp. 1075–1080
4. Series, R. W. (1989) Czochralski Qrowth of Silicon under an Axial Magnetic Field, *J. Cryst. Growth*, **97** pp. 85–91
5. Tkacheva,T. M. and Petrov, G. N. (1994) Defects Formation in MCZ Si during Heat Treatment and Irradiation in HV (1 Mev) TEM, *Proc. 4th Scien. Busin. Conf. "Silicon-94", Roznov pod Radhostem, Czech Republic*, TECON Scientific, Roznov pod Radhostem, pp. 239–250

Index